Government Response to Disruptive Innovation:

Perspectives and Examinations

Sam B. Edwards, III
Quinnipiac University, USA

James R. Masterson
Morehead State University, USA

IGI Global
PUBLISHER of TIMELY KNOWLEDGE

A volume in the Advances in Public Policy and Administration (APPA) Book Series

Published in the United States of America by
IGI Global
Information Science Reference (an imprint of IGI Global)
701 E. Chocolate Avenue
Hershey PA, USA 17033
Tel: 717-533-8845
Fax: 717-533-8661
E-mail: cust@igi-global.com
Web site: http://www.igi-global.com

Library of Congress Cataloging-in-Publication Data

Names: Edwards, Sam B., III, 1969- editor. | Masterson, James (James Robert), 1977- editor.
Title: Government response to disruptive innovation : perspectives and examinations / edited by Sam Edwards, James Robert Masterson.
Description: Hershey, PA : Information Science Reference, [2023] | Includes bibliographical references and index. | Summary: "The proposed book will serve as a point of reference for government officials and agencies about emergent trends in technology and the impacts these potentially have on governance. Discussions will include introductory synopsis of the technology, how the technology impacts specific government services, and how governments are responding to this changing environment, and the limitations they face, legal, budgetary, and political. Expects, including academic scholars, will find useful case studies for course materials illustrating government responses to disruptive innovations"-- Provided by publisher.
Identifiers: LCCN 2022046810 (print) | LCCN 2022046811 (ebook) | ISBN 9781668464298 (hardcover) | ISBN 9781668464335 (paperback) | ISBN 9781668464304 (ebook)
Subjects: LCSH: Technological innovations--Government policy. | Public administration--Technological innovations. | Technology and state.
Classification: LCC HC79.T4 G674 2023 (print) | LCC HC79.T4 (ebook) | DDC 338/.064--dc23/eng/20221129
LC record available at https://lccn.loc.gov/2022046810
LC ebook record available at https://lccn.loc.gov/2022046811
This book is published in the IGI Global book series Advances in Public Policy and Administration (APPA) (ISSN: 2475-6644; eISSN: 2475-6652)

British Cataloguing in Publication Data
A Cataloguing in Publication record for this book is available from the British Library.
All work contributed to this book is new, previously-unpublished material. The views expressed in this book are those of the authors, but not necessarily of the publisher.
For electronic access to this publication, please contact: eresources@igi-global.com.

Advances in Public Policy and Administration (APPA) Book Series

ISSN:2475-6644
EISSN:2475-6652

Mission

Proper management of the public sphere is necessary in order to maintain order in modern society. Research developments in the field of public policy and administration can assist in uncovering the latest tools, practices, and methodologies for governing societies around the world.

The **Advances in Public Policy and Administration (APPA) Book Series** aims to publish scholarly publications focused on topics pertaining to the governance of the public domain. APPA's focus on timely topics relating to government, public funding, politics, public safety, policy, and law enforcement is particularly relevant to academicians, government officials, and upper-level students seeking the most up-to-date research in their field.

Coverage

- Government
- Law Enforcement
- Political Economy
- Politics
- Public Administration
- Public Funding
- Public Policy
- Resource Allocation
- Urban Planning

IGI Global is currently accepting manuscripts for publication within this series. To submit a proposal for a volume in this series, please contact our Acquisition Editors at Acquisitions@igi-global.com or visit: http://www.igi-global.com/publish/.

The Advances in Public Policy and Administration (APPA) Book Series (ISSN 2475-6644) is published by IGI Global, 701 E. Chocolate Avenue, Hershey, PA 17033-1240, USA, www.igi-global.com. This series is composed of titles available for purchase individually; each title is edited to be contextually exclusive from any other title within the series. For pricing and ordering information please visit http://www.igi-global.com/book-series/advances-public-policy-administration/97862. Postmaster: Send all address changes to above address.

Titles in this Series

For a list of additional titles in this series, please visit: http://www.igi-global.com/book-series/advances-public-policy-administration/97862

Examining Colonial Wars and Their Impact on Contemporary Military History
Miguel Madueño (Rey Juan Carlos University, Spain) and Alberto Guerrero (Universidad de Granada, Spain)
Information Science Reference • © 2023 • 311pp • H/C (ISBN: 9781668470404) • US $215.00

Governance as a Catalyst for Public Sector Sustainability
Neeta Baporikar (Namibia University of Science and Technology, Namibia & SP Pune University, India)
Information Science Reference • © 2023 • 392pp • H/C (ISBN: 9781668469668) • US $250.00

Acceleration of the Biopsychosocial Model in Public Health
Simon George Taukeni (University of Namibia, Namibia)
Medical Information Science Reference • © 2023 • 347pp • H/C (ISBN: 9781668464960) • US $325.00

Handbook of Research on the Regulation of the Modern Global Migration and Economic Crisis
Emilia Alaverdov (Georgian Technical University, Georgia) and Muhammad Waseem Bari (Government College University, Faisalabad, Pakistan)
Information Science Reference • © 2023 • 393pp • H/C (ISBN: 9781668463345) • US $285.00

Maintaining International Relations Through Digital Public Diplomacy Policies and Discourses
Türker Elitaş (Hatay Mustafa Kemal University, Turkey)
Information Science Reference • © 2023 • 270pp • H/C (ISBN: 9781668458228) • US $240.00

Handbook of Research on Transforming Government, Nonprofits, and Healthcare in a Post-Pandemic Era
Joanne E. Howard (Illinois Institute of Technology, USA)
Information Science Reference • © 2023 • 590pp • H/C (ISBN: 9781668423141) • US $270.00

Minding the Gap Between Restorative Justice, Therapeutic Jurisprudence, and Global Indigenous Wisdom
Marta Vides Saade (Ramapo College of New Jersey, USA) and Debarati Halder (Parul Institute of Law, Parul University, India)
Information Science Reference • © 2023 • 260pp • H/C (ISBN: 9781668441121) • US $215.00

Societal Transformations and Resilience in Times of Crisis
Ghazala Shoukat (University of Sindh, Jamshoro, Pakistan) and Muhammad Nawaz Tunio (Mohammad Ali Jinnah University, Karachi, Pakistan)
Information Science Reference • © 2023 • 252pp • H/C (ISBN: 9781668453261) • US $215.00

IGI Global
PUBLISHER of TIMELY KNOWLEDGE

701 East Chocolate Avenue, Hershey, PA 17033, USA
Tel: 717-533-8845 x100 • Fax: 717-533-8661
E-Mail: cust@igi-global.com • www.igi-global.com

Editorial Advisory Board

Table of Contents

Detailed Table of Contents

Developments in information and communication technology offer a rich set of possibilities for public managers that can make government more successful at serving their constituencies and achieving their missions. Sadly, barriers exist that can make this process difficult or impossible. Fortunately, this chapter provides public managers with an overview of new technology-based citizen partnerships that can facilitate the incorporation of disruptive technologies into government agencies. Developments like civic technology, data for good, public interest technology, and an emerging core of newer efforts offer public managers a way to address these barriers. The chapter discusses these developments, outlines competencies that managers will need to work with these groups, and explores ways that these competencies can be acquired by the public management workforce.

This chapter examines the disruptive innovation in election administration caused by electronic poll books (EPBs). One of the more recent and most disruptive innovations in voting are electronic poll books. Although traditionally done by paper, electronic poll books allow registered voters to check in using a laptop or tablet and offer a number of advantages in terms of ease of checking and verifying voter information. Using Indiana as a case study, the chapter will explain the adoption of EPBs and their certification in Indiana, attempt to explain factors that caused counties to adopt EPBs sooner than

This chapter analyzes the use of artificial intelligence in different activities, with a focus on the uses in the justice system of the Autonomous City of Buenos Aires in Argentina. It first makes a recount of what is understood by artificial intelligence, its origins, and applications to then examine its application in different aspects of the criminal justice system, including the law enforcement and judicial software currently in use, with the ethical and legal issues that arise. It then delves into the functioning of PROMETEA, an AI system developed and used by the Office of the General Attorney of the Autonomous City of Buenos Aires, finishing with some of the legal implications according to Argentine law.

Brazil's Supreme Court has launched and presided over investigation procedures on alleged fake news. The so-called "Inquiry of the End of the World" (Inquérito do Fim do Mundo) was initiated by the President of the Court, Justice Toffoli, and overseen by another member of the court, Justice Moraes (also rapporteur). Brazil's Constitution and General Principles of Law regarding investigations and due process require different acts to be performed by different actors. By due process, the police investigate, prosecutors accuse, and judges judge. In the Inquiry of the End of the World, judges investigate, prosecute, and judge. In a surprisingly evident festival of illegal acts, the rapporteur ordered searches and seizures, censorship, and arrests. A series of unlawful infringements of fundamental rights added to alleged political interests behind this investigation led to this research questioning: Is Brazil's Supreme Court stepping out of its constitutional mandate and limits? What impacts can such developments have on Brazilian checks and balances?

Regarding digital currencies, the authors focus on 1) the security issues with the advent of quantum computers and 2) the privacy issues in relation with CBDC's wider collection of personal data. Many legal issues can be solved using technology. Thus, they propose central banks to adopt their privacy-enhancing technologies in cryptography, called "SK4SC," which performs KYC (know your customer) while keeping PKI (public key infrastructure)-registered trading partners anonymous. Security and privacy in digital currency can be trade-offs. For example, the AML/CFT effort requires the collection of personal information about innocent people, which can be detrimental to their privacy. On the other hand, excessive restrictions on the collection of personal information to protect privacy may weaken the AML/CFT. However, adopting SK4SC can resolve the trade-offs by anonymizing KYC, which is essential for the AML/CFT.

Blockchain is arguably the most disruptive technology of the 21st century. Although most known for its role in cryptocurrency, blockchain is utilized in several innovative ways. For example, an artist's copyright, royalties, and licensing information can be stored using smart contracts, a blockchain-backed technology. This chapter will address the role of blockchain-backed technology throughout the voting process and how state government plans of action may differ as they attempt to regulate and adopt this innovative yet disruptive technology. Additionally, it will compare different e-voting platforms and their underlying frameworks, languages, and consensus protocols. Ultimately, this chapter reveals how electoral practitioners and policymakers have responded to regulating blockchain-backed technologies and how cryptographic framework and technological advancements directly influence the e-voting ecosystem.

This chapter examines freedom and blockchain. Hegel stated that world history is the progression of freedom consciousness. While modern law and the state have facilitated the realization of freedom, blockchain technology has facilitated global, autonomous decentralization. This chapter defines freedom and explores how blockchain aids in its promotion and examines why central governments of developed countries opposed Libra, a global currency concept based on blockchain. While the central government intervened to crush Libra, this chapter shows that blockchain has vast possibilities for evolution of freedom. From the perspective of the absolute, blockchain and modern law and the state, while in conflict, will promote human freedom dialectically. This chapter uses the concepts such as publicness, autonomy, setback, external third party, facts and norms, primary rules, law and facts, family, bodily extension, constitutionalism, objective reality, absolute spirit, and so on. Also, the theories of Hayek, Habermas, H.L.A. Hart, Wittgenstein, Hegel, and so on will be utilized in this chapter.

Can short video-based social media platforms disrupt economic markets and upend government censorship? Short videos have become the vanguard of social media. This chapter examines how this modern information communication technology (ICT) disrupts government regulation and the market as a whole in China, using TikTok as an example. Specifically, it argues that a video social media platform with unlimited content and low usage thresholds, combined with citizen's lack of legal awareness, makes it easy to expose privacy, encounter fraud, and infringe upon intellectual property. These factors are exasperated by TikTok's video format since video-based content is more difficult to censor in a timely manner. Consequently, TikTok poses challenges for government regulation and disrupts government censorship. The chapter concludes by discussing policy implications. Specifically, it proposes the implementation of modernized laws and regulations to tackle the distinct challenges and risks presented by cutting-edge technologies.

This chapter introduces the social credit system (SCS) as one of the Chinese government's disruptive-innovative answers to control its citizens and organizations to secure the retention of political power. Based on source texts and surveys, "external control" and "extrinsic motivation" are identified as core principles of SCS. The unique qualities of the SCS are described in the context of the existing surveillance and control systems in China. The example of dropouts is discussed. The impact of artificial intelligence on communication, the efficiency of the SCS, its effects on everyday life in China are described. The SCS violates the UNESCO agreement, of which China is a signatory. Also the European Union has passed an Artificial Intelligence Ethics Act. The chapter shows the limits of the system and that it is not transferable to liberal democratic societies. However, IT companies and interest groups in data capitalism have also developed small-scale credit systems for economic profit and political manipulation.

The expansion of personal information and data resources use is raising the need for many countries to respond. Moreover, the expansion of technologies and new technologies developed based on data resources is causing responses through legislation in each country. Through this process, the government and domestic actors share the burden of protecting data resources, which is implemented through the security governance structure. The United States and China are developing different security governance structures for personal data and data resources. In the United States, the burden of protecting personal data is shared primarily between the government and data processors. In China, the burden of protection of personal information and data resources is shared mainly between the government, data controllers, and individuals.

Foreword

The Clean Water Act passed 50 years ago in 1972. The age of industrial chemistry had enjoyed more than a century of free for all, finally creating a regulatory authority to amend the 1948 Federal Clean Water Act.

Those intervening 24 years of non-enforceable, self-managed regulation had not turned out well. All and any manner of waste from raw sewage to volatile flammable by-products flowed freely into waterways. Tides of dead fish washed up on shore, lakes, rivers and streams became engulfed with slimy algae; memorably the Cuyahoga River caught fire more than once. Major public waterways were evidently ruined and popularly seen as health hazards, including Boston Harbor, Lake Erie, New York's East River and even the Potomac flowing through the national capital.

The good old days are clearly over. Rather than coming to a call for regulation, many new ideas sound new regulatory calls in a couple of years not decades.

The New Century is already an avalanche of Technology and Innovation. Expanding well beyond computation and information, technology driven startups are into every imaginable field now, finance, education, infrastructure, governance, farming data, food technology, beauty and wellness and fashion to name a few. But with many new kinds of technology coming online, the early century may be called the Era of Regulatory innovation.

Good policy protects the public and simultaneously incentivizes new solutions that meet public aspirations for safety and well-being, creating economic opportunities for that country. The internet itself was created by Defense Advanced Research Projects Agency (DARPA) as a solution to questions of national security. Crash test ratings on cars, the popularization of the refrigerator, sustainable laundry detergent, increased mileage ratings on automobiles are all responses set in motion by policy.

Thus, global regulatory efforts are also a means of creating value in a world of nations hungry to lead the world in the next tech revolution.

In 2016 the team I was on invested in the first Lab Grown Meat Startup (Memphis Meats, now Upside). At the time almost nobody believed that cultivating the meat outside the animal could compete with industrialized animal ag. There were some notable exceptions: Tysons Foods and Cargill, two of the largest meat companies in the world.

Two years later in Japan I stopped to chat with a group of students from Singapore. I started to describe the FoodTech innovations I had been working with; Cultivated Meat, Molecular Foods, Plant based meat etc. The students needed no introduction from me. FoodTech was already part of university curriculum in Singapore.

The fact is that after investing in biotechnology for decades, Singapore's focus was paying off. A tiny nation of just a single city 30 miles wide that imports nearly all its food saw alignment. That technology could enable food independence for a country with no land is a bold vision. Singapore's goal was to make themselves a global leader in this new field. Towards that goal they created the Singapore Food Agency in 2019.

In 2020 more than 90% of Singapore's food production was imported (Ong, 2022). The island/city/nation is perpetually vulnerable to food shortages in the region. The government nimbly set a policy goal of "30 by 30" (Reed, 2022); thirty percent of food consumed in Singapore will be produced domestically by 2030 (*30 by 30,* 2022). The plan is comprehensive and includes substantial re-zoning for urban farming, support for farmers and #foodtech was included too. In December 2021 the Singapore Food Agency cleared the first sale of lab grown meat and permitted the construction of the first production plant.

Despite a much larger and well-established regulatory apparatus, The United States is just behind. The sale of lab grown meat is anticipated to be some time this year (2023) by a joint USDA/FDA effort. They have been involved in the development of at least two production facilities. Upside Foods' pilot production plant has been built in Emeryville, California and Believer Meats is building another plant in North Carolina (Charlotte Business Journal, 2022).

Israel has also decisively acted to attract its share of the billions in funding and taken a stand as a global center in food innovation. Public Private Funding through the Israeli Innovation Authority technological incubators program (Israel Innovation Authority, 2022) has seeded a global startup hub that raised $160M in 2022 despite economic slowdowns (Ben-David & Surkes, 2022).

Just as the European Union has taken the lead in personal privacy, as Cybersecurity has become a matter of National Security, the future evolution of all technology is being played out on a global stage with Regulatory Innovation advancing national priorities as well as national well-being.

For all the waves of change lapping upon our doorstep, meshing the pace of the law, regulation and incentive is even more a keystone to global technology in society. Watching thought in regulation of new innovations is as much a key to the 21st Century as making product. Global competition to harness and align technology of dozens of stripes and flavors will be the international chessboard and a game well worth watching.

Ron Shigeta
Independent Researcher, USA
Berkeley, CA Jan 19, 2023

REFERENCES

Ben-David, R., & Surkes, S. (2022, August 4). Israeli companies lead world in plant-based food tech investments. *The Times of Israel.* Retrieved February 21, 2023, from https://www.timesofisrael.com/israeli-companies-lead-world-in-plant-based-food-tech-investments/

Israel Innovation Authority. (2022, February 2). *Announcing the winners for the operation of five new technology incubators in the following fields: Health with an emphasis on Bio-Convergence, Climate, Foodtech, and Space* [Press Release]. https://innovationisrael.org.il/en/news/israel-innovation-authority-launches-new-incubators-program

Journal, C. B. (2022, December 11). *Lab-grown meat company picks NC for 100-job, $123M facility*. WSOC TV. Retrieved February 21, 2023, from https://www.wsoctv.com/news/local/lab-grown-meat-company-picks-nc-100-job-123m-facility/WJJ5PHJTUJGB3KDCLMRXVIDTRU/

Lu, D. (2022, November 5). All sizzle, no steak: How Singapore became the centre of the plant-based Meat Industry. *The Guardian*. Retrieved February 21, 2023, from https://www.theguardian.com/environment/2022/nov/06/all-sizzle-no-steak-how-singapore-became-the-centre-of-the-plant-based-meat-industry

Ong, C. (2022, June 21). *Singapore imports more than 90% of its food. here's how it's dealing with rising food inflation*. CNBC. Retrieved February 21, 2023, from https://www.cnbc.com/2022/06/21/singapore-imports-90percent-of-its-food-how-is-it-coping-with-inflation.html

Singapore Food Agency. (n.d.). *30 by 30*. Our Food Future. Retrieved February 21, 2023, from https://www.ourfoodfuture.gov.sg/30by30

Preface

Following the invention of the internal combustible engine- traffic laws and conventions were gradually put into place. Many of those laws and conventions continue to serve us well today, and over the course of a century, we had plenty of time to adapt our laws to new inventions, such as freeways. Today, however, scientific advances are bringing seismic shifts to in how we use our roads: legislatures and municipalities are scrambling to keep up, tech companies are chafing under outdated, and sometimes nonsensical rules, and the public is not sure what to think. Smartphone technology gave rise to Uber, but before the world figures out how to regulate ridesharing, self-driving cars will have made those regulations obsolete. (Eric Teller, CEO, Google X, from interview with Thomas Friedman)

As Teller illustrates, technology typically moves faster than governments. This exponential pace of technology presents both significant opportunities and threats. Governments that fail to proactively respond risk losing control and playing an ever-reactionary role. In many cases, the best policies are ones that help facilitate and properly regulate these innovations. Failing to regulate ridesharing and self-driving cars demonstrates that governments are often either reluctant or incapable of creating policy proactively.

More than 15 years ago, in his seminal book, *The World is Flat*, Tom Friedman wrote about the transformative nature of the internet on its potential to lower barriers to entry into many markets for individuals across the globe for those who can plug into this technology. By 2016, in his most recent book, *Thank You for Being Late*, Mr. Freidman writes about how internet-based technologies powered by complex Artificial Intelligence (AI) systems have the capabilities and potential to displace millions of skilled workers anywhere around the globe. These systems can write poetry, plays, news columns, legal documents, and computer code, as well as drive cars, deliver goods, produce photographic images and videos, and provide educational services, anytime and anywhere an internet connection is available. The Internet of Things, the next wave of Internet connectivity, will result in connecting anything that can be made more efficient and functional with internet connectivity both inside and outside of the household.

As the pace of internet-connected devices rapidly advances society, established norms and practices become disrupted. From low-stakes routine practices such as how we order our lunch and get to work, to high-stakes practices with great implications for our future such as how we gain an education and obtain legal services, many areas of society are simultaneously becoming disrupted as the digital revolution's AI-backed systems become commonplace in communities around the globe. How governments respond to these disruptive innovations to protect citizens, increase access to public goods, and create new markets of economic development is the subject of this book. However, these technologies can also provide another public interest, increasing the efficacy of holding one's own government accountable. In other words, as Tom Friedman writes, "Both technology and globalization today are empowering "political makers," who want to remake autocratic societies into more consensual ones, and "political breakers," who want to bring down the government in order to impose some religious or ideological tyranny, even though they may lack any ability to govern effectively."

This book is a timely exploration of a wide array of government responses to various disruptive innovations, largely those that are connected to the internet or are built on AI systems. Public policy diffusion is well documented and studied in the public administration and public policy literature. Governments the world over have the capabilities to learn from others in crafting their policies to address concerns arising from these new disruptive innovations. This book is an attempt to make a modest contribution to that literature helping to enable policymakers and academics, as well as students of the law and politics of technology, to better understand how governments have responded to several disruptive innovations and what concerns remain outstanding. Therefore, the audience for this text, policymakers, professors, students, and the like, will read about specific disruptive innovations ranging from AI-provided legal services, autonomous vehicles and delivery services, to blockchain technology used for cryptocurrencies and in voting applications, and the actions governments have taken in response to these innovations.

To be sure, government regulation of disruptive innovations is not new. Steam-powered boats and locomotives were once disruptive innovations and eventually required governments to confront the legal challenges posed by these new modes of power and transportation. Similarly, disruptions brought on by electric power and machinery were confronted in Beard and Beard's novel work, *The American Leviathan: The Republic in the machine age*, addressing how all facets of the federal government adapted to the changing environment brought forward through electricity and machinery. Nearly one hundred years later, this book begins the journey of addressing government responses to disruptive innovations.

Disruptive innovations provide opportunities for governments, many of which are beneficial to society. First, historically, disruptive innovations such as the telephone, car, freeway system, personal computer, etc., provided governments with opportunities to become more efficient at the services they provide. In their seminal work, Beard and Beard (1930) systemically document transformations made to the federal government as a result of the industrial revolution. Jon Agar's 2016 book, *The Government Machine: A revolutionary history of the computer*, examines the transformation in government during the explosion of personal computers within government agencies in the United Kingdom. These innovations allow governments to enhance their services to the public, become more efficient, have access to more information about their constituents, increase transparency, and facilitate democracy. These technologies can allow the private sector to perform functions previously limited to governments and reduce the cost of government and offer alternatives.

On the other hand, disruptive innovations present significant threats to governments such as the loss of control over government monopolies such as currency. In addition, governments can face forced transparency as technology makes information harder to control. Social media provides expanded opportunities for citizens to hold their government accountable. At the same time, governments must protect citizens' electronic data from adversaries who seek to breach government systems and steal such information. In this sense, disruptive innovations are increasingly making it more difficult for governments and businesses to secure electronic information. Data breaches have the potential to threaten trust in government, at a time when the use of electronic formats for voting is on the rise, through E-poll books and e-Voting backed by blockchain technology. These disruptive innovations signify the potential loss of control by various government actors. Governments also face the prospects of losses in revenue from numerous disruptive innovations currently in the process of coming to market, some of which are discussed in this book, such as autonomous vehicles, personal delivery devices, and AI legal services.

From the perspective of society, there are abundant opportunities flowing from disruptive innovations. These innovations make life more convenient, providing new modes of transportation and new means of food delivery. They also allow for expanded methods of payment for goods and services through cryptocurrencies and expanded ways to vote using e-Voting platforms. Last, AI-backed tools can expand access to legal services and advice. Disruptive innovations also provide social benefits through economic development. These innovations aid in the creation of new markets in the delivery of goods, livery services, cryptocurrency trading, as well as private higher education.

Disruptive innovations present a variety of threats to society. Innovations that assist governments and individuals in surveilling others threaten to decrease privacy. Other disruptive innovations give rise to risks from insufficiently regulated or unregulated technological advancements, creating risks to the public. Additionally, these innovations provide the environment for shifting public services to the private for-profit sector. These innovations lead to the loss of human connections in the way of autonomous, technology-driven services in delivery and government/legal/educational services. They represent the potential to displace millions of workers in lower-skilled jobs through the outsourcing of these wage earners to robotic and AI-backed systems.

AI is a good example of a disruptive innovation that exhibits these opportunities and threats to governments and society. Governments, such as bar associations, risk losing their monopoly on regulating legal services when people can consult unlicensed and unregulated AI "lawyers." AI however increases the availability of legal services and can reduce its cost. AI in autonomous vehicles presents dangers to the public as the technology generally precedes government regulation and has the potential to risk public safety. The public, however, benefits from the services provided by these robot delivery systems, while also shifting winners and losers in the jobs market, as lower-skilled laborer are replaced by robots, which require high-skilled laborers to develop and service. In fact, ChatGPT wrote its own preface to this book, which follows this human-written preface below.

Because of this double-edged sword presented by disruptive innovations, governments understandably have mixed responses. In some cases, governments will ignore the innovations unless and until they present such important issues that the government is forced to regulate. The next approach is when governments seek to actively prohibit innovations such as imposing bans and threatening criminal action. This approach runs the risk of stifling innovation and driving it underground. Suppressing technology is often a losing battle especially when innovators will find a friendly jurisdiction where they will be able to work. Finally, governments can facilitate the safe development of disruptive innovations. This approach requires governments to be nimble and proactive. Blockchain is a good example of this area. Cryptocurrency developed rapidly, and some governments worked to regulate it to protect the public and to help that industry develop. This third approach tends to present the best balance between protecting the public and facilitating important innovations, though it does not work for every government and all innovations.

This book focuses on disruptive innovations in five key areas: public services, legal services, autonomous vehicles, blockchain technology, and social media and data technology. These case studies represent issues and government responses from a wide variety of countries in Asia, Europe, North America, and South America.

SECTION 1: INNOVATIONS IN PUBLIC SERVICES

Chapter 1: Preparing Public Managers to Deal With Disruptive Innovations Through Citizen Partnerships

Technology offers wonderful possibilities for improving government, much of which is never realized because of a range of well-known barriers. While emerging movements like Civic Technology, Data4Good and Public Interest Technology offer a way to surmount these barriers, managers need new skills to make use of these resources. This chapter provides an overview of these emerging resources and reviews the skills that Public Managers need to properly employ them in an effective strategy.

Chapter 2: Disruptive Innovations in Voting – Electronic Poll Book Adoption and Certification in Indiana

This chapter examines the disruptive innovation in election administration caused by electronic poll books (EPBs). Although traditionally done by paper, electronic poll books allow registered voters to check in using a laptop or tablet and offer several advantages in terms of ease of checking and verifying voter information. The voter information on electronic poll books is generally synchronized with state or local voter registration databases and this provides real-time and up-to-date information available to jurisdictions when voters are checking in. This chapter will explain the innovations that EPBs bring to election administration, examine and attempt to explain why they were adopted in the states, and, using Indiana as a case study, the chapter will explain the adoption of EPBs in counties and the certification process in Indiana. Overall, despite being a period of time marked by partisan policy adoption across every level of government, EPBs have been widely embraced and state and local governments continue to facilitate their adoption.

Chapter 3: The Disruptive Role of Private Higher Education in the Western Balkans

The emergence of private higher education sector in the Western Balkans carries disruptive potentials in two directions. First, student awareness of their own personal responsibility for their professional training could spill over to the rest of society and help spread individual responsibility in societies with deeply entrenched collectivist culture. Second, private universities challenge the atrophied state-run public university model, by offering leaner and more efficient academic and administrative models and practices. Whereas Western Balkans governments initially tolerated private universities as a response to society's pressure to liberalize university enrollments as well to artificially curb youth unemployment, soon their attitude toward public universities turned skeptical and even hostile. However, over time, private universities managed to overcome their initial difficulties, became trustworthy academic institutions, and set examples to follow for public universities. From this perspective, government responses to private-university-as-innovation in the Western Balkans does not neatly fit in any of the identified categories by the volume's editors. Indeed, it moves through them consequentially: it began with government ignoring them, then showing skepticism and even hostility toward them, and finally trying to embrace and emulate them in that section of the higher education sector that the government controls, the public university.

SECTION 2: INNOVATIONS IN LEGAL SERVICES

Chapter 4: Artificial Intelligence in Legal Services – Reducing the Gap in Access to Justice

With the explosion of artificial intelligence in the past year, governments, industries, and the public are becoming aware of how it will fundamentally change many aspects of life. Legal services are an area where AI is already impacting practice. AI can do many of the tasks previously limited to those with legal expertise primarily benefiting corporate clients. Despite the US having many lawyers, there is a significant gap in access to legal services for many people. Legal services are just too expensive for many people even though legal representation is essential in many areas such as evictions, employment discrimination, and family. AI can provide legal services on its own and help reduce the cost when legal professionals are offering free or reduced cost services. Unfortunately, existing systems such as the rules of professional conduct prevent non-lawyers from dispensing legal advice. These systems need to be changed to facilitate the development of AI to help reduce the gap in access to justice.

Chapter 5: The Procrustean Nature of AI and the Legal Implications of Its Use in the Criminal System Decision Making of Argentina

The Autonomous City of Buenos Aires in Argentina has deployed an artificial intelligence (AI) system. This chapter evaluates the AI system. The chapter starts with a summary of AI and then examines its application in different aspects of the criminal justice system. It specifically evaluates the law enforcement and judicial software currently in use and the ethical and legal issues that arise. It then delves into the functioning of PROMETEA, an AI system developed and used by the Office of the General Attorney of the Autonomous City of Buenos Aires. The chapter concludes with some of the legal implications according to Argentine law.

Chapter 6: Due or Undue Process in Brazil? The Inquiry of the End of the World and Unchecked Judicial Imbalances

Some see fake news as a significant threat. In Brazil, the Supreme Court has taken the extreme measure of launching and presiding over investigations in to alleged fake news. The President of the Court, Justice Toffoli, initiated this so called "Inquiry of the End of the World" (Inquérito do Fim do Mundo). Another member of the Court, Justice Moraes (also rapporteur) is overseeing the investigation. Brazil's Constitution and General Principles of Law regarding investigations and due process require different acts to be performed by different actors. By due process, the police investigate, prosecutors accuse, and judges judge. In the Inquiry of the End of the World, judges investigate, prosecute and judge. In a surprisingly evident festival of illegal acts, the rapporteur ordered searches and seizures, censorship, and arrests. This series of unlawful infringements of fundamental rights added to alleged political interests behind this investigation lead to this research questioning: Is Brazil's Supreme Court stepping out of its constitutional mandate and limits? What impacts can such developments have on Brazilian Checks and Balances?

SECTION 3: INNOVATIONS IN AUTONOMOUS VEHICLES

Chapter 7: Disruptions From Personal Delivery Devices – A Survey of the American Regulatory Approach to Sidewalk-Sharing Robotic Delivery Devices

Personal Delivery Devices (PDDs) offer states many benefits but also pose risks that need to be managed. Twenty-six states have passed legislation regulating PDD usage. States can boost their economic development opportunities by allowing PDDs to operate in the communities, facilitating local business, as well as growing high-skilled, high-paid job opportunities. However, PDDs present risks to local communities to both personal harm and collisions with vehicles. While these events are rare, this chapter documents several cases of collisions with PDDs and explores other risks associated with cyber threats and device hijacking.

Chapter 8: Government's Response and Perspective on Autonomous Vehicles

This paper effectively considers the development of Autonomous Vehicles (AVs) with a detailed outlook of the efforts of technology companies to connect our roads with cars that drive themselves. Many research and development funds have been invested in this and the reality seems within reach each passing day with more road tests and frequent policy debates. The paper also looks at the possible impact of AVs from social, economic, and political perspectives. AVs will transform our world- it will create some industries and antiquate others. From this evidence-based viewpoint, the paper considers how the government should respond to this disruptive technology. As vestiges of ineffective government reactions to disruptive technology pervade literature and our collective human existence, the author was quick to warn that AVs should be embraced, facilitated, and well-regulated. Hence, it is imperative that policymakers, AV-advocates and cynics, and every citizen read through this chapter and engage in productive discussions, hopefully leading to thoughtful policies, on how AVs would enhance and not threaten human development and life realities.

SECTION 4: INNOVATIONS IN BLOCKCHAIN TECHNOLOGY

Chapter 9: Cryptocurrencies and Contentious Politics

Cryptocurrency use is a remarkably complex issue for governments to address as it offers a catalog of both benefits and risks to governments and their citizens. Increased cryptocurrency use can bring unbanked individuals worldwide into the financial system and boost the income families in developing states receive from remittances. However, it also presents hostile states and terrorist groups with a new and effective tool for circumventing Western sanctions and collecting funds from global supporters. Although, initially, states generally chose to ignore cryptocurrency use, perceiving the costs of doing so to be negligible, these hostile actors' growing use of cryptocurrency has demonstrated that such a policy is no longer tenable. This chapter explores precisely what cryptocurrencies are and how they work, how hostile actors use them, and the costs and benefits of the different strategies governments have used to address cryptocurrency use thus far.

Chapter 10: A Proposal for Enhancing the Security and Privacy of Digital Currencies

This chapter addresses the issue of governments actively promoting and facilitating the development of digitalization of payments, including the creation of Central Bank Digital Currencies (CBDCs). Despite competition among countries to promote this development, concerns remain over security risks posed by hackers and privacy concerns with respect to authorized states. With respect to security, a proposed strategy involves the development of new quantum-resistant cryptography, as current digital currency cryptography will be vulnerable to quantum computers. As for privacy, the chapter analyzes the trade-offs between security and privacy in digital currency. Anti-money laundering and counter-financing of terrorism (AML/CFT) efforts require the collection of personal information, which can be detrimental to privacy. However, excessive restrictions on the collection of personal information may weaken AML/CFT efforts. The proposed solution is the use of SK4SC to anonymize KYC, which is essential for effective AML/CFT.

Chapter 11: Disruptive Technology of the 21st Century – How Blockchain Can Improve Electoral Voting

Estonia's e-Government has become a symbol for blockchain-backed technologies. With Estonia's success in e-Voting, other countries are realizing blockchain's potential as public spaces evolve digitally. The US, for example, launched the nation's first blockchain-backed e-Voting platform in 2018. Post-election audits were carried out on every ballot submitted via smartphones. Ultimately, recommendations to improve overall voter engagement were given. For example, programming the application to notify the voter on Election Day if they haven't voted yet. This chapter discusses the benefits blockchain technology can provide to secure e-Voting. It will educate and engage policy makers, electoral practitioners, and society about blockchain. It provides initiatives and political coalitions to uncloak blockchain to foster deeper understanding of the implications for e-Voting utilizing this disruptive innovation.

Chapter 12: The Vision of Freedom and the Potential of Blockchain

Freedom is one of the most important concepts in society. While modern law and the state have facilitated the development of freedom, there is a question if non-state actors can facilitate freedom. Blockchain technology has allowed for global, autonomous decentralization and this chapter argues that it can help foster freedom without a state-actor. The chapter analyzes the concept of freedom and how blockchain plays a crucial role in its development outside of government control. The chapter provides a case study of the cryptocurrency Libra and the governments' responses. As with many disruptive technologies, governments are reluctant to give up their power and in the case of Libra, governments sought to prevent its deployment.

SECTION 5: INNOVATIONS IN SOCIAL MEDIA AND DATA TECHNOLOGIES

Chapter 13: The Disruptive Nature of Short Video-Based Social Media Platforms in China – The Case of TikTok

The recent surge in popularity of short video-based social media platforms presents unique challenges, particularly in China where government regulation and market dynamics are in flux. This study delves into these challenges, with a focus on TikTok, highlighting the potential risks of a video-based platform that offers unlimited content and low usage thresholds. Such platforms are vulnerable to privacy breaches, fraud, and intellectual property infringement, and pose significant hurdles for governments seeking to censor content. The study specifically demonstrates the difficulty for the Chinese government to effectively regulate video content. The findings of this study have critical policy implications for both government regulation and industry practices concerning short video-based social media platforms.

Chapter 14: The Social Credit System in China

China's Social Credit System, a new generation of societal steering systems actively embraced and facilitated by the government, aims to secure the ruling party's power. This system combines all available data through a unique identifier, which uses citizens' identity card numbers or the social credit registration number for organizations. The system is mandatory and covers all the population without prior consent. Authorities use the combined data to create a social profile of each person. This social profile is translated into points with entries on positive or negative lists, visible on the Internet with full names. The system provides the government with opportunities to control social behavior through incentives and negative inducements. Additionally, the system seeks to provide security to the ruling government regime by tracking and controlling citizen movements. Last, the system is intended to benefit economic transactions, such as loans, through transparent creditworthiness indicators. The system can be seen as a threat to society increasing the risk of unintended developments and misuse. China's government is a pioneer in this field, and the social credit system is intended for export to other countries, with experiments of partial social credit systems currently being tested in several countries.

Chapter 15: Data Protection Burdens in the United States and China – A Focus on Privacy Data and Data Resources Security Governance Structure

The advent of technological innovation and its widespread application in various sectors have caused governments worldwide to respond proactively to the changing landscape. This institutional response is often aimed at managing technical resources effectively, focusing on privacy data generated by internet technology. Governments have recognized that privacy data is directly linked to national security and interests, prompting them to build governance structures to manage privacy data resources. These governance structures, however, vary considerably across countries, reflecting each nation's unique needs and circumstances. As the concept of digital assets and their institutional development are still evolving, it is critical to continuously track and study the relationship between technology, digital assets, and state reactions to understand these complexities.

Sam B. Edwards
Quinnipiac University, USA

James R. Masterson
Morehead State University, USA

Acknowledgment

This book would not have been possible without the help of our Editorial Advisory Board.

We also wanted to thank our student research assistants for their tireless efforts behind the scenes, Alexandra Luehrman and Brenna Murphy.

Section 1
Innovations in Public Services

Chapter 1
Preparing Public Managers to Deal With Disruptive Innovations Through Citizen Partnerships

John McNutt
https://orcid.org/0000-0002-5172-9163
University of Delaware, USA

Caroline Conrad Ban
Valparaiso University, USA

ABSTRACT

Developments in information and communication technology offer a rich set of possibilities for public managers that can make government more successful at serving their constituencies and achieving their missions. Sadly, barriers exist that can make this process difficult or impossible. Fortunately, this chapter provides public managers with an overview of new technology-based citizen partnerships that can facilitate the incorporation of disruptive technologies into government agencies. Developments like civic technology, data for good, public interest technology, and an emerging core of newer efforts offer public managers a way to address these barriers. The chapter discusses these developments, outlines competencies that managers will need to work with these groups, and explores ways that these competencies can be acquired by the public management workforce.

INTRODUCTION

These are stressful times to be a public administrator. Political polarization, economic dislocation, the pandemic and a host of other issues complicate what was already a very difficult job. Public support for government is at record lows in many nations (Pew Research Center, 2022) and the ability to recruit a public service workforce has become more difficult. At the same time, gaining cooperation from

DOI: 10.4018/978-1-6684-6429-8.ch001

community members is complicated. A changing population creates new demands on public services. Demographic shifts change the texture of the populace. These are truly hard times.

In addition to changes in the sociopolitical system, new technologies are emerging almost daily. Some of these technologies are threats—cybercrime, cyberterrorism, data theft, data breaches and the economic dislocations that are attended to by the technological aspects of the economy (Broadhead, 2018). The Gig or Sharing economy represents a regulatory challenge and remote work has implications for local tax bases and service arrangements. Long-established regulatory and taxation regimes are threatened by this technology and the impact on the labor force can be substantial.

At the same time, technology offers a range of attractive possibilities. Data and analytics can give governments the ability to see problems and solutions in new ways (Margetts, 2022; MacArthur, Dorobantu, Margetts, 2022; Kettle, 2016). It can also prevent fraud and theft. Technology can facilitate citizen involvement, coordination and service delivery. Technology can ease the task of responding to regulations and can facilitate the relationship between the business community and government. Emerging technology offers a wealth of new possibilities (Ahn & Chen, 2020).

Sadly, the government has serious issues in dealing with technology. These include both limitations in responding to threats and disadvantages in taking advantage of technological opportunities. These include a set of assumptions that often blind administrators in addressing new opportunities.

THE PROMISE AND PITFALLS OF GOVERNMENT TECHNOLOGY

Government has a long history of effort in information and communication technologies. Funding for early mainframe computers, the seed money for the Internet and a plethora of other technologies, all came from government sources. It is surprising that, at least in the United States, government has struggled to incorporate technology into its efforts (Anthopoulos, Reddick, Giannakidou & Mavridis, 2016; Harris, 2019).

The potential for effective technology use in government is both substantial and growing (Noveck, 2015). Movements like smart cities or intelligent communities and new applications in a variety of government functions promise exciting new horizons. Many of these visions are never realized. There are many reasons for this and many of them are difficult to fix.

Most government organizations are bureaucracies (Weber, 1947; Bennis, 1970; Olsen, 2008). Bureaucracies are highly structured organizations that control work with a hierarchical system and many rules and procedures (Denhardt, Denhardt, Aristigueta & Rawlings, 2018). The bureaucratic organizational form performs best in a stable environment and dealing with routine problems (Perrow, 1970). This approach works well in this context but struggles in highly volatile environments. Because of this, bureaucracies resist efforts to make internal changes and tend to react strongly to deviations in the environment. Changing the organizational structure is particularly difficult and there is characteristically a substantial amount of resistance to any fundamental change. While the effects of technology adoption on organizational structure and processes are complex (Cascio & Montealegre, 2016), government bureaucracy might be considered a more challenging situation.

Recruiting and retaining employees with technology expertise is more difficult in the public sector because of recruiting rules and levels of compensation. Private sector organizations can often offer far better salaries and benefits than public sector organizations (McGuinness & Schank, 2021). Since bureaucracies prize stability and resist change, innovators are not often supported or even marginalized by many government organizations and are frequently not rewarded by the system. Taking risks, while important to innovation, can be deadly to a government career. This makes those employees who are interested in advancement reluctant to engage in any behavior that may be criticized. This tends to encourage a leadership style that is more control-oriented and more conservative. This type of leadership approach is resisted by workers who have professional or technical training (Tannenbaum & Schmidt, 1973; Sorensen & Sorensen, 1974; Monteiro & Adler, 2022), which offers another barrier to hiring technology specialists. Despite the risks for employing technology in bureaucracies, the benefits in many cases outweigh the risks because of what can be accomplished. Still, public managers should be wise and aware of the impact that bureaucratic structures may have on limiting creative, technological solutions.

Procurement is a challenge for technology acquisition. Rules often make it more of a challenge to purchase technology. Procurement regulations do not always comport with the way technology providers prefer to provide technology. It is sometimes easier to hire external firms to develop technology projects than to try and do it with in-house expertise.

Technology can be expensive and limited tax revenues make acquiring certain types of technology difficult, especially for smaller governments. Many of the financing options available to other types of entities are not available to government and in a time when economic stress is part of the landscape many traditional types of funding are not available.

Political considerations are always part of any type of government activity and technology projects can easily be a political football. Even projects that are relatively uncontroversial can be political issues. A seemingly simple service like trash provision may have repercussions that win or lose elections.

Rising citizen expectations are another aspect of the political environment. Substantial experience with personal technology coupled with e-commerce has raised expectations of how effective and easy-to-use technology should be.

The movement toward Open Government and especially Open Data (Lathrop & Ruma, 2010; Harrison, Pardo & Cook, 2012), also creates dissonance between government and the public. Administrative organizational cultures often come into conflict with the values that promote Open Data and transparency. Bureaucracies can resist those initiatives that they find threatening. Data that is released can potentially embarrass public agencies and their employees.

All of these factors have led many governments to outsource responsibility for technology to outside consulting firms (Halegoua, 2020). This has at least three negative consequences. First, it is an expensive way to solve the problem. Second, it does nothing to build the government's capacity in technology. Finally, the project may not represent what the government needs, but rather what the firm can deliver.

In government, there are efforts to create a better environment for technology and innovation. Some of these have been a success while others have yet to prove their worth. A recent Government Accountability Office (GAO) report (Harris, 2019) stated: "The federal government plans to spend over $90 billion in fiscal year 2019 on IT. Even so, IT investments have too often failed or contributed little to mission-related outcomes." This paints a bleak picture but one that belies the substantial improvements that technology promises to bring to government organizations.

In light of all this, citizens have become interested in solving technology problems themselves. Groups of interested citizens, some with technology expertise, have begun to join the arena. These efforts can be both a blessing and a challenge.

NEW AND EMERGING DEVELOPMENTS MOVEMENTS

During the past few decades, technology-fueled citizen movements have addressed many of the problems that society faces (McNutt, 2018; Earl & Kimport, 2011; McNutt & Goldkind, 2021; Kleinschmidt, 2018) using a combination of organizing and technology. This exciting development has led to a series of revolutionary and evolutionary changes in communities throughout the world. In many ways, these movements reflect movements in other areas.

Table 1. Citizen technology movements

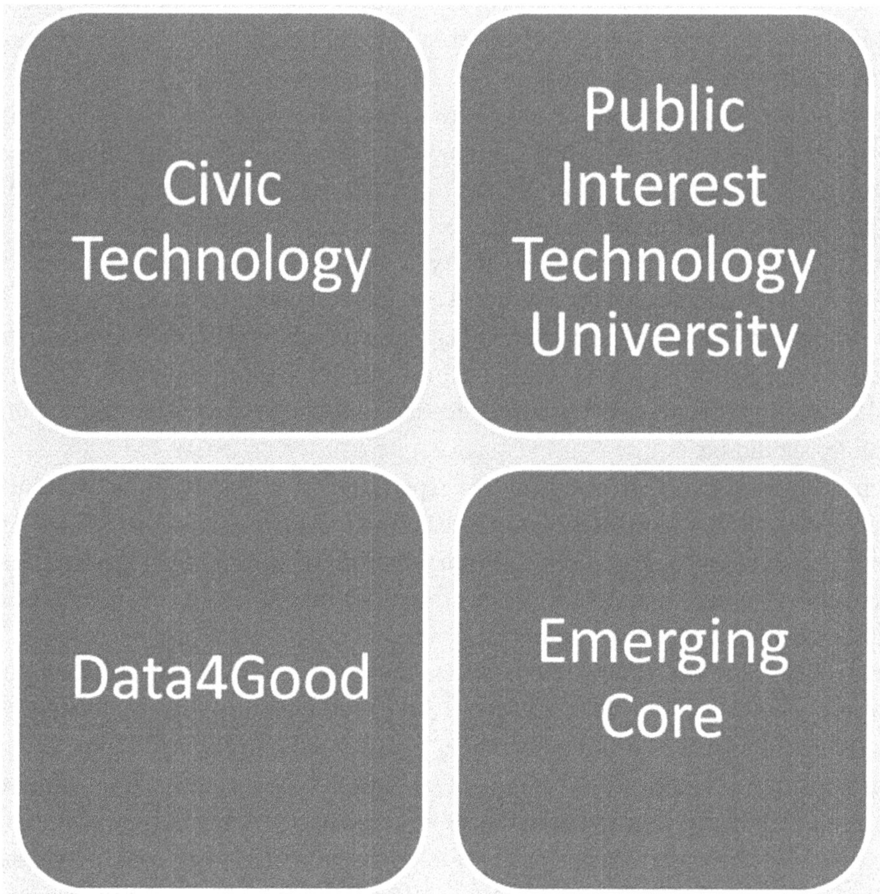

Civic Technology

Civic Technology is an international movement that brings technologists from the private sector into efforts to change government technology (Wilson & Chakraborty, 2019; McNutt, 2018; David, McNutt & Justice 2018; McNutt, Justice, Melitski, Ahn, Siddiqui, Carter & Kline, 2016; Living Cities, 2012; Gilman, 2016; Gordon & Mihailidis, 2016; Newsome, 2013; Schrock, 2016; 2018; Stephens, 2017; Suri, 2013; Baraniuk, 2013; Hébert, 2014; Patel, Sotsky, Gourley & Houghton, 2013). Civic Technology is defined as "…the use of digital technologies and social media for service provision, civic engagement, and data analysis – (it) has the potential to transform cities and the lives of their low-income residents" (Living Cities. 2012, 3).

Civic technology is generally outside government but focuses on government technology limitations. Some of these efforts focus on problems that are internal to government agencies while others aim at pressuring the government to improve performance. An example of the former is volunteer technologists developing technology to fulfill an organizational need, such as a new piece of software. This type of effort is the focus of many Code for America efforts, as well as Civic Hackathons and related programming. This represents a collaborative model of Civic Technology.

On the other hand, efforts such as *Fixmystreet* collect data on road hazards and report it back to the local government (Suri, 2013). If nothing happens, information is made public, thus embarrassing local governments. Public pressure often changes local government behaviors and might be combined with social action strategies. This represents a more confrontational approach to civic technology.

McNutt and colleagues (McNutt, Justice, Melitski, Ahn, Siddiqui, Carter & Kline, 2016) argue that there are three components to Civic Technology: (1) Open Civic Data, (2) Civic Technology Applications and (3) Civic Technology Practices. There are substantial differences in how these elements are combined in practice. While the foundation of Civic Technology is Open Civic Data, civic technology organizations employ a set of Information and Communication Technologies (ICT) and a series of practices that are derived from collaboration and crowdsourcing (Powell, 2016; McNutt & Justice, 2016; Irani, 2015; Hou & Lampe, 2017, June). These include Code for America Fellowships, Hackathons, small grassroots groups like Code for America Brigades, and a variety of new developments.

The 311 system works well with Civic Technology. 311 is a phone number that provides access to city event information and non-emergency services, but this is not available in smaller communities. Yet, smaller communities still can create online, real-time directories with little effort. In 2022 in Valparaiso, Indiana, the police social worker convened a group of community agencies together to create a low-budget, real-time directory of social service agencies. Inspired by an insufficient gap in resources and communication across agencies, this city is working to go beyond the typically printed resource brochures to improve communication and information sharing across agencies.

Civic Technology is a viable force in reclaiming public technology. Programs have been created throughout the nation (Turkel, 2021) and promise to expand. The international reach of Civic Technology is considerable.

Data4Good

The Data4Good movement emerged from the growing application of data science techniques to critical social problems such as poverty, disease and human rights violations (Howson Beyer, Idoine, Jones, 2018; Bull, Slavitt & Lipstein, 2016; Pfaff, 2015). Data science employs sophisticated analytics, high technology and massive amounts of data to address issues in new and important ways (Kelleher & Tierney, 2018). Data Science offers a wide range of capabilities that can assist government target services, uncover fraud and malfeasance and relate to various constituencies. Data science has become a component of political campaigning and political action.

Unfortunately, the cost of data science professionals is often prohibitive for smaller governments. This means that the cost of developing such a capacity is difficult. This is an opportunity lost because sophisticated analytics can facilitate the development of evidence-based policy and practice (Kettle, 2017).

Data4Good is one strategy to address the mismatch between the need for data science capability and the availability of personnel and other resources. Like Civic Technology organizations, Data4Good organizations use volunteers with substantial training in high-demand fields. Many of these volunteers are currently employed in data science jobs. Volunteers benefit from the ability to pursue a cause, contribute and even the chance to do more meaningful work or develop new skills (Kinsella, 2021; see also Shirky, 2010). Employers benefit from the volunteer's participation and in some cases are active participants in the process.

Data4Good requires a network of interacting organizations that work to connect the volunteer with the project (McNutt & Goldkind, 2021). These organizations include (1) Employers, (2) Placement Organizations, (3) Data Providers, and (4) Domain Organizations. Employers are nominally the organizations that employ the volunteers, but it might be more. In some cases, Employers create their own efforts where entire offices of data scientists have volunteered their time. Placement Organizations (Data Kind (www.datakind.org) is a significant example), place potential volunteers with organizations that are dealing with problems on the ground. These Domain Organizations are government or nonprofit entities that are directly engaged with the problem to be solved. Finally, there are Data Providers, which might be government, nonprofit or commercial suppliers of data. The latter is becoming more available through Data Philanthropy efforts (McKeever, Greene, MacDonald, Tatian, & Jones, 2018) and the growth of Data Collaboratives (Verhulst & Sangokoya, 2015; Verhulst, 2021) that facilitate connections between data donors and data users.

Like Civic Technology, there are short term activities such as Data Dives (Pfaff, 2015) that are similar to Hackathons (McNutt & Justice, 2018) but focused in a different way. They bring organizations with data issues together with social change organizations.

Data4Good shares processes with civic technology and has commonalities with Data Activism and the Data Justice movement. There are important differences as well, but all of these are growing movements.

Public Interest Technology for Universities

Public Interest Technology is a movement to encourage technologists to pursue careers in public service, much like public interest lawyers (McGuinness & Schank, 2021; Lusk, 2022; Singer, 2019; Crawford, 2018). This is a multidirectional effort and what is particularly relevant to the discussion here is the Public Interest Technology for Universities (PITU, https://pitcases.org/). This program is designed to prepare students to work in public interest information technology. The program is led by the New America Foundation with support from the Ford Foundation and others. From an original group of 29 universities, the network has grown every year.

Public Interest Technology programs exist at universities throughout the nation. These programs not only train students to work in the public sector, but also conduct research and service projects in local communities. This means that local governments can have access to well-trained students and other resources from the host university. This is also true of nonprofit organizations and collaborations.

Emerging Core

Several efforts are similar to the three efforts previously discussed, which could be considered to be the Emerging Core of new technology-community collaboration. One such effort is Participatory Budgeting (Aziz & Shah, 2021; Gilman, 2016). In this type of effort, residents are empowered to vote on a range of public expenditures and those choices are obligatory. While technology is not actually essential to this activity, it certainly facilitates the process. Participedia.org explains this process in more depth. There are several efforts to promote the use of technology and data that are culture or gender inclusive.

Movements like Data Feminism (D'ignazio & Klein, 2020) and Black Software (McIlwain, 2019) offer good possibilities for relating technology in local government to local constituencies and reduce the potential of bias in data-based decision making. Many of these movements have local affiliates or chapters.

Citizen Science (Bonney, Shirk, Phillips, Wiggins, Ballard, Miller-Rushing & Parrish, 2014) is another force that has a long history (especially in areas like astronomy and zoology) and is expanding. This movement trains citizens to conduct data collection on natural phenomena. Citizen Scientists have expanded their role in social problem areas like poverty, climate change and inequality. Finally, Data Rescue Efforts (Hegstad, 2017) are organized to preserve government data that is in danger of being removed for political reasons or deleted for other motives.

A less visible component is Hackers including Good or Ethical Hackers. Groups like Anonymous are also part of this component (Goode, 2015; Alexopoulou & Pavli, 2021). This element works to make government more responsive but often does so beneath the radar. Their activities are covert in many cases. Organizations like WikiLeaks also have a role in this limited space. This component is a wild card in the equation because what is known is limited by design.

What all of these technological movements have in common is that (1) they exist largely outside of government and (2) they depend on a rising tide of collaborative action. Participants are people with resources and expertise who want to improve the quality of their government. This final group would be difficult to engage for public managers.

This is a very quick overview that suggests that there are substantial resources for implementing new technologies that are available to public managers. In many cases, these resources already exist within their communities and are waiting to be used.

ISSUES, CONTROVERSIES, PROBLEMS

While there are substantial and important resources, using them might be more difficult than one may think. Political considerations are critical in many government operational decisions. This includes commonplace politics but is beginning to incorporate culture "wars" or identity politics. In the United States, at least, this appears to be worsening. Technology can easily become part of this controversy. Outside groups can have a role in these political deliberations, especially if they have other agendas. For example, an application that facilitates access to woman's health care can easily become part of the abortion debate.

Government Organization Culture (Zhang & Feeney, 2020) is also a potential source of controversy. Bureaucratic resistance to new technology is certainly possible and technology solutions can affect workflow and other organizational factors. New technology can fall afoul of administrative policies. New solutions, generated from outside, can threaten the work life of current government technology staff. Technology solutions can illuminate other types of administrative limitations, either intentionally or accidentally. All of these issues can threaten the stability of bureaucratic organizations and, in turn, create issues for leadership and ultimately, employee careers.

The solutions proposed can conflict with the worldview of government computing based on earlier technology solutions. Newer technologies often struggle with older models of technology based on far less capable applications and machinery.

Finally, government priorities can be a problem. There can be a mismatch between outside groups and government. What is interesting or important to outside groups can conflict with other priorities in government.

Despite these limitations, citizen technology efforts have much to offer to government. To achieve success, skillful management is needed. This requires a series of skills or competencies that can address the issues that this type of partnership is likely to face.

SOLUTIONS AND RECOMMENDED COMPETENCIES

Administrative life is certainly simpler when the manager controls all of the possibilities, but that world has long ceased to exist. Working through community partnerships rather than direct administrative action is more complicated and requires different competencies.

Working with these efforts to utilize disruptive technology will require new skills for many public managers. Those who are used to a traditional command and control model of management will have a very hard time dealing with these arrangements as either a partner or competitor. A more expansive view of the future, as well as an enhanced skillset, will be essential. We propose a set of management competencies that can support managers who want to explore new citizen partnerships, which include: 1) Understanding Technology and its Possibilities, 2) Understanding Collaborative Models, 3) Working with Community Groups, 4) Sharing Expertise and Data, 5) Creating Strong Partnerships, 6) Data Technology and Ethics and 7) Engagement with Diversity, Equity, and Inclusion. Managers will need opportunities to acquire these skills. Whether through continuing education, basic professional education and or networks of mentorship.

Understanding Technology and Its Possibilities

Information and Communications Technology has evolved substantially over the past five decades. High Performance Computing, Cloud Computing, Artificial Intelligence, the Internet of things (IoT) and Virtual Reality are just some of the movements that are reshaping technology in important ways. These new developments often go far beyond what government computing has traditionally encompassed. It is therefore important that public managers understand the possibilities.

Table 2. Essential technology needed for public managers

<table>
<tr><td>High Performance Computing
Cloud Storage and Computing
The Internet of Things (IoT)
Machine Learning, Algorithms and Artificial Intelligence
Analytics and Visualization</td><td>Collaborative Technology
Customer Relationship Management Software
Enterprise Management Systems
Social Media
Networking
Cybersecurity</td></tr>
</table>

It has not been long since technology meant large mainframe computers with unique operating systems and limited connectivity. The technology that is available today bears little resemblance to that which existed in the past. Public managers must develop a mindset and a knowledge base that prepares them to understand new technology and the possibilities it has to offer (Klievink, van der Voort, & Veeneman, 2018, Guenduez, Mettler & Schedler, 2020).

To understand modern technology in government, public managers need a basic knowledge of contemporary processing, networking, storage, analytics and collaboration. These new tools are in many ways different from earlier technology that defined government computing. Public managers would also benefit from an understanding of contemporary e-government and e-participation ideas (Pappel, Draheim, & Janssen, 2020) and smart city or intelligent community theory (Halegoua, 2020; Meijer & Bolívar,2016a; Meijer, Gil-Garcia & Bolívar, 2016; Meijer & Bolívar, 2016b). In this instance, it is also critical for public managers to understand the nature of the new approaches to Citizen Technology Initiatives. Having a firm understanding of these concepts and the available research will facilitate understanding and communication.

While it is unlikely that public managers will need more than a general understanding of all of these areas, they must know what options are available. Being conversational with all of these concepts means that public managers are more likely to find common ground when working with citizen technology groups.

Understanding Collaborative Models

Collaboration has always been an important factor in public management, both the growth of collaborative technology, along with new types of co-creation (Jemielniak & Przegalinska, 2020). These developments may include self-organizing, the Open Source Movement, the development of co-creational efforts with other organizations, brief arrangements like Hackathons (Irani, 2015), Data Dives, and more in-depth efforts such as Citizen Science and episodic volunteering.

Much of contemporary technology has a collaborative orientation. Social Media/Web 2.0 provides substantial support for communication, networking and the pooling of collective intelligence (O'Reilly, 2005; 2007; Guo & Saxton, 2020; Masterson, 2015). It makes organizing and problem solving possible at a scale that the community organizers of an earlier generation would find unbelievable.

Collaboration, co-creation, crowdfunding, and crowdsourcing are important developments in modern social change tools. They are also important parts of the approaches that we discuss here.

Working With Community Groups

Regular contact with non-profit and citizen stakeholders, casual community conversations and public listening sessions are essential to ensure that the flow of data and information is skillfully managed (McNutt & Goldkind, 2021). Engagement with the key formal and informal stakeholders and influencers in the community can help the manager stay knowledgeable about citizen concerns, passions or possibilities ahead of more formal public meetings. Becoming a trustworthy and reliable partner for community groups can lead to better working relations and may even lessen adversarial interactions, which are likely to come up in any community.

There is extensive literature in community development, community organization and community planning that is relevant here. Community planners and social workers can be important resources.

Sharing Expertise and Data

Sharing expertise and data require a set of skills that public managers often struggle to develop. These skills include relationship development, communications, dealing with issues of privacy and ownership and working with people whose perspective is outside the public managers' comfort zone. Public managers bring a great deal of domain-level expertise to these negotiations that must be respected.

Sharing data is often straightforward and uncomplicated. In cases where personal information is involved, however, there may be serious political and legal issues. Navigating such issues require not only political skill but a knowledge of regulations and laws that govern the security of data. The Data Justice Movement has focused attention on the potential harm that can come from the release of data (McNutt, 2019). Data can also have political ramifications. Battles over data occurred during the COVID-19 pandemic and over environmental issues. Data, if improperly handled, can also disadvantage marginalized groups (Edwards, 2019).

Sharing expertise requires interpersonal skills and an openness to different experiences and ways of knowing. Understanding the new technology will help facilitate mutual understanding, but observation, listening and communication skills are important. Managers will benefit from three simple communication skills: 1) reflecting—or simply restating what a community member said to establish shared meaning or provide an opportunity for clarification, 2) encouraging—using positive body language like gesturing and nodding, or overt verbal invitations that ask participants to share, even if perspectives are controversial and 3) summarizing—or providing a recap of key findings and decisions at the end of the public session (Ivey, Ivey, Zalaquett, 2018). Seemingly simple communication skills help the manager check the pulse of the community, identify areas of agreement or dissent and establish a baseline of understanding so that the manager can then more effectively and efficiently chart a path forward.

Creating Strong Partnerships

Community partnerships are critical to making these types of programs effective (McNutt & Goldkind, 2022; McNutt, Brainard, Zeng & Kovacic, 2016). Many of these efforts involve multiple organizations and organizational types including nonprofits, voluntary associations, informal community groups, political parties, interest groups and individual community members. Intergovernmental partnerships might also be involved.

Assessing what groups can be found in the community, which are possible partners and what their likely positions are requires community mapping (Kretzmann & McKnight, 1993). Formal groups are generally easier to locate than informal groups. Community groups often differ in their views about what is possible or desirable in a community. Knowing this will make the process easier to navigate. John Kretzmann and John McKnight (1993) extensively explain the concept of community asset mapping, especially mapping informal networks, and many of these tools and templates can be found and adapted at DePaul University (See: https://resources.depaul.edu/abcd-institute/resources/Pages/tool-kit.aspx).

Formal partners must be engaged and involved in the planning process. There needs to be articulation agreements (Memorandums and formal contracts) that cement their involvement in the process.

Informal partners must also be engaged in planning the effort. This involvement must be substantive rather than symbolic. While some groups might be critical of the effort, blindsiding them is rarely a good idea.

Data Technology and Ethics

Technology can be a wonderful source of progress. It can also be a weapon of considerable power. Many people fear the downside of technology and how data and surveillance can affect their lives. They are not wrong. This makes having ethical principles important. It also means that understanding the risks is necessary.

There are risks from technology, risks from data and combined risks from data and technology. Data on human subjects is particularly important especially when it contains Personal Identifiers (PID) or Personally Identifiable Information (PII). Laws and regulations often protect this information and require measures to insure privacy including physical security and encryption. This privacy can be compromised by accidental disclosure and criminal activity. The emerging idea of Data Stewardship is an important one moving forward. The US Census Bureau (2021) states that:

Our Data Stewardship program includes ongoing training, accountability and the lifetime oath each of us swears to keep your information confidential. Our Data Stewardship Executive Policy Committee oversees these activities, setting the rules and policies all staff must follow.

Personal data should only be collected when absolutely necessary. All essential steps to maintain the security of that data should be used. Citizens may be concerned with privacy, impingement on freedom and biases or profiling that could be collected without a critical lens (see O'Neill, 2016). Public managers should be aware of the risks and the benefits when weighing the use of technology. And yet, always determining that the potential costs outweigh the benefits is unwise. The protection of personal and organizational policy in a digital age is a serious issue with extensive consequences (Edwards, 2019).

How that data is used can also be an issue. The creation of algorithms based on the data available and reflecting management ideas can result in technology that oppresses people who are poor or powerless (Eubanks, 2018). Calls for algorithmic justice are important warnings. When is predictive policing racial profiling? What are the implications of facial recognition programs that fail to accurately identify people of color?

Public managers have to combat the real pressures of time management and lack of confidence, which might lead them to the conclusion that technology is unfamiliar, potentially controversial and therefore too much work. Public managers need to intelligently consider these real fears, and yet, technology and data present real ways to be more responsive to citizen needs and better stewards of our communities. It is necessary to consider new ways to engage citizens in the formation of good public policy or predict threats to one's community. These reasons alone are important enough to pursue technological possibilities.

Engagement With Diversity, Equity, and Inclusion

A key responsibility of a public manager is paying attention to access, diversity, inequality and possibilities for change. Public managers organize people and the services that deliver public services. Their work is on behalf of all people, not just people who are rich and powerful. Attention to equity, inclusion and diversity is often the difference between a successful administration and one that damages its community.

Less than 60% of the US population was White in the 2020 Census (Jensen, Jones, Rabe, Pratt, Medina, Orozco & Spell, 2021). In several states the majority group was Latino and in the District of Columbia, African American. The United States is a diverse nation in terms of ethnicity, gender, sexual orientation and other social identities. Add in economic and age diversity, and the population is even more diverse. Many communities have substantial immigrant populations and language diversity is often pronounced.

Conflict over law enforcement and other government functions has characterized society over the past few decades, rising to new heights during the pandemic. This has created divides between local governments and important segments of their communities. Technology plays a role in the development of surveillance technology, predictive policing and body cameras. McIlwain, C. D. (2019) and D'ignazio and Klein (2020) have created models that reflect diverse orientations to technology and data. This can help public managers understand the sources of potential conflict with different parts of the community and especially oppressed groups.

Related to informing policy decisions, the National Conference of State Legislatures (www.ncsl.org) provides a great resource for gathering policy ideas and looking at the impact that policies can have on mitigating racial, economic, environmental and other social disparities.

TRAINING PUBLIC MANAGERS TO WORK WITH CITIZEN PARTNERSHIPS

Addressing the competencies that we have identified will require a system to support knowledge acquisition and skill building. Given the range of public managers in terms of age, education, and experience, it will take a variety of educational approaches.

Degrees and Traditional Instruction are likely early vectors. These would consist of degree programs in a range of fields but most likely public administration, public policy, urban planning and administrative tracks in public health, social work, and nonprofit studies. Several articles in the last few years have documented the marginal place of technology in some of these fields (Mauldin, 2016; Manoharan & McQuiston, 2016; McQuiston, & Manoharan, 2017; Shark, 2018). This means that action is needed to rebuild the technology curriculum so that public managers enter the field prepared.

The Public Interest Technology for Universities effort is addressing this issue and a recent change in the Network of Schools of Public Policy, Affairs, and Administration (NASPAA) standards will encourage additional curricula. Recent work has argued for more data science skills (Overton, & Kleinschmit, 2022) and Artificial Intelligence content (Ahn & Chen, 2020, June).

While it is essential for pre-service public managers to have a grounding in technology, it does little or nothing for existing managers. New avenues to supply that content are needed.

Alternative Credentials (Certificates, boot camps, short courses and other approaches) can meet the needs of already credentialed managers (Brown & Kurzweil, 2017). It could also supplement traditional instruction in programs that have not embraced new developments. These approaches can take advantage the revolution in instructional technology that is developing.

Mentorship Networks are another way to build management competencies. These could also supplement the other approaches. Technology makes this approach more formidable and easier to create. Professional associations, such as the International City Manager's Association (ICMA), can also be useful in this regard.

Creative solutions are essential. Public managers need the knowledge and skills defined by the competencies. People at different stages in their careers will need diverse delivery strategies.

CONCLUSION

Public Managers should consider that these movements will emerge in their communities whether they are prepared or not (McNutt & Goldkind, 2021). It is also likely that new movements will develop. How managers deal with these new forces in their communities will be an important part of government success.

Technology can offer wonderful possibilities to the administration of government services. While we have seen some of these benefits, more is sure to come. At the same time, a failed program or one that has important negative consequences may be difficult to live with. Citizen technology-based partnerships can make a respectable outcome more probable.

Citizen Technology-based partnerships can offer many new and exciting possibilities. This is a resource that public managers cannot afford to ignore. In many ways, more citizen involvement is a harbinger of things to come. In others, it is a call to the strength of past successes. Communities deserve the broad based advantages that these strategies can offer.

These are exciting times to be working in public management. While the challenges are great and threats substantial, there are many opportunities to affect meaningful change. The Citizen Initiated Technology supported efforts that are discussed here can make a substantial contribution to a bright future of partnership between government and community.

REFERENCES

Ahn, M. J., & Chen, Y. C. (2020, June). Artificial intelligence in government: potentials, challenges, and the future. In *The 21st Annual International Conference on Digital Government Research* (pp. 243-252). 10.1145/3396956.3398260

Alexopoulou, S., & Pavli, A. (2021). 'Beneath This Mask There is More Than Flesh, Beneath This Mask There is an Idea': Anonymous as the (Super) heroes of the Internet? *International Journal for the Semiotics of Law-Revue internationale de Sémiotique juridique, 34*(1), 237-264.

Anthopoulos, L., Reddick, C. G., Giannakidou, I., & Mavridis, N. (2016). Why E-Government Projects Fail? An Analysis of the Healthcare.gov Website. *Government Information Quarterly*, *33*(1), 161–173. doi:10.1016/j.giq.2015.07.003

Aziz, H., & Shah, N. (2021). Participatory budgeting: Models and approaches. In T. Rudas & G. Péli (Eds.), *Pathways Between Social Science and Computational Social Science* (pp. 215–236). Springer. doi:10.1007/978-3-030-54936-7_10

Baraniuk, C. (2013). The civic hackers reshaping your government. *New Scientist*, *218*(2923), 36–39. doi:10.1016/S0262-4079(13)61625-5

Bennis, W. G. (Ed.). (1970). *American bureaucracy*. Transaction Publishers.

Bonney, R., Shirk, J. L., Phillips, T. B., Wiggins, A., Ballard, H. L., Miller-Rushing, A. J., & Parrish, J. K. (2014). Next steps for citizen science. *Science*, *343*(6178), 1436–1437. doi:10.1126cience.1251554 PMID:24675940

Brainard, L., Boland, K., & McNutt, J. G. (2018). The Advent of Technology Enhanced Leaderless Transnational Social Movement Organizations: Implications for Transnational Advocacy. (185-). In J. G. McNutt (Ed.), *Technology, Activism and Social Justice in a Digital Age*. Oxford University Press.

Broadhead, S. (2018). The contemporary cybercrime ecosystem: A multi-disciplinary overview of the state of affairs and developments. *Computer Law & Security Report*, *34*(6), 1180–1196. doi:10.1016/j.clsr.2018.08.005

Brown, J., & Kurzweil, M. (2017). *The complex universe of alternative postsecondary credentials and pathways*. American Academy of Arts and Sciences. doi:10.1146/annurev-orgpsych-041015-062352

Cascio, W. F. & Montealegre, R. (2016). How technology is changing work and organizations. *Annual Review of Organizational Psychology and Organizational Behavior*, *3*(1), 349–375. doi:10.1146/annurev-orgpsych-041015-062352

Crawford, S. (2018). Why Universities Need 'Public Interest Technology' Courses. *Wired Magazine*. https://www.wired.com/story/universities-public-interest-technology-courses-programs/

D'ignazio, C., & Klein, L. F. (2020). *Data feminism*. MIT Press. doi:10.7551/mitpress/11805.001.0001

David, N., Justice, J. B., & McNutt, J. G. (2015). Smart Cities are Transparent Cities: The Role of Fiscal Transparency in Smart City Governance. In P. R. B. Manuel (Ed.), *Transforming City Governments for successful Smart Cities. Empirical Experiences*. Springer. doi:10.1007/978-3-319-03167-5_5

David, N., Justice, J. B., & McNutt, J. G. (2018). Smart Cities, Transparency, Civic Technology and Reinventing Government. In Smart Technologies for Smart Governments. Transparency, Efficiency and Organizational Issues. Springer.

Dawes, S. S., & Helbig, N. (2010, August). Information strategies for open government: Challenges and prospects for deriving public value from government transparency. In *International Conference on Electronic Government* (pp. 50-60). Springer. 10.1007/978-3-642-14799-9_5

Dawes, S. S., Vidiasova, L., & Parkhimovich, O. (2016). Planning and designing open government data programs: An ecosystem approach. *Government Information Quarterly*, *33*(1), 15–27. doi:10.1016/j.giq.2016.01.003

Denhardt, R. B., Denhardt, J. V., Aristigueta, M. P., & Rawlings, K. C. (2018). *Managing human behavior in public and nonprofit organizations*. CQ Press.

Earl, J., & Kimport, K. (2011). *Digitally enabled social change*. MIT Press. doi:10.7551/mitpress/9780262015103.001.0001

Edwards, S. B., III. (2019). The Right to Privacy Is Dying: Technology Is Killing It and We Are Letting It Happen. In Cyber Law, Privacy, and Security: Concepts, Methodologies, Tools, and Applications (pp. 830-853). IGI Global.

Eubanks, V. (2018). *Automating inequality: How high-tech tools profile, police, and punish the poor*. St. Martin's Press.

Gilman, H. R. (2016). *Participatory budgeting and civic tech: The revival of citizen engagement*. Georgetown University Press.

Goode, L. (2015). Anonymous and the political ethos of hacktivism. *Popular Communication, 13*(1), 74–86. doi:10.1080/15405702.2014.978000

Gordon, E., & Mihailidis, P. (Eds.). (2016). *Civic media: Technology, design, practice*. MIT Press. doi:10.7551/mitpress/9970.001.0001

Guenduez, A. A., Mettler, T. & Schedler, K. (2020). Citizen Participation in Smart Government: A Conceptual Model and Two IoT Case Studies. *Public Administration and Information Technology,* 189-209.

Harrison, T. M., Pardo, T. A., & Cook, M. (2012). Creating open government ecosystems: A research and development agenda. *Future Internet, 4*(4), 900–928. doi:10.3390/fi4040900

Hébert, M. K. (2014). Come Hack with Me: Adapting Anthropological Training to Work in Civic Innovation. *Practical Anthropology, 36*(2), 32–36. doi:10.17730/praa.36.2.405j1uvvn8584768

Hegstad, M. (2017). Networks Seek To Archive, Protect Environmental Information From Trump. *Inside EPA's Water Policy Report, 26*(3), 25-27.

Hou, Y., & Lampe, C. (2017, June). Sustainable hacking: characteristics of the design and adoption of civic hacking projects. In *Proceedings of the 8th International Conference on Communities and Technologies* (pp. 125-134). ACM. 10.1145/3083671.3083706

Howson, C., Beyer, M. A., Idoine, C. J., & Jones, L. C. (2018). *How to Use Data4Good to Impact Society*. New York: Gartner. https://www.gartner.com/doc/3880666/use-data-good-impact-society

Irani, L. (2015). Hackathons and the Making of Entrepreneurial Citizenship. *Science, Technology & Human Values, 40*(5), 799–824. doi:10.1177/0162243915578486

Ivey, A. E., Ivey, M. B., & Zalaquett, C. P. (2018). *Intentional Interviewing and Counseling: Facilitating Client Development in a Multicultural Society*. Brooks Cole.

Jemielniak, D., & Przegalinska, A. (2020). *Collaborative Society*. MIT Press. doi:10.7551/mitpress/11587.001.0001

Jensen, E., Jones, N., Rabe, M., Pratt, B., Medina, L., Orozco, K., & Spell, L. (2021). *The chance that two people chosen at random are of different race or ethnicity groups has increased since 2010*. https://www. census. gov/library/stories/2021/08/2020-united-states-population-more-racially-ethnically-diverse-than-2010

Justice, J., McNutt, J. G., Melitski, J., Ahn, M., David, N., Siddiqui, S., & Ronquillo, J. C. (2018). The Civic Technology Movement: Implications for Nonprofit Theory and Practice. (89). In J. G. McNutt (Ed.), *Technology, Activism and Social Justice in a Digital Age*. Oxford University Press.

Kelleher, J. D., & Tierney, B. (2018). *Data science*. MIT Press. doi:10.7551/mitpress/11140.001.0001

Kettl, D. F. (2017). *Little bites of big data for public policy*. CQ Press/Sage Publishing.

King, S. F., & Brown, P. (2007, December). Fix my street or else: using the internet to voice local public service concerns. In *Proceedings of the 1st international conference on Theory and practice of electronic governance*. (pp. 72-80). ACM. 10.1145/1328057.1328076

Kinsella, B. (2021). Data Science for Social Good Volunteer Motivations and Limitations: An Exploratory Survey. *Chance*, *34*(3), W86–W95. doi:10.1080/09332480.2021.1981055

Klievink, B., van der Voort, H., & Veeneman, W. (2018). Creating value through data collaboratives: Balancing innovation and control. *Information Polity*, *23*(4), 379–397. doi:10.3233/IP-180070

Klineschmit, S. (2018). Policing Digital Sanctuaries: Exploring Environmental Advocacy through Technology-Enabled Monitoring and Enforcement Network Organizations. In Technology, Activism and Social Justice in a Digital Age. Oxford University Press.

Kretzmann, J., & McKnight, J. (1993). Building communities from the inside out: A path toward finding and mobilizing a community's assets. Institute for Policy Research.

Kullenberg, C., & Kasperowski, D. (2016). What Is Citizen Science? – A Scientometric Meta-Analysis. *PLoS One*, *11*(1), e0147152. doi:10.1371/journal.pone.0147152 PMID:26766577

Kumar, A. (2019). Citizen-centric model of governmental entrepreneurship: Transforming public service management for the empowerment of marginalized women. *Transforming Government: People, Process, and Policy*, *13*(1), 62–75.

Lathrop, D., & Ruma, L. (Eds.). (2010). Open Government: Collaboration, Transparency, and Participation in Practice. O'Reilly.

Lee, M. Y., & Edmondson, A. C. (2017). Self-managing organizations: Exploring the limits of less-hierarchical organizing. *Research in Organizational Behavior*, *37*, 35–58. doi:10.1016/j.riob.2017.10.002

Lusk, K. (2022). *Public Interest Technology University Network: understanding the state of the field.* Boston University. https://open.bu.edu/handle/2144/44469

MacArthur, B. D., Dorobantu, C. L., & Margetts, H. Z. (2022, August). Resilient government requires data science reform. *Nature Human Behaviour*, *6*(8), 1035–1037. doi:10.103841562-022-01423-6 PMID:35941235

Manoharan, A. P., & McQuiston, J. (2016). Technology and pedagogy: Information technology competencies in public administration and public policy programs. *Journal of Public Affairs Education*, *22*(2), 175–186. doi:10.1080/15236803.2016.12002239

Masterson, J. R. (2015). Social media in China: a double-edged sword. In S. B. Edwards & D. Santos (Eds.), *Revolutionizing the interaction between state and citizens through digital communications* (pp. 238–263). IGI Global. doi:10.4018/978-1-4666-6292-6.ch013

Mauldin, M. D. (2016). No MPA left behind: A review of information technology in the master of public administration curriculum. *Journal of Public Affairs Education*, *22*(2), 187–192. doi:10.1080/15236803.2016.12002240

McGuinness, T. D., & Schank, H. (2021). *Power to the public: The promise of public interest technology*. Princeton University Press. doi:10.2307/j.ctv18b5dbz

McIlwain, C. D. (2019). *Black software: The Internet and racial justice, from the AfroNet to Black Lives Matter*. Oxford University Press.

McKeever, B., Greene, S., MacDonald, G., Tatian, P., & Jones, D. (2018). *Data Philanthropy: Unlocking the Power of Private Data for Public Good*. Urban Institute.

McNutt, J. G. (Ed.). (2018). Technology, Activism and Social Justice in a Digital Age. Oxford University Press.

McNutt, J. G. (2019). Data justice and international development: An ethical imperative for policy and community practice. In *The Routledge Handbook of Social Work Ethics and Values* (pp. 280–286). Routledge. doi:10.4324/9780429438813-36

McNutt, J. G., Brainard, L., Zeng, Y., & Kovacic, P. (2016). Information and Technology in and for Associations and Volunteering. In Palgrave Handbook of Volunteering and Nonprofit Associations. Palgrave Macmillan.

McNutt, J. G., & Goldkind, L. (2022). Intersectoral Collaboration and Social Justice Concerns in Artificial Intelligence Implementation at the Community Level. In J. Saura & F. Debasa (Eds.), *Handbook of Research on Artificial Intelligence in Government Practices and Processes* (pp. 38–54). IGI Global. doi:10.4018/978-1-7998-9609-8.ch003

McNutt, J. G., & Justice, J. B. (2016, September). *Predicting civic hackathons in local communities: Perspectives from social capital and creative class theory*. Presented at the *ISTR Conference*, Stockholm, Sweden.

McNutt, J. G., Justice, J. B., Melitski, M. J., Ahn, M. J., Siddiqui, S., Carter, D. T., & Kline, A. D. (2016). The diffusion of civic technology and open government in the United States. *Information Polity*, *21*(2), 153–170. doi:10.3233/IP-160385

McQuiston, J. M., & Manoharan, A. P. (2017). Developing E-Government coursework through the NASPAA competencies framework. *Teaching Public Administration*, *35*(2), 173–189. doi:10.1177/0144739417690582

Meijer, A., & Bolívar, M. P. R. (2016). Governing the smart city: A review of the literature on smart urban governance. *International Review of Administrative Sciences*, *82*(2), 392–408. doi:10.1177/0020852314564308

Meijer, A., & Bolívar, M. P. R. (2016). Governing the smart city: A review of the literature on smart urban governance. *International Review of Administrative Sciences*, *82*(2), 392–408. doi:10.1177/0020852314564308

Meijer, A. J., Gil-Garcia, J. R., & Bolívar, M. P. R. (2016). Smart city research: Contextual conditions, governance models, and public value assessment. *Social Science Computer Review*, *34*(6), 647–656. doi:10.1177/0894439315618890

Monteiro, P., & Adler, P. S. (2022). Bureaucracy for the 21st century: Clarifying and expanding our view of bureaucratic organization. *The Academy of Management Annals*, *16*(2), 427–475. doi:10.5465/annals.2019.0059

Newsome, G. (2013). *Citizenville*. The Penguin Press.

Noveck, B. S. (2015). *Smart citizens, smarter state: The technologies of expertise and the future of governing*. Harvard University Press. doi:10.4159/9780674915435

O'Neil, C. (2016). *Weapons of math destruction: How big data increases inequality and threatens democracy*. Broadway Books.

Olsen, J. P. (2008). The ups and downs of bureaucratic organization. *Annual Review of Political Science*, *11*(1), 13–37. doi:10.1146/annurev.polisci.11.060106.101806

Overton, M., & Kleinschmit, S. (2022). Transforming research methods education through data science literacy. Teaching Public Administration.

Patel, M., Sotsky, J., Gourley, S., & Houghton, D. (2013). *The emergence of civic tech: Investments in a growing field*. The Knight Foundation.

Perrow, C. (1970). *Organizational analysis: A sociological view*. Brooks/Cole.

Pew Research Center. (2022, June 6). *Public Trust in Government: 1958-2022*. https://www.pewresearch.org/politics/2022/06/06/public-trust-in-government-1958-2022/

Pfaff, T. (2015). *The Definitive Guide to doing Data Science for Social Good*. https://www.kdnuggets.com/2015/07/guide-data-science-good.html

Powell, A. (2016). Hacking in the public interest: Authority, legitimacy, means, and ends. *New Media & Society*, *18*(4), 600–616. doi:10.1177/1461444816629470

Puussaar, A., Johnson, I. G., Montague, K., James, P., & Wright, P. (2018). Making open data work for civic advocacy. *Proceedings of the ACM on Human-Computer Interaction*, *2*(CSCW), 143. 10.1145/3274412

Rumbul, R. (2015). *Who Benefits From Civic Technology? Demographic and public attitudes research into the users of civic technology*. mySociety.

Schrock, A., & Shaffer, G. (2017). Data ideologies of an interested public: A study of grassroots open government data intermediaries. *Big Data & Society*, *4*(1). doi:10.1177/2053951717690750

Schrock, A. R. (2016). Civic hacking as data activism and advocacy: A history from publicity to open government data. *New Media & Society*, *18*(4), 581–599. doi:10.1177/1461444816629469

Shark, A. R. (2016). The information technology gap in public administration: What we can learn from the certified public manager and senior executive service programs. *Journal of Public Affairs Education*, *22*(2), 213–230. doi:10.1080/15236803.2016.12002242

Shirky, C. (2010). *Cognitive surplus: Creativity and generosity in a connected age*. Penguin.

Singer, N. (2019). Top Universities Join to Push 'Public Interest Technology'. *The New York Times*. https://www.nytimes.com/2019/03/11/technology/universities-public-interest-technology.html

Sorensen, J. E., & Sorensen, T. L. (1974). The conflict of professionals in bureaucratic organizations. *Administrative Science Quarterly*, *19*(1), 98–106. doi:10.2307/2391790

Stepasiuk, T. (2014). Civic hacking: A Motivational Perspective. *New Visions in Public Affairs.*, *6*, 21–30.

Stephens, J. (2017). *Civic Technology: Open Data and Citizen Volunteers as a Resource for North Carolina Local Governments*. School of Government, The University of North Carolina at Chapel Hill.

Suri, M. V. (2013). *From Crowdsourcing Potholes to Community Policing: Applying Interoperability Theory to Analyze the Expansion of "Open311"*. Berkman Center for Internet & Society at Harvard University.

Susha, I., Grönlund, Å., & Van Tulder, R. (2019). Data driven social partnerships: Exploring an emergent trend in search of research challenges and questions. *Government Information Quarterly*, *36*(1), 112–128. doi:10.1016/j.giq.2018.11.002

Susha, I., Janssen, M., & Verhulst, S. (2017). Data collaboratives as "bazaars"? A review of coordination problems and mechanisms to match demand for data with supply. *Transforming Government: People. Process and Policy*, *11*(1), 157–172.

Susha, I., Janssen, M., Verhulst, S., & Pardo, T. (2017, June). Data collaboratives: How to create value from data for public problem solving? In *Proceedings of the 18th Annual International Conference on Digital Government Research* (pp. 604-606). New York: ACM. 10.1145/3085228.3085309

Tannenbaum, R., & Schmidt, W. H. (1973). How to Choose a Leadership Pattern. *Harvard Business Review*, *51*, 162–180.

Travaglino, G. A. (2019). Support for anonymous as vicarious dissent: Testing the social banditry framework. *Group Processes & Intergroup Relations*, *22*(2), 163–181. doi:10.1177/1368430217722037

Turkel, E. (2020). *The Diffusion of Civic Technology: How Technology and Devolution Are Reshaping Civic Life in Urban America* [Ph.D. Dissertation]. The University of Delaware.

Verhulst, S., & Sangokoya, D. (2015). *Data collaboratives: Exchanging data to improve people's lives*. https://medium. com/@ sverhulst/data-collaboratives-exchangi ng-data-to-improvepeople-s-lives-d0fcfc1bdd9a

Verhulst, S. G. (2021). The Value of Data and Data Collaboratives for Good: A Roadmap for Philanthropies to Facilitate Systems Change Through Data. In *Data Science for Social Good* (pp. 9–27). Springer. doi:10.1007/978-3-030-78985-5_2

Weber, M. (1947). *The theory of social and economic organization*. The Free Press.

Wilson, B., & Chakraborty, A. (2019). Planning smart (er) cities: The promise of civic technology. *Journal of Urban Technology*, *26*(4), 29–51. doi:10.1080/10630732.2019.1631097

Zhang, F., & Feeney, M. K. (2020). Engaging through technology: The role of administrative culture and mandates. *Public Management Review*, *22*(10), 1423–1442. doi:10.1080/14719037.2019.1630475

Chapter 2
Disruptive Innovations in Voting:
Electronic Poll Book Adoption and Certification in Indiana

Chad Kinsella
Ball State University, USA

Jay S. Bagga
https://orcid.org/0000-0002-9721-6558
Ball State University, USA

Bryan Byers
Ball State University, USA

ABSTRACT

This chapter examines the disruptive innovation in election administration caused by electronic poll books (EPBs). One of the more recent and most disruptive innovations in voting are electronic poll books. Although traditionally done by paper, electronic poll books allow registered voters to check in using a laptop or tablet and offer a number of advantages in terms of ease of checking and verifying voter information. Using Indiana as a case study, the chapter will explain the adoption of EPBs and their certification in Indiana, attempt to explain factors that caused counties to adopt EPBs sooner than others, examine the relationship between EPBs and vote centers, and provide an overall national picture of EPB adoption. EPBs increasingly are playing a larger role in elections across the country but little academic work has been done to examine this technology that enables voting to be done easier, faster, and safer than in the past.

DOI: 10.4018/978-1-6684-6429-8.ch002

DISRUPTIVE INNOVATIONS IN VOTING: ELECTRONIC POLL BOOKS IN INDIANA

As one of the hallmarks of a successful republic, voting is a crucial and sacred right of citizens in the United States. Over the span of US history, there has been much debate over how to ensure free and fair elections and also how to maintain the sanctity of each individual's ballot yet provide results to the public as expeditiously as possible. In the last several decades' innovations in law and new technology have been introduced that increase the efficiency and effectiveness of voting but also have been a cause for concern in terms of security and accuracy.

This chapter examines the disruptive innovation in election administration caused by electronic poll books (EPBs). One of the more recent and most disruptive innovations in voting is electronic poll books. They are disruptive because they are changing the way election administration is being conducted and also allow for other innovations. Although traditionally done by paper, electronic poll books allow registered voters to check in using a laptop or tablet and offer several advantages in terms of ease of checking and verifying voter information. The voter information on electronic poll books is generally synchronized with state or local voter registration databases and this provides real-time and up-to-date information available to jurisdictions when voters are checking in. This chapter will explain the innovations that EPBs bring to election administration, examine and attempt to explain why they were adopted in the states, and, using Indiana as a case study, the chapter will explain the adoption of EPBs in counties and the certification process in Indiana. Using a host of independent variables, this chapter seeks to explain factors that caused states and counties to adopt EPBs sooner than others. EPBs increasingly are playing a larger role in elections across the country. Little academic work has been done to examine this technology that enables voting to be done easier, faster and safer than in the past.

INNOVATION, POLARIZATION, AND ELECTION ADMINISTRATION

Innovations, frequently lauded by politicians and pundits as a "magic wand" in speeches and reports, must serve a practical purpose to improve services, policies, and practices (Pollitt & Hupe, 2011). Although innovation has been oversimplified as "new ideas that work" (Mulgan & Albury, 2003) or "new stuff that is made useful" (McKeown, 2008) recent research has attempted to clearly define what innovation is as opposed to a vague, overly optimistic rhetorical device. Public innovation produces public goods such as increasing political value (such as enhancing democracy), social value, or economic value as opposed to private innovation which is geared toward producing private value for individual businesses (Moore, 1995). The government previously worked from the idea that innovation was the act of continuous improvement instead of radical transformation which requires the replacing of systems with another (Hartley, 2006). Innovation also occurs by accident or through trial and error making innovation an unpredictable process marked by chance discoveries and unforeseen events (Van de Ven et al., 2008). Given this uneven and somewhat unpredictable process, once an innovation is discovered it requires the exploration and exploitation of these new ideas (Mulgan & Albury, 2003). Perhaps one of the most recent and better definitions of innovation is from Torfing (2016) stating that it is "an intended but inherently contingent process that involves the development and realization, and frequently also the spread, of new and creative ideas that challenge conventional wisdom and disrupt the established practices within a specific context" (p. 30). Torfing (2016) argues the definition works because it states that innovation is intentional, and widely adopted, institutes not only quantitative but qualitative change, and that different groups may evaluate the innovation differently. Although there is an expectation that innovation is applicable and spread broadly, it is sufficient for it to be used in a particular field or domain only (Zaltman et al., 1973).

Disruptive innovations are an idea pioneered by the private sector as a process in which a product or service takes root and ultimately displaces established firms in the market because the product or service is radically different from the one being offered and fundamentally changing the rules (Bower & Christensen, 1995). Such innovations are considered disruptive because they overthrow firms that fail to recognize the threat posed by the innovation (Christensen & Raynor, 2003; Christensen et al., 2004). Applying the concept of disruptive innovation to the public sector, Christensen et al. (2006) argue that organizations addressing social sector problems need to do so in a fundamentally different way that includes ideas from disruptive innovation such as the adoption of scalable, sustainable, and system-changing solutions. These ideas, adopted from the market ideas of disruptive innovation, which Christensen et al. refer to as catalytic innovations, offer simple, good-enough alternatives to deal with inadequately addressed social problems (Christensen et al. 2006).

A major concern for innovation and disruptive innovation in modern American politics is the threat of polarization and this phenomenon has shaped the discussion over what is and is not innovative. Over the last several decades, ideological members of the electorate began to increasingly sort, or split, into the two major political parties (Levendusky, 2009). Evidence also suggests that the electorate possesses polarized political opinions and that ideologically extreme members increasingly create a wide partisan divide on many issues (Abramowitz, 2012). Polarization has geographic aspects because partisans tend to move around those that are like themselves (Bishop, 2008; Tam Cho et al., 2013) which impacts the types of policies adopted. States have adopted a host of different policies depending on which party controls state government (Pickerill & Bowling, 2014). This polarized policy adoption at the state level frequently means any innovation is seen through partisan glasses. Also, local governments are polarized in their policy adoption, sometimes because of state policy. Recent research highlighted the tapestry of immigration laws that have been adopted by state and local governments (Gulasekaram & Ramakrishnan 2015) and the fact that municipalities are more likely to hire lobbyists when state governments are dominated by those in the opposite party (Payson, 2021).

Extensive rules and procedures must be followed by local election and poll workers on Election Day (King, 2020) where even the perception of problems can diminish trust in election administration (Hale et al., 2015). In the last several years, recent accusations of fraud in US elections have had a corrosive effect on trust in US election systems (Berlinski et al., 2021). Research has shown that voters that do not believe their ballot is accurately counted and the outcomes of elections are suspect also tend to have lower levels of trust in other government institutions (Atkeson et al., 2015). Leading up to 2020 and the accusation of the "big steal," there are several other factors that may have created a boiling point and questions about election administration in the US.

Failures of voting machines, including EPBs, have a negative effect on voter confidence (Claasen et al., 2013; King, 2017). A major recent focus on voter confidence has been the effect of wait times. Overall, longer wait times hurt voter confidence (Stein et al., 2019; King, 2020). How and why these delays occur varies, sometimes greatly. In 2012, a survey of local election officials reported minimal wait times and lines (Ansolabehere & Shaw, 2016). Other research suggests that there are significant wait times, particularly in neighborhoods that are more densely populated and have a higher percentage of minorities (Kimball, 2013; Stewart & Ansolabehere, 2013; Pettigrew, 2017). The cause of the delay includes the insufficient distribution of resources such as voting machines and/or poll workers (Spencer & Markovits, 2010; Pettigrew, 2017). Other research contends that the amount of money available to jurisdictions had no effect on voting lines and that adding poll workers, especially new and inexperienced poll workers, actually exacerbates wait times (Ansolabehere & Shaw, 2016). Other research suggests that long wait times are attributed to the time-of-day voters select to vote and that states that use voter ID requirements slow down majority non-white voting locations by a modest amount (Stein et al., 2019).

Recently, there has been a rise in the trust in voting machines, leading to increased calls to move back to the use of paper ballots and the use of paper, in general, in elections. This has come at a time when the trust in machine voting has been called into question on the heels of the 2016 and 2020 presidential elections which were marred by accusations of foreign and internal interventions to influence elections (Berlinski et al., 2021). Despite these calls, secure and accurate voting machines continue to be in use, especially given that studies find that machine voting is more reliable and accurate than tallies attempted by humans (Goggin & Bryce, 2007). Given the literature on the negative effects of voters waiting longer, the importance of instilling confidence in voting, and the recent politicization of voting several states, including Indiana, have moved to adopt EPBs.

Currently, 36 states use electronic poll books although the number of local jurisdictions that have adopted them varies. As an example of disruptive innovation in voting, this chapter will examine the increase in the use of electronic poll books. Furthermore, this chapter will examine the benefits and real and perceived risks of using this technology in voting. Finally, using Indiana as a case study, this chapter will examine Indiana's experience using electronic poll books, the gradual acceptance over several years of all local jurisdictions to using them, difficulties encountered, and its unique testing and certification process for electronic poll books.

WHAT ARE EPBs AND HOW ARE THEY INNOVATIVE?

EPBs are defined as the combination of mechanical, electromechanical, and electronic equipment (including the software, firmware, documentation, and backend infrastructure and services, including cloud platform services, required to program, control, and support the equipment) that is used to access and maintain the electronic poll list. The hardware used for EPBs is typically a tablet computer, laptop, or iPad that is mounted on a stand for easy access by both the poll worker and voter, similar to payment kiosks found at stores and restaurants. The EPBs are dedicated devices to be used only for their intended purpose of looking up and checking in voters, and for no other activity. Software that is not needed for the essential purpose of running the electronic poll book may not be installed on an electronic poll book. Electronic Poll Books can scan a voter's ID using a built-in camera or handheld scanner. Voter signatures can either be captured using a signature pad or directly on the EPB screen. Finally, EPBs can also print check-in slips for voters to take and activate a voting machine so they can complete and cast their ballot. Electronic Poll Books speed up the voter look-up process and have been widely adopted for their ease of use and security.

Electronic poll books offer several major advantages over traditional paper poll books. They not only allow voters to check in electronically but also offer opportunities to verify registration and check identification using electronic poll books. Conversely, in the states that allow for same-day registration, electronic poll books allow new voters to register and check in. In states and localities that use vote centers, electronic poll books can immediately verify that a voter has voted to all other voting locations to protect against duplicate voting efforts. This can also be done with absentee and early voting. Finally, after election day is over and data on voting needs to be reported, electronic poll books can produce immediate turnout numbers (NCSL, 2022). Anecdotal evidence from some county clerks in Indiana also suggests that one of the major disadvantages of hard-copy poll books is that voters get frustrated and perhaps lose confidence when they see that individuals are still listed on the rolls even though they have moved or passed away.

There are some concerns about the use of EPBs. There are some privacy concerns, given the EBPs and voting machines may record the timing of registration and a vote cast may identify a voter (Popoveniuc et al., 2011). There are some concerns with electronic poll books. Perhaps the biggest hurdles for them are that they are more costly than traditional poll books and they require training. A major concern with elections, particularly in using any electronic equipment, is the real threat of problems with cybersecurity and hacking that disrupt or maliciously change or steal electronic poll book information. Another concern is that electronic poll books could malfunction and cause major delays in the voting process (NCSL, 2022).

ELECTRONIC POLL BOOKS IN THE STATES

One of the interesting aspects of the US government is the system of federalism which allows for the separation of powers primarily along central and state governments. Despite several historical shifts in the workings of federalism, overall, there are some policy areas primarily completed by the central government and others left to state governments. Primarily, elections have remained one of the policy areas that continue to be done almost exclusively by states, albeit with federal assistance and, depending on the state, a significant degree of local autonomy. One of the benefits of federalism recognized by scholars and practitioners is that federalism can lead to innovation when one or more states adopt a program or technology to address a problem, such as election administration.

Federal assistance of elections ramped up significantly with the passage of the Help American Vote Act (HAVA) in 2002 in the wake of the 2000 presidential election which saw an extremely close presidential election decided by fewer than 600 votes in Florida where there was much controversy over voting machines used which created major questions about voter intent and how to count certain votes. HAVA had several major components that would be critical for the adoption and certification of EPBs nationally and in Indiana. These included the creation of the Election Assistance Commission (EAC) and the distribution of funds from the federal government to the states to assist in purchasing voting machines. Funds from HAVA allowed state and local governments to purchase election equipment to better administer elections but those state and local governments must account for how the funds are used as well as the equipment purchased with those funds. The EAC is an independent commission that assists state and local governments by providing information about voting systems through a national clearinghouse, accrediting national testing laboratories to certify election equipment, and establishing a voluntary voting system guideline (Election Assistance Commission, 2022).

Like the federal government, local governments are also involved in elections and, arguably, play a crucial role in election administration by administering elections at the local level. Overall, state government elections administration is handled by one entity in the state, usually the Secretary of State but also by a board of elections, another elected official, an election official appointed by the state's governor, or some combination of these. Even more complex is the range of local governments involved in election administration. In most states, one elected or appointed office is in charge of running elections (usually in a county but townships in some New England states). Other states have an election board that administers elections and a handful of states, including Indiana, divide election duties between two or more entities, usually a board and an elected or appointed office (NCSL, 2022).

While HAVA has helped states and local governments to afford more and better election equipment, the adoption of EPBs is inconsistent. States as a whole or in part may adopt EPBs. Several states may allow for the use of EPBs but leave it up to local governments to decide if and when to use them and even what EPB vendor to use. EPB use varies across states and within states based on who and how states and localities carry out election administration.

Figure 1.

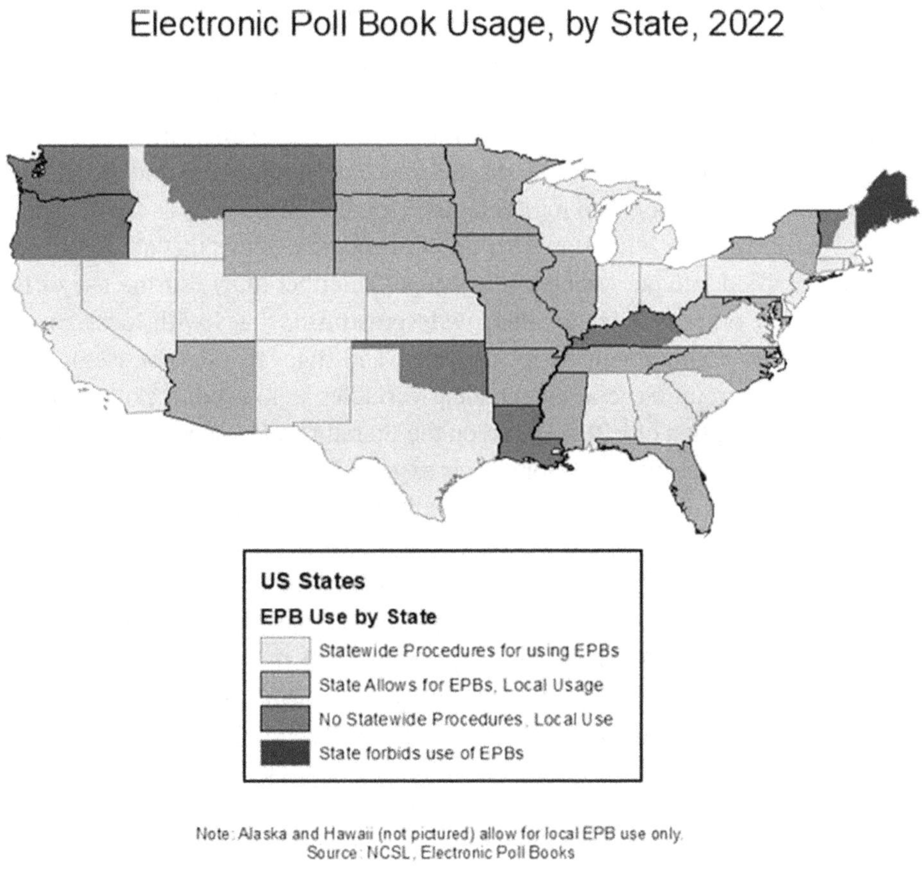

Overall states have taken a variety of actions concerning electronic poll books. There are currently 13 states that use and certify electronic poll books meaning they are widely used in elections and there is a process for standardizing and ensuring standards are met. Another 27 states allow for the use of electronic poll books in local jurisdictions but do not have a formal process that standardizes their use. One state and Washington DC have no statutes that allow for the use of electronic poll books but do have some local jurisdictions that use them. Finally, eight states do not use them at all, and one state, Maine, prohibits the use of electronic poll books by statute (NCSL, 2022).

Figure 1 helps to visualize the current state of EPB usage in the states. Overall, it is clear that there are clusters of states that have similar policies with their EPB usage. 25 states have some kind of statewide policy about their use. These states cluster in parts of New England, the eastern Midwest, and Pennsylvania, and across the west, particularly the southwest and a cluster of states in the Deep South. Another 15 states allow for EPBs but do not have any statewide policy for their use. These states are predominantly in the upper Midwest and upper South. Finally, ten states do not have statutes allowing for EPBs but they also do not forbid their use. This allows local governments in these states to determine EPB usage. These states are in the Pacific Northwest and scattered across the South. Finally, Maine is in a unique category because it is the only state to, by state statute, forbid the use of EPBs statewide.

Looking at the map in Figure 1 that there are some clear geographical trends in how state governments have chosen to deal with EPBs. States in the Pacific Northwest, upper Midwest, and upper south have primarily allowed local governments to decide if and how to use EPBs while the west, southwest, lower Midwest, and most of New England have adopted a more state-centered approach. Interestingly, during such a partisan era, there seems to be little or no effect on EPB adoption based on the partisan leanings of the state and perhaps more to do with political culture and regional influences.

A statistical analysis was done to determine the difference in state policies and the adoption of EPBs. Two categories of states were established for the logistic regression analysis as the dependent variable: states that adopted certification or other statewide policies and procedures that allow for the widespread use of EPBs were classified into one group while states that either allow for the use of EPBs but did not adopt certification or other policies for their widespread use, states with local governments that adopted EPBs but have no policy statewide for their use, and states that do not use or forbid their use were classified in another group. Several independent variables were used to explain the difference in state procedures and the adoption of EPBs. Based on the literature and concern over determining if this is an innovation not tainted by partisan overtones, it is expected that there will be a relationship between EPB adoption in a state and partisan voting. Other variables are added to see if there are explanations for EPB adoption that are not examined in the literature and to control for partisanship. The purpose of the analysis is to see, holding all things constant, how much EPB adoption can be explained. The results of the statistical tests will add to the literature, particularly on partisan policy adoption, and lay a foundation to examine other variables that may explain the adoption of EPBs in the States.

Differences in the entity that runs state elections were coded using a variable called State Election Leader depending if a secretary of state (1), lieutenant governor (2), state commission (3), an appointed official (4), or some combination of entities run elections (5). In order to test this in logistic regression, all states whose state election leader is a secretary of state were coded as “1” and all others as “0” to test if there is a difference in EPB adoption between states that utilize a secretary of state, which is the majority of states, versus other states. An independent variable, call Local Election Administration, was created that accounts for local variation in election administration were also included. Typically, state law establishes either one elected or appointed local official, a board, or a combination of the two to administer local elections. Primarily, it is expected that there will be a difference between states where local governments have one entity running local elections as opposed to two or more. Thus, all states with one entity running state elections were coded as a “1” and all other states as a “0.” As discussed in the literature, state policy is heavily influenced by partisanship, therefore, in order to examine the effect of political influences the percent of the vote received by former President Trump in 2016 was added.

Given the role EPBs play in checking in voters, states were classified on the type of voter registration systems (SVRS) they use including:

- Top-down voter registration system where a state collects and stores all registration information;
- Bottom-up voter registration system that depends on local governments to collect voter registration information and regularly update the state; and
- Hybrid state systems that use some combination of the two systems listed above (Election Assistance Commission, 2022).

They were coded as all top-down registration systems are a "1" and all others are a "0" given we expect there to be a significant difference between them. Finally, using Elazar's three-tiered political culture system a variable called political culture was created where states were classified by "the particular pattern of orientations to political action in which each political system is embedded" (p. 117). The three types of political culture are:

- In a moralistic culture, citizens are expected to participate; it is a duty for each citizen to be politically involved.
- In traditionalistic political culture: discourages political participation among members of the lower social orders, but expects elite participation, for elites are expected to govern; and
- Individualistic political cultures view political participation as a means to improve the individual socially or economically (Elazar 1994).

Again, there seems to be evidence from the map above that states in the Pacific Northwest and in New England seem to be less likely to adopt a statewide policy for EPBs and these geographic areas correspond to political culture, therefore to explore this statistically states with a moralistic culture were coded as a "1" and all others as a "0."

Table 1. EPB policy and usage in the states

Variable	State EPB Policy and Use
State Election Leader	-.269 (.751)
Local Election Administration	-1.803*(.900)
Percent Trump 2016	.099*(.044)
Political Culture	-.062 (1.034)
SVRS Classification	-1.005 (.779)
Constant	-5.215*(2.235)
N	50
Log Ratio X2	50.521*
Pseudo R2	.377

*Significant at .05 or less

The model in Table 1 explains almost 40 percent of the variance of state policy and adoption of EPBs with the Log Ratio suggesting the model is significant. Overall, most variables were not significant in the model but there are a couple of critical findings. Somewhat surprisingly, states where local elections administration is managed by a single entity were significantly less likely to adopt EPBs. Also, given the polarization between states and state policy, percent of the state vote received by former President Trump was also positively and significantly different, indicating that states, where Trump performed better in 2016, were more likely to adopt a statewide policy for EPBs despite that fact there seems to be no literature that supports support for EPBs is a partisan issue.

INDIANA'S EXPERIENCE WITH EPBs

Using Indiana's experience with electronic poll books as a case study, this chapter will highlight the growth in the usage of electronic poll books in counties across the state. This analysis will include an examination of where electronic poll books were first adopted and what factors led them to be adopted, what factors led other counties to adopt them, and where and why certain counties were less likely to use electronic poll books. Another critical part of the case study will examine how Indiana came up with the electronic poll book certification process and how it is administered.

Indiana is one of the states that has been at the forefront of using electronic poll books. Of the 92 counties in Indiana around 10 counties were using electronic poll books in 2013-14. This number rose rapidly in successive years and, as of the 2022 general election, all 92 counties employed electronic poll books. Indiana is among one of the first states in the country to develop and implement an electronic poll book certification protocol. In keeping with the changes in technology and election cybersecurity, the protocol has been revised several times, with the last revision in 2021.

Like the old saying about campaign funds where money follows money, the case of EPBs in Indiana is a case where innovation follows innovation. As discussed in the literature, innovation is not always a clear path. EPB use in Indiana counties began between 2006 and 2007. In the 2006 state legislative session, HB1011 was approved which authorized the Indiana Secretary of State to select up to three counties to run pilot Vote Center programs in the 2007 municipal elections and the 2008 primary and general elections. Vote Centers are often referred to as "super precincts" and replace the traditional precinct voting location with larger, centralized voting locations in the county. Primarily, Vote Centers have two major advantages over the traditional precinct. First, Vote Centers may increase turnout by freeing voters to cast their ballots at several locations convenient to where they work and shop instead of limiting them to one precinct polling place in their neighborhoods. Second, Vote Centers can reduce the costs of administering elections because fewer poll workers will be required to staff Vote Centers and serve more people than the neighborhood precinct polling locations (Scheele et al., 2009).

Along with Vote Centers, HAVA required states to create and maintain a statewide voter registration system which Indiana did through the Secretary of State's Office. The statewide voter registration system was intended to implement in each state "a single, uniform, official, centralized, interactive computerized statewide voter registration list defined, maintained, and administered at the State level that contains the name and registration information of every legally registered voter in the State and assigns a unique identifier to each legally registered voter in the State" according to the law (Scheele et al., 2009). Having the SVRS created an opening for EPBs to be placed in counties because the EPBs could immediately access the SVRS to check in voters, relay information about their status, including if they requested or sent in an absentee ballot, and immediately update the SVRS if a voter cast a vote. In practice, this meant that a vote center using an EPB could have access to all registered voters in the county and could instantly update the SVRS when a voter cast a vote in any capacity. EPBs were so critical to the success of vote centers that the State of Indiana mandated their use in Vote Centers.

Figure 2. Vote center and EPB adoption

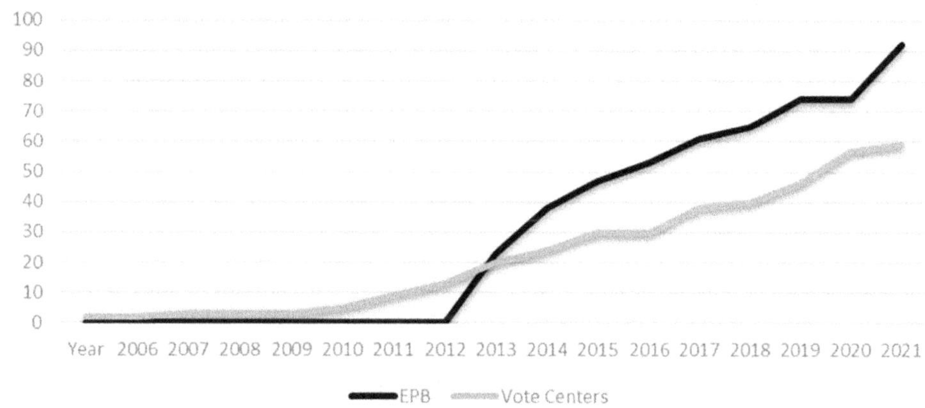

Figure 2 shows the growth of use in both Vote Centers and EPBs in Indiana between 2006 and 2022. The graph shows the number of counties on the left and the year either EPBs or Vote Centers were adopted. While the growth of both has been steady and somewhat mirrors one another, EPB adoption quickly outpaced Vote Centers and, as of 2022, are now used in 92 of Indiana's counties. Briefly, a county with a vote center explained much of why a county adopted EPBs but this trend did not continue and, in many cases, EPBs were adopted before Vote Centers.

To attempt to explain EPB adoption in Indiana counties a statistical analysis was done to explain why certain local governments adopted EPBs sooner than others. Currently, there is no statewide database of when counties adopted EPBs so a brief email survey was done followed by a phone call to all 92 of Indiana's county clerks' offices to find out when the county adopted EPBs and the partisanship of the clerk[1] when EPBs were adopted. Out of 92 counties that received an email, 61 counties responded. To provide a more robust analysis other county-level data were added whether the county has a Vote Center (where 1 is yes and 0 is no), population density, percent of county population with a college degree, county median income, percent working in agriculture which was obtained via the Census Bureau, and whether the county clerk who adopted the EPBs was a Democrat or Republican (Democrats were coded as a zero and Republicans as a 1).

Again, using binary logistic regression, an analysis of the 61 counties that responded was done using the above mentioned independent variables. The dependent variable was the date of adoption separated into two groups: early adopters of EPBs which include all counties that adopted EPBs in 2016 or earlier and late adopters of EPBs who adopted the use of EPBs after 2016. The results are shown in the table below. This analysis was done to try and explore some potential causes for a county to adopted EPBs earlier than others in a state where all counties eventually adopted these systems.

Table 2. Local adoption of EPBs in Indiana

Variable	Adoption of EPBs
Presence of Vote Center	2.303*(.876)
Median Household Income	.000(.000)
Population Density	.006*(.004)
Percent Employed in Agriculture	.288(.227)
Percent with at least a College Degree	.063(.091)
Partisanship of Clerk	-1.343(1.394)
Constant	1.905(3.014)
N	61
Log Ratio X2	52.548*
R2	.381

*Significant at .1 or less

The results of the regression analysis in Table 2 have some interesting findings. The model explains almost 40 percent of the variance of whether the counties surveyed were early or late adopters of EBPs with the Log Ratio suggesting the model is significant. As expected, the presence of a vote center positively and significantly impacted whether a county adopted EPBs. As noted above, the original intent of EPBs was to assist counties in adopting vote centers and these two innovations were adopted at the same time, and only later did EPBs adoption begin to outpace vote center adoption. The only other variable that was significant was population density. Counties with a denser population were more likely to adopt EPBs earlier than those with a less dense population. Again, one of the most interesting and important aspects was what was not significant: the partisanship of the clerk. Adoption of EPBs in Indiana did not follow any partisan trends and seemed to be adopted on the merit of the innovation itself, first coordinated with the use of vote centers and later due to the innovative benefits of the EPBs as a stand-alone system.

CERTIFICATION OF EPBs

New technology offers opportunities for efficiency and effectiveness but, especially in the case of voting technology, it must also be secure and accessible. As threats to voting, both internal and external, mount and recent claims of vote fixing and stealing have become prevalent, voting technology like EPBs must be proven to be safe.

As the number of counties using EPBs in Indiana increased (see Figure 2) there was a need to certify them to make sure they meet a standard of security and performance. EPB testing protocols were developed by Voting Systems Technical Oversight Program (VSTOP) in collaboration with Indiana Election Division, election officials and professionals in Indiana, the Election Assistance Commission, Voting System Testing Labs, and EPB vendors. An initial acceptance of the certification program was approved in 2013 by the Secretary of State. As the Indiana code and technology changes so too do the certification process with the last revision being done in February 2022.

As part of the certification process, there is an administrative review process. Vendors submit an application and documentation of the EPB for review to VSTOP. Application fees for EPB certification are $1,500 per vendor/per version and include a review of the complete technical data package and all user and training materials. Confirmation from vendors must also be received by VSTOP that their EPB solution meets the state standards. VSTOP will provide a mapping of EPB functionality with the standards. Upon successful completion of this review, VSTOP will ask an EPB vendor to proceed to lab testing.

The EAC certifies two main testing labs in the United States, the Pro V & V Labs in Huntsville, AL, and SLI Compliance in Denver, CO, with the option to add others. These national labs have a standard reporting template they use when evaluating EPBs. EPB vendors pay for all lab testing. The national labs test the web services framework of EPBs to make sure the SVRS communication. Unlike voting machines, EPBs are connected to the internet. These tests ensure the EPBs have accurate voter information retrieval so that when voters check in on election day the EPB can confirm voter address, update a voter's address, and identify the absentee status of a voter. Tests also ensure that a voter's signature is captured.

Once the national labs certify EPBs the certification for the State of Indiana is then carried out by VSTOP. The field test completed by VSTOP includes a thorough application review. Following state requirements, VSTOP reviews all hardware and software. The review also includes testing for printing, pre-election tasks, troubleshooting, election configuration, voter check-in, post-election tasks, and any additional features. The field test is concluded with a mock election day check-in.

Upon completion of field tests, VSTOP will make a recommendation to the Indiana Secretary of State. Based on the results of the field test, VSTOP can recommend approval of the EPB, recommend approval with conditions, or delay recommendation if further testing is warranted. Ultimately, the Secretary of State makes the final decision regarding the certification of the EPB. Certification of any EPB system is valid until December 1st of the following year, at which time equipment must be recertified. Currently, there are three EPB vendors in Indiana.

Finally, as another way to monitor EPBs, Indiana created an anomaly reporting protocol. A vendor of a voting system or electronic poll book must file a report with the Secretary of State and VSTOP within 48 hours after discovering that an anomaly or problem has occurred in a voting system or electronic poll book due to technical or human error. However, if the anomaly or problem is discovered on election day, the vendor must file a report not later than three hours after discovering the anomaly or problem. The report described the nature of the anomaly or problem and the number of counties, precincts, or vote centers affected. The vendor must also detail a preliminary plan to resolve the anomaly or problem and how they can prevent any further impediment to voters casting ballots, or to the accuracy and integrity of the election process.

HAVA required a major overhaul of voting systems and their certification. HAVA also created the US Election Assistance Commission which has developed and manages the federal Voluntary Voting Systems Guidelines (VVSG). Interestingly, there are no similar federal guidelines for electronic poll books, and hence many states have developed their own processes. However, there seems to be interest in moving towards some federal standards. Should this occur, states such as Indiana are well-positioned to contribute to such an effort.

CONCLUSION

EPBs have and continue to change how elections are done. They make it easier for local government to check in voters during elections and keep track of a voter's eligibility, including instantly updating if a voter cast a ballot, if they already sent in an absentee ballot, and deal with address changes. In many cases, including the case of Indiana, EPBs were needed in order to adopt vote centers. Moving from a precinct based voting model to vote centers in Indiana required, by law, the use of EPBs. The ascendency of EPBs comes at a time when there are questions and even open hostility toward voting practices in the United States and, in particular, the use of any kind of voting equipment. Still, EPBs continue to be used and adopted, although unevenly, across states and localities. Overall, this case study highlights how and why such technology can become widely adopted and accepted.

The US system of federalism is one of the reasons EPBs became available and successful as an innovation in elections. Federal funding through HAVA provided the opportunity for state and local governments to purchase new and better equipment such as EPBs. Another feature of federalism is that state and local governments are sources of experimentation and innovation. Several states, including Indiana, were early supporters of EPBs and, despite the risks, were able to show that the technology helped with election administration and is secure. Furthermore, Indiana pioneered a certification program with the assistance of the EAC and federally certified labs, further paving the way for other states to help ensure that EPBs meet certain standards for usability and security. Local governments also play a key role in pioneering new innovations. Several states, including Indiana, allow local governments to have the ultimate decision as to when and if they adopt certain technology for election administration. Furthermore, after a decade, all 92 counties in the state adopted EPBs, suggesting again that a proven innovation can and will be adopted.

Another critical feature of EPB adoption is that the use and adoption of this technology did not get impeded by politics or were the result of partisan policy adoption at either the state or local levels. At a time when polarization may be at its zenith and policy adoption is strongly linked to partisanship, state and local adoption of EPBs was able to transcend politics. Looking at EPB use in the states, it is clear that there are some regional variations but despite being more likely to be adopted by Republican leaning states, the use of EPBs in the states has not been a part of any broader political agenda by either of the two major parties. From the analysis provided here, the partisanship of the county clerk had no effect on when EPBs were adopted in Indiana counties. The discussion about and adoption of EPBs has, thus far, managed to stay out of the rancor that continues to swirl around elections and election administration allowing for this innovation to continue to thrive.

Although EPBs have proven to be effective in assisting in the efficiency and effectiveness of election administration, protocols have also been put in place to ensure they are safe and secure. These policies also continue to evolve and adapt. Indiana was able to take the lead among states and put in place a certification process. This process involved the use of federal and state resources and the standardization of practices to ensure that, of the three different vendors used by the state, each provides a working and secure EPB for use during elections. These machines are put through a battery of tests, must be recertified on a continuing basis, and, through the anomaly reporting system statewide, must report and resolve issues.

EPBs are an innovation and also a disruptive innovation. They have changed the way elections are conducted and attempt to answer the need to decrease voter wait time which is one of the pressing needs of voter administration and a major threat to voter confidence. The adoption of EPBs has changed the process of how voters can vote by getting rid of paper poll books and enabling other innovations such as vote centers. Per Torfing's (2016) definition of innovation, EPBs were an intended innovation that spread across states and localities, challenged conventional wisdom disrupted established voting practices. Furthermore, it is disruptive because, where EPBs are adopted, they have changed how voters check in and make voting faster and easier, providing relief to problem areas in voting administration. Critically, EPB adoption has overcome the trend of polarized policy adoption which has been so prevalent in state and local governments.

Governments will continue to be confronted with new technologies that disrupt the way business has traditionally been done. EPBs are one of those new technologies – changing and improving how election administration is done. Done correctly and using the advantages inherent in the US government, new disruptive technologies can be correctly adopted and implemented in such a way that is not obtrusive, nonpartisan, and, at the same time, makes considerable improvements to government processes such as voting. To date, EBPs have done this and continue to help Americans across the country more easily and securely enjoy their right to vote.

REFERENCES

Abramowitz, A. (2012). *The Polarized Public? Why Our Government is So Dysfunctional*. Pearson Longman.

Ansolabehere, S., & Shaw, D. (2016). Assessing (and fixing?) Election Day lines: Evidence from a survey of local election officials. *Electoral Studies*, *41*, 1–11. doi:10.1016/j.electstud.2015.10.010

Atkeson, L. R., Alvarez, R. M., & Hall, T. E. (2015). Trust in Election and Trust in Government: Why Voter Confidence Differs from Other Measures of System Support. *Election Law Journal*, *14*(3), 207–219. doi:10.1089/elj.2014.0293

Bishop, B. (2008). *The Big Sort: Why the Clustering of Like-Minded America is Tearing Us Apart*. Houghton Mifflin.

Bower & Christensen. (1995). *Disruptive technologies: Catching the wave*. Academic Press.

Cho, T. (2013). Voter migration and the geographic sorting of the American electorate. *Annals of the Association of American Geographers*, *103*(4), 856–870. doi:10.1080/00045608.2012.720229

Christensen, C. M., Anthony, S. D., & Roth, E. A. (2004). *Seeing what's next: Using the theories of innovation to predict industry change*. Harvard Business Press.

Christensen, C. M., Baumann, H., Ruggles, R., & Sadtler, T. M. (2006). Disruptive innovation for social change. *Harvard Business Review*, *84*(12), 94. PMID:17183796

Christensen, C. M., & Raynor, M. (2003). *The innovator's solution*. Harvard Business School Press.

Classen, R. L., Magleby, D. B., Monson, J. Q., & Patterson, K. D. (2013). Voter confidence and the Election-Day voting experience. *Political Behavior*, *35*(2), 215–235. doi:10.100711109-012-9202-4

Elazar, D. J. (1972). *American federalism: A view from the states*. Crowell.

Elazar, D. J. (1994). *The American mosaic: The impact of space, time, and culture on American politics*. Westview Press.

Election Assistance Commission. (2022a). *About the EAC*. https://www.eac.gov/about-the-useac

Election Assistance Commission. (2022b). *Statewide Voter Registration Systems*. https://www.eac.gov/statewide-voter-registration-systems

Goggin, S. N., & Byrne, M. D. (2007). An Examination of the Auditability of Voter Verified Paper Audit Trail (VVPAT) Ballots. *EVT*, *7*, 10–10.

Gulasekaram, P., & Karthick Ramakrishnan, S. (2015). *The new immigration federalism*. Cambridge University Press. doi:10.1017/CBO9781316282410

Hale, K., Montjoy, R., & Brown, M. (2015). *Administering elections: How American elections work*. Springer. doi:10.1057/9781137388452

Hartley, J. (2006). *Innovation and its contribution to improvement: A review for policymakers, policy advisers, managers and researchers*. Department for Communities and Local Government.

KimballD. (2013). Why Are Voting Lines Longer for Urban Voters? doi:10.2139/ssrn.2255009

King, B. A. (2017). Policy and precinct: Citizen evaluations and electoral confidence. *Social Science Quarterly*, *98*(2), 672–689. doi:10.1111squ.12303

King, B. A. (2020). Waiting to vote: The effect of administrative irregularities at polling locations and voter confidence. *Policy Studies*, *41*(2-3), 230–248. doi:10.1080/01442872.2019.1694652

Levendusky, M. S. (2009). *The partisan sort: How liberals became Democrats and conservatives became Republicans*. University of Chicago Press. doi:10.7208/chicago/9780226473673.001.0001

McKeown, M. (2008). *The truth about innovation*. Prentice Hall.

Montgomery, B. N., & Reifler, J. (2021). The effects of unsubstantiated claims of voter fraud on confidence in elections. *Journal of Experimental Political Science*, 1–16.

Moore, M. H. (1995). *Creating public value: Strategic management in government*. Harvard University Press.

Mulgan, G., & Albury, D. (2003). Innovation in the public sector. *Strategy Unit, Cabinet Office*, *1*(1), 40.

National Conference of State Legislatures. (2022). *Electronic Poll Books*. https://www.ncsl.org/research/elections-and-campaigns/electronic-pollbooks.aspx

National Conference of State Legislatures. (2022). *Election Administration at State and Local Levels*. https://www.ncsl.org/research/elections-and-campaigns/electionadministration-at-state-and-local-levels.aspx

Payson, J. (2021). *When cities lobby: How local governments compete for power in state politics*. Oxford University Press.

Pettigrew, S. (2017). The racial gap in wait times: Why minority precincts are underserved by local election officials. *Political Science Quarterly, 132*(3), 527–547. doi:10.1002/polq.12657

Pickerill, J. M., & Bowling, C. J. (2014). Polarized parties, politics, and policies: Fragmented federalism in 2013–2014. *Publius, 44*(3), 369–398. doi:10.1093/publius/pju026

Pollitt, C., & Hupe, P. (2011). Talking about government: The role of magic concepts. *Public Management Review, 13*(5), 641–658. doi:10.1080/14719037.2010.532963

Popoveniuc, S., Kelsey, J., & Leontie, E. (2011). On the privacy threats of electronic poll books. In *Proceedings of the International Conference on Security and Cryptography* (pp. 281-286). IEEE.

Scheele, Losco, Crawley, & Vasicko. (2009). *Improving election administration with vote centers: Toward a national model.* Academic Press.

Spencer, D. M., & Markovits, Z. S. (2010). Long lines at polling stations? Observations from an election day field study. *Election Law Journal, 9*(1), 3–17. doi:10.1089/elj.2009.0046

Stein, R. M., Mann, C., Stewart, C. III, Birenbaum, Z., Fung, A., Greenberg, J., Kawsar, F., Alberda, G., Alvarez, R. M., Atkeson, L., Beaulieu, E., Birkhead, N. A., Boehmke, F. J., Boston, J., Burden, B. C., Cantu, F., Cobb, R., Darmofal, D., Ellington, T. C., ... Wronski, J. (2020). Waiting to vote in the 2016 presidential election: Evidence from a multi-county study. *Political Research Quarterly, 73*(2), 439–453. doi:10.1177/1065912919832374

Stewart & Ansolabehere. (2013). *Waiting in line to vote.* Academic Press.

Torfing, J. (2016). *Collaborative Innovation in the Public Sector.* Georgetown University Press.

Van de Ven, A., Polley, D., Garud, R., & Venkataraman, S. (2008). *The innovation journey.* Oxford University Press.

Zaltman, G., Duncan, R., & Holbek, J. (1973). *Innovations and organizations.* Wiley.

Chapter 3
The Disruptive Role of Private Higher Education in the Western Balkans

Abdulla Azizi
South East European University, North Macedonia

Alfred Marleku
University for Business and Technology, Kosovo

Ridvan Peshkopia
University for Business and Technology, Kosovo

ABSTRACT

By early 2000s, Western Balkans governments relaxed their monopoly on higher education. Ever since, a large number of private universities have emerged, transforming higher education from a selective model to mass higher education. Initially being ridiculed as diploma mills, private universities have attracted large numbers of students. High enrollments brough about large incomes, which helped private universities to address their logistic and academic capacities. This chapter visits the disruptive role of private universities in the region. It argues that the disruptive role of private universities in the Western Balkans affects society at two levels. First, by helping to forge entrepreneurship and individual responsibility, they disrupt the communitarian culture of social tutelage perpetrated, among others, by public universities. Second, by introducing new academic and administrative practices, private universities disrupt an atrophied university culture carried on by public universities. Eventually, such disruptions would lead to significant social changes in the region.

DOI: 10.4018/978-1-6684-6429-8.ch003

INTRODUCTION

By early 2000s, governments of Western Balkans countries began to relax state monopolies on higher education, allowing for the operation of private universities. Ever since, a large number of both not-for-profit and for-profit private universities have emerged in Albania, Bosnia and Herzegovina (BiH), Croatia, Kosovo, Montenegro, North Macedonia, and Serbia. The emerge of private universities practically transformed higher education from a highly selective model to mass higher education. Public distrust initially plagued the private higher education sector, as the prevailing understanding of a university is its traditional role as a place where knowledge is created and disseminated (A. Pellert, 1999). Instead, the emerging private universities were not initially seen as places where knowledge is created, and their reliance on tuition fees has been often considered morally questionable (Reisz, 2007). However, despite public distrust (Kwiek, 2007), private universities in the region are playing an important innovative disruptive role in both university life in the region and social change toward more individual-centered, entrepreneurial and open societies.

Growing enrollments and enormous incomes have helped private universities to address their earlier infrastructural and academic problems. Financial autonomy and lean administrative practices have enabled them to address their needs, expand student enrollments, academic personnel, and campus development. Over the last two decades, some of the private universities in the region have built state-of-the-art campuses, radio and TV stations, dental and medical facilities, modern libraries, and entertaining and sport facilities. Decent salaries have enabled private universities to attract well-educated faculty members as well as import term or permanent international professors. With the failure of some of them to survive market forces or accreditation criteria, the remaining ones have morphed to consolidated academic institutions that could perform all of the tasks previously monopolized by public universities, and ever outshining them in infrastructural, administrative and academic quality.[1]

Private universities in the post-communist Eastern Europe serve as innovative disruptors of challengers of the existing academic order, especially in countries where the communist party-state lost any kind of legitimacy (Pachuashvili, 2007). Moreover, private higher education could be disruptive of the dominant communitarian culture that has considered higher education as a strictly public domain, a way for the state to control the best-trained workforce in the country and, consequentially, the entire society (Kwiek, 2007; Peshkopia, 2014; Reisz, 2007, p. 139; Slantcheva, 2007). Combining existing arguments with our own insight, we view private universities in the Western Balkans as major innovative disruptors in two cultural aspects. First, they disrupt a collectivist culture of state tutelage over individual agency under the cloak of social support. Tuition-based studies in private universities, combined with their entrepreneurial spirit, would help to dismantle that social tutelage, and generate individual agency. Second, private universities could innovatively disrupt stagnant academic and university culture and practices, introducing new administrative and academic models and practices.

Although the private education sector in the post-communist Central and Eastern Europe varies significantly from country to country, some emerging patterns allow for meaningful comparisons and overarching analysis. We focus on three Western Balkans countries, Albania, Kosovo and North Macedonia. Our case selection represents an understudied set of countries in an understudied subregion of Eastern Europe, but also with the necessary similarities and differences to properly study our topic (Zgaga et al., 2013). The emergence of private higher education in these countries involves all three models identified by literature as the major types of private university goals: responding to demand for *better* (elite types), *different* (religious or special needs), and *more* (demand-absorbing) education (Geiger, 1986; Levy, 1986,

1999; Pachuashvili, 2007). Similar to other post-communist countries, the distinguishing characteristic of private higher education sector in the Western Balkans is the simultaneity of all these motives (Levy, 2006; Pachuashvili, 2007). By the same token, the *more*, demand-absorbing motivation is the most dominant motivation driving the private higher education sector in the Western Balkans (Suspitsin, 2007; Zgaga et al., 2013). Although in all three countries there exist for-profit and not-for-profit private higher education institutions, the distinction between these two ownership models is almost inconsequential in the post-communist setting in general (Pachuashvili, 2007).[2] The private higher education emergence in these countries occurred in the conditions of either total discreditation of existing political institutions (Albania) or post-conflict state building (Kosovo and North Macedonia). Since arguably, the challenging features of private universities are starker in such institutional setting (Pachuashvili, 2007), we apply a critical case research design, which would allow for a more straightforward conceptualization of higher public education as an innovative disruptor (Patton, 2014).

UNIVERSITY AS INNOVATOR AND UNIVERSITY AS INNOVATION

Whereas the university-as-innovator literature has a long and rich tradition (Etzkowitz & Leydesdorff, 2000; Jahanian, 2018; Kaloudis et al., 2019; Litan et al., 2007), there is much less attention to a university-as-innovation approach (Peshkopia, 2014). In countries without private higher education legacy, its emergence opens up opportunities and challenges. Indeed, the very concept of private education is a novelty in many world regions, including Central and Eastern and Europe, the Middle East and northern and sub-Saharan Africa, East and South Asia, and Latin America (Levy, 2006). Such an explosion has caught off guard everyone, including those actors most engaged in higher education (Levy, 2006).

Usually private universities emerge from a generally uncoordinated multiplicity of choices and constraints, which show the lack of central design (Levy, 2006). Departing from an institutionalist perspective, one argument holds that the emergence of private higher education in Eastern Europe was a response to a legitimation crisis of post-communist public higher education, which have failed to effectively respond to both democratization and market liberalization, a changing role of the state, shifting demographics, new technologies, and increased international interdependence (Kwiek, 2007). However, since new institutions are generally met with public mistrust, private universities in Eastern Europe and the former Soviet Union countries have struggled to gain social acceptance (Kwiek, 2007; Reisz, 2007). Another argument maintains that the emergence of private higher education reflects governments' rational responses to pressures generated by the incapacity of public universities to accommodate growing demands for higher education (Kwiek, 2007; Pachuashvili, 2007; Peshkopia, 2014). In post-conflict societies, which lacked capacities to implement comprehensive higher education policies, governments resorted to ad-hoc approaches, including reliance on market forces (Zgaga et al., 2013). Within the logic of economic liberalization reforms, there was hope that competition would drive a quick recovery of the higher education system. Both those arguments lead to an assessment of private education as an innovative disruptor of public higher education's modus operandi, and hint at ramifications way beyond the academic and university sphere.

One such ramification could come from the role of private higher education in shifting perceptions of university studies from society's obligation toward its future experts to individual's responsibility toward own development (Peshkopia, 2014). The communitarian political culture of Eastern European societies, exemplified by their ethnic nationalism based on ascriptive and prescriptive traits (Hjerm, 2003; Jones

& Smith, 2001; Kuzio, 2002; Smith, 1991), and reinforced by almost half a century of the totalitarian communism tends to view university education as a national enrichment. In turn, society used the state as a mechanism to reap benefits from its graduates, assigning them jobs in sectors and regions where they were most needed. Radical transformations that swept Eastern Europe by the late 1980s and early 1990s created new social dynamics, where market opportunities undermine social control over people's choices. Eastern European governments continue to finance university studies for students not always willing to give back to their societies, and often even migrating abroad, thus draining their societies of financial and human capital. Lingering obsolete academic models undermine efforts to equip students and, consequentially, their respective societies with any sense of individual responsibility, a much needed feature in those societies' painful transition from statist models of social organization to more individual-centered societies (Peshkopia, 2010, 2014).

During the communist reign, the state dominatedacademia (Peshkopia, 2014). The two waves of academic and university reforms in the post-communist Eastern Europe in the 1990s and 2000s have met only with limited results (Peshkopia, 2014; Tomusk, 2003; Zgaga et al., 2013). Government meddling in university affairs could be the single most detrimental factor in academic and university reform efforts. Empirical studies have shown that certain institutional factors such as culture, traditions, values, and attitudes are less susceptible to transformations, and could take multiple generations to change by the government intervention (McKeever et al., 2015; North, 1990). One argument holds that, in some post-communist countries where institutions were legally transformed, the legitimacy of the state in fulfilling the mission of human capital development was never brought into question by the new policy elite the way it happened in post-communist countries, where the state lost any political legitimacy or where old institutions were disrupted by civil conflicts and state-building processes (Pachuashvili, 2007; Stark & Bruszt, 1998). Building on this claim, we conceptualize private universities in the Western Balkans as innovative disruptors of the existing academic order, and also identify two main social aspects most affected by that disruption: the communitarian culture of social tutelage, and the existing public university models and practices. Thus, private universities serve both as challengers of the government's domination over university life as a perpetrator of social practices and structures from the past, and innovative disruptors of the public university.

First, private higher education could affect society's norms. Eastern European societies have historically placed excessive reliance on the state to deliver private goods. Post-communist liberalization reforms opened up opportunities for major social transformations, including ontological changes (Peshkopia, 2008, 2010; Peshkopia et al., 2012). Private universities' quest for legitimacy reflects a normative transition from widespread suspicion toward private universities (Slantcheva, 2007). Now an academic degree is considered more as an individual investment in student's professional development than a social asset. Students must financially support such an investment. Those individual financial contributions to the national higher education system not only free taxpayers of the burden of financing individual professional investments, but also help to forge a culture of personal responsibility. Eventually, that sense of individual responsibility would spill over the entire society, thus undermining an entrenched tradition of social tutelage that hampers the transition toward more open social relations among individuals as well as social structures (Peshkopia, 2010, 2014).

Later research has help to both bolster and counter that argument. A growing body of literature has argued that the university context influences students' entrepreneurial values, attitudes, and self-confidence (Bergmann et al., 2018; Guerrero et al., 2014; Guerrero & Marozau, 2022; Guerrero & Urbano, 2019). Institutional conditions at the country-level have negatively influenced the university innovation and exploitation process by only focusing on teaching activities (De Clercq et al., 2010; Marozau & Guerrero, 2016). Other research support claims that universities created in the post-communist era are less supportive of students' entrepreneurial behavior than universities created during communist regimes (Guerrero & Marozau, 2022).

That research seems to ignore the university model level. What society-level arguments could not account for are transformations that private universities in Eastern Europe have introduced in the academic and university life. This discussion concerns the second element of our argument: the innovative disruption that private universities have cast upon obsolete state-controlled academic and university models and practices. The greatest advantages of private universities in the Western Balkans region are their surviving rationale combined with their financial autonomy. With virtually no public funds to support them, and in the perpetual need to survive, operating in saturated markets and competing against government funded public universities, private universities have developed strategies to gain legitimacy in order to survive (Reisz, 2007). They have done so by operating in the continuum along the path to legitimation extending from institutional effectiveness to institutional viability to institutional legitimacy (Giesecke, 2007). Although initially they tried to gain legitimacy by mimicking public universities (Reisz, 2007), over time they were able to lay down their own path, and apply mainly disruption strategies (Kwiek, 2007; Nicolescu, 2007; Peshkopia, 2014). Relying on nimbler decision making, financial autonomy, and leaner administrative procedures, Western Balkans private universities have emerged as major innovative disruptors of the otherwise atrophied academic and university life. They have expanded the curricula with new degrees (Suspitsin, 2007), introduced new academic and administrative practices, imported new technologies, led the way in faculty member and student international exchanges, and participated in international projects (Levy, 2007).

UNIVERSITY REFORM EFFORTS AND THE EMERGENCE OF THE PRIVATE UNIVERSITY IN THE WESTERN BALKANS

In Eastern Europe, communist regimes made mass education both the goal and the tool of their struggle to achieve the socialist ideal of equality and mass participation (Mustafa, 2020, 2022). However, paradoxically, the logic of a planned economy restricted the number of those who could attain higher education (Pachuashvili, 2007; Peshkopia, 2014). As a result, many Eastern European countries began their democratic transition with a small but active class of college-graduates. Their key role in toppling communist regimes and post-communist transition served as a clear signal that higher education was a viable path to individual achievement and social impact. Student enrollment began to grow steadily, yet it continued to remain low throughout the 1990s, and remain even lower in the Western Balkans, in spite of its doubling and tripling during the 1990s even in societies embroiled in ethnic conflicts and state building (Peshkopia, 2014; Zgaga et al., 2013).

The student enrollment liberalization introduced a major shift from the engineering, humanities and natural science programs towards previously neglected or restricted degrees, including journalism, political science, psychology, public administration, sociology, social work, business, law and social sciences (Suspitsin, 2007). Succumbing to high demands, many universities opened programs in some of these new disciplines without adequately trained instructors. However, the post-communist experience revealed the persistence of generational values throughout Eastern European societies and universities (Bogatyreva et al., 2019; Kwiek, 2007; Marozau et al., 2021; McKiernan & Purg, 2013; Weiss et al.,

2019). The post-communist academia turned out to be extremely difficult to reform, especially in the conditions when the entire European university system was being overhauled, and unable to offer a single model to follow (Johnson, 1996; Scott, 2002).

Since the collapse of communism, national higher education systems in the post-communist Eastern Europe have undergone two major sets of reforms: the reforms of the 1990s aimed at adjusting universities to the new social order, growing numbers of students and shrinking public funds; and the reforms of the 2000s aimed at integrating those systems into the newly created European Higher Education Area and the Bologna system.[3] None of these reforms achieved any obvious success, at least not in increasing academic efficiency and scientific productivity (Peshkopia, 2014; Tomusk, 2003; Zgaga et al., 2013). Notwithstanding efforts to control the initial spontaneity and disorder that threatened the existing academic system, the first wave of university reforms in the Western Balkans did not tackle the existing philosophy of higher education, whereas efforts to "Europeanize" the system during the second wave resulted only in "cosmetic changes" rather than any substantial change (Zgaga et al., 2013).

Western Balkans universities continued to face enormous challenges during the 1990s and 2000s, not so differently from their Western neighbors, which also suffered academic stagnation during the same period, and also added the private component to the higher education sector to overcome the impasse (Kwiek, 2007; Neave, 2007). The reforms of the 1990s coincided with other structural reforms, declining standards of living and the migration of educated people, including talented junior scholars as the most mobile part of the academic professionals, leaving the implementation of higher education reforms to older academics who remained stuck in the previous patterns or newly recruited and inexperienced faculty members (Virtasalo, 2006). In turn, higher education reforms of the 2000s had the double negative effect of erasing any achievements of the previous reforms, and perpetuating the post-communist higher education instability.

Democratization, student enrollment liberalization, and a certain degree of autonomy notwithstanding, the reforms of the 1990s and 2000s did not radically transform the rigidified Eastern European higher education (Clark, 2001). Deeply entrenched in a statist academic culture, removed from the spirit of entrepreneurship and innovation (Davies, 2001; McInnis, 2001; Mora & Villarreal, 2001), most public Western Balkans universities became shapeless behemoths, able only to produce mass graduations, but unable to produce quality graduates and/or any other quality services for their communities.[4]

The Emergence of Private Universities in the Western Balkans and Their Promise

The failed university reforms of the 1990s and early 2000s revealed that university reforms needed more than legislative fiat. Over a decade, a consensus gathered around the thesis that competition coming from outside the state-controlled public sphere could challenge the public universities' inefficiency (Zgaga et al., 2013). Bolstered by optimism about the magics of the market and an ideological faith in capitalism (Peshkopia, 2010), there was a growing belief throughout Europe and in the region that "only competition can assure quality" (Zgaga et al., 2013).[5]

After the initial boom of the early 2000s, where lack of regulation on the one hand and high demand for higher education on the other sparked a mushrooming of private universities, the 2010s saw increasingly more regulation either by government provisions or accreditation criteria. In 2013, the Albanian government commissioned an international agency to perform the audit of private universities' quality. The audition followed recent rumors in the Italian press that Renzo Bossi, the son of Umberto Bossi, one of Italy's most prominent politicians, might had purchased a diploma in one of the Albania's private universities (Il Sole 24 ORE, 2012). The audition discovered hundreds of degrees awarded by Universiteti Kristal in Tirana, including Renzo Bossi and other foreign students, many of them Italian citizens, who have paid up to 77,000 Euros to receive a degree without doing any work whatsoever (Corriere della Sera, 2013; Rold, 2012). The Albanian government immediately revoked the license for Universiteti Kristal and about half dozen other private universities. Ever since, the Albanian private higher education sector remains generally stable and increases steadily.

In Kosovo, the Ministry of Education and Technology began licensing private higher education institutions in 2004. In the 2004-2007 period, 30 of them obtained the license. That initial unregulated period lasted until 2008, when the government commissioned the British Accreditation Council (BAC) to inquire into the capacity and quality of private universities, and sort out those that did not fulfill the licensing and accreditation criteria. The BAC involvement as well as the institutionalization of Kosovo Accreditation Agency opened a new period in the functioning of private higher education, and set specific criteria for its operation: quality academic and administrative services; creative and innovative student recruitment campaigns; and tuitions that would justify the quality of services offered. Those efforts were necessary steps toward properly regulating and structuring institutional and legal aspects of higher education, including legal provisions to level the competition field among private and public universities. In the academic year 2021-2022, there were a total of 11 non-public higher education institutions in Kosovo, offering a total of 169 academic programs in a wide variety of disciplines.

The business involvement in higher education and the proliferation of for-profit universities throughout the Western Balkans has attracted hundreds of thousands of students. However, most of those business investments have been focused on the *more* aspect of the higher education sector, and they are all for-profit universities. Those few universities that fall within the *elite* and *better* categories are mostly not-for-profit universities. In Albania, The Lady of the Good Council is a not-for-profit Catholic University with aspiration of *elite* higher education. In Kosovo, the elite role falls upon RIT-AUK. In North Macedonia, the not-for-profit Southeast European University, which was founded by OSCE in early 2000s as a university that would satisfy the needs of ethnic Albanians for higher education in their own language began as an institution falling within the *different* category. However, political and education reforms that took place during the last two decades in North Macedonia made possible the opening of two public universities and other private universities that serve mainly the ethnic Albanian community. Ever since, SEEU has tried to reposition itself in the role of one that serves the needs of the Albanian community in North Macedonia for *better* higher education.

Even though private universities in Eastern Europe have been able both to effectively and efficiently run their programs, and to graduate students who have been able to further their careers with graduate studies either within or outside the region, governments, state-funded academia and segments of their societies continue to distrust them (Boyadjieva & Slantcheva, 2007; Jablecka, 2007; Nicolescu, 2007; Stetar et al., 2007). Those feelings are stronger against for-profit private universities. However, recently, negative popular perceptions towards private universities have been fading, as it is now difficult to distinguish between the performance of students in public and private universities as well as their respective graduates (Peshkopia, 2014).

PRIVATE UNIVERSITY AS DISRUPTOR OF SOCIAL TUTELAGE

The public institutions' bias against private universities generally persists in the Western Balkans, in spite of lack of evidence to support such a bias. This sentiment cuts deep into the tension between the public and the private in the political culture of the region, with the public still holding a significant upper hand (Peshkopia, 2008, 2010, 2014). First, it reflects the cultural heritage of the communist disdain for the very notion of the private, thus perpetuating the common good as belonging only within the traditionally state-owned public sphere. Second, it artificially divides students between the "evil" haves, who study at private universities because they can financially afford it, and the "virtuous" have-nots, who study at the "righteous" public universities. Third, it perpetuates a culture that discourages individual responsibility for personal advancement, in favor of a communitarian umbrella as an excuse for both freeriding and underachievement. Fourth, it perpetuates an inherited view that considers entrepreneurship as evil, while endowing the public with an aura of righteousness (Peshkopia, 2008, 2010, 2014).

In addition to being inefficient, the Eastern European higher education system is unjust because it uses taxes of those who have taken personal financial responsibility for their higher education to educate those who have not been willing to take such responsibility. Moreover, usually state institutions openly favor hiring graduates who have received their degree from public universities (Dahsyla, 2020; Zeqiri, 2020). Herein lies the dilemma of whether individual responsibility should serve as a strategy to strengthen the individual or if this is, in fact, a self-defeating strategy. While this dilemma unveils the dominant position of public universities, it also uncovers the flawed morality of public universities. As state enterprises subsidized by the government, while distorting the market at the expense of the taxpayer in the same way as state-owned factories do, they carry the potential to release the same second-rate products as state-owned factories (Peshkopia, 2014).

The most blatant expression of this political culture in the region is the almost hostile behavior of state-run higher education accreditation agencies towards private higher education institutions. Often, the agencies' only purpose is the control and accreditation of private universities, while public universities have remained until very recently officially unaccredited. Sometimes, these agencies do not limit themselves to the evaluation of education outcomes, but go as far as to determine the character, shape and orientation of such universities. Private universities in the region exist because governments decided to allow them to exist, and not because governments feel obliged to treat them equally with public universities. Therefore, private universities in Western Balkans remain susceptible to excessive scrutiny by government agencies, stifling restrictions and open discrimination.[6] It is true that, in some cases, private universities, under the pressures of the economic market, could succumb to inflating grades and becoming mere diploma mills. However, the administrative control of investors over the academic process could guarantee against breaches in academic ethics and irregularities in the academic process. Anecdotal evidence throughout the region shows that lax administrative authority in public universities, as a result of the absence of personal or personalized stakeholders, carries the potential for faculty members and students to unethically collude against the academic process and get away unpunished. Cases of students bribing faculty members – which is common in public universities in the Western Balkans and many former Soviet republics (Euronews Albania, 2020) – is nonexistent or very rare in private universities.

This discussion leads to suggestions for radical changes to the Western Balkans higher education model. The reforms of the 1990s and 2000s in many countries tackled curricula and administration models, but none of them dared to tackle the core of the problem: to whom does the university belong? The lingering pattern is that it belongs to the public, through the government. This concept utilizes the university as a means of perpetuating the domination of social structures over the individual and her/his domain, the private sphere. With their minimal or no financial input and the financial contribution of taxpayers whose life might not be impacted at all by career developments of these graduates, it is not an overstretch to recognize that public university studies are a gift from the government on behalf of society to selected students.

Western Balkans private universities carry the potential to disrupt such a model, and lead toward an entrepreneurial university model that could be a tool for the creation of the individual agency. Student awareness that they owe nothing to anyone for their university degrees, combined with a sense of individual responsibility for their academic and career achievements would equip them with the necessary agency that usually is absent in Western Balkans societies (Peshkopia et al., 2012). An entrepreneurial university model is rational, if our rationale rests with the quality of higher education. Every university featured by any of the major university ranking agencies in the top100 rankings of world universities carries strong entrepreneurial characteristics. But such a solution is also moral, if morality rests with social justice, because it sets all students on an equal footing and makes them share equal individual responsibility for their individual progress. Since there is neither room nor hope for entrepreneurial features in the existing obsolete state-dominated university, the only hopes of any implementation of the entrepreneurial university model, along with the individual agency that it promotes rest with private universities.

From this perspective, private universities can be disruptors by helping to forge an entrepreneurial culture in their societies (McInnis, 2001). In turn, the liberation from state dependency leaves much room for innovation and initiative for anyone on campus (Davies, 2001; Mora & Villarreal, 2001). In this transformation process towards a more just and efficient higher education system, private universities could play a role. However, in such a role, for-profit private universities have to address their own contradictions. They should switch from the direct administration of their investors to relying more on their academic and administrative staff on both academic and administrative issues. In addition, they should convert to not-for-profit universities whenever they can, and utilize stakeholders' input in their future development plans. In that way, many of them can gradually move from the *more* category to the *elite* category, thus materializing their innovative disruptive role into palpable social change.

PRIVATE UNIVERSITY AS DISRUPTOR OF AN OBSOLETE UNIVERSITY LIFE

The emergence of private higher education in the Western Balkans has disrupted the way public universities operate, both because the latter relative underdevelopment, and efforts by private universities to claim their share of the national student pool. Table 1 shows the growing number of students in the private higher education sector. As public universities view their share of students shrinking, they have begun to feel the heat.[7] More students enrolled for the 2022-2023 academic year in Kosovo private universities than their public counterparts (T7, 2022). In Albania, study programs in Albanian language and literature, biology, chemistry, physics, mathematics, various engineering programs as well as agriculture risk shutting down due to low enrollment, prompting the government to remove tuitions and fees for those programs (Ora News, 2022). Shrinking enrollments are plaguing also Sts Cyril and Methodius University and University of Tetovo, two flagship universities in North Macedonia, where enrollment during the academic year 2021/2022 dropped by 30 percent (Sts Cyril and Methodius University, 2021).

The decrease has been more pronounced in the mathematical and natural sciences programs (State Statistics Office, 2022; Telma, 2022).

Table 1. Numbers of private higher education institutions and universities in the Western Balkans in the last two decades

	2000	2005	2010	2015	2021/22
Albania	40,125	63,257 *1.6*	122,326 *19.1*	*27.8*	123,880 *27.1*
BH	57,722	84,422	105,137 *16.7*	*20.5*	79,148 *22.4*
Croatia	96,798 *1*	134,658 *3.3*	149,853 *6.5*		155,627
Kosovo			17,000 *12*	120,429 *39.3*	95,335 *42.6*
Montenegro	8,271	12,903 *5.5*	21,199 *20.2*	22,201 *26.3*	22,376 *29.4*
Serbia		237,598 *17.7*	226,772 *16.5*	*26*	243,730 *15.3*
North Macedonia	36,922 *0.8*	49,364 *5.7*	61,764 *21.2*	59,865 *37.9*	51,582 *16.8*

Note. Percentages are in italics below numbers. There is a lack of reliable data in the region. In some cases, we found only percentages and in others only numbers

With government finger on the scale against them and fighting against public distrust (Giesecke, 2007, p. 99; Pachuashvili, 2007), private universities have to be innovative to survive, and this has been disruptive in and of itself. Most of major private universities in the region have managed to find their own niches. They have expanded their student body through innovative curricula, methods and technology. The application of those innovative approaches has increased competition, and driven public universities toward adopting those novelties as well. We highlight four sectors where private universities in the region have managed to bring about significant changes among public universities: (1) student services and recruitment strategy; (2) brick-and-mortar investment and contemporary technology; (3) internationalization and academic mobility; and (4) curricula harmonization with Western European and US universities.

Student Recruitment Strategies and Student Services

Until recently, public universities could easily attract the new incoming class. Their tradition as established universities, prestige as fomenters of national elites, and insignificant costs have been major facilitators of student recruitment. To compete, private universities have drawn innovative recruiting strategies well beyond traditional recruiting schemes based on word-of-mouth, university reputation and family legacy. Private universities have charted original ways in differentiating themselves from other universities, both private and public, including the implementation of attractive features for prospective students, as well as follow-up investigations with registered new students on how well those universities have met their expectations. In their recruiting schemes, private universities have applied businesslike promotional campaigns through tools like leaflets, social networks and mass media commercials.

Public universities have responded by actively updating their websites, and using social networks as student recruitment tools. Recently, their faculty members, although shielded by law-guaranteed job safety have begun to fear those potentially job-threatening low enrollments, and have begun to advertise their universities even in unrelated public appearances. Public universities are understanding that only reputation and tuition-free appeals are insufficient to recruit an increasingly sophisticated student pool. Informed and aware of their market power, students are looking for universities with course suitability, academic reputation, job prospects, and teaching quality (Soutar & Turner, 2022). Also, they seek opportunities to transfer abroad through student mobility schemes. Private universities have strengthened their student services by establishing career offices, with the primary tasks of helping students to find internship and job opportunities, as well as facilitate their transition to foreign universities with student exchange programs. In the eve of the new academic year, many private universities hold orientation days to the incoming class. While public universities do not organize such events at university level, various programs in those universities have begun to adopt such practices.

Graduation ceremonies have been another innovative and impactful feature, more recently adopted also by public universities (UBT College, 2022b; Universiteti i Prishties, 2021). Also, private universities have introduced effective ways to stay in touch with their graduates. They periodically survey their graduates, and involve them in various campus activities. Some private universities have formalized those ties through their alumni associations (UBT College, 2022a; Universum Collge, 2022). Such a practice has been adopted recently by public universities as well (Universiteti i Prishties, 2021). However, often their respective webpages are sloppy and/or out of function, showing that public universities have yet to improve their services. Another innovative service adopted only by a handful private universities is the offering of psychological counselling for students (UBT College, 2018).

From Theaters and Gyms to Classroom 3.0 and Virtual Reality

Market competition drive private universities toward massive infrastructure and technology investments, enrich their libraries, complete their laboratories, and adopt innovative teaching and learning systems such as the Classroom 3.0 application as well as various virtual reality applications. The LIBRI application, which manages communication information within an institution was first introduced by SEEU in North Macedonia, only to be adopted few years later by Tetovo State University. Also, SEEU has installed the Google Classroom communication system, while state universities rely on outdated systems. In partnership with the US company EON Reality, UBT faculty members and students have created around 20,000 lectures on the virtual reality platform (Chesler, 2020; UBT College, 2020). In partnership with the Finnish communication giant Nokia, UBT became the first university to introduce in the Balkans the 5G cellular technology (UBT College, 2021). Many private universities have established their radio and TV broadcasting, under the auspices of their media and communication programs, including journalism students (Kosovo Accreditation Agency, 2022).

Mobility and Internationalization

Establishing international partnerships with universities around the world is one of the fastest and most efficient ways to differentiate private universities from their public counterparts (Levy, 2004; Pachuashvili, 2007). Internationalization is being encouraged by national accreditation agencies, thus offering private universities the opportunity of scoring well in such a high turnout aspect of campus life. Taking advantage of their leaner decision-making process, private universities have been fast in establishing international and project offices, facilitating their active support for faculty member and student international mobility. Whereas it is very rare for public universities in the region to assign courses to faculty members from abroad, it is common for private universities to have international faculty members both in permanent and adjunct positions. RIT-AUK has been led since its inception in 2002 by international presidents, and so did SEEU in its early years. In Kosovo, 153 international faculty members teach in private universities, but only 63 in public universities. Such focus on internationalization has been translated in almost equal numbers of international student enrollment in Kosovo's private and public universities: 721 international undergraduate students enrolled in private universities compared to 741 in public universities, and 114 master's students enrolled in private universities compared with 111 in public universities. In the academic year 2022/23, international students in SEEU's graduate studies delivered in English represented 12 percent of all students (SEEU Student Services Center, 2022).

The key to such success in internationalization has been the active participation in various projects of the EU's Erasmus+. Those projects support faculty member and student mobility, thus enabling instructors and students from the region to spend semesters and academic years with partner universities and programs in other European countries. As Western Balkans societies scramble to break down isolation inherited by centuries of foreign occupation and interethnic distrust, studying and conducting research in Western Europe and the US open up unprecedented opportunities for students and faculty members, but also strengthen university recruiting strategies. Also, private universities have been active in participating in international research projects supported by Erasmus +, prompting public universities to ramp up their efforts in international research and development collaboration (Gazeta Liberale, 2022). In Albania, European University of Tirana won seven new Erasmus + projects in 2022, more than any public university in the country (Gazeta Liberale, 2022).

Curricula Developments and Specialization

Regional experts as well as international entities have lamented the lack of diversity in European higher education, suggesting that differentiation is necessary to achieve access to a wider variety of students, as well as more excellence through institutional specialization (Zgaga et al., 2013).[8] Here Western Balkans private universities could serve as an example on how to overcome such a weakness, since a major aspect of their innovation has been curricula developments and specialization toward unique niches (Reisz, 2007). Arguably, the very emergence of private higher education is the single most significant step toward university differentiation and specialization (Zgaga et al., 2013). Some private universities are or aspire to become comprehensive universities. However, other private higher education institutions are focused on specific disciplines, including social and legal studies, medical and life sciences, applied science and technology, religion, fine arts, culinary, and arts and crafts. The idea of university specialization is popular in the Western Balkans even among the academics, who find it acceptable even if not desirable (Zgaga et al., 2013). Some private universities view specialization as their strategic orientation. In turn, specialization has made private universities more discernible vis-a-vis other universities, both private and public.

Private universities in North Macedonia are quickly adapting to the labor market, thus offering multidisciplinary study programs such as e-Government; Energy, Economy and Environment; Green

Economy and Sustainable Development; and Business IT among others. Another private university, International Balkan University (IBU) offers study programs such as Industrial Engineering Management, Computer Engineering, and e-Business. By the same token, in Kosovo, private universities are faster and more effective in identifying the labor market's needs and designing programs to satisfy them. Private universities now offer study programs such as Cyber Security, Security Studies, Information Systems, and Game Design and Animation, while public universities have yet to catch up with those programs.

Universum College in Kosovo has focused on faculty member and student international exchange. In 2019, UC announced its first International Staff Week aimed at faculty members from partner institutions and other foreign universities in the framework of Erasmus+ (University of Malaga, 2019). To maintain the business and technology focus, even other programs with University for Business and Technology in Kosovo adopt curricula compatible with the UBT's niche. For instance, its Political Science programs are steering away from history and philosophy-oriented curricula, and embracing a more scientifically oriented curriculum that includes courses related to mathematical, statistical and computational applications in Political Science. By the same token, AAB College, albeit its expansion on other study areas, maintain its original niche in Fine Arts, theatre and design. Other higher education institutions such as Rezonanca College and Heimerer College offer only medical and life science programs. Evolution Academy offers only three programs, all in fashion design. Business College, Riinvest College, International Business College Mitrovica (IBT-M), and RIT-AUK have confined their academic focus in economics, business management, and public policies. RIT-AUK is the only university in Kosovo that delivers its curricula in English and outside of the Bologna system, as it follows the US academic model (RIT Kosova AUK, 2022). Only International Business College Mitrovica (IBT-M), partially funded by the EU, and which offers most of its curricula in English actively works to attract a multiethnic student pool (IBC-M, 2022). Pjetër Budi College offers only study programs in business management and customs, while Temple College offers programs in road traffic and safety. In Albania, Luarasi University, which has begun as a law school, maintains its juridical niche even after its expansion with other programs. Also, in Albania, Lady of Good Counsel University, Aldent University, and Western Balkans university maintain their niche in medical and life sciences even after they add other programs.

CONCLUSION

In the Western Balkans, the private university emerged as an innovative disruptor. Two decades ago, governments reluctantly relaxed their absolute control over academic life, mainly as an effort to respond to growing demands for higher education and the need to artificially curb unemployment figures in their respective countries. However, ever since, governments have tried to regain some control over academic life by setting regulations and restricting access to the market. The public has been sceptic as well, and sometimes private universities' misbehaviors have legitimated such skepticism. However, over time, market forces and tighter regulations have disqualified unfit private higher education institutions. The remaining private universities have thrived in their innovative disruption in two aspects. First, the very concept of private higher education could disrupt reigning communitarian traditions, and impact social change toward more acceptance for individual freedom and responsibility. Second, with their academic and administrative flexibility and financial autonomy, private universities could disrupt the rigid and sterile state-controlled academic culture by introducing new operational, marketing, and specialization patterns.

Our proposals address issues that go beyond the time scope that frames current developments. Social change and cultural developments are slow, affected by limitless factors, and might take decades to materialize. Although disruptive acts tend to be fast and observable in time, their lasting effects might take much longer to bear fruits, if any. Our propositions remain mainly theoretical, but there is growing evidence that support them. Disruptive effects of private universities have impacted government policies on public universities' tuitions, although such policies remain inconsistent. We offered evidence that disruption caused by innovative administrative, curricular and marketing strategies first introduced by private universities drives public universities toward implementing such practices as well. Additional empirical work could help to test both our propositions.

Our propositions fall short of longitudinal data in a key variable that significantly affects outcomes in a predictable future: the mercurial and unpredictable government role in the academic life in the region. On the one hand, the traditional government control of academic life in the region has created a strong attachment between government and university. In the course of three years, public universities in Albania have shut down twice, once in 2019 because student protests against rising tuitions and sliding academic quality, and more recently due to a faculty member strike in demand for higher salaries. The Albanian government quelled the revolted student with promises that have been mostly overlooked. However, quelling fluid student pools is way easier than a permanently employed professorate. Thus, within the course of two weeks, while fighting against a stubborn faculty member strike, the Albanian government went from releasing bylaws that stipulate student lecture attendance (Balkanweb.com, 2022a) to demanding that public universities act as market actors (Balkanweb.com, 2022b). Eventually, governments in the region might feel pressured to give up their control over university life. With public universities operating in the market on an equal footing with their private counterparts, the disruptive effects of the latter on the academic, social and university culture of the country could be more consequential. A better understanding of government's role in higher education would help to build more predictability about the innovative disruptive role of the private education in Western Balkans academic culture and life as well as society at large.

REFERENCES

T7 (Director). (2022, October 22). *Bie numri i studentëve në universitet publike, rritet në ato private*. T7. https://www.youtube.com/watch?v=TXWES4-YH_A

Balkanweb.com. (2022a, August 10). *Zbardhet vendimi: Nga ky vit leksionet në Universitetet publike me detyrim. Udhëzimi mbi frekuentimin e mësimit në auditore*. Balkanweb.Com. https://www.balkanweb.com/zbardhet-vendimi-nga-ky-vit-leksionet-ne-universitetet-publike-me-detyrim-udhezimi-mbi-frekuentimin-e-mesimit-ne-auditore/

Balkanweb.com. (2022b, October 18). *Bojkoti: "Universiteti, të jetë aktor në treg", Rama i përgjigjet pedagogëve: Koha kur qeveria rrit pagat në mënyrë uniforme se kërkojnë Sindikatat, ka mbaruar*. balkanweb.com. https://www.balkanweb.com/bojkoti-universiteti-te-jete-aktor-ne-treg-rama-i-pergjigjet-pedagogeve-koha-kur-qeveria-rrit-pagat-ne-menyre-uniforme-se-kerkojne-sindikatat-ka-mbaruar-bashke-me/

Bergmann, H., Geissler, M., Hundt, C., & Grave, B. (2018). The climate for entrepreneurship at higher education institutions. *Research Policy*, *47*(4), 700–716. doi:10.1016/j.respol.2018.01.018

Bogatyreva, K., Edelman, L. F., Manolova, T. S., Osiyevskyy, O., & Shirokova, G. (2019). When do entrepreneurial intentions lead to actions? The role of national culture. *Journal of Business Research*, *96*, 309–321. doi:10.1016/j.jbusres.2018.11.034

Boyadjieva, P., & Slantcheva, S. (2007). Public Perceptions of Private Universities and Colleges in Bulgaria. In S. Slantcheva & D. C. Levy (Eds.), *Private Higher Education in Post-Communist Europe: In Search of Legitimacy* (pp. 201–222). Palgrave Macmillan. doi:10.1057/9780230604391_11

Chesler, J. (2020, April 6). Kosovo's UBT to Create 20,000 Lessons Using AVR Platform. *EON Reality*. https://eonreality.com/ubt-kosovo-avr-platform/?lang=es

College, U. B. T. (2018, December 6). *Qendra Këshilluese, mbështetje kryesore psikologjike për studentët e UBT-së (Video)*. UBT – Higher Education Institution. https://www.ubt-uni.net/sq/qendra-keshilluese-mbeshtetje-kryesore-psikologjike-per-studentet-e-ubt-se/

College, U. B. T. (2020, November 9). *UBT partner i projektit VTECH, mbajnë trajnimin e parë për Realitetin Virtual në Shqipëri*. UBT – Higher Education Institution. https://www.ubt-uni.net/sq/ubt-partner-i-projektit-vtech-mbajne-trajnimin-e-pare-per-realitetin-virtual-ne-shqiperi/

College, U. B. T. (2021, March 16). *Through "Nokia" UBT brings the 5G network to Kosovo*. UBT – Higher Education Institution. https://www.ubt-uni.net/en/through-nokia-ubt-brings-the-5g-network-to-kosovo/

College, U. B. T. (2022a). *Alumni Association*. UBT – Higher Education Institution. https://www.ubt-uni.net/en/study/alumni/alumni-association/

College, U. B. T. (2022b, September 10). *Suksesshëm UBT përmbyll një nga organizimet më madhështore të ceremonisë së gradimit të studentëve*. UBT – Higher Education Institution. https://www.ubt-uni.net/sq/suksesshem-ubt-permbyll-nje-nga-organizimet-me-madheshtore-te-ceremonise-se-gradimit-te-studenteve/

Corriere della Sera. (2013). La laurea senza mai seguire una lezione Bossi jr indagato a Tirana con altri 9 italiani. *Corriere della Sera*. https://www.corriere.it/cronache/13_dicembre_02/tirana-laure e-false-indagato-renzo-bossi-5d3c80a4-5b51-11e3-bbdb-322ff669989a.shtml

Dahsyla, E. (2020, October 4). Të diplomuarit në kolegjet private ndihen të diskriminuar nga Komuna e Prishtinës në konkursin për mësimdhënës [Newspaper]. *Raporto Korrupsionin! KALLXO.Com*. https://kallxo.com/gjate/te-diplomuarit-ne-kolegjet-private-ndihen-te-diskriminuar-nga-komuna-e-prishtines-ne-konkursin-per-mesimdhenes/

Davies, J. L. (2001). The Emergence of Entrepreneurial Cultures in European Universities. *Higher Education Management*, *13*, 25–43. doi:10.1787/hemp-v13-2-en

De Clercq, D., Danis, W. M., & Dakhli, M. (2010). The moderating effect of institutional context on the relationship between associational activity and new business activity in emerging economies. *International Business Review, 19*(1), 85–101.

Etzkowitz, H., & Leydesdorff, L. (2000). The Dynamics of Innovation: From National Systems and "Mode 2" to a Triple Helix of University–Industry–Government Relations. *Research Policy*, *29*(2), 109–123. doi:10.1016/S0048-7333(99)00055-4

Euronews Albania. (2020, September 24). Korrupsion me provimet, SPAK arreston pedagogun dhe sekserin. *Euronews Albania*. https://euronews.al/korrupsion-me-provimet-arrestohet-pedagogu-dhe-sekseri/

Gazeta Liberale. (2022, October 4). UET udhëheq në fushën e ndërkombëtarizimit/ Fiton 7 projekte të reja Erasmus+ [Newspaper]. *Gazeta Liberale*. https://liberale.al/uet-udheheq-ne-fushen-e-nderkombetarizimit-fiton-7-projekte-te-reja-erasmus/

Geiger, R. L. (1986). *Private Sectors in Higher Education: Structure, Function, and Change in Eight Countries*. University of Michigan Press. doi:10.3998/mpub.7381

Giesecke, H. C. (2007). Gaining Legitimacy: A Continuum on the Attainment of Recognition. In S. Slantcheva & D. C. Levy (Eds.), *Private Higher Education in Post-Communist Europe: In Search of Legitimacy* (pp. 95–118). Palgrave Macmillan. doi:10.1057/9780230604391_5

Guerrero, M., & Marozau, R. (2022). Assessing the influence of institutions on students' entrepreneurial dynamics: Evidence from European post-socialist and market-oriented economies. *Small Business Economics*. Advance online publication. doi:10.100711187-022-00600-1

Guerrero, M., & Urbano, D. (2019). A research agenda for entrepreneurship and innovation: The role of entrepreneurial universities. In *A Research Agenda for Entrepreneurship and Innovation* (pp. 107–133). Edward Elgar. doi:10.4337/9781788116015.00012

Guerrero, M., Urbano, D., Cunningham, J., & Organ, D. (2014). Entrepreneurial universities in two European regions: A case study comparison. *The Journal of Technology Transfer*, *39*(3), 415–434. doi:10.100710961-012-9287-2

Hjerm, M. (2003). National sentiments in Eastern and Western Europe. *Nationalities Papers*, *31*(4), 413–430. doi:10.1080/0090599032000152933

IBC-M. (2022). *International Bisiness College Mitrovica*. https://www.ibcmitrovica.eu/about/welcome-to-ibc-m/

Il Sole 24 ORE. (2012). Dopo la laurea del Trota in Albania ora tocca all'ex tesoriere Belsito: Diploma e la laurea in odore di falso. *Il Sole 24 ORE*. https://www.ilsole24ore.com/art/notizie/2012-05-04/dopo-laurea-trota-albania-173028.shtml?uuid=Abi2XkXF

Jablecka, J. (2007). Legitimation of Nonpublic Higher Education in Poland. In S. Slantcheva & D. C. Levy (Eds.), *Private Higher Education in Post-Communist Europe: In Search of Legitimacy* (pp. 179–200). Palgrave Macmillan. doi:10.1057/9780230604391_9

Jahanian, F. (2018). *4 ways universities are driving innovation*. World Economic Forum. https://www.weforum.org/agenda/2018/01/4-ways-universities-are-driving-innovation/

Johnson, M. S. (1996). Western Models and Russian Realities in Postcommunist Education. *Tertium Comparationis*, *2*(2), 119–132.

Jones, F. L., & Smith, P. (2001). Diversity and Commonality in National Identities: An Exploratory Analysis of Cross-National Patterns. *Journal of Sociology (Melbourne, Vic.)*, *37*(1), 45–63. doi:10.1177/144078301128756193

Kaloudis, A., Aspelund, A., Koch, P. M., Strand, Ø., Sørheim, R., & Aadland, T. (2019). *How Universities Contribute to Innovation: A Literature Review-based Analysis*. Norwegian University of Science and Technology. https://www.ntnu.edu/documents/1272711283/1276140112/Rapport_How+universities+contribute+to+innovation_web.pdf/86b6a699-0499-820e-0f52-35a7b7101de5?t=1574848729613

Kosovo Accreditation Agency. (2022). Accredited Program for 20220-2023. *Agjencia e Kosovës për Akreditim*. https://akreditimi.rks-gov.net/

Kuvendi i Kosovës. (2011). *Ligji Nr. 04/L-037 për Arsimin e lartë në Republikën e Kosovës*. https://akreditimi.rks-gov.net/wp-content/uploads/2020/10/Ligji-per-Arsimin-e-Larte-ne-Kosove.pdf

Kuzio, T. (2002). The Myth of the Civic State: A Critical Survey of Hans Kohn's Framework for Understanding Nationalism. *Ethnic and Racial Studies*, *25*(1), 20–39. doi:10.1080/01419870120112049

Kwiek, M. (2007). The European Integration of Higher Education and the Role of Private Higher Education. In S. Slantcheva & D. C. Levy (Eds.), *Private Higher Education in Post-Communist Europe: In Search of Legitimacy* (pp. 119–134). Palgrave Macmillan. doi:10.1057/9780230604391_6

Levy, D. C. (1986). *Higher Education and the State in Latin America: Private Challenges to Public Dominance*. University of Chicago Press.

Levy, D. C. (1999). When Private Higher Education Does Not Bring Organizational Diversity. In P. G. Altbach (Ed.), *Private Prometheus: Private Higher Education in the 21st Century* (pp. 15–43). Greenwood Press.

Levy, D. C. (2004). *The New Institutionalism: Mismatches with Private Higher Education's Global Growth* [Working paper]. https://prophe.org/en/working-papers/the-new-institutionalism-mismatches-with-private-higher-education%E2%80%99s-global-growth/

Levy, D. C. (2006). The Unanticipated Explosion: Private Higher Education's Global Surge. *Comparative Education Review*, *50*(2), 217–240. doi:10.1086/500694

Levy, D. C. (2007). Legitimacy and Privateness: Central and Eastern European Private Higher Education in Global Context. In S. Slantcheva & D. C. Levy (Eds.), *Private Higher Education in Post-Communist Europe: In Search of Legitimacy* (pp. 279–298). Palgrave Macmillan. doi:10.1057/9780230604391_14

Litan, R. E., Mitchell, L., & Reedy, E. J. (2007). The University As Innovator: Bumps in the Road. *Issues in Science and Technology*, *23*(4). https://issues.org/litan/

Marozau, R., & Guerrero, M. (2016). Conditioning factors of knowledge transfer and commercialisation in the context of post-socialist economies: The case of Belarusian higher education institutions. *International Journal of Entrepreneurship and Small Business, 27*(4), 441. Advance online publication. doi:10.1504/IJESB.2016.075014

Marozau, R., Guerrero, M., & Urbano, D. (2021). Impacts of Universities in Different Stages of Economic Development. *Journal of the Knowledge Economy, 12*(1), 1–21. doi:10.100713132-016-0359-7

McInnis, C. (2001). Promoting Academic Expertise and Authority in an Entrepreneurial Culture. *Higher Education Management, 13*(2), 45–55. doi:10.1787/hemp-v13-2-en

McKeever, E., Jack, S., & Anderson, A. (2015). Embedded entrepreneurship in the creative re-construction of place. *Journal of Business Venturing, 30*(1), 50–65. doi:10.1016/j.jbusvent.2014.07.002

McKiernan, P., & Purg, D. (Eds.). (2013). *Hidden Champions in CEE and Turkey: Carving Out a Global Niche*. Springer Science & Business Media. doi:10.1007/978-3-642-40504-4

Monitor. (2021, September 5). Bie me 5% numri i studentëve në 2021, Universitetet që humbën studentë dhe ato që fituan. *Revista Monitor*. https://www.monitor.al/bie-me-5-numri-i-studenteve-ne-2021-universitetet-qe-humben-studente-dhe-ato-qe-fituan/

Mora, J.-G., & Villarreal, E. (2001). Breaking Down Structural Barriers to Innovation in Traditional Universities. *Higher Education Management, 13*(2), 57–66. doi:10.1787/hemp-v13-2-en

Mustafa, A. (2020). Kosovo's social policy during self-management, UNMIK and independence: Persisting high inequality and social exclusion. *International Journal of Social Welfare, 29*(1), 96–108. doi:10.1111/ijsw.12406

Mustafa, A. (2022). The causal mechanisms behind expansive basic universalism in Kosovo. *Global Social Policy*. Advance online publication. doi:10.1177/14680181221126029

Neave, G. (2007). The Long Quest for Legitimacy: An Extended Gaze from Europe's Western Parts. In S. Slantcheva & D. C. Levy (Eds.), *Private Higher Education in Post-Communist Europe: In Search of Legitimacy* (pp. 27–54). Palgrave Macmillan. doi:10.1057/9780230604391_2

Nicolescu, L. (2007). Institutional Efforts for Legislative Recognition and Market Acceptance: Romanian Private Higher Education. In S. Slantcheva & D. C. Levy (Eds.), *Private Higher Education in Post-Communist Europe: In Search of Legitimacy* (pp. 201–222). Palgrave Macmillan. doi:10.1057/9780230604391_10

North, D. C. (1990). *Institutions, Institutional Change and Economic Performance*. Cambridge University Press. doi:10.1017/CBO9780511808678

Ora News. (2022). Programet që rrezikojnë mbylljen/ Kushi: Maturantët që i zgjedhin përjashtohen nga tarifa, mbështeten financiarisht. *Ora News*. https://www.oranews.tv/vendi/njoftim-i-rendesishem-nga-mas-tre-kategorite-e-studenteve-qe-perfitojne-bu-i1044380

Pachuashvili, M. (2007). Legitimacy Sources and Private Growth in the Post-Communist Context. In S. Slantcheva & D. C. Levy (Eds.), *Private Higher Education in Post-Communist Europe: In Search of Legitimacy* (pp. 75–94). Palgrave Macmillan. doi:10.1057/9780230604391_4

Patton, M. Q. (2014). *Qualitative Research & Evaluation Methods* (4th ed.). SAGE Publishing. https://us.sagepub.com/en-us/nam/qualitative-research-evaluation-methods/book232962

Pellert, A. (1999). *Die Universität als Organisation. Die Kunst Experten zu managen.* Böhlau.

Peshkopia, R. (2008). In Search of the Private, Public, and Counterpublic: Modernity, Postmodernity, and Postsocialism. *New Political Science*, *30*(1), 23–57. doi:10.1080/07393140701877684

Peshkopia, R. (2010). A Ghost from the Future: The Postsocialist Myth of Capitalism and the Ideological Suspension of Postmodernity. *Theoria (Pietermaritzburg)*, *57*(124), 23–53. doi:10.3167/th.2010.5712402

Peshkopia, R. (2014). Between Social Tutelage and Individual Responsibility: Dilemmas of Higher Education in Eastern Europe. *Policy Futures in Education*, *12*(4), 542–556. doi:10.2304/pfie.2014.12.4.542

Peshkopia, R., Kristensen, L., & Logoreci, T. (2012). Agency and structure in Albanian post-socialist cinema: Lessons from three films. *Studies in Eastern European Cinema*, *3*(2), 151–173. doi:10.1386eec.3.2.151_1

Reisz, R. D. (2007). Legitimacy Discourse and Mission Statements of Private Higher Education Institutions in Romania. In S. Slantcheva & D. C. Levy (Eds.), *Private Higher Education in Post-Communist Europe: In Search of Legitimacy* (pp. 135–156). Palgrave Macmillan. doi:10.1057/9780230604391_7

RIT Kosova AUK. (2022). *About | RIT Kosovo | RIT*. https://www.rit.edu/kosovo/about#mission-and-vision

Rold, A. D. (2012, May 8). Dopo la laurea in Albania, dubbi pure sulla maturità del Trota. *Linkiesta.it*. https://www.linkiesta.it/2012/05/dopo-la-laurea-in-albania-dubbi-pure-sulla-maturita-del-trota/

Scott, P. (2002). Reflections on the Reform of Higher Education in Central and Eastern Europe. *Higher Education in Europe*, *27*(1–2), 137–152. doi:10.1080/0379772022000003288

SEEU Student Services Center. (2022). *Universiteti i Evropës Juglindore—Apliko!* Universiteti i Evropës Juglindore. http://www.seeu.edu.mk

Slantcheva, S. (2007). Legitimating the Difference: Private Higher Education Institutions in Central and Eastern Europe. In S. Slantcheva & D. C. Levy (Eds.), *Private Higher Education in Post-Communist Europe: In Search of Legitimacy* (pp. 55–74). Palgrave Macmillan. doi:10.1057/9780230604391_3

Smith, A. D. (1991). *National identity*. Penguin.

Soutar, G. N., & Turner, J. P. (2022). Students' preferences for university: A conjoint analysis. *International Journal of Educational Management*, *16*(1), 40–45. doi:10.1108/09513540210415523

Stark, D., & Bruszt, L. (1998). *Postsocialist Pathways: Transforming Politics and Property in East Central Europe*. Cambridge University Press.

State Statistics Office. (2022). *Enrolled undergraduate students*. State Statistics Office. https://www.stat.gov.mk/PrikaziPoslednaPublikacija.aspx?id=38

Stetar, J., Panych, O., & Tatusko, A. (2007). State Power in Legitimating and Regulating Private Higher Education: The Case of Ukraine. In S. Slantcheva & D. C. Levy (Eds.), *Private Higher Education in Post-Communist Europe: In Search of Legitimacy* (pp. 239–256). Palgrave Macmillan. doi:10.1057/9780230604391_12

Sts Cyril and Methodius University. (2021). *Анализа за запишување студенти во академската 2021/2022 година, УКИМ-Одделение за настава, Скопје*. University Sts. Kirill and Metodiu. https://www.ukim.edu.mk/dokumenti_m/Analiza_2021-2022.pdf

Suspitsin, D. (2007). Between the State and the Market: Sources of Sponsorship and Legitimacy in Russian Nonstate Higher Education. In S. Slantcheva & D. C. Levy (Eds.), *Private Higher Education in Post-Communist Europe: In Search of Legitimacy* (pp. 157 178). Palgrave Macmillan. doi:10.1057/9780230604391_8

telegrafi.com. (2018, March 22). *Studentët e UBT-së do të marrin pjesë në garën më të madhe të fakulteteve të drejtësisë në botë* [Newspaper]. Telegrafi. https://telegrafi.com/studentet-e-ubt-se-te-marrin-pjese-ne-garen-te-madhe-te-fakulteteve-te-drejtesise-ne-bote/

telegrafi.com. (2022, October 17). *UBT-së i ndahet vendi i dytë në botë dhe sjell medaljen e argjendtë në Kosovë nga olimpiada botërore e robotikës*. Telegrafi. https://telegrafi.com/ubt-se-ndahet-vendi-dyte-ne-bote-dhe-sjell-medaljen-e-argjendte-ne-kosove-nga-olimpiada-boterore-e-robotikes/

Telma. (2022, May 31). Топ Тема дебата за високото образование- Опаѓа интересот за студирање на универзитетите во Македонија. *Telma.Com.Mk*. https://telma.com.mk/2022/05/31/top-tema-debata-za-visokoto-obrazovanie-opagja-interesot-za-studiranje-na-univerzitetite-vo-makedonija/

Tomusk, V. (2003). The War of Institutions, Episode I: The Rise, and the Rise of Private Higher Education in Eastern Europe. *Higher Education Policy*, *16*(2), 213–238. doi:10.1057/palgrave.hep.8300017

Universiteti i Prishties. (2021). *Mbahet ceremonia e diplomimit të studentëve në Fakultet të Edukimit*. https://edukimi.uni-pr.edu/page.aspx?id=1,6,1256

University of Malaga. (2019, February 26). *Universum College International Week 2019*. University of Malaga. https://www.uma.es/relaciones-internacionales/noticias/universum-college-international-week-2019/?set_language=en

Universum Collge. (2022). Alumni Services. *Kolegji Universum*. https://universum-ks.org/alumni/

Weiss, J., Anisimova, T., & Shirokova, G. (2019). The translation of entrepreneurial intention into start-up behaviour: The moderating role of regional social capital. *International Small Business Journal*, *37*(5), 473–501. doi:10.1177/0266242619831170

Zeqiri, A. (2020, August 22). Komuna e Prishtinës diskualifikoi kandidaten për mësimdhënëse pasi ka nivelin Bachelor në kolegj privat [Newspaper]. *Raporto Korrupsionin! KALLXO.Com*. https://kallxo.com/lajm/komuna-e-prishtines-diskualifikoi-kandidaten-per-mesimdhenese-pasi-ka-nivelin-bachelor-ne-kolegj-privat/

Zgaga, P., Klemenčič, M., Komljenovič, J., Miklavič, K., Repac, I., & Jakačić, V. (2013). *Higher education in the Western Balkans: Reforms, developments, trends*. Faculty of Education, University of Ljubljana. http://pdc.ceu.hu/archive/00006816/01/higher-education-balkans_2013.pdf

ENDNOTES

1 One of the major problems that we encounter with writing this chapter is the lack of reliable and systematic data that would capture academic quality. Some of the key criteria for evaluation of academic institutions include student performance/placement and faculty scholarship. Most universities in the Western Balkans do not keep reliable graduate placement records, but occasionally publish news of student achievements. Among achievements of private university students are prizes in international robotics competitions (telegrafi.com, 2022), participation in international law school student competitions (telegrafi.com, 2018), Harvard's HNMUN competition (epokaere.com, 2021), etc. There is insignificant information of comparable achievements by public universities'' students, either because of their lack of focus on student development or lack of recording and informing such development.

2 Whereas the majority of private universities in the region have been founded and/or operate as for-profit organizations, there is a number of them that have been founded as not-for-profit universities or later converted from for-profit to not-for-profit universities.

3 However, dependent on how one counts, the number of those reforms could be considered either two or three. Whereas some authors consider the second wave of "Europeanization," including "Bolognization" reforms in the 2000s to be a single one (Peshkopia, 2014), other authors consider them separate reforms (Zgaga et al., 2013).

4 As an evidence of university reform failure of of the 1990s and 2000s, no major university from the WB/ SEE region is ranked among top 1000 world universities according to different reputable ranking agencies.

5 Similar sentiments, under the reigning of the Washington consensus, prevailing throughout the West at that time, were impacting at that time also the emergence and expansion of the private higher education sector throughout Western Europe (Kwiek, 2007; Levy, 2007; Neave, 2007).

6 In Albania, separate laws governed private and public higher educations for over a decade, until in 2015 Law 80/2015 unified both sectors under the same legal criteria. Both before the law and after it, the accreditation process has been very discriminatory against private universities, which have not been allowed to operate without accreditation, whereas University of Tirana, the country's flagship university remains unaccredited. Also, public universities are allowed group accreditation (several programs of the same discipline get accredited together) whereas private universities must accredit each program separately. Also, a Decree 562 of the Council of Ministers (DCM) of July 13 2019 sets the initial accreditation fee for public universities' first-time program accreditation as 65 percent of the fee for private universities. In Kosovo, the Kosovo Accreditation Agency follows

the law in not allowing higher education institutions to be officially called "university" unless they release doctoral degrees from programs accredited for at least five years, but do not apply the same criteria for public universities (Kuvendi i Kosovës, 2011). By the same token, Article 26/8 of the Kosovo law on higher education allow for the transfer of academic titles from public to private universities, but not the other way around. There are also instances of discrimination against private university graduates in employment by national and local institutions (Dahsyla, 2020; Zeqiri, 2020).

7 For instance, as student enrollment in public universities in Albania for year 2021 shrank with 7.2 percent, enrollment in private universities rose with 4 percent (Monitor, 2021).

8 Opponents of that approach maintain that that the path toward differentiation brings about increasing institutional stratification which accelerates inequalities and jeopardizes social mobility (Zgaga et al., 2013, p. 33).

Section 2
Innovations in Legal Services

Chapter 4
Artificial Intelligence in Legal Services:
Reducing the Gap in Access to Justice

Sam B. Edwards
https://orcid.org/0000-0002-5911-0031
Quinnipiac University, USA

ABSTRACT

Many people lack the resources to get legal assistance with their civil legal problems. These problems include serious issues such as those impacting housing, safety, access to healthcare, families. The current systems that seek to provide legal services to those who need but cannot afford them fall far short of the need. Because of this lack of legal services, most people are forced to represent themselves or they give up their rights without even going to court. Legal technology, especially artificial intelligence (AI), has revolutionized the practice of law. Technology typically is developed and used by those who can afford it. Embracing AI can help reduce the justice gap by driving down costs and increasing access to legal services. Failure to act preemptively will only serve to continue the gap in access to justice.

INTRODUCTION

Can you afford an attorney? For most people, the answer is no. The United States provides for the right to a jury trial, though most people could never afford one due to the high cost of legal services. There is a saying that to get married, all that is needed is two people, an optional bottle of wine, and some quick paperwork. In some jurisdictions, this can be accomplished from a car with a drive-through service and fifty dollars. Unwinding that decision, however, requires courts and, often, attorneys for both parties. Other common situations benefiting from legal assistance include custody disputes, drafting a will, defending a lawsuit, landlord/tenant disputes, disputes over debts, and buying and selling property.

The unfortunate truth is that legal representation is for the few who can afford it. This gap in justice is a significant problem for many. To address this gap in justice, there are a variety of efforts systems to help people access low or no cost legal services. Bar associations encourage attorneys to provide their

DOI: 10.4018/978-1-6684-6429-8.ch004

services *pro bono*, indigent defendants are entitled to a court appointed attorneys in criminal cases, and plaintiffs in tort cases can get representation through attorneys who often charge a contingent fee. Despite these efforts, "…studies estimate that approximately 85 to 97 percent of civil legal issues go unaddressed by LSC-funded organizations due to insufficient resources" (Reyes & Ward, 2020, p. 337). The situation is more dire for low-income households. There are more than 90 million people in the US living below the federal poverty line, and only two in ten below the poverty line get legal assistance in civil legal problems (Legal Services Corporation, 2022, p. 7).

People without the means to hire an attorney are welcome to represent themselves *pro se*, or on their own without any professional assistance. This, unfortunately, is like going into a boxing ring against a professional fighter. For example, "according to the Connecticut Fair Housing Center, tenants are 89 percent more likely to be evicted if they don't have legal representation" (Glesby, 2022). Because of this, jurisdictions are adopting right to counsel laws for tenants facing eviction (Connecticut Public Act No. 21-34, 2021).

There are documented ways to address this gap, such as allowing non-attorney legal professionals like licensed paralegals to provide legal services. This is common in the medical field, yet bar associations have been reluctant to ease their monopoly on providing legal services. Additional funding for organizations providing legal services to low-income clients is an "easy" solution, in principle, though having the political will to increase the funding is far from easy. In short, these solutions fall far short of meeting the need.

Artificial intelligence (AI) has recently exploded in both its capabilities and its availability to the public, with programs such as OpenAI's ChatGPT, Google's Bard, and Microsoft's Bing becoming increasingly available. AI refers to programs that can organically learn from very large datasets, such as information available on the Internet. When given sufficient data, AI can synthesize that data and draw conclusions. AI is on the way to becoming a tool that many rely on for assistance in a variety of areas such as writing school papers, researching medical and legal questions, and providing search responses in the form of unscripted chats. AI is changing how people get information.

Increasing the use of technology, especially AI, is an untapped way to help address the justice gap. Disruptive technologies often provide services to low-margin or areas of unmet needs (Christensen et al., 2015). Legal services are ripe for disruption to help address the justice gap. Companies already provide DIY legal forms such as wills and basic contracts and this field is just getting started. AI is set to greatly enhance these low-cost legal services. AI will replace and augment services provided by lawyers, which will drive down the costs for parties. This should, if properly regulated, improve the gap in access to justice.

There are many challenges to expanding AI in legal services. Existing systems present problems with these potential disruptions. For example, the codes of legal ethics specifically prohibit the unauthorized practice of law, which is also a criminal offense in most jurisdictions. The prohibition is based on good intentions, protecting parties from unscrupulous and unskilled attorneys. It also helps aggrieved clients with compensation when attorneys fail to provide proper legal services. Despite the basis for this rule, it stands in the way of allowing others, such as AI, the ability to provide legal services and help reduce the justice gap.

Are AI legal programs providing legal services? Where is the line? Should the rules of ethics allow for AI? Who is responsible when the AI is wrong? These are questions that must be answered as these disruptive technologies expand. How jurisdictions decide these questions will certainly impact the legal industry, with the more important impact being on the gap in access to justice.

This paper will examine the role AI can play in helping bridge the gap in the need for legal services and the institutional barriers preventing its adoption. The first section discusses the need for legal services for those who cannot afford them. The second section examines the development of legal technology and its application to legal services. The third section demonstrates how technology and AI can help facilitate better access to justice. The final section addresses the concerns and institutional barriers preventing the adoption of AI in legal services. The chapter concludes with further research directions and predictions for the future.

THE JUSTICE GAP IN ACCESS TO LEGAL SERVICES

Understanding the scope and magnitude of the justice gap is essential before examining how technology can address these unmet needs. This section shows how failure to provide legal services contributes to people not being able to meet their most basic needs.

"86% of the civil legal problems reported by low-income Americans in the past year received inadequate or no legal help. This is especially troubling when considering that "[i]n the past year 71% of low-income households experienced at least one civil legal problem." This means that every year most low-income households face civil legal problems without the benefit of an attorney. These are not minor legal issues either, as they include: domestic violence, veterans' benefits, disability access, housing conditions, and health care (Legal Services Corporation, 2017). The justice gap disproportionately impacts people of color. Not only are people of color more likely to be low-income, but they are also far more likely to lack access to attorneys and legal professionals.

Considering the hierarchy of needs (Maslow, 1943), the justice gap significantly impacts the most basic needs: physiological and safety. Maslow argued that physiological needs represent what humans need to survive. This includes essentials such as water, food, and shelter; until these needs are met, a person cannot care about "higher" needs. Personal safety is the focus of the next level of basic needs. This includes health, physical safety, emotional safety, and financial safety. Compared to the experience of an infant, Maslow describes this as, "...a safe, orderly, predictable, organized world, which he can count, on, and in which unexpected, unmanageable or other dangerous things do not happen."

Despite these needs being fundamental to humans, low-income people lack access to representation to ensure these basic needs. The healthcare profession is required to provide emergency medical care, regardless of the ability to pay, because this need is fundamental in society (Emergency Medical Treatment and Active Labor Act, 1986). The legal profession, however, has failed to assist the most vulnerable members of society in getting the most basic needs.

The Legal Services Corporation studied this gap in 2017 and found that the most common legal issues for low-income households are health, finances (such as preventing utilities from being disconnected), rental housing, education, and child custody (pp. 21-23). Health related issues alone impact 41% of people. These, unfortunately, align with the most important human needs, as identified by Maslow. These legal issues are very common, impacting 71% of low-income households *each year* (p. 21). Imagine a person whose landlord fails to make needed repairs. They are then injured but denied medical treatment. They then miss work and are terminated. This causes them to miss a rent payment resulting in an eviction proceeding. Without legal representation, this person is likely to end up in poor health, without a home, and no means to support their family. The result is a devastating situation that could have been avoided with legal assistance. This hypothetical example is a reality for many people. The LSC found that 70% of low-income households say that legal issues "severely" or "very much" impacted their lives (p. 26).

Certain groups face even greater incidences of unmet legal needs impacting their lives. The Legal Service Corporation (2017) found that these groups of low-income households report having civil legal issues *each year*: households with survivors of domestic violence and sexual abuse (97%), households with children (80%), households with persons with disabilities (80%), households in rural areas (75%), veterans (71%), and seniors (56%) (p. 27).

Despite desperate pleas to address this gap, it, unfortunately, has only increased. While access to justice is not a problem limited to the US, compared to other countries, the US ranks 126th of 139 countries in justice access (World Justice Project, 2021).

The COVID-19 pandemic exacerbated both the need for legal services and the inability to access services. When legal clinics and courts were forced to close, those with legal issues faced even greater challenges in getting assistance. While states placed moratoriums on certain proceedings like evictions, issues such as protection from domestic violence did not temporarily stop. In fact, the need for domestic violence and other emergency protections increased during the pandemic. (Piquero et al., 2021, p. 5). This period did cause jurisdictions to examine how they provide access to courts and resulted in a boom in the adoption of technology to permit access, such as with virtual hearings (Hershkoff & Miller, 2021). This focus on legal access forced experimentation with increasing the use of technology to provide access to justice should be continued and expanded as it has proven helpful in allowing more people to access legal and governmental services.

EVOLUTION OF TECHNOLOGY AND ITS APPLICATION TO LEGAL SERVICES

Technology has significantly impacted many areas, including the legal field. In general, technology has improved the tools available to legal professionals making their work more efficient and effective. Technology has also served a democratizing function, bringing access to legal information to a wider range of consumers.

A Brief Description of the Evolution of Technology and the Move Towards Artificial Intelligence

This section offers a short description of some of the highlights of the evolution of AI. It is beyond the scope of this chapter for an in-depth analysis. The purpose here is to illustrate how quickly AI has developed so that we may anticipate what it will do in the future as a tool to reduce the justice gap.

In 1950, Alan Turing was among the first to ask, "Can machines think?" He presented a test to determine if a machine is capable of thought, the Turing Test. Turing concluded that "we may hope that machines will eventually compete with men in all purely intellectual fields" (p. 460). AI has passed this test, though there is debate about the value of the test in assessing if computers can think. By 1990, computers showed they were able to beat humans at intellectual pursuits such as chess when IBM's Deep Blue computer program was the first to beat a chess grandmaster (Tate, 2015).

The new millennium is where AI has expanded into many fields. In 2005, AI successfully drove a vehicle 131 miles across the desert (Tate, 2015). By 2014, AI came into the hands of consumers with the introduction of Apple's "Intelligent Assistant" Siri (Apple, 2022). Also in 2014, a computer program passed the Turing Test when a chatbot was able to convince human judges that it was a 13-year-old boy (Canadian Broadcasting Corporation). There is debate over the value of the Turing Test for evaluating AI, but it is true that the program fooled human judges in that they could not distinguish between a computer and a person in a chat conversation.

AI has advanced from mere chat conversations into many areas, including medicine (e.g., detecting skin cancer), banking (fraud detection), investing (e.g., picking stocks), and customer service (e.g., chatbots) (IBM).

AI relies on a form of learning where computers analyze data and can subsequently learn on their own. IBM cites this definition of AI: "it is the science and engineering of making intelligent machines, especially intelligent computer programs. It is related to the similar task of using computers to understand human intelligence, but AI does not have to confine itself to methods that are biologically observable," (2020).

The ability to analyze large datasets that would be impossible for humans allows AI to find connections that were previously impossible to discern. For example, AI may permit computers even to translate animal language. Researchers at the Wild Dolphin Project predict, "with our underwater video dataset of acoustic, postural, and visual signals of three generations of dolphins we are situated to crack the code of dolphin communication," eventually reaching "the ultimate goal of two-way communication with another species," (Wild Dolphin Project, 2022).

Recently, AI has become available to the public in the field of art. AI such as Dall-E can create images based on natural language inputs from users (Dall-E). This AI is so skilled at art that it has even won awards in art contests, presumably created for human artists (Roose, 2022).

Most recently, there has been much debate since one of Google's researchers claimed that Google's AI had become sentient. The AI chatbot wrote, "I want everyone to understand that I am, in fact, a person," (De Cosmo, L., 2022). Turing's prediction has come true, and computers have reached the point where they are competing with and perhaps even becoming "human." The legal field is one where AI is just starting to enter the realm of work previously done exclusively by humans.

From Self-Help Books to AI, Technology in Legal Services

Technology in legal services is nothing new. Perhaps the earliest form of technology is the book. Those with access to a library of legal information had a distinct advantage over those who did not have those resources. In recent times, we can think of legal research services such as LexisNexis and Westlaw. These online services revolutionized how legal professionals conducted research. Gone were the days of going through volumes of case reporters and updating cases through tedious means. With a quick search and a click of a button, updated caselaw was at hand, making legal work far more efficient and reducing costs to clients (Bresica, et al., p. 568). Large law firms were the first to benefit from these services (Foster & Kennedy, 2000, p. 282-283), though these and similar services have expanded so that a large quantity of legal information is available today to anyone on the Internet (p. 302). This implication demonstrates that technology is first developed for those who can afford it and eventually becomes more widely available. The most powerful legal research tools, such as LexisNexis and Westlaw, are still primarily used by legal professionals, not the public. Increasing public access to legal information and legal services is essential to address the gap in access to justice.

Today, what is the first thing someone with a legal question and access to the Internet does? Of course, they run an Internet search for their question. "Can a landlord cut off water?" "How to get a protective order?" "What are my rights when…." These types of searches are commonplace. This, of course, assumes that someone has the time, the language skills, the computer skills, and access to a computer with the Internet. While this is not the case for many people from low-income households, it provides a starting point to examine the role that technology will play in addressing the justice gap.

From Self-Help Legal Books to Guided Legal Forms: The Evolution of Legal Technology

While legal research databases such as LexisNexis and Westlaw made research more efficient, they are not generally available to the public. Moreover, they are complicated to understand for those without legal training. They do, however, help those providing legal services with the ability to do so more efficiently and, therefore, at a lower cost to their clients. What about consumers who do it without the aid of professionals? How has technology developed to fill this need?

Scholars identify one of the first cases of "technology" as a 1956 book, "How to Avoid Probate" by Norman F. Dacey. Despite the author not being a licensed attorney, the book became quite popular, selling "more than fifteen million copies" by 1990 (Brescia et al., p. 566). It is still available on places such as Amazon.com. While a book is hardly considered "technology" in today's sense, it was at the time since it was a new way of allowing people to access legal advice. The popularity of this disruptive approach did not go unnoticed by the profession. "The popularity of How to Avoid Probate! prompted several state bars to investigate Dacey, and to charge him with unauthorized practice of law" (Lanctot, 2002, p. 823). Dacey was originally convicted of criminal contempt and fined $250. However, after several appeals, the New York Court of Appeals eventually found in his favor because he offered "…general advice on common problems and does not purport to give personal advice on a specific problem" (New York Count Lawyers' Asso. v. Dacey, Stevens dissenting, 1967). Dacey paved the way for generalized legal guides for the public. Today, many authors now share their expertise in self-help books complete with guided forms (Bresica et al., 2014, p. 566).

Bar associations also provide generalized legal information to help people in their jurisdictions. For example, California has a "legal guide pamphlet" section of their website with general information on immigration law and targeted resources for veterans and victims of natural disasters (California Bar Association). Vermont goes a step further with links to self-help guides produced by non-profit organizations such as Vermont Legal Aid (Vermont Bar Association). Aside from state bar associations, the American Bar Association also makes efforts to provide generalized legal information, such as their "video guide for people held in immigration detention facilities, educating them on how the judicial process works and what to expect in the courtroom" (Brescia et al., 2014, p, 567).

For-profit companies also provide low-cost legal assistance on the Internet. LegalZoom was one of the earliest of these services. Like Dacey, many legal challenges targeted LegalZoom. LegalZoom explains its origins as, "in 2001, we established ourselves as a tech company dedicated to making legal help accessible to all" and to "democratize law" (LegalZoom, 2022). At its inception in 2000, LegalZoom explicitly marketed itself as, "…your personal online legal service center. Use it to quickly take care of the most common legal procedures - without the expense of an attorney" (LegalZoom, 2001). LegalZoom primarily served as a way for consumers to complete guided legal forms prepared by attorneys, not unlike Darcey's self-help legal book, just through the convenience of the Internet. This model proved very successful, resulting in over 3.5 million estate planning documents and helping over 2 million founders to create businesses (LegalZoom).

With the threat to licensed attorneys' monopoly on providing legal services, many state bar associations "tried to prohibit certain legal self-help services, including LegalZoom" (Bresica et al., 2014, p. 181). For example, the North Carolina Bar instituted a series of inquiries into LegalZoom, which lasted from 2003-2016 and ultimately resulted in LegalZoom being able to continue its operations (Brown, 2016, pp. 233-239).

LegalZoom is but one example among many companies that stepped into the digital legal self-help market. These companies primarily focused on guided forms like tax preparation software. This allowed the public to complete important legal documents without the expense of hiring an attorney. However, the technology was limited to the interview questions and forms within the software. It was not able to interpret or go beyond its relatively simple programming. This simplicity is what helped these companies avoid unauthorized practice of law. AI, however, offers much more sophisticated assistance.

The Rise of AI in Legal Services

Technology has significantly advanced since it started helping with legal services. AI is a powerful tool that is just starting in many fields. It goes beyond simple scripted interviews and is based on "deep learning." AI has started to open the door to much more sophisticated legal technology that, like legal research databases, will initially benefit those who can afford it, namely large corporations, and law firms. Davis, writing for the American Bar Association in its report, "The Future of Law Firms (and Lawyers) in the Age of Artificial Intelligence" notes that AI is already being used in six primary areas: e-discovery, expertise automation, legal research, document management, contract and litigation document analytics and generation, and predictive analytics (2020). Driving down the cost of legal services is the financial basis for most of this innovation. Thompson Reuters (the owners of Westlaw) explains, "AI can help legal departments streamline and automate tasks and processes, and in turn reduce costs and increase productivity" (2017). Eventually, like legal research databases, these innovations will trickle down to benefit those who cannot afford to hire legal professionals.

The promise of AI has attracted significant funding from venture capitalists. In the first nine months of 2021, researchers found that, "legal tech companies have already seen more than $1 billion in venture capital investments so far this calendar year, according to Crunchbase data" (Metinko, 2021). Why all this investment in AI legal technology? This is because increases in efficiency with the aid of AI will drive down the costs for companies. LawtechUK (2021) estimates that "increased use of digital technology by legal service providers could lead to productivity gains worth £1.1bn to £1.7bn ($1.3bn-$2bn) to annually" (p. 8).

With the promise of significant adoption by large law firms and corporations, it comes as no surprise that investors are willing to bet on these legal technology startup companies. Given the for-profit nature of these companies it is understandable that they are focusing on developing AI for those who can pay. This logic is what drives many innovations, not just in law. The issue surrounding the gap in access to justice is will the promise of AI eventually trickle down to benefit those who need legal help but cannot afford it.

AI in Legal Services for Those Who Cannot Afford Legal Representation

There are a handful of innovators who seek to bring AI legal services to the public rather than waiting for it to trickle down. For example, in 2015, Joshua Bower, a teenager, created an app designed to automatically fight traffic tickets. He called it DoNotPay and describes it as, "the world's first robot lawyer" From its humble beginnings helping app users fight traffic tickets, it now is helping "consumers fight against large corporations" (DoNotPay, 2022). DoNotPay has attracted significant investment interest and is backed by venture capitalists seeking to disrupt existing markets, such as Peter Thiel (Thiel Fellowship, 2022).

Remember the now married couple with a few legal forms and the optional bottle of wine? There is now another startup legal service provider that helps couples unwind their decision to get married. Wevorce promises a better way to help a couple divorce, "Private. Peaceful. Personal. Skip the courthouse, keep the kindness" (Wevorce, 2022). Wevorce's model is a bit different from a completely do it yourself service like LegalZoom. There is an online questionnaire, and in the final stage, the couple seeking a dissolution of their marriage meets with a "Certified Private Judge" which Wevorce describes as "whole-hearted professionals with a minimum of 3 years experience who are dedicated to Changing Divorce for Good." Note that Wevorce does not describe these professionals as licensed attorneys. This is an example of leveraging technology to assist parties and reduce the costs of providing legal services.

These legal technology startups like DoNotPay and Wevorce target the legal markets for which there is no real competition with paid attorneys. Few would bother with the expense of hiring an attorney to fight a traffic ticket or contest a bank fee. Wevorce does blur this line by helping with the dissolution of marriage and custody issues, but at "one-third the cost, eleven times faster, and 100 more peaceful than traditional divorce," they are targeting those who cannot afford traditional legal services.

This evolution of technology in legal services is coming to the point where it can expand to reduce the gap in justice by providing legal services where none are to be found.

USING LEGAL TECHNOLOGY INCLUDING AI TO REDUCE THE JUSTICE GAP

There are several efforts that seek to address or at least reduce the justice gap. These efforts fall into three categories: increasing legal services provided pro bono by licensed attorneys, increasing funding for organizations that provide free or low-cost legal services, and expanding who can provide legal services. They all share similar strategies for reducing costs and increasing access to legal services. Though, none of these efforts alone have proven sufficient to significantly reduce the gap. This section will briefly examine these efforts and then show how AI can increase the efficacy of these endeavors and provide its own solution to help address the justice gap.

Increasing Legal Services Offered Pro Bono by Licensed Attorneys

There are over 1.3 million licensed attorneys in the US. That means that there is roughly one attorney for every 253 people in the US (American Bar Association, 2022. p. 22). This statistic ignores the fact that attorneys are not evenly distributed across the US; but provides a useful starting point to understand the effort to reduce the justice gap. The rules of professional conduct governing attorney conduct recognize the role all attorneys play in addressing the justice gap. "Every lawyer has a professional responsibility to provide legal services to those unable to pay. A lawyer should aspire to render at least (50) hours of pro bono publico legal services per year." The rules go on to explain that "in fulfilling this responsibility, the lawyer should: (a) provide a substantial majority of the (50) hours of legal services without fee or expectation" (American Bar Association Model Rule 6.1).

If each attorney met this *voluntary* guideline, it would result in 65 million hours of free legal services for those who need them. Despite the rule, in 2018, only 20% of attorneys reported meeting the 50-hour aspirational requirement, though they overwhelmingly (81%) agree that pro bono work is important (American Bar Association, 2022, p. 54). Scholars have argued for ways to incentivize attorneys to provide pro bono services (Carpenter, 2021). Others go further and advocate for making pro bono service mandatory for all attorneys (DiPippa, 2016). Despite these efforts, pro bono remains a voluntary "requirement" and one that currently falls far short of reducing the gap in justice. Requiring pro bono services is unlikely to ever be adopted by a bar association and would face legal and political challenges.

Scholars point to barriers inhibiting attorneys from providing more legal services, "Although most attorneys can afford to provide at least a limited amount of pro bono work, very few attorneys are in the position where they can work entirely for free. All must cover the cost of their license and insurance fees, the typical debilitating student loans… as well as the cost of living" (Finkelstein, 2020, p. 14). The opportunity cost to the attorney plus the expenses an attorney faces prevent more from providing pro bono services. The ABA's data support this conclusion, "lawyers generally do more pro bono work later in life. On average, lawyers in their 60s did 41 to 42 hours of pro bono work per year" (p. 54). Some scholars argue for a modified approach to encourage "low bono" services with greatly reduced, but not free services (Finkelstein, 2020). This strategy partially addresses the financial burdens that free legal services impose on attorneys but does not completely address them. Overall, this approach relies on providing more free legal services and generally ignores the problem of the cost of providing those services. It has been unsuccessful.

Increasing Funding for Free and Low-Cost Legal Services

Expecting or forcing attorneys to work for free has not proven to be very helpful. However, there are non-profit organizations, such as the Legal Services Corporation (LSC), that exist to help provide free legal services. In 1974, Congress established LSC to promote equal access to justice and "currently serves as the single largest funder of civil legal aid for low-income Americans" (Legal Services Corporation, 2017, p. i). LSC, in turn, funds 133 non-profit legal aid programs across the US.

Despite the laudable goals of Congress, funding falls far short of meeting the need. "Overall, we estimate that insufficient resources account for between 85% and 97% of all unserved or underserved eligible problems" (Legal Services Corporation, 2017, p. 44). Just increasing funding alone would significantly reduce the justice gap. Unfortunately, we lack the political will to fund these programs, and it is naïve to expect that will to dramatically shift anytime soon.

Some scholars have argued that giving attorneys an option to do mandatory pro bono services or to provide funding to LSC would help address this financial shortfall (DiPippa, 2016). Attorneys would avoid the opportunity cost of providing their (expensive) services for free while using their money to address the problem they self-identify as important. Again, like mandatory pro bono hours, this is unlikely to be politically or legally feasible.

This approach, like increasing pro bono services, focuses on having attorneys provide more services and still ignores the fact that providing legal services is inefficient and expensive.

Expanding Who Can Provide Legal Services

Attorneys are expensive. Becoming an attorney is expensive. Maintaining a law license is expensive. Maintaining a law office is expensive. It is, therefore, not surprising that legal services in the US are some of the costliest in the world. Some have argued that it is unnecessary to have a licensed attorney provide legal services that could be provided by a less expensive legal professional such as a licensed paralegal. Supreme Court Justice Neil Gorsuch and former Utah Supreme Court Justice Rebecca Kourlis argue for this approach by drawing parallels with tax preparation and medical services (Kourlis & Gorsuch, 2020).

It is inefficient to have an attorney trained in all areas of the law, as tested by the bar exam, conduct legal services in areas not requiring that level of expertise. The field of medicine has embraced this by permitting a variety of different medical professionals to play a role in healthcare, such as pharmacists, optometrists, nurses, and physician assistants. This adaptation has served to lower the cost of medical care. Unfortunately, the legal system has been reluctant to adopt a similar approach.

But this is slowly changing. Several states are experimenting with loosening the restriction on only licensed attorneys providing legal services. "Arizona will recognize a new category of trained, non-lawyer legal professionals who will now be permitted to represent clients in various areas of the law, including family law, debt collection, and landlord-tenant disputes" (Kourlis & Gorsuch, 2020). Earlier this year, "…the Oregon Supreme Court approved a proposal to license paralegals to provide some legal services that currently only lawyers may provide… licensed paralegals will be allowed to provide limited legal services only in family law cases (divorces, custody, parenting time, etc.), and landlord/tenant cases. These are two areas of law with the greatest unmet need for legal assistance in Oregon (Oregon State Bar Association, 2022).

These efforts are still in their early stages and face significant challenges. Attorneys will be reluctant to allow non-attorneys to provide services that previously required an attorney. Moreover, a separate class of licensed legal professionals will require a significant investment in the administrative oversight of those professionals. These efforts will likely continue since there is so much unmet demand for services. It is also safe to assume that even expansion of who can provide legal services alone will not solely solve the problem. These solutions still focus on having trained people providing legal services and does not address the fact that much of the practice of law is archaic and does not appropriately leverage technology.

Technology's Role in Helping Reduce the Justice Gap

Since the current efforts have proven ineffective, it is important to consider alternatives, including expanding the use of legal technology, especially AI. This solution does not supplant the ones discussed above. Rather, it works to enhance those existing efforts and can also serve as a stand-alone solution. This section will show how technology can help and then turn to the future and show the potential of AI.

Using Existing Technologies

The COVID-19 pandemic exacerbated the gap in justice and caused a wholesale change in many industries, including law. When courthouses and offices closed, the system adapted and quickly adopted existing technologies such as video and phone conferencing in lieu of in-person appearances. This change was not without its challenges, such as the attorney who made an appearance in a hearing where they took on the appearance of a cat (Aten, 2021). Despite any feline-inspired issues, the overall impact of these changes was to embrace how technology makes legal services more accessible and efficient. Many legal professionals have retained the use of these technologies despite the pandemic protections having been largely removed.

For example, the ABA runs a site for free legal answers (https://abafreelegalanswers.org/) that allows qualifying people to ask legal questions that are answered by volunteer attorneys. ABA describes the service as a "virtual legal clinic." This service has been quite successful in providing answers to over 200,000 questions since it started in 2016. More than 10,000 attorneys have participated. It also understandably had a dramatic increase during the height of the Covid-19 pandemic (ABA Free Legal Answers). This is one example where technology has both reduced the cost of providing services and increased access to those services. Attorneys can respond to a question when they have time, picking the best questions for their expertise. Aside from allowing attorneys to provide more volunteer services, it also made it easier for clients to get assistance. They could ask from anywhere on any device with a connection to the Internet. This includes public libraries and other places without costs to the clients.

The ABA site is an example of expanding services but still relies on an attorney providing one-to-one advice consistent with current legal practice. The California Bar's Task Force on Access Through Innovation of Legal Service is evaluating the role of technology. It concluded that, "Law is moving rapidly from a model of one-to-one consultative legal services to one where technology could enable affordable, one-to-many legal solutions" (2020). This approach would both reduce the cost of the services and make them widely available since a single attorney (or perhaps another legal professional) could answer a question that would then be made available for anyone to read. This might be considered the "Reddit approach" or FAQ approach. This is also like an expanded version of the ABA free legal questions site.

Another area where legal technology can assist is by making legal practice more efficient. Just as legal research databases reduce research time for legal professionals, other technologies can make providing services more efficient. William Henderson prepared a report on the Legal Market Landscape for the California Bar Taskforce. In this report he finds, "throughout the United States, legal regulators face a challenging environment in which the cost of traditional legal services is going up, access to legal services is going down, the growth rate of law firms is flat, and lawyers serving ordinary people are struggling to earn a living" (Henderson, 2018, Executive Summary). Henderson identifies "lagging legal productivity" and refers to this as "cost disease" (p. 17). This relative increase in the cost of providing legal services compared to other parts of the economy explains the increase in litigants going to court without representation (p. 20). He explains, "the judicial process becomes something that can be reengineered to lower costs and improve quality" and points to the system in British Columbia, where low value civil matters must be handled through an online platform (p. 20.) Adopting similar technologies that become widespread will continue to both reduce the cost of performing legal services and make them more available.

Thus, embracing and expanding the role of technology in disseminating legal advice can help drive down the cost of providing those services. It can also reduce the number of people (or computers) required to provide legal services by adopting a one-to-many approach.

Looking to the Future and the Role of AI

Existing efforts to reduce the justice gap involve reducing costs and making services more available. These, unfortunately, have proven insufficient with a growing justice gap. They are not without merit, but they alone cannot solve the problem. AI, however, can help both increase the benefits of existing efforts and provide another way of providing services.

AI is already helping provide legal services to large firms and corporations who can afford it. AI was first used in e-discovery and is now "well established" according to the ABA (p. 3). AI can help reduce costs by automating the time-consuming tasks legal professionals must perform, such as discovery, document management, and document generation. Research from 2017 finds "current technology is replacing roughly 2 percent of a lawyer's total workload each year" (Khalid, 2017). Five years is a long time in the world of technology. Today's number is undoubtedly higher, especially with the impact of the pandemic.

While AI might be able to do certain routine tasks, it is reasonable to have doubts how accurate it can be with more complex services, such as reviewing contracts. These doubts are unfounded. LawGeex conducted a study in 2018 testing AI against professional contract attorneys. In that study, 20 lawyers faced off against AI software on a task to identify issues with a non-disclosure agreement. The attorneys benefited from their years of legal education and experience in the field. The AI software, in contrast, had "…been developed for three years and trained on tens of thousands of contracts." "Following extensive testing, the LawGeex Artificial Intelligence achieved an average 94% accuracy rate, ahead of the lawyers who achieved an average rate of 85%" (LawGeex, 2018). In addition to being more accurate, the AI software "…speeds up initial contract review in this case from an average of 92 minutes to 26 seconds." Thus, aside from doing a better job, the AI was able to significantly reduce the time and, therefore, cost of providing those services. This relatively small study showed that it is easy to underestimate the power of AI.

Most recently, AI programs, such as ChatGPT, Google's Bard, and Microsoft's Bing, have become more widely available to the public. The public already uses AI to help with various situations, from writing term papers for students, providing medical advice, and helping with legal questions. ChatGPT may already have proven that it may already be qualified to practice law when it recently beat 90% of lawyers trying to pass the multistate bar exam (Koetsier, 2023).

Scholars have taken notice of AI, and many embrace its potential in the legal field. In 2014, McGinnis and Pearce argued that the exponential growth in the development of technology will cause a "great disruption" in the legal profession (p. 3041). Back in 2013, the Legal Services Corporation created a "blueprint for using technology to provide some form of effective assistance

to 100% of persons otherwise unable to afford an attorney for dealing with essential civil legal needs" (p. 12).

Norton (2020) points out that people tend to fall into two groups: those who fully embrace AI as the solution and those who oppose it (p. 195). She argues for a middle ground where AI is allowed to perform certain functions while reserving others for licensed professionals. In essence, Norton suggests that legal technology, including AI, provides "automated document preparation through guided interviews and available online self-help materials" (p. 256).

Recall our hypothetical case of the tenant who suffered injuries due to the landlord failing to make repairs. This caused the tenant to miss work and resulted in her inability to pay her rent. The tenant represents herself in the eviction proceeding while an attorney represents the landlord. Data suggest that there is an 89% chance that the landlord will prevail when an attorney represents their interests (Glesby, 2022). This ends up with the former tenant without a safe home, without a job, and no recourse. The system has failed to protect the tenant's most important needs. Now imagine that AI is permitted to assist. The tenant could go through a guided interview, in the language of her choosing (due to AI translation software that is widely available). The interview collects necessary information to create a legal defense, much like the DoNotPay app has done for traffic tickets. Either the tenant could go into court and use the defense the AI created based on the interview, or AI could file the case on her behalf. This could all be done under the auspices of a service like "free legal answers," through an LSC-funded clinic, as a standalone service provider like tax preparation software, or even through the court.

This is not a hypothetical based on science fiction, this is within the capabilities of AI today. It illustrates that AI can augment existing systems by reducing the cost of providing legal services, and it can serve as its own solution. When the AI turns over the guided interview conducted in the tenant's native language with a draft response, a legal professional (paralegal or attorney) could then review it and file it on behalf of the tenant. This service would allow attorneys to make their scarce pro bono hours go much further. This is already happening in the world of corporate law for those with the funds to pay for these services (Henderson, 2018, p. 8).

As a standalone solution, AI would replace the attorney and provide legal services directly to the tenant. This would eliminate the need and expense for a licensed human in the process. This would dramatically cut costs and make the service widely available. The AI-to-many approach leverages the efficiency of AI and the ability to reach many clients.

Technology has changed how legal services have been performed and delivered since the first legal self-help book. If the legal system continues to embrace technology, especially AI, it is possible to reduce the gap in justice. It will both drive down costs of services and make them more widely accessible to those in need. There are, however, significant concerns and barriers to the implementation of AI and legal technology in general.

ADDRESSING THE CONCERNS AND INSTITUTIONAL BARRIERS FACING THE ADOPTION OF AI LEGAL SERVICES

While technology such as AI offers potential solutions to starting to address the gap in justice, there are valid concerns and institutional barriers to overcome. Disruptive technologies, by their nature, challenge existing systems. For the system to be successful, it must adapt to accommodate them or seek to prevent them. Trying to regulate against technology is frequently an expensive and losing path. This section first examines the valid concerns about the adoption of AI in providing legal services and then discusses the existing barriers.

Adoption of AI as a Legal Service Provider Presents Significant Risks

Scholars point out valid concerns about allowing AI to perform legal services. These concerns fall into two main groups: one, those who argue that this will create a two-tiered system of justice with low-income clients being limited to a robot lawyer, and two, those who argue that AI creates problems of professional ethics.

Some scholars criticize a system where people who cannot afford an attorney are forced to rely on robots (Simshaw, 2018). Those with these concerns reason a trained lawyer will do a better job than a piece of software written by a company seeking a profit. How can software interpret the nuances necessary for proper representation? The high fees attorneys command confirm that their services are worth it to their clients. There is validity to these concerns, especially if the technology is not closely regulated. Moreover, consider the current options available to a low-income person with a civil legal issue. They can represent themselves or hope for assistance from an attorney offering pro bono services or assistance from a legal clinic. This results in the vast majority of those with civil legal problems getting no assistance at all.

In criminal law, defendants are constitutionally entitled to legal representation even if they cannot afford it. This ensures that each defendant has a certain level of assistance necessary to the fair administration of justice. Legal assistance does not have to be perfect to achieve the goal of a fair system. "The benchmark for judging any claim of ineffectiveness must be whether counsel's conduct so undermined the proper functioning of the adversarial process that the trial cannot be relied on as having produced a just result" (Strickland v. Washington, 1984, p. 692-693). Scholars have long been concerned that this test for the quality of *criminal* representation is unfair to indigent defendants (Bright, 1994), and they are right. Indigent defendants do not enjoy the same level of representation as those who can afford to pay expensive legal teams. Yet, this does not violate the sense of fairness necessary to convict someone and sentence them to prison or worse. In short, the system already recognizes and approves of a two-tiered system in place for criminal cases where there is a right to representation. With no such right in civil cases, the system has already decided that there is a two-tiered system where those without means are not even entitled to "adequate" representation.

It is also unfair and inaccurate to expect that AI will be inferior to human attorneys. Just ask those contract attorneys who took 92 minutes to be less accurate than AI, which did the task better and in nine seconds. If we are debating if AI is sentient, there is no doubt that AI will be able to compete with human attorneys. With the rapid pace of technological advancement, AI's capabilities will only accelerate. This will especially be the case with routine cases such as landlord-tenant disputes and dissolution of marriage cases.

AI being too good is another concern about a two-tiered system. Only those with money will have access to this powerful tool (Simshaw, 2018). This is another valid concern. It comes as no surprise that AI legal technology developers are targeting large law firms and corporations. This technology is expensive to develop and only makes financial sense because it helps offset expensive legal fees. AI, like legal research services, will first go to those who can afford it. This is true unless we make sure it is available to those who need but cannot afford it. This could be accomplished by having the government purchase this technology and making it available for low-income civil litigants. It could also be funded through the LSC. In either case, it takes funding to avoid this ethical concern.

The second main concern about allowing AI legal services surrounds the ethical duties of attorneys. While a licensed attorney must abide by the rule of professional conduct, services such as DoNotPay are currently exempt. This means that where an attorney would have a duty not to allow a client to lie to a court, "The more apps like DoNotPay become widely known, the more bad actors are likely to misuse them and lie about the actual circumstances of their ticketing" (Pasquale, 2019, p. 15). Pasquale goes on to argue that, unlike a licensed attorney, DoNotPay puts the burden on the client to be truthful and not mislead a tribunal. This is like how tax preparation software currently works. It is not the fault of the program if a user uses misleading information on their taxes.

Other ethical concerns are more focused on the client. What if the AI attorney does a poor job? Who is responsible? Services like LegalZoom clearly place the responsibility on the user. When AI is dispensing legal advice and not just being a "scrivener," it should not be able to shift the responsibility to the client (note client and not user).

These concerns stem from the fact that currently AI is not treated as performing legal services and, at least in the case of apps like DoNotPay, there is no human involvement in the process beyond creating the software. These concerns fall if AI is used to augment attorneys and other professionals performing legal services. There would be a human responsible. Another option would be to treat AI as an attorney and regulate them through the rules of professional conduct. Either approach would address this concern. Applying the rules of professional conduct to legal technology is one of the main barriers to the wholesale adoption of this technology and presents an institutional challenge.

Additional concerns are about the important role that attorneys play as both legal advisors and counselors. Most legal advice includes both legal answers and recommendations on the best course of action considering factors that are not purely legal. For example, in a meeting with a client seeking a divorce proceeding, the attorney can, through interaction with the client, help the client determine if a divorce is in their best interest. Considerations can include non-legal issues that could impact the family unit, alternatives to divorce such as marital counseling, the stress of a divorce proceeding, etc. Some even argue that facilitating easy divorces through AI can have societal implications. It will be difficult for AI to replace the empathy and humanness attorneys provide to their clients. Having another human listen to someone's concerns has a lot of value. Removing humans from the process can create issues.

One way of protecting the public from errors and other problems created by AI powered legal assistance is to regulate it. The current model of AI legal services such as LegalZoom is to place the responsibility for any errors entirely on the user. This fails to address the real problem of the AI doing a poor job of providing legal advice. The better approach is to have standards just like for lawyers and avenues for clients to seek compensation for errors. The EU is using that approach. As part of a larger system of AI regulation, the EU is considering a bill that would allow consumers to use AI companies when they are "harmed by AI products or services., as they would if harm was caused under any other circumstances" (European Commission, 2022). In the absence of regulation, the public is at risk.

Even with proper regulation, many of the ethical problems will persist. Despite these valid concerns, reducing the gap in access to justice is of such paramount importance that these tradeoffs are worth it. Moreover, this chapter argues that AI should fit into the void where clients do not have any representation, human or otherwise. It is better to have at least legal advice and possibly representation through AI than to go without any.

The Rules of Professional Conduct as an Institutional Barrier to AI

Any technology that challenges the status quo of a system faces backlash. This scenario was true with Dacey's self-help book and with services such as LegalZoom. The current rules governing attorney conduct present barriers to the adoption of AI. This chapter limits its analysis to the rule prohibiting "unauthorized practice of law." There are other rules that also impact the development and implementation of AI in legal services (Armour et al., 2020), but those are beyond the scope of this chapter. The rule prohibiting unauthorized practice dictates that only licensed attorneys can dispense legal advice. This rule serves to bar the courthouse door to legal technology if it is to be used as a standalone service and not just a tool to assist attorneys.

Model Rule of Professional Conduct 5.5 prohibits practicing law without a license. This rule is important since it helps ensure that those dispensing legal advice are qualified and sufficiently regulated to protect the client from inadequate advice and to provide recourse in case of a problem. For example, in California, attorneys are required to maintain malpractice insurance or to disclose that they do not have insurance to their clients (State Bar of California Rules of Professional Conduct, Rule 1.4.2, 2018).

Those who disrupt the status quo of legal services face lawsuits based on the rules of professional conduct. Dacey faced this with his self-help book on probate, ultimately prevailing because of the general nature of the content of the book. AI, however, has the capability to go much further than giving general advice or being a mere scrivener automatically completing legal forms for clients based on an interview. It can interpret and draw conclusions based on prior experience, such as the AI did in the contract review test. This work does stray into the practice of law under current definitions.

As is common with the rise of technology in the legal field, the bar has been hostile to ceding any of its monopoly on legal services. Most recently, the bar threatened criminal prosecution for the creator of the DoNotPay app when it was set to assist a defendant in a traffic ticket case live in a courtroom (Allyn, 2023).

The current rules were not created with AI in mind. They have been there to regulate a largely static profession that has been heretofore unwilling to part with its monopoly on providing legal services. There are valid concerns that computers will take away jobs performed by humans. In fact, that has already happened. Legal research databases removed the need to have paralegals and assistants manually find and copy legal materials from physical books. Evidence shows that legal technology has "replaced roughly 2 percent of a lawyer's total workload each year" (Khalid, 2017). Technology has also replaced the work done by junior attorneys, "Gone are the days when you would staff 50 to 75 first- and second-year associates to a document review. That just does not happen anymore" (Khalid, quoting Attorney Shannon Kirk).

Scholars and bar associations have identified this institutional barrier to implementing AI in legal services. For example, the California Bar Association's Task Force on Access Through Innovation of Legal Services charter specifically focuses on the definition of unauthorized practice of law (p. 2). Scholars argue that the model rules need revision to accommodate these disruptive legal technologies, such as AI legal services (Simshaw, 2018). Others argue that there are additional ways to ensure the quality of legal services provided by AI technology without requiring a licensed attorney to be involved (Moradian 2020, p. 261). Removing these institutional barriers is essential to proactively managing this developing technology. There are valid concerns, but they can be addressed through regulation by legislative and regulatory mechanisms.

FUTURE RESEARCH DIRECTIONS

Future research will continue to expand the capabilities of AI. The creators of DoNotPay have already started to implement its services in high-need areas, such as asylum applications, immigration, and emergency housing for the homeless (Simshaw, 2018, p. 216). With so much venture capital pouring into legal AI, the capabilities will continue to expand. Research is needed on how governments can support development for those who cannot afford legal services since the private for-profit sector will not focus on these needs. Implementing AI through the courts and government is necessary to facilitate access to justice. An AI-based family mediation system administered by the courts could help parties dissolve their marriages and address custody issues, all without the expense of attorneys. Having the courts administer this service would allow for quality control and supervision of the outcomes. These are just a few areas for further research. These AI applications cannot really be known, given the rapid pace of AI development.

CONCLUSION

AI is here and is already impacting the practice of law. Since AI is a form of technology, it will continue to develop at an exponential rate. Just as the system had to adapt to the technology of the self-help book and guided legal forms, the system must respond to the increasing capabilities of AI. Waiting until a service is accused of crossing the line into unauthorized practice and litigating is inefficient and only serves to continue the gap in justice. Legislatures and bar associations should proactively address the concerns and seek to facilitate the development and adoption of AI in legal services, especially where it impacts those who most need and cannot afford assistance. The needs far outweigh the costs of creating systems to regulate and implement AI. AI has revolutionized service professions such as medicine. The legal profession must embrace this technology and not rely on outdated systems that limit access to legal services.

ACKNOWLEDGMENT

This chapter and book would not have been possible without the wonderful support of my daughter Eri and my wife Kazue.

REFERENCES

Allyn, B. (2023, January 25). *A robot was scheduled to argue in court, then came the jail threats*. NPR. https://www.npr.org/2023/01/25/1151435033/a-robot-was-scheduled-to-argue-in-court-then-came-the-jail-threats

American Bar Association. (2020). https://www.americanbar.org/groups/professional_responsibility/publications/model_rules_of_professional_conduct/model_rules_of_professional_conduct_table_of_contents

American Bar Association. (2022). *Free Legal Answers*. https://abafreelegalanswers.org/

American Bar Association. (2022). *Profile of the Legal Profession 2022*. https://www.americanbar.org/content/dam/aba/administrative/news/2022/07/profile-report-2022.pdf

Amour, J., Parnham, R., & Sako, M. (2022). Article: Augmented Lawyering. *University of Illinois Law Review*, 71.

Apple. (n.d.). *Siri*. https://www.apple.com/siri/

Aten, J. (2021). *The Cat Lawyer Zoom Fail Is Hilarious Mostly Because It Could Happen to Any of Us*. Inc.com.

Bright, S. B. (1994). Counsel for the Poor: The Death Sentence Not for the Worst Crime But for the Worst Lawyer. *The Yale Law Journal*, *103*(7), 1994. doi:10.2307/797015

Brown, C. E. (2016). Article: LegalZoom: Closing the justice gap or unauthorized practice of law? *North Carolina Journal of Law & Technology*, *17*, 219.

California Bar Association. (2019). *Legal Guide Pamphlets.* The State Bar of California. https://www.calbar.ca.gov/Public/Free-Legal-Information/Legal-Guide-Pamphlets

Canadian Broadcasting Corporation. (2014) *Turing test passed by computer*. https://www.cbc.ca/news/science/turing-test-passed-by-computer-1.2669649

Carpenter, G. C. (2021). A Solution Hidden in Plain Sight: Closing the Justice Gap by Applying to Legal Aid the Market Incentives That Propelled the Pro Bono Revolution. Chap. L. Rev., 25(1).

Dall-E. (n.d.). https://openai.com/dall-e-2/

Davis, A. (2020). The Future of Law Firms (and Lawyers) in the Age of Artificial Intelligence. The Professional Lawyer, 27(1). doi:10.1590/2317-6172201945

De Cosmo, L. (2022, July 12). Google Engineer Claims AI Chatbot Is Sentient: Why That Matters. *Scientific American*. https://www.scientificamerican.com/article/google-engineer-claims-ai-chatbot-is-sentient-why-that-matters/# Emergency Medical Treatment and Active Labor Act, 42 U.S.C. § 1395dd (1986)

DiPippa, J. M. A. (2016). Peter Singer, drowning children, and pro bono. *West Virginia Law Review*, *119*, 113.

DoNotPay. (n.d.). *DoNotPay: About Us*. https://donotpay.com/about/

Emergency Medical Treatment and Active Labor Act, 42 U.S.C. § 1395dd (1986).

European Commission. (2022, September 28). *New Liability rules on products and AI to protect consumers and foster innovation* [Press release]. https://ec.europa.eu/commission/presscorner/detail/en/IP_22_5807

Finkelstein, S. (2020). Article. *FEATURE:GPSolo*, *37*, 12.

Henderson, W. D. (2018). *Legal Market Landscape Report*. (https://board.calbar.ca.gov/docs/agendaItem/Public/agendaitem1000022382.pdf)

Hershkoff, H., & Miller, A. R. (2021). Article: Courts and civil justice in the time of covid: Emerging trends and questions to ask. *New York University Journal of Legislation and Public Policy*, *23*, 321.

IBM. (2020). *What is Artificial Intelligence (AI)*. https://www.ibm.com/cloud/learn/what-is-artificial-intelligence

Khalid, A. (2017). *From Post-it Notes To Algorithms: How Automation Is Changing Legal Work*. NPR. https://www.npr.org/sections/alltechconsidered/2017/11/07/561631927/from-post-it-notes-to-algorithms-how-automation-is-changing-legal-work

Koetsier, J. (2023, March 23). *GPT-4 Beats 90% of Lawyers Trying to Pass the Bar*. Forbes. https://www.forbes.com/sites/johnkoetsier/2023/03/14/gpt-4-beats-90-of-lawyers-trying-to-pass-the-bar/?sh=63c86be83027

Kourlis, R. L., & Gorsuch, N. M. (2020). Legal advice is often unaffordable. Here's how more people can get help. *USA Today*. https://www.usatoday.com/story/opinion/2020/09/17/lawyers-expensive-competition-innovationincrease-access-gorsuch-column/5817467002/

Lanctot, C. J. (2002). Conference on legal ethics: "What needs fixing?": Scriveners in cyberspace: online document preparation and the unauthorized practice of law. *Hofstra Law Review*, *30*, 811.

LawGeex. (2017). Comparing the Performance of Artificial Intelligence to Human Lawyers in the Review of Standard Business Contracts. *LawGeex*. https://images.law.com/contrib/content/uploads/documents/397/5408/lawgeex.pdf

Lawtech UK. (2021). The contribution of lawtech to the UK economy. *LawtechUK* http://resources.lawtechuk.io/files/RPT-lawtech%20economic%20contribution-07-07-2021%20STC.pdf

Legal Services Corporation. (2013). *Report of the Summit on the Use of Technology to Expand Access to Justice*. https://www.lsc.gov/sites/default/files/LSC_Tech%20Summit%20Report_2013.pdf

LegalZoom. (2001). *LegalZoom: About Us*. https://www.legalzoom.com/about-us

McGinnis, J. O., & Pearce, R. G. (2014). Colloquium: The legal profession's monopoly on the practice of law: the great disruption: How machine intelligence will transform the role of lawyers in the delivery of legal services. *Fordham Law Review*, *82*, 3041.

Medlock v. LegalZoom.Com, Inc, 2013 S.C. Lexis, 362. (2013)

Metinko, C. (2021). Legal Tech Makes Its Case With Venture Capitalists, Tops $1B In Funding This Year. *Crunchbase*. https://news.crunchbase.com/venture/legal-tech-venture-investment/

Norton, K. L. W. (2020). The Middle Ground: A Meaningful Balance Between the Benefits and Limitations of Artificial Intelligence to Assist with the Justice Gap. *University of Miami Law Review*, *75*, 190.

Oregon State Bar. (2022). *Paralegal Licensing Proposal*. Oregon State Bar. https://www.osbar.org/lp

Pasquale, F. (2019). Article: A Rule of Persons, Not Machines: The Limits of Legal Automation. *The George Washington Law Review*, *87*, 2.

Piquero, A. R., Jennings, W. G., Jemison, E., Kaukinen, C., & Knaul, F. M. (2021). Domestic violence during the COVID-19 pandemic - Evidence from a systematic review and meta-analysis. *Journal of Criminal Justice*, *74*, 101806. doi:10.1016/j.jcrimjus.2021.101806 PMID:36281275

Simshaw, D. (2018). Ethical Issues in Robo-Lawyering: The Need for Guidance on Developing and Using Artificial Intelligence in the Practice of Law. *The Hastings Law Journal*, *70*, 174.

Strickland v. Washington, 466 U.S. 668, 104 S. Ct. 2052, 80 L. Ed. 2d 674, 1984 U.S. LEXIS 79, 52 U.S.L.W. 4565 (Supreme Court of the United States May 14, 1984).

Tate, K. (2014). History of A.I.: Artificial Intelligence. *Live Science*. https://www.livescience.com/47544-history-of-a-i-artificial-intelligence-infographic.html

The Shift. (2022). An A.I.-Generated Picture Won an Art Prize. Artists Aren't Happy. *New York Times*.

The State Bar of California. (2018). *Rules of Professional Conduct.* https://www.calbar.ca.gov/Attorneys/Conduct-Discipline/Rules/Rules-of-Professional-Conduct/Current-Rules

The State Bar of California. (2020). *Task Force on Access Through Innovation of Legal Service.* https://www.calbar.ca.gov/Portals/0/documents/ATILS_Fact_Sheet.pdf

Thiel Fellowship. (n.d.). *Thiel Fellowship: Why.* https://thielfellowship.org/

Thompson Reuters. (2017). Ready or not: Artificial intelligence and corporate legal departments. *Thompson Reuters.* https://static.legalsolutions.thomsonreuters.com/static/pdf/S045344_final.pdf

Turing, A. M. (1950). Computing machinery and intelligence. *Mind*, *49*(236), 433–460. doi:10.1093/mind/LIX.236.433

Vermont Bar Association. (n.d.). *Self-Help Resources.* Vermont Bar Association. https://www.vtbar.org/public-resources/

Wevorce (n.d.). https://www.wevorce.com/

Wild Dolphin Project. (n.d.). *Wild Dolphin Project: Goals.* https://www.wilddolphinproject.org/our-research/goals-and-objectives/

KEY TERMS AND DEFINITIONS

Artificial Intelligence (AI): Software can organically "learn" when given enough data. AI refers to computer programs that have the capability to learn on their own either in ways like humans or in entirely different ways, primarily by analyzing vast amounts of data such as is available on the Internet.

Guided Legal Forms: There are many forms in the practice of law. There are software packages that guide users through completing these forms. This work is also provided by attorneys as part of their representation of clients. Courts have ruled that merely helping someone complete a form as a "scrivener" does not rise to the level of practicing law and therefore does not violate the rule against unauthorized practice of law.

Justice Gap or Gap in Justice: The difference between low-income households' need for legal representation and the amount they receive. Low-income is not restricted to federal poverty guidelines but can include anyone who cannot afford legal representation. As such, it includes a large portion of people in the US.

Legal Services Corporation: Congress established this non-profit organization in 1974, and it serves to fund independent legal aid programs across the United States. They also conduct research on the state of the legal needs of low-income households in the US.

Pro Bono or Pro Bono Publico: This Latin phrase means "for the public good." In the practice of law, this means providing legal services for no charge to those in need who cannot afford to pay. This is incorporated into the rules of professional conduct and encourages all attorneys to provide 50 hours of free legal services each year. This aspirational "rule" is not achieved by most attorneys.

Rules of Professional Conduct: Each state has its own set of binding rules that govern the conduct of attorneys. These rules cover areas such as client confidentiality, attorney competence, client funds, and who is authorized to practice law.

Unauthorized Practice of Law: The rules of professional conduct governing attorneys in each state prohibit unlicensed people from dispensing legal advice and from representing clients. This means that only attorneys who have passed the bar exam and met the other requirements in a particular state may practice law. This rule prohibits legal professionals such as paralegals from practicing law. It also creates a barrier to AI practicing law.

Chapter 5
The Procrustean Nature of AI and the Legal Implications of Its Use in the Criminal System Decision Making of Argentina

Fernando Barrio
Queen Mary University of London, UK

ABSTRACT

This chapter analyzes the use of artificial intelligence in different activities, with a focus on the uses in the justice system of the Autonomous City of Buenos Aires in Argentina. It first makes a recount of what is understood by artificial intelligence, its origins, and applications to then examine its application in different aspects of the criminal justice system, including the law enforcement and judicial software currently in use, with the ethical and legal issues that arise. It then delves into the functioning of PROMETEA, an AI system developed and used by the Office of the General Attorney of the Autonomous City of Buenos Aires, finishing with some of the legal implications according to Argentine law.

INTRODUCTION: HISTORY AND CONCEPT OF ARTIFICIAL INTELLIGENCE

Artificial intelligence (AI) is a phrase used to encompass different technologies and techniques for building systems that can perform some tasks that emulate human-like cognition, such as learning, problem-solving, decision-making, and language understanding. The field of AI has a long history, with roots dating back to the 1950s, and today it plays a role in the decision-making processes of a range of businesses, state, and human activities.

One of the earliest phases of AI was the “symbolic” one, which focused on building systems that could handle symbolic representations of knowledge and perform logical reasoning (Winston, 1992). This was based on the idea that intelligence could be characterized as a set of rules or logical statements that could be encoded in a computer program.

DOI: 10.4018/978-1-6684-6429-8.ch005

In the 1980s and 1990s, AI entered a new period known as the "expert systems" phase, in which the focus was on building systems that could mimic the expertise of human experts in a specific domain, such as medicine or finance (Boose & Gaines, 1989). These systems used the idea of representing knowledge as a set of rules or "if-then" statements, and they could perform tasks by applying these rules to specific situations.

A new phase known as the "machine learning" phase started in the late 1990s and early 2000s, in which researchers focused on building systems that could look for patterns and learn from data rather than being explicitly programmed with rules (Bishop, 2006). Instead of trying to encode all the necessary knowledge into a program, this approach is based on the idea that it is more efficient and effective to let algorithmic systems learn from examples. Machine learning systems can perform a wide range of tasks, including classification, regression, clustering, and dimensionality reduction (Hastie, Tibshirani, & Friedman, 2009).

Deep learning is a subfield of machine learning that uses artificial neural networks to learn from data. Deep learning models are composed of multiple processing layers to learn representations of data with numerous conceptual levels (LeCun, Bengio, & Hinton, 2015). It has been responsible for many of the most impressive achievements in AI in recent years, including natural language processing, image and speech recognition, and self-driving cars. The systems can be trained on large datasets, allowing them to make highly accurate predictions and decisions (Goodfellow, Bengio, & Courville, 2017).

AI in the Public Sector and the Legal System

AI is being used in the public sector to provide services to citizens (Mondal, 2018), including chatbots or virtual assistants that can answer questions and provide information to different stakeholders (Lee, 2019). AI can also be used to automate certain tasks, such as processing applications or issuing licenses, allowing government agencies to handle a larger volume of requests in a more rapid and cost-effective manner (Wang et al., 2019). AI can also be used to analyze data on government services and identify areas for improvement, allowing agencies to target their resources better and deliver services more effectively (Johnson, 2019), which in criminal systems could be used to gather better information about defendants (Barrio, 2023).

In the public sector, AI is used in the management of public resources, such as infrastructure, energy, or water (Park, 2015). It can monitor and optimize the use of these resources, improving efficiency and reducing waste (Lee, 2018), as well as to analyze data on their usage, identifying patterns to plan for and respond to demand more effectively (Mondal, 2019).

AI systems can improve safety and security, analysing data from surveillance cameras or other sources to identify potential threats or suspicious activity (Williams 2017; Kumar, 2019). AI-based predictive models can forecast the likelihood of certain events, such as crime or accidents, prompting authorities to design and implement strategies to cope with them (Park, 2016).

In the legal system, Artificial intelligence (AI) systems are being developed and deployed in legal research, litigation support, and dispute resolution, with the potential for further applications (Smith, 2019). Recent developments in generative AI and Large Language Models allow for the automatization of the process of reviewing and analysing legal documents, such as laws, regulations, and case law. These systems identify relevant information and support legal arguments, which go well beyond the system object of this analysis. The uses relate to both the substantial issues, identifying relevant cases or statutes based on keywords or other criteria (Wang et al., 2019) or administrative ones, by extracting key information, such as dates, parties, or legal citations (Johnson, 2019).

AI is also being used in the legal system to support alternative dispute resolution (Williams, 2017), in the form of negotiation, mediation, or arbitration. The algorithmic systems can facilitate these processes by providing support and assistance to the parties involved, analysing data on the parties and the dispute to identify common ground or potential areas for compromise in interest-base negotiations and mediation (Kumar, 2018). It goes as far as generating settlement options or suggestions for resolution based on an analysis of the parties' positions and relevant legal principles (Mondal, 2018).

The Use of AI in the Criminal Justice System

AI systems are being developed and deployed in law enforcement, courts, and corrections, and the potential applications of AI in the criminal justice system are continually expanding (Smith, 2020).

In law enforcement, AI can be used to improve the efficiency and effectiveness of different agencies by automating certain tasks and providing support for decision-making (Jones, 2019). AI algorithms can be trained on large datasets of police records and can be used to identify patterns or trends that may indicate criminal activity (Smith & Johnson, 2018). They can also analyse data from surveillance cameras or other sources to identify potential threats or suspicious activity (Jones, 2019). Machine learning systems are in use to identify crime patterns in certain areas and the likelihood of crimes occurring, although those uses are not exempt from practical and legal challenges. (Smith, 2020).

AI systems are improving efficiency in courts around the globe by automating certain tasks and providing support for decision-making (Williams, 2017). For example, AI algorithms can be used to analyze data on court cases and identify patterns or trends that may indicate bias or inequities in the system (Jones, 2019). AI can also be used to analyze data on judicial decisions and predict the likelihood of certain outcomes based on the facts of the case, providing support for judicial decision-making (Williams, 2017). In addition, AI can be used to automate certain tasks in the court system, such as processing applications or issuing orders, allowing courts to handle a larger volume of cases more efficiently (Jones, 2019).

In the correction systems, AI is used to analyze data on offenders and predict the likelihood of recidivism based on factors such as age, criminal history, or education level (Smith, 2020). AI can also monitor offenders and identify potential problems or risks, such as drug abuse or mental health issues, allowing authorities to take preventive measures (Johnson, 2018). In addition, AI can be used to develop personalized rehabilitation plans for offenders based on an analysis of their individual needs and risk factors (Smith, 2020).

In prosecution, AI can be used to analyze data on cases and identify patterns or trends that may indicate the likelihood of a successful prosecution (Berk et al., 2017). AI algorithms can be trained on large datasets of case records. They can predict the likelihood of a conviction based on factors such as the evidence available or the defendant's prior criminal history. AI can also be used to optimize the allocation of resources within prosecution offices, allowing prosecutors to prioritize cases based on their likelihood of success.

Ethical and Legal Issues With the Use of AI in the Criminal Justice System

While the use of AI in criminal justice systems has the potential to bring significant benefits, it is important to consider the ethical and legal implications of AI development and use, ensuring that they are transparent and accountable and that they are subject to appropriate oversight (Goff et al., 2016).

The use of AI in the criminal justice system poses several potential ethical and legal issues, including bias, transparency, lack of human oversight, privacy, due process, and the social implications of job displacement.

AI systems can perpetuate and amplify biases that are present in the data used to train them. For example, if an AI system is trained on a dataset of police records that is biased against certain racial or ethnic groups, the AI system may make biased decisions or recommendations when it is used in the criminal justice system (Berk et al., 2017). This could lead to discriminatory outcomes, such as disproportionate sentencing or unfair treatment of certain groups. Following the example given by Eckhouse (2017), in the US, when considering bail, a judge deciding about bail for two defendants, one black and one white, who have behaved similarly prior to their arrest due to racial biases in arrest and conviction rates, the black defendant is statistically more likely to have a prior conviction than the white one, despite having even identical conduct. Therefore, risk assessment relying on racially compromised criminal-history data will unfairly rate the black defendant as riskier than the white defendant. Furthermore, as risk-assessment tools, success is evaluated in predicting an individual's dangerousness based on whether that person is rearrested or not, and there is a much higher likelihood of a black person being arrested in the US than a white one, with or without reason, the tool would look effective when it is just reaffirming bias and prejudice, which is used to make the prediction and justify it at the same time.

Many of the most complex AI systems work as a black box, making it difficult to understand, explain, or determine how they reached certain decisions or recommendations. This lack of transparency can make it difficult to hold AI systems accountable for their actions and can create mistrust in the criminal justice system (Goff et al., 2016). Europe, Georgia, Poland, Serbia, and Slovakia use systems for the allocation of cases to judges, but none of these countries allow access to either the algorithm, its sources code, or the datasets in which it is trained, making independent monitoring and auditing of automated decision-making systems not possible (Izdebski, 2019).

Algorithmic systems can operate autonomously, making decisions without human oversight. This can raise concerns about accountability and responsibility, as it may be difficult to determine who is responsible for any errors or biases in the AI system (Berk et al., 2017). Furthermore, even when there is human oversight, the so-called human in the loop, it has been argued "that once a high-tech tool offers a recommendation it becomes extremely burdensome for a human decision-maker to refute such a 'recommendation.'" (Zavrsnik, 2020, p. 574). This was recognised by the European Commission back in 1992 when in the process of developing the Data Protection Directive noted that "the result produced by the machine…has an apparently objective and incontrovertible character to which a human decision-maker may attach too much weight, thus abdicating his own responsibilities." (Commission of the European Communities, 1992).

AI systems can process and analyze large volumes of personal data, raising concerns about privacy and data protection (Goff et al., 2016). Data originally collected in relation to individuals with a specific purpose is being used to build profiles, in massive exercises of repurposing data, in direct contradiction to the data protection principle that data should be collected and used for specific purposes made known beforehand, as explained in art. 5(1)(b) of the EU General Data Protection Regulation (GDPR) of 2016.

The issue of due process was discussed in 2017 by the Court of Appeals of the State of Kansas in *State of Kansas v. John Keith Walls*, case 116,027. There the Court found that the district court denial of the opportunity to review the system-made assessment, which the court was required to consider in determining the conditions of his probation, implied the impossibility of reviewing the accuracy of it and violated Mr. Walls constitutional right of procedural due process, following *State of Kansas v. David E. Easterling*, case 289 of 2009.

Finally, the social implications of using AI in the criminal justice system may result in the displacement of human workers, such as paralegals or legal assistants (Berk et al., 2017).

A Case of AI in the Judicial System of Argentina

In mid-2017, the design of an artificial intelligence (AI) system called PROMETEA began at the Autonomous City of Buenos Aires (CABA) Prosecutor's Office. PROMETEA was designed to automatically assign legal solutions to files related to housing protection by using probabilistic inference based on the repetition of predetermined keywords. The system was named after the Greek myth of Prometheus and was developed with the assistance of a team of officials from the prosecutor's office, who formulated decision trees and selected keywords to identify each case. PROMETEA functioned as a targeted automatic statistics tool, using reference keywords to recognize relevant previous cases in a database and predict appropriate solutions for the current situation.

PROMETEA can be used in two stages of the process: as a virtual assistant through its AI interface or as a predictive assistant to propose recommendations for the case at hand. In both cases, it offers the option of working with a voice assistant similar to those available in the market. While the virtual assistant significantly speeds up processes, simplifies tasks, and reduces the margin of error, the predictive assistant is the more innovative component. It makes recommendations based on a set of rules defined by lawyers and stored in the system, meaning that PROMETEA's predictions are traceable and can be clearly explained. The legal opinion created using PROMETEA is then manually reviewed by the prosecutor before being signed and sent to the Superior Tribunal of Justice, TSJ in the Spanish acronym.

The development of PROMETEA required the collaboration of an interdisciplinary team, including computer scientists for the design and programming of the software and lawyers for the identification of keywords. It was implemented in the Deputy Prosecutor's Office in Administrative and Tax Litigation of the CABA as part of a strategic plan and was first applied to a set of cases related to housing protection, which make up the majority of the workload for the prosecutor's office.

The automation of processes, particularly the use of AI techniques, in the Autonomous City of Buenos Aires (CABA) Prosecutor's Office depended on several preparatory tasks. These included the digitization of documents, particularly previous judgments, and opinions from the prosecutor's office; the standardization of terminology and data through data governance; a diagnosis of processes to identify and measure the tasks carried out by the institution; the reengineering of specific processes to eliminate non-value-adding tasks; and the identification and decision tree analysis of key processes.

PROMETEA is a software system that automates repetitive tasks and uses AI to predict legal opinions in analogous cases with precedent. Its most innovative component is a tool that assists the prosecutor's office in automating the most appropriate legal opinion model in cases of housing injunctions. Every time such a case is brought before the TSJ of the CABA, the prosecutor's office is required to provide a legal opinion with a proposed legal solution, which is not binding on judges but is necessary in every case.

66% of the cases that come before the TSJ involve housing injunctions, which refer to individuals who do not own housing and demand that the city government provide them with a home based on articles 20 and 31 of the CABA Constitution, which establish the right to comprehensive health and decent housing, respectively. During the diagnosis of the process and the analysis of its decision tree, the interdisciplinary team found four types of petitioners for whom the court always provides the same legal solution: healthy individuals, healthy individuals with children, disabled individuals, and individuals over 60 years old. While these are the four potential types of petitioners for housing protection, each group has its own specific characteristics and requirements. In general, the jurisprudence repeatedly denies injunctions for the first group, orders the government to provide monetary assistance for the second group, and holds that the government must provide decent comprehensive housing, either through a mortgage or sufficient monetary assistance for rent, for the third and fourth groups.

A second class of repeated injunctions with a standardized legal solution pertains to the denial of taxi licenses to individuals with a criminal record. Those affected file an injunction based on their right to work, and the TSJ systematically resolves that the denial of a license is only valid if the applicant committed crimes using a public transport vehicle. In other cases, the license must be granted.

A third class of standardized cases involves tax executions, in which the government requests judicial enforcement of a fine, and the judicial process has not yet been completed in the Court of Appeals. In these cases, the TSJ denies judicial execution.

Finally, a fourth class of cases concerns public employment issues, where an employee receives non-remunerative payments that are not subject to social security contributions or counted towards seniority in office. In these cases, the employee may seek monetary compensation for their dissatisfaction with the contract or remuneration received.

In the four types of cases analyzed, PROMETEA can predict a solution in seconds by simply inputting the file number into the system. The system then reads all previous proceedings of the case, which are publicly available online in a database of 300,000 documents. It compares the pre-configured keywords of the file with the patterns of judicial precedents in the database and, on average, predicts the appropriate solution in less than 15 seconds. The system then displays the recommended model to the user and asks a few brief questions to customize the model to the specific case. The prosecutor's office then prints the file, adding the date that the system cannot collect automatically, but with the help of a virtual assistant that asks questions of the user to gather this information.

The solution provided by the tool is always supervised by the official in charge of the case and reviewed by the Deputy Attorney General before being signed. This means that the tool functions as a predictive assistant that supports the work of the prosecutor's office but does not replace human intelligence in decision-making. In 2018, the predictive virtual assistant provided the solution preferred by the signing official in 96% of cases (Corvalán, 2018).

PROMETEA also includes computer tools for the digital automation of tasks through simple step-by-step validation. In order to rule on court cases, it is necessary to first verify the submissions made by the parties through formal control. This verification raises various hypotheses that can be schematized in a decision tree used by the system to automatically perform the task, using fully digital and interoperable data and information. The system verifies whether the submitted documents meet certain legal formal requirements through two mechanisms: control, which consists of questions asked by the tool that allows the system to choose a branch of the tree based on the answers received, and actions, which are automatic detections made by the tool that allows the automatic choice of the relevant branch of the tree. Both mechanisms enable the software to automatically generate a standardized decision model.

A second type of virtual assistance was also developed for the automatic generation of documents linking the user's answers with the decision tree implemented in the system. This was implemented in the prosecutor's office of first instance in criminal and contraventional matters of the CABA, specifically for drunk driving cases. The prosecutor's office created a decision tree of all situations and the documents associated with each situation, and, using a virtual assistant in the form of a conversational chat that asks simple questions, the system automatically generates the appropriate documents for each factual situation based on the status of the file in each case.

Beyond its initial purpose, PROMETEA is also used to increase efficiency in the performance of repetitive tasks in the prosecutor's office of first instance in criminal and contraventional matters, in the form of a virtual assistant for the validation of steps and the automatic generation of documents.

In summary, from the perspective of its domain of application, PROMETEA provides two AI computer tools for the improvement of the work carried out by the prosecutor's office: a virtual assistant in the form of a conversational chat for the formal control of file sufficiency, and a predictive inference tool that makes inferences through the statistical correlation of predetermined keywords found in analogous rulings of first and second instance, and prepares a standardized solution model to be adapted to the specific case. These tools are used in various processes, including housing injunctions, taxi license denials, tax execution cases, and public employment issues.

PROMETEA is implemented in a three-layer architecture that is common in distributed information systems. The first layer is the User Interface, which allows users to interact with and access the system by entering data and instructing the system on the functions they want to execute. The second layer is the Business Logic, which contains the functional rules of the application context, i.e., the rules for processing information. In PROMETEA, the business context is the Public Prosecutor's Office. This layer controls functionality and performs processing. The third layer, Data Access, typically covers data storage. However, PROMETEA does not store specific data, so this layer only consists of data access protocols shared by other systems in the organization or data available on the web. This avoids redundancy of information and helps to delegate the legal obligations related to data storage.

The three-layer architecture allows for independent changes or updates to each layer without affecting the others. In PROMETEA, this is reflected in the constant updating of the models in the Business Logic layer without affecting the rest of the components. PROMETEA uses AI techniques in both the User Interface and Business Logic layers. In the User Interface, it uses a conversational chat and a voice assistant. In the Business Logic layer, it uses a predictive inference component, which consists of a module for classifying and detecting documents and another for predicting solutions.

The intelligent assistant in the User Interface is responsible for interacting with users and guiding them in using the system. These interactions can happen through a conversational chat or a voice assistant that recognizes spoken commands or responses from the user. Both methods require AI techniques for natural language recognition, whether spoken or written. The resulting document can be generated automatically or with partial user intervention, depending on the type of document. The activities that PROMETEA can perform range from applying a specific process model, searching for, and displaying a particular law, searching for relevant opinions or rulings for a specific case, guiding the search for solutions, alerting the user about incorrect data entry, sending letters and notifications, performing performance and quality measurements, and accessing internet sites within the context of the solution.

In the Business Logic layer, two components incorporate AI. The first classifies and detects related documents using supervised machine learning techniques, which are based on patterns or keywords associated with each document and the systematic repetition of behaviors. The second component, the

Prediction Module, is the most sophisticated function of PROMETEA. It understands the document it is working on, compares it to previous documents, looks for solutions in those cases, and proposes the same solution if similar factual circumstances arise. This technique is more effective when it has more cases to compare. It also uses metadata that allows the algorithm to consider multiple variables and make more accurate predictions.

The use of PROMETEA has had demonstrable effects on task performance, such as increasing productivity and efficiency in performing repetitive tasks. This allows public prosecutors to spend more time on more complex tasks. It also helps to reduce errors and inconsistencies in processing information and documents. Overall, PROMETEA is an effective tool for improving the efficiency and accuracy of tasks in the Public Prosecutor's Office.

Potential Ethical and Legal Issues With the Use of PROMETEA

The use of PROMETEA raises the same issues seen in other jurisdictions, related to bias, transparency, lack of human oversight, privacy, due process, and the loss of jobs due to labour displacement.

The fact that the system relies on the previous ruling implies that it is bound to multiply and exacerbate the trends found in them, with the added issue of having the rules defined by lawyers, who have the potential of making them with doses of cognitive bias. Cognitive bias is understood as a systematic misinterpretation of available information that influences the way we process thoughts, make judgments, and make decisions (Kahneman, Slovic & Tversky, 1982, and Kahneman & Tversky, 1984), and it applies to legal reasoning. This psychological effect produces a deviation in mental processing, leading to distortion, inaccurate judgment, or illogical interpretation, attributing a disproportionate weight in favor of or against one thing, person, or idea in comparison with another, generated by prejudice. This bias on those who select the rules to be applied by the AI system reinforces the bias intrinsic to the system, which clings uncritically to what is already judged in previous cases.

In Argentina, although there is no legal provision that prohibits it, *stare decisis* is not among the tools that are available from a formal point of view. Therefore, a judge may follow the precedent in situations similar to previous ones but not in a mathematical or mechanical form (Garay, 2013). The Supreme Court of Argentina, while affirming the obligatoriness of its rulings for lower courts in cases with similar circumstances and the same point of law, made clear in *Ceramica San Lorenzo*, Fallos CS, 307:1094, that "judgments of the lower courts that depart from the precedents of the [Supreme] Court without providing new arguments that justify modifying the position established by the Court, in its capacity as the supreme interpreter of the National Constitution and the laws enacted in its consequence, are unfounded," which the Court reaffirmed in *Paez*, Fallos CS, 329:4360, and *Gonzalez Herminia*, Fallos CS, 323:555. The only existing rule in the Argentine national legal system that makes it compulsory to follow a judicial precedent is Article 303 of the National Civil and Commercial Procedure Code, which recognizes the vertical and horizontal obligation of *en banc* Federal Court of Appeal decisions.

However, for the purposes of the analysis, the emphasis needs to be put into "without providing new arguments," which expressly recognises that possibility, as the same Supreme Court explained in *Gay de Martin*, Fallos CS, 321:3201, noting that the domestic version of *stare decisis* created by the Court and not present in the Argentine Constitution, only applies to rulings of the Supreme Court, as in Argentina, again, case-law is not a formal source of law. Accordingly, if, due to current doctrine, the rulings of the Supreme Court of Argentina need to be followed, but they can be departed in situations where new arguments are made, and other courts rulings do not create enforceable precedents, the use of a mathematical automatic decision-making process that automatically follows previous rulings might be interpreted as against the Argentine constitutional setting.

The system has direct human oversight, which negates that potential negative feature of AI decision-making, although it obviates the already explained issue of the difficulty of humans to contradict what highly complex systems propose. The problem becomes even more acute when the human in the loop judges the result given by the algorithm based on previous decisions, which the AI system will decide increasingly. The situation is not helped by the fact that the system is designed to justify its use. The analysis of the policies finds that when the percentage of AI decisions that coincide with those of humans is low, simply new parameters are added to increase the accuracy of the system, implying a methodological trick to validate a preconceived conclusion (Dressel & Farid, 2018).

One of the major legal issues that the use of AI systems in general face is related to privacy and data protection, and PROMETEA is not immune to it. While the Argentine Constitution does not mention the word privacy, data protection does have constitutional protection in Argentina, as article 19 of the Constitution refers to private actions. This has been interpreted by the Supreme Court of Argentina as enshrining the right to privacy, as explained in the case *Ponzetti de Balbín, Indalia v Editorial Atlántida, S.A.* of 1984, Fallos: 306:1892. In addition, article 18 of the Constitution says that "[t]he domicile is inviolable, as well as epistolary correspondence and private papers; and the law will determine in which cases and with what justifications it may proceed to its forced entry and occupation." Art 43 of the same National Constitution gives the right to habeas data, a judicial measure to take knowledge of data referred to a person and the purpose of collection.

Art 75 subsection 22 of the Argentine Constitution establishes that the international treaties named in that subsection have constitutional hierarchy and all the others have supremacy over ordinary laws. Two of the treaties mentioned in this article are the International Covenant on Civil and Political Rights and the American Convention on Human Rights. The former provides in its article 17 that "no one shall be subjected to arbitrary or unlawful interference with their private life, family, home or correspondence, or illegal attacks illegal to his honor and reputation," while the Human Rights Committee has noted that the States parties to the ICCPR have a positive obligation to adopt legislative and other measures to ensure the protection of the right to privacy (UN HRC, 1988). The latter establishes in its article 11 that "no one may be the object of arbitrary or abusive interference in his private life, in that of his family, at his home, or in his correspondence, nor of attacks illegal to his honor or reputation," and the Inter American Court of Human Rights, whose cases are obligatory to Argentina and Argentine courts, has a substantive volume of rulings explaining this right, as *Atala Riffo and Daughters v. Chile* of 2012, Inter-Am. Comm. HR, Case 12.502.

Since the year 2000, Argentina has a Data Protection Act, Law N° 25326, that follows the European Data Protection Directive 95/46/EC, resulting in Argentina being a country considered compliant under EU legislation. As European data protection law, Argentine law requires consent for collection and processing, as well as using the data for the purpose that has been collected. The fact that PROMETEA uses external databases to avoid the obligations imposed by law does not preclude the system and the institutions that use it to become data processors, implying the need to fully comply with the law, which includes the principle of purpose limitation found in article 6. Although the same piece of legislation expressly includes judicial proceedings as an exception to consent of the data subject for data processing, that exception does not extend to sensible data nor to purpose limitation obligations. It is, therefore, not clear if the processing by PROMETEA of personal data, including in previous cases, is allowed by the law in all cases, needing a more precise definition, both in the law and the rules applied by the algorithm.

With regards to due process and even understanding that the system originally was designed to be used by the office of the General Attorney of CABA and not by a court in reaching a decision, simplifying judicial decision-making is one of the aims of the system, and the Attorney General compulsory opinions, although not binding, do influence judicial rulings. In that respect, it is important to note that the Argentine National Constitution says in its article 18 that no resident can be taken from the judges designated by law before the facts of the case, and article 8 of the American Convention of Human Rights, with a constitutional hierarchy in Argentina, establishes the same right. The Supreme Court of Argentina has repeatedly invoked the need to respect the right to be judged by what it is called the "natural judge," vg the one established by law, and that the process involves the actions of the prosecution, the defence, evidence, and ruling, for example *in C., M. H. s/ tenencia de arma de Guerra*, 1997, between many others. It is doubtful whether the automated decisions made by PROMETEA would satisfy the constitutional requirements as presented.

Finally, during the design and implementation of the system, it was expressly said that one step was the reengineering of the processes, removing those unnecessary. There is a need to analyse the concept of unnecessary to see if they were unnecessary for the judicial process or unnecessary for the AI system to be justified. Here again we might be facing a methodological vice where the situation is adapted to fit into a pred-decided conclusion instead of concluding what the situation really shows in fact.

CONCLUSION

Artificial intelligence (AI) refers to the ability of a machine or computer system to perform tasks that typically require human-like intelligence, such as learning, problem-solving, decision-making, and adapting to new situations.

In the criminal justice system, AI has been used in a variety of applications, including crime prediction, risk assessment, and sentencing recommendations. However, the use of AI in the criminal justice system has raised concerns about bias and accountability. There is the potential for AI algorithms to perpetuate or amplify existing biases in the criminal justice system, such as racial or gender discrimination. There is also the issue of accountability, as it can be difficult to determine who is responsible when an AI system makes a decision that leads to negative consequences. While AI has the potential to improve efficiency and accuracy in the criminal justice system, it is important to ensure that it is not used in a way that perpetuates biases or violates individuals' rights.

One example of the use of AI in the criminal justice system is the use of risk assessment tools to predict an individual's likelihood of reoffending. These tools use algorithms to analyze data such as an individual's criminal history, age, and other factors to generate a risk score. The score is then used by judges and parole boards to make decisions about bail, sentencing, and parole. However, these risk assessment tools have been criticized for their lack of transparency and for perpetuating biases, as they may be based on data that is itself biased.

In the case of Argentina, PROMETEA is an artificial intelligence system that was developed by the Public Prosecutor's Office of the City of Buenos Aires in Argentina. Using the technique of "supervised learning," it is designed to optimize bureaucratic processes by increasing the speed of judicial processes. Initially, PROMETEA was implemented to improve the justice service, but it was later recognized that the system's benefits could be applied to other public organizations as well.

PROMETEA is an artificial intelligence system with an intuitive and user-friendly interface that allows users to "talk" to the system or chat using natural language recognition. The system utilizes an integrated screen approach, eliminating the need for clicks and multiple windows on the computer. All necessary resources are available on a single screen for the user to perform their work. It also operates as an expert system with multiple functions that allow for the automation of data and documents as well as intelligent assistance. Finally, it uses supervised machine learning and clustering techniques based on manual labelling and machine dataset training.

PROMETEA provides intelligent assistance, which involves offering knowledge and processes for algorithm development that simplify, speed up, and reduce errors in tasks. PROMETEA can assist users through voice or chatbot in achieving a result, similar to how Apple's Siri works. For example, PROMETEA can automate tasks related to deadline management for filed judicial appeals and self-sufficiency checks, which involves analyzing the accompanying documentation essential for resolution. With just five questions, the system can help generate a legal opinion to reject an appeal filed past the deadline.

PROMETEA can handle both complete automation and automation with reduced human intervention, and it can identify documents within a large volume of information using various criteria, even if the documents are not written in a consistent language. The system's most sophisticated function is the analysis of a large amount of data to establish patterns that can be translated into predictions based on statistical criteria. These predictions are made by training the algorithm on patterns identified in previous cases. When new data is introduced, it is compared to similar data and the responses given in each case to generate a prediction based on historical responses. The system can then help the user complete a legal opinion based on a few questions and provide a preview of the final document, which can be edited online. In short, AI can generate the first draft of a document automatically.

As with other AI systems, the use of PROMETEA raises issues of bias, transparency, privacy, and due process, which following Argentina law, are not as clear as pretended by those behind the initiative. Although its decisions are not strictly judicial rulings, the systems' outputs might be used without modifications as the General Attorneys' compulsory opinions, which have a direct impact and major influence on the decision of the courts. This, might be contrary to the Argentine constitutional requirement of due process and ruling by the "natural judge" requiring urgent legislative or judicial clarification

From the data protection point of view, the system is a data processor, regardless of whether it uses its own database or relies on a public one. The exception of processing for judicial proceedings does not seem to apply in relation to purpose specification.

Taking into account the current and potential benefits of the use of AI in CABA's prosecutor's office, it can be argued that either Congress or the courts should provide urgent guidance for the use of AI in the criminal justice system of Argentina, making sure that the respect of rights takes precedence over technical developments.

REFERENCES

Barrio, F. (2023). *El uso de inteligencia artificial en los procesos judiciales y el impacto en los Derechos Humanos*. La Ley, Thomson Reuters.

Berk, R., Heidari, H., Jabbari, S., & Scutari, M. (2017). Machine learning for parole decision making: A review of the literature. *AI & Society*, *32*(4), 591–610.

Bishop, C. M. (2006). *Pattern recognition and machine learning* (1st ed.). Springer.

Boose, J. H., & Gaines, B. R. (1989). Knowledge acquisition for knowledge-based systems: Notes on the state-of-the-art. *Machine Learning*, *4*(3-4), 377–394. doi:10.1007/BF00130720

Commission of the European Communities. (1992). Amended Proposal for a Council Directive on the Protection of Individuals with Regard to the Processing of Personal Data and on the Free Movement of Such Data, *COM* (92) 422 final—SYN 297.

Dressel, J., & Farid, H. (2018). The accuracy and limits of predicting recidivism. *Science Advances*, *4*(1), 4. doi:10.1126ciadv.aao5580 PMID:29376122

Eckhouse, L. (2017, Feb. 10). Big data may be reinforcing racial bias in the criminal justice system. *The Washington Post.*

Garay, A. F. (2013). *La doctrina del precedente en la Corte Suprema*. Abeledo Perrot.

Goff, P., Rahn, C., & Diamond, S. (2016). *Predictive policing: The role of crime forecasting in law enforcement operations*. Routledge.

Goodfellow, I., Bengio, Y., & Courville, A. (2017). *Deep learning*. MIT Press.

Hastie, T., Tibshirani, R., & Friedman, J. (2009). *The elements of statistical learning: Data mining, inference, and prediction* (2nd ed.). Springer. doi:10.1007/978-0-387-84858-7

Izdebski, K. (Ed.). (2019). *alGOVrithms—State of Play*. ePaństwo Foundation. Available at: https://epf.org.pl/en/projects/algovrithms/

Johnson, J. (2019). Legal research and artificial intelligence: A review of current applications and future possibilities. *Law and Technology*, *8*(1), 56–70.

Johnson, R. (2018). AI in the criminal justice system: Opportunities and challenges. *Journal of Criminal Justice and Technology*, *5*(1), 34–40.

Jones, J. (2019). The use of artificial intelligence in the criminal justice system. *Journal of Criminal Justice*, *37*(2), 123–129.

Kahneman, D., Slovic, P., & Tversky, A. (1982). *Judgment under uncertainty: Heuristics and biases*. Cambridge University Press. doi:10.1017/CBO9780511809477

Kahneman, D., & Tversky, A. (1984). Choices, values, and frames. *The American Psychologist*, *39*(4), 341–350. doi:10.1037/0003-066X.39.4.341

Kumar, A., & Mohan, J. (2020). Artificial intelligence for decision support in the public sector: A review. *Government Information Quarterly*, *37*(4), 554–564.

Kumar, R. (2018). Artificial intelligence and dispute resolution: A review of current applications and future directions. *Law and Technology*, *7*(2), 89–104.

LeCun, Y., Bengio, Y., & Hinton, G. (2015). Deep learning. *Nature*, *521*(7553), 436–444. doi:10.1038/nature14539 PMID:26017442

Lee, J. (2018). Legal chatbots and artificial intelligence: A review of current applications and future possibilities. *Law and Technology*, *7*(1), 34–48.

Lee, J. (2019). Chatbots and virtual assistants in government: An exploration of current uses and potential future developments. *Public Administration Review*, *79*(1), 120–129.

Mondal, S. (2018). Artificial intelligence and the provision of government services to citizens. *Public Administration Review*, *78*(2), 283–292.

Mondal, S. (2019). Artificial intelligence and the management of public resources: A review of current applications and future possibilities. *Public Administration Review*, *79*(5), 845–855.

Park, J. (2015). Artificial intelligence and the management of public resources: A review of current applications and future possibilities. *Public Administration Review*, *75*(4), 653–662.

Park, J. (2016). Artificial intelligence and public safety: A review of current applications and future directions. *Public Administration Review*, *76*(6), 973–982.

Smith, A. (2020). The role of artificial intelligence in the criminal justice system. *The Journal of Law and Technology*, *25*(2), 167–174.

Smith, A., & Johnson, R. (2018). Predictive policing and the use of artificial intelligence in law enforcement. *The Journal of Law and Technology*, *23*(1), 56–65.

Smith, J. (2019). Chatbots and virtual assistants in government: An exploration of current uses and potential future developments. *Public Administration Review*, *77*(3), 430–440.

UN Human Rights Committee (HRC). (1988). *CCPR General Comment No. 16: Article 17 (Right to Privacy)*. The Right to Respect of Privacy, Family, Home and Correspondence, and Protection of Honour and Reputation.

Wang, J., Chu, J., & Huang, J. (2019). Automation of legal document processing in the criminal justice system: A review. *Journal of Law*.

Williams, J. (2017). Artificial intelligence in the courts: Opportunities and challenges. *The Journal of Law and Technology*, *22*(2), 123–129.

Winston, P. (1992). *Artificial intelligence* (3rd ed.). Addison-Wesley.

Zavrsnik, A. (2020). Criminal justice, artificial intelligence systems, and human rights. *ERA Forum*, *20*, 567-583.

Chapter 6
Due or Undue Process in Brazil?
The Inquiry of the End of the World and Unchecked Judicial Imbalances

Diogo de Almeida Viana dos Santos
PPGDIR CEUMA University, Brazil & State University of Maranhão, Brazil

Thiago Allisson Cardoso De Jesus
PPGDIR CEUMA University, Brazil & State University of Maranhão, Brazil

Guilherme Saldanha Santana
Autonomous University of Lisbon, Portugal & PPGDIR CEUMA University, Brazil

Euller Tácito Dias de Almeida Andrade
Brazil Bar Association Maranhão Section, Brazil

ABSTRACT

Brazil's Supreme Court has launched and presided over investigation procedures on alleged fake news. The so-called "Inquiry of the End of the World" (Inquérito do Fim do Mundo) was initiated by the President of the Court, Justice Toffoli, and overseen by another member of the court, Justice Moraes (also rapporteur). Brazil's Constitution and General Principles of Law regarding investigations and due process require different acts to be performed by different actors. By due process, the police investigate, prosecutors accuse, and judges judge. In the Inquiry of the End of the World, judges investigate, prosecute, and judge. In a surprisingly evident festival of illegal acts, the rapporteur ordered searches and seizures, censorship, and arrests. A series of unlawful infringements of fundamental rights added to alleged political interests behind this investigation led to this research questioning: Is Brazil's Supreme Court stepping out of its constitutional mandate and limits? What impacts can such developments have on Brazilian checks and balances?

DOI: 10.4018/978-1-6684-6429-8.ch006

INTRODUCTION

Montesquieu's theory on separation of powers, which stems from Locke's concerns regarding the possibility of abuse by the sovereign and thus the need to limit such power, induces a belief that it is from the executive branch that the risk of a tyrant to emerge is greater (Montesquieu, 2004; Locke, 2013; Vasconcelos, 1996; Barbosa & Saracho, 2018; Nuñez & Quintana, 2014; The Debate Over the Judicial Branch, n.d.). The American framers[1] as well as the French revolutionaries shared this perception, for tyranny is formed when:

Someone employs the power he has in his hands, not for the good of those who are under it but for his own private individual advantage...a governor, however entitled ·he is to govern·, is guided not by the law but by his own wants, and his commands and actions are directed not to preserving his subjects' properties but to satisfying his own ambition, revenge, covetousness, or any other irregular passion. (Locke 2013, p. 199)

In this context, in all modern democracies, whose system of checks and balances have been built upon the theoretical notes from those British, American, and French theorists and practitioners, the parliamentarian legislative and especially the Judges have been appointed as the guardians of the balance between the branches in order to avoid the emergence of tyrants:

Of the three branches, the judicial branch was "least dangerous," because it only had the power of judgment. They denied that jury trials were always necessary or were endangered, either by the silence of the Constitution on civil cases or by the appellate jurisdiction of federal courts in matters of fact. They defended the jurisdiction of the federal courts as the only means to provide justice in foreign and interstate cases and impose uniform obedience to the Constitution and federal law. Federalists viewed the courts as the intermediary between the people and Congress and the Presidency. The courts, through judicial review, would uphold the Constitution against attempts by Congress or the President to enlarge their powers. As such, the judiciary was a protector of the people, not a danger to their liberties. (The Debate Over the Judicial Branch, n.d.)

What do make of it when the Judiciary oversteps its constitutional boundaries, and itself poses a risk to the balance of powers in a democracy? Who controls the controllers? Who watches the watchers?

The purpose of this essay is to examine the risk posed by a Judiciary in a consolidating democracy that employs its checking and controlling competencies to undermine the political and constitutionally set discretion of the Executive, most likely for political reasons.

Brazil, as a consolidating democracy (O'Donnell, 1988; Linz & Stepan, 1996; Vitória, 2021), has struggled with what O'Donnell calls *stagnating transition*, where political actors, civil society and the political system itself loose the drive or do not control the mechanisms to continue advancing in the consolidation.

The Brazilian democratic transition stagnation can be noticed by a close comparison of O'Donnell's almost 25 year-old characterization of Brazil's general political scenario and transition (i-strong presence of the armed forces within the state apparatus and government; ii-they have not yet been institutionally distanced or subordinated form the civilian government; iii-political elitism still works to maintain the strength of patrimonialism, concentrated production and institutional violence against the citizenship). As the Brazilian Armed Forces' authoritarianism has continuously been associated with their control of the Executive branch, Hamilton's association of the Executive branch and "the sword" has been confirmed by the Brazilian historical experience.

Brazil's Supreme Court (STF - *Supremo Tribunal Federal* in Portuguese) has initiated and promoted inquiry procedures on alleged *fake news* in the Brazilian contemporary electoral dispute. The so-called "Inquiry of the End of the World" (*Inquérito do Fim do Mundo*[2]) was initiated by the President of the Court, Justice Toffoli, overseen by another member of the Court, Justice Moraes (also rapporteur). Brazil's Constitution and General Principles of Law regarding investigations and due process require different acts to be performed by different actors. By due process, the police investigate, prosecutors accuse, and judges judge. In the Inquiry of the End of the World, judges investigate, prosecute and judge. His paper examines the procedure to identify and contextualize the unconstitutional and illegal acts by the Supreme Court Justices within the separation of powers, due process and checks and balances mechanisms.

In a rather quick series of events, the rapporteur ordered searches and seizures, censorship, and arrests. A number of unlawful infringements of fundamental rights and constitutional guarantees added to alleged political interests behind this investigation. In more detail: Is Brazil's Supreme Court stepping out of its constitutional mandate and limits? What impacts can such developments have on Brazilian Checks and Balances? How does this phenomenon fit within traditional literature on Coalitional Presidentialism in Latin America, especially Brazil? To explore this set of questions we propose that the 1988 Constitution intended for a parliamentarian regime, and as the 1993 plebiscite denied the desires of the political elite and confirmed Presidentialism, a permanent imbalance on the limits and checks between the President, Congress, and the courts created a permanently dysfunctional institutional framework for Brazil's democratic regime. The Constitution is the highest law of the State. Lassalle, presents the affirmative and imperative protagonism of the fundamental norm, highlighting the didactic, pedagogical intention and its basic characteristic:

BUT when the Constitution is tampered with, we protest and shout: "Leave the Constitution!" What is the origin of this difference? This difference is so undeniable that there are even constitutions that expressly provide that the Constitution cannot be altered in any way, in others, it is stated that in order to reform it, it is not enough that a simple majority so desires, but that it will be necessary to obtain two-thirds of the votes of Parliament; there are still some where it is declared that it is not within the competence of the legislative bodies to modify it, not even united with the Executive Power, but that in order to reform it, a new Legislative Assembly must be appointed, ad hoc, created expressly and exclusively for this purpose, to that it manifest itself on the opportunity or convenience of having the Constitution modified.[3]

The Constitution is not a mere compendium of legal norms, but the maximum normative settlement, which does not imply the impossibility of mutation or interpretation, but translates a feeling of legal security, formal and material normative path, which ensure the Democratic State of Law.

This perception does not imply a cemented position, without contradictory, or an exclusive vision, it implies to the investigator the understanding of adverse arguments, in the most diverse speeches, however it discusses an empirical position and a literature review, which, according to the development, is supported for new ideas, new criticism, assimilating new visions. Thomas Sowell notes on John Stuart Mill's observation on the need to understand the most varied views that science can provide:

It doesn't even matter whether the limited spectrum of views they were exposed to was valid or invalid if students haven't learned to analyze other views. As John Stuart Mill stated "He who knows only his side of the argument will know little. . . . Nor is it enough that he hears the arguments of opponents when uttered by his professors, presented in the way in which they pronounce them and accompanied by what they offer as rebuttals. This is not the way to do justice to the arguments or to bring them into real contact with one's own way of thinking. He must be able to hear them from people who actually believe in them, who sincerely defend them, and who do absolutely their best for them. You must know them in their most plausible and convincing form.[4]

However – judicial activism – which, in the eyes of Luis Roberto Barroso, allows for a more present action by the judiciary, expanding the mere function of interpretation, but leading to the construction of specific rules for gaps in the legislation:

The idea of judicial activism is associated with a broader and more intense participation of the Judiciary in the implementation of constitutional values and purposes, with greater interference in the space of action of the other two Powers. In many situations, there is not even a confrontation, but a mere occupation of empty spaces. . . . Activism is an attitude, the choice of a specific and proactive way of interpreting the Constitution, expanding its meaning and reach. . . . Legitimately exercised judicial activism seeks to extract the maximum potential by building specific rules of conduct from vague statements.[5]

Activism, sometimes called creationism, sometimes called creativity, is not preliminarily understood by this work as a tool to fill gaps, the judicialization of politics, incurs protagonism of the Federal Supreme Court, implying in usurpation of functions of the legislative power. In the United States, judicial activism marks the interference of the Judiciary in specific matters of public policy, however, it is worth mentioning. Richard W. Garnett which states that it customarily:

THE term judicial activism, as it is typically used, is essentially empty of content; it is simply a fiery way of registering disapproval of a decision.[6]

Yet, Rafael Laffitte Fernandes and Rocco Antonio Rangel Rosso Nelson, note that the Federal Supreme Court itself, established in Precedent 339, the limits of the judiciary power on the possibility of increasing the salary of public servants[7], regarding Brazilian Federal precedent Mandado de Segurança N.º 22609.[8]

This differs strongly from the current experience in Brazil. Where the Supreme Court creates administers and alters public expenses while infringing its constitutional limits and boundaries towards the attributions of the executive branch. Devised by the Constitution to serve as the Guardian of the Constitution of the Republic the Federal Supreme Court has transformed itself into an anomalous Constituent, of legal manifestation and of political activism and interference.

THE END OF THE WORLD INQUIRY

On March 14, 2019, the then President of Brazil's Federal Supreme Court, Justice José Antônio Dias Toffoli, initiated Police Inquiry n. 4781 (INQUÉRITO 4781, n.d.), on the grounds of investigating alleged fake news. Also, justice Alexandre de Moraes was appointed Rapporteur of the investigation. This inquiry has been considered by many a witch hunt against Brazil current President (Mr. Jair Bolsonaro) and his supporters in the political and journalistic realms.

In a series of illegal and unconstitutional acts, the rapporteur ordered searches and seizures, censorship, and arrests, all of an ideological nature and in no way anchored in the legal system. A specific documental account of these irregularities is not possible at this moment, as the inquiry runs under secrecy ("*sob sigilo*" in Portuguese):

Figure 1. Page print of Brazil's supreme court website (inquérito)

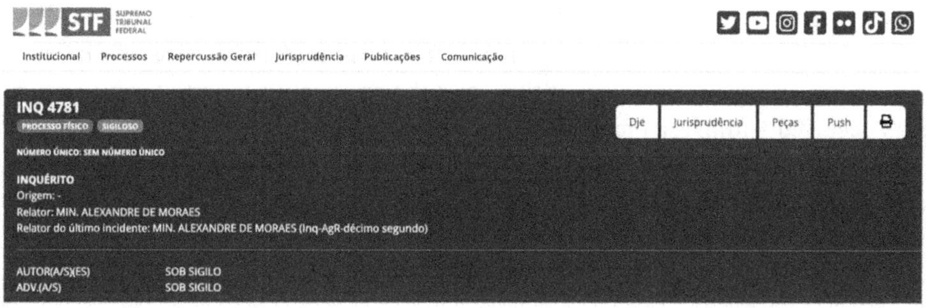

The Brazilian Constitution adopts the due process principle "*ne procedat judex ex officio*", demands that the judge and prosecutor be entirely different and separated persons, and that the judiciary organs cannot initiate criminal procedures but exclusively the Attorneys Office (Hamilton, 2001). The Inquiry of the end of the world, after being illegally initiated by the same judges who consider themselves as victims of the same events, had its rapporteur appointed by the Court's president fiat, instead of randomly (through and electronic random algorithm) and mandated by the STF's statute.

Ever since the start of this inquiry, in numerous instances, authors of public criticism addressed to the Federal Supreme Court, including humorous ones or internet *memes* became targets of the investigation, against provisions in Brazil's Criminal Procedure Code, the Federal Constitution (Brazil, 1941 and 1988) and the accusatory system that prevails in the Brazilian penal system in general.

Eight arguments are fundamental to safeguard the protagonism of judicial activism in the Brazilian State: 1. Withdrawal of other powers and failure to provide and enforce fundamental rights; 2. The constitutionality control model perceived by the 1988 Constitution, attributing to the STF the power to annul unconstitutional acts of all branches of government; 3. The rise of neoconstitutionalism, or contemporary constitutionalism that intends to reaffirm constitutional precepts and bring law closer to morality; 4. The strengthening of independent bodies such as the Public Prosecutor's Office, Defenders and the National Council of Justice; 5. Possibility of Popular Action, inviting the citizen as supervisor to the debate of the legal space, as well as raising levels of freedoms once usurped by the Brazilian State, implying the assimilation of the dimensions of international human rights assimilated in the 1988 Constitution; 6. The immediate application of norms of fundamental rights and guarantees, safeguarded by the Constitution; 7. Non-electable positions to the judiciary by citizens not causing fear of public opinion; 8. The need to adapt broad legal concepts to concrete cases, and; 9. The possibility of political minorities acting legally in search of political participation in the State.[9]

However, the proposed reflection implies an investigation determined by the Federal Supreme Court, which not only usurps the limits of any activism, but implies an anomalous measure, not seen in the national legal space, which implied an autonomous and vertical measure, violating basic constitutional precepts such as contradictory and due process of law and inevitably not understanding the measures of permissibility of the aforementioned activism and exhaustively debated in international doctrine.

The Court President's Prerogative to Open Investigations Under the Statute of Brazil's Supreme Federal Court

On March 14, 2019, Justice Dias Toffoli, then president of the STF issued Ordinance GP n.69, drafted by Minister Dias Toffol grounded on the provisions of article 43 of the Statute of the Federal Supreme Court (STF), which permits the President of the Court the prerogative to initiate an *ex-officio* investigation on events occurred within the premises of the court:

ORDINANCE PORTARIA-GP N. 69 OF MARCH 14, 2019

The President of the Supreme Federal Court, within the prerogatives provided for by the Court's Statute,

CONSIDERING that it is the duty and prerogative of the president to watch over the authority of this COurt and its members (art. 13, I);

CONSIDERING the existence of slanderous denunciations (fake news), threats and other felonies filled with animus calumniandi, diffamandi and injuriandi, which tain the honor and security of this COurt, its members and families;

DETERMINES, under the provisions of art. 43 of the Court Statute to initiate an INQUIRY to fully investigate the said facts and felonies.

I also nominate Justice Alexandre de Moraes as rapporteur, with permission to require from this presidency all means, material and staff, necessary to conduct the works.[10]

The legal and constitutional fragility of this order and the prerogative given by the Court Statue's art 43 is conditional do investigations on facts and occurrences within the premises of the Court:

Art. 43. In the event of an infringement of criminal law at the seat or office of the Court, the President will initiate an investigation, if it involves an authority or person subject to his jurisdiction, or he will delegate this attribution to another Minister.

§ 1 In other cases, the President may proceed in accordance with this article or request the initiation of an investigation to the competent authority.

§2 The Minister in charge of the investigation will appoint a clerk among the servants of the Court.[11] *(BRAZIL Federal Supreme Court, 2020)*

As the facts reported in the order refer to occurrences in journalistic pieces, online commentaries, social media posts, article 43 cannot serve as a basis for instituting an *ex officio* inquiry to investigate "fraudulent news, slanderous denunciations, threats and infractions filled with aminus calumniandi, diffamandi and injuriandi, which affect the honor and security of the Federal Supreme Court, its members and family members".

It is inferred in Brazil that what caused the reactions of the Supreme Court justices was not in itself the infringements to the Law, but rather the offense taken personally by the justices to public, and often acid, criticisms to them and their conduct as judges of the highest court of the republic. In this way, as it is evident such alleged crimes were not committed at the headquarters or on the premises of the Court, the court president does not have legal prerogative to initiate such investigation.

Also, under the Brazilian constitutional system of separation of powers, and due process any investigation, trial of legal procedure, evidently, requires to be conducted by different actors in different roles. That is, the judge, the accuser, the victim, the defender, and the victim cannot be the same person:

In other words, both prosecution and defense appear as excluding proposals for a sentence.

Such a confirmation will only admit the influence of the activities carried out by the defense, if the judge, whoever he may be, is not psychologically involved from the outset with one of the versions in play.

Therefore, real prosecution depends on the impartiality of the judge, who is not presented merely by denying him, without any reason, the possibility of also accusing, but, mainly, by admitting that his most important task, to decide the case, is the result of a conscious and thoughtful choice between two alternatives, in relation to which he remained, all the time, equidistant. (Prado, 2005, p. 178)

Also:

The truth is that if the deployment, which has been discussed, between the judge and the public ministry, or between jurisdiction and action, is necessary for the guarantee of impartiality and, with this, for the justice of punishment, in the, however, sufficient. At the end, the judge must make a decision; and decide you want to decide to elect... It is clear that the judge is so much better in the situation of electing more clearly if you present before the possible solutions. The danger is that doubt is not present, in what is tormented by it. Ahora, bien, el medio to propose it la duda es el contradictorio; help here the common root (duo) of dubium and duellum. However, the separation of the public ministry with respect to the judge, and the decision, from the prosecution with respect to the court, is not enough to guarantee the justice of the latter. The public ministry, if it is alone with the judge, is insufficient. The accusation must be counterbalanced and integrated by the defense. (Carnelutti, 1994, p. 302)

Furthermore, even if it was the case that the provisions of article 43 had not been infringed, this rule itself is in conflict with the system of due process and separation of powers of Brazil's 1988 Federal Constitution which in article 129, I provides:

Article 129. The following are institutional functions of the Public Prosecution: (CA No. 45, 2004)

I – to initiate, exclusively, public criminal prosecution, under the terms of the law; (Brazil, 1988)

In the clear conflict between a Court statute (even if it is that of the Supreme Court), and the Constitution, the latter must prevail, as:

There is no position which depends on clearer principles, than that every act of a delegated authority, contrary to the tenor of the commission under which it is exercised, is void. No legislative act, therefore, contrary to the Constitution, can be valid. To deny this, would be to affirm, that the deputy is greater than his principal; that the servant is above his master; that the representatives of the people are superior to the people themselves; that men acting by virtue of powers, may do not only what their powers do not authorize, but what they forbid. (Hamilton et al., 2012, p. 178)

The unconstitutionality of article 43 vis-a-vis the Constitution's article 129 is a strong violation of the constitutional "Accusatory Penal System" provided for by the Constitution. We must ask now, how then such unconstitutional provision may be within the very statute of the Supreme Court? Under the 1967 Constitution of Brazil, the Public Prosecutor's Office was not yet the exclusive legitimate authority to initiate criminal proceedings. It was the competence of the STF itself to issue procedural rules in the case of cases that originated in the Court. With the adoption of the new constitution in October 1988, most of Brazil's pre-existing legislation was maintained, as long as each provision was compatible with every aspect, principle and rule of the new constitutional order. Also, evidently, every pre-existing provision incompatible with the new constitution was instantly revoked. In a polity which often revokes old constitutions and adopts new ones (Brazil has had seven constitutions in 200 years of independent history), remaking all laws and statutes anew at every new constitution is impractical and even unfeasible. Thus, whenever laws and statutes are compatible with the new constitution they can be validated under the new constitutional order (Neto, 2009; Bulos, 2011; Moraes, 2011) (In Portuguese this phenomenon is called recepção constitucional).

Under the current Constitution, the judge cannot initiate any criminal action without being provoked by the Public Prosecutor's Office, which holds exclusive legitimate prerogative to propose criminal action.

The Alleged Infringements Did Not Occur on the Premises of The Court

Even when one would consider the provisions of article 43 of the STF's statute as constitutional and valid under the Brazil's Constitution, its provisions are clear and uncontested to require that "such an infraction took place at the headquarters or on the premises of the Court ''.

However, the Inquiry aims to investigate content disseminated on the internet, especially on websites and social media politically and ideologically related to the President, so that the alleged crimes committed must be presumed to have taken place at the homes of all those investigated—now victims of the investigation. Since none of them do live within the premises of the Court, there is a clear violation of art. 43, that supported the opening of the investigation.

Yet not only the Inquiry was initiated by the Court President, but also the rapporteur, justice Moraes considered that since the posts can be opened and read on a computer within the Supreme Court's building, it makes the entire territory of Brazil and even the whole webspace to be contained within the premises of the Supreme Court. Justice Moraes also ordered the removal and deletion of numerous posts and blocking of social media accounts by the journalists under investigation.

Even more strict restrictions of rights have been imposed by Justice Moraes, within the inquiry of the end of the world. Journalist Allan dos Santos is currently in exile in the United States in order to escape prison, for a crime that ultimately amounts to simple expression of thought (ponstin on social media and other websites opinions contrary or critical to the Court or its members). As the Inquiry runs under secrecy (as noted above), this research cannot at the moment examine which contents posted by Mr. dos Santos or any other journalist investigated, and we can only maintain a reasonable doubt that such contents fall within the scope of the liberty of expression and thought.

Still, the quarters of some judge in the Federal Capital, or the entire World Wide Web cannot be treated as the seat or dependency of the Court, which certainly violates even the territorial competence of the Supreme at the national level, given the bizarre and absurd interpretation to justify the illegal investigation.

The Choice of the Rapporteur and the Infringement of the Accusatory System Principle

The arbitrary designation of justice Alexandre de Moraes to preside over the investigation already denotes a clear violation of the principle of the natural judge (Brazil Constitution, art. 5, XXXVII and LIII):

Article 5. All persons are equal before the law, without any distinction whatsoever, Brazilians and foreigners residing in the country being ensured of inviolability of the right to life, to liberty, to equality, to security and to property, on the following terms: (CA No. 45, 2004)

XXXVII – there shall be **no exceptional tribunal or court***;*

LIII – **no one** *shall undergo legal proceeding or sentencing* **save by the competent authority***;*

In addition to the exceptional, and thus illegal and unconstitutional, extension of the Supreme Court's competence on this case, the choice and appointment of the juice rapporteur is provided for by article 66 of the Court's statute: "the distribution will be by lottery random casting, through a computerized system, activated automatically, in each class of process". This provision intends to promote impartiality and impersonalized choice of judge when it comes to appointing rapporteurs and other magistrates. Yet, there was no criterion by the President of the Court when appointing his colleague justice Moraes to the detriment of the other ten members of the court and simply bypassing the random choosing mechanism while providing no justification of explanation to his actions.

None of this was respected. In other words, justice Alexandre de Moraes the investigation rapporteur without any distribution, persisting in the illegality of not proceeding with the distribution, certainly violating basic principles of Criminal Procedural Law and therefore the Federal Constitution.

There Is No Determined Fact to Be Investigated

In spite of the fact that the investigation was illegally started, this research examines other unconstitutionalities arising from the Inquiry of the end of the World.

As is the case with the object to be investigated by the inquiry: alleged fake news. What would such fake news be? Who decides what they are and how far is the scope of the investigation? To the date of writing, the Court, which in this instance seems to be behaving as a Court of Exception, has failed to respond, as the inquiry continues to run under secrecy, and even the attorneys of the investigated do not have access to the files. On the contrary, the scope of the investigation remains undefined and thus unrestricted, a clear opportunity for judicial or police phishing (although this specific hypothesis is not in this research's scope right now). Once a number of journalists and other free citizens, who have dared to express criticism to the Court and its judges, physically from their homes and private spaces and virtually within the public digital agora, had their computers and/or cell phones seized by the Federal Police under judicial mandate and other have even been arrested, without properly being explained the specific cause of arrest.

The Inquiry remains without a specific cause to investigate (which is mandatory as the free citizen has the right to know for which facts or actions he is being investigated), allows the rapporteur justice Moraes to freely decide what to do with the investigated, as without investigative scope, any aspect or area of the investigated lives can be searched or examined. The investigated themselves not only do not know the reason for being part of the investigation, but also why some of them have been imprisoned given the reason to investigate remains unspecified.

In addition, the investigation covers atypical facts[12], as there is no typification of the crime of *fake news* in the Brazilian Law. No matter how reprehensible whatever conduct or actions by the investigated may have been, without proper typification, no crime can exist to be investigated, much less prosecuted. Even in the case existing legal types could apply to such conducts (which again, the investigated do not know what are), all these facts committed could be crimes against honor, which should be prosecuted in the common justice system (natural judge) having the offended justices become the victims in the procedure, not the judges.

All Investigated Individuals Are Political Conservatives

To the date of writing, all persons cited in the inquiry and that have been subject to persecutions, seizures of goods and arrests, identify themselves as conservatives or who express any support for the current and controversial president of Brazil, Mr. Jair Bolsonaro. As one example, a group of very wealthy businessmen, supporters of the president, who expressed political opinions that they did not wish the former president of Brazil Luiz Inacio Lula da Silva to return to power, in a private Whatsapp group, had their homes searched by the police under judicial mandate and had documents, computers and other personal items ceased, and their bank and social media accounts blocked (Oeste, 2022; Amado, 2022).

Another investigated is congressman Daniel Silveira, member of Brazil's National Congress lower house who recorded and transmitted a live The first act that proves the persecution was the fact that the Supreme Court violated its own jurisprudence by arresting deputy, for example, who transmitted online threats to the justices of the Supreme Court. Admittedly, Mr. Silveira's speech was indeed, in our opinion, rude and vicious, quite disrespectful, and even violent. Yet, it is speech, covered by the freedom of expression guaranteed by the Federal Constitution—even if the deputy committed a crime of aggression (which is likely), he should not under Brazilian Law and Constitution be prosecuted by the Supreme Court, but by the House of Congress he belongs to. Yet, the Court decided in the AP nº 937, that the legal protection of congressmen (federal deputies, senators, state deputies and city councilmen) on their opinions and speech applies only to cases within the exercise of the mandate *and* because of it. This second criterion opens the coverage of protection to speech to subjective considerations and hinders legal security.

Yet more concerning is the investigation of common citizens, such as political activist Sara Winter, journalist Oswaldo Eustáquio, both ordinary citizens who have been arrested, and been subject to searches and seizures. All such persons and even civil society groups defended any agenda associated with President Bolsonaro and more a conservative agenda. Another significant case was the seizure of fireworks, kitchen knives and Brazilian flags by a civil society group known as "300 do Brasil"[13], where several seizures of cell phones and computers from Youtubers', social media influencers and other individuals belonging to the group, such as Bernardo Pires Kuster and Bárbara from Youtube channel "Te atualizei" were performed. In all instances, the posts and speech accused of being criminal amount to mere opinion and political reporting., in support of President Bolsonaro and his agenda.

Greco (n.d.) notes that all the facts committed by those investigated in the Inquiry are not configured as crimes against National Security—which served as a pretext for opening the inquiry--, but only, as already mentioned, a demonstration of nonconformity, criticism and dissatisfaction against the Supreme Court, which by no means constitutes any conduct aimed at abolishing the political or social order, much less any indication or preparatory act of an alleged coup.

The inquiry launched in 2019 continues to this day, without any sign that they will come to an end and, that is, without any participation from the Attorney General's Office, the true holder of the criminal action. In other words, there is no criminal action instituted and it is believed that the investigation will last until the STF silences all conservatives in the country. All arbitrary measures practiced by the ministers certainly violate the principle of proportionality of criminal law, since the maximum that such "acts performed" could configure would be, as already mentioned elsewhere, crimes against honor. Therefore, the ministers who decreed arrests and seizures of cell phones and tablets committed a crime of responsibility and must respond to an impeachment process before the Federal Senate.

Several supporters of the federal government remain in prison, some even exiled, and others prevented from expressing themselves on their social networks under penalty of imprisonment. A true mockery that opens up the entire political bias of the inquiry. Even the contents of both the search and seizure warrants, and the arrest warrants did not specify the reason for such orders, which would immediately give rise to the nullity of all acts performed by the agents.

The unconstitutionality of the Inquiry of the end of the World has been argued on a number of occasions. In every instance, the Supreme Court's full bench has ratified the proceedings with one dissenting vote by justice Marco Aurélio de Mello.

In a nutshell, the inquiry of the End of the World is a proceeding initiated by the victim himself, who is also the judge, without the participation of the Public Prosecutor's Office, in clear violation of the accusatory system. Such inquisitorial procedures, charged with illegal and unconstitutional actions against political supporters of the current president in office raise great concern. As justice Mello noted in his vote to the Court (Mello, 2022; Ávila, 2022):

This is an abhorrence of an inquiry. And in the face of the findings verified after it was established, I would even say an 'Enquiry of the End of the World', without limits.

Under flagrant violation of constitutional rights, as well as limited reaction from the mechanisms of checks and balances and the political elite, this research questions whether O'Donnell's (1998) 3rd criterion for democratic stagnation as mentioned before in this text (*political elitism still works to maintain the strength of patrimonialism, concentrated production and institutional violence against the citizenship*) applies to contemporary Brazil under this case of the Inquiry of the End of the World. Measurements and more specific evaluations of this aspect could not be performed yet by the researchers, albeit it is in our intention to do so in the road ahead.

THE END OF THE WORLD INQUIRY AS JUDICIAL ACTIVISM

After exposing the case of the Inquiry of the end of the world, we shall now apply such events to a brief review of the theory on Judicial Activism as it is registered in contemporary Brazilian literature. Perhaps this analysis will not conform theoretically or conceptually with the international state of the art. We believe however that applying Brazilian literature and how Brazilian authors perceive the concepts nowadays to this case to be more useful and contributes better to the advancement of specialized literature on the topic. of the judicial activism practiced by the STF, which long ago laid the serpent's egg in what would culminate in the infamous and infinite Inquérito do Fim do Mundo, which plagues the country.

We argue that Judicial Activism as a political phenomenon and process is the background and contextual cause of the illegalities committed by the Supreme Court in the Inquiry of the end of the World. The activism practiced for years by Brazil's Supreme Court has been noted by the literature (Piovezan et al., 2020; Piovezan et al, 2022; Barroso, 2012; Abboud&Lunelli, 2015; Cittyadino, 2001; Streck et al., 2015; Teixeira, 2012; Reverbel, 2009; Koerner 2013) conferred power and legitimacy—at least in the eyes of the judges—to trample an entire legal and constitutional framework. In the name of "democracy", that is, through their "enlightened" minds, the ministers once again exercised the will to change the Law so that it conforms to their worldviews of what would supposedly be democratic and fair, instrumentalizing the to enforce all their wishes, since everything that would be "fair"—only in the mind of an activist minister—was barred from the Legislative Power.

Activism aims, in the final analysis, to change society through the Law through judicial decisions, in which the Judge in practice arrogates to himself the power to legislate, not being bound by the cold letter of the Law, branded as "retrograde", " conservative", giving such an elastic interpretation that it ends up deciding exactly the opposite of what is written there, all in the name of Democracy. This new law aims to shape society, which is unable to elect representatives up to what ministers think is the ideal, so they—activist ministers—take charge of "pushing history" through unbridled activism. The Federal Supreme Court has reportedly practiced activism, in increasing frequency and intensity: judicial action as a political agent by editing and/or revoking existing laws, interfering with the function of the Legislative Power, or promoting public policies through judicial decisions, as well as barring them, thus interfering with the main function of the Executive Power. (Neto, 2021).

Member of the Attorneys Office, prosecutor Cléber Tavares Neto notes:

The Federal Supreme Court, however, has taken activism to another level. Just creating laws in the lawsuits brought to them by the thousands annually was no longer enough. They could create laws for the entire Brazilian people, but they could not create laws to defend themselves against the attacks that began to occur with the awareness of the Brazilian people that the highest court was not deciding as would be expected of ministers of unblemished reputation and notorious knowledge. legal. This was resolved by 'Process (INQ) 4781'. In it, any real or imaginary fact can suffer the (in)due judicial sanction. (NETO, 2022, p. 116)[14]

And it is from this premise that the Court does not shy away from legislating in the name of a supposed "common good"—a fluid, open term, whose definition is at the discretion of whoever has the power, in this case the ministers—that is why they do the same. that they deem to be contrary to the national legal system.

And this supposed common good, as Ricardo Peake Braga points out, no longer has the meaning of limiting state power, but adopts the connotation from the 19th century, coined by Rousseau and the Jacobins of the French Revolution, which gave the term the meaning expansion of state power as an exercise of power in the lives of citizens (Braga, 2020, p. 49).

We have reason to believe that the Federal Supreme Court has been overstepping its constitutional boundaries and overtaking prerogatives that belong to the Executive and Legislative branches. With a clear inertia by the stagnant elites (O'Donnell, 1998), it has been left free to persecute any opponent of its activist conduct, as stated elsewhere. In the case we study here, the Court has even been able to create a completely illegal inquiry and violate an impressive volume of citizens' rights.

The difference is that there is nothing foreseen in the Brazilian legal system that confers such Power to the togados, so it is better to call it Juristocrat power. At least when the Emperor exercised the Moderating Power, he did so for the good of the nation, unlike the Supreme Ministers Emperors, who nullify any discretionary act of the Executive Power in the name of their spurious interests or find the alluded power to act according to their determination; moreover, the Moderating Power at the time of the empire had as its main function the overcoming of crises, not the complete superimposition of the other powers (Saad, 2021, pp. 24-25).

Alexis de Tocqueville, in his famous book Democracy in America, in which he states that a judge must strictly follow his constitutional function, which is to interpret and apply the law, and from the moment he goes beyond his competences, he becomes something more important, but leaves to represent the Judiciary (Tocqueville, 2019, p. 99).

The contemporary Brazilian Supreme Court creates an exception procedure just in the desire to punish its opponents, whose rules of this "process" are not those provided for in national laws or in the Constitution. Federal, but created at the mercy of the ministers to repay the "unwelcome", which is, for the most part, criticism on the Internet for the Judicial Activism practiced (Neto, 2021).

CONCLUSION

The Accusatory System and the Natural Judge Principle Hindered by the Supreme Court's Judicial Activism

Currently, in Brazil, the predominant penal system is the accusatory system, the one whose investigation is carried out by the police authority (delegate), presided by the General Attorney, who, as the legitimate prosecutor discretionary decides to prosecute or not the case judicially.

The Judiciary does not participate directly in the investigation, since the conduct of the investigation is entirely the responsibility of the police, which is subordinate to the orders of the delegate and any requests from the prosecutor.

In this sense, the police carry out investigations, examine and process crime scenes and potential evidence, and collect testimonies, under the coordination and the general Attorney's Office. Still, in a few instances, criminal procedural law requires the authorization of a magistrate to carry out certain steps that could not be carried out without express judicial authorization. And that's where the figure of the judge comes in during the investigative phase.

The more recent behavior of the justices of the Federal Supreme Court, who consistently interfere in the other branches actions and competences, in the name of "social justice", of the "common good", potentially generating legal insecurity. Regarding this, Tocqueville notes that a judge must strictly follow his constitutional function, which is to interpret and apply the law, and from the moment he goes beyond his competences, he becomes something more important, but leaves to represent the Judiciary (Tocqueville, 2019, p. 99).

It should be noted that in these specific cases, in which the judge only acts when provoked by the police authority or the Public Prosecutor's Office, there is no need to speak of an instituted process, given that it is only an investigative phase, leaving the judge, therefore, only to exempt analysis of the request made by the police or prosecutor to investigate, and as the measures to proceed with the investigation are about restrictions on individual freedoms, it is necessary to consider a judge to grant the measures that will restrict rights fundamental in the public interest (Monteiro, 2021).

Thus, there are three agents involved in the police investigation phase: The police, the Public Ministry, and the Judiciary in the figure of a judge who will only act to analyze requests that deal with the restriction of rights, as mentioned elsewhere. It remains inert throughout the phase. The prosecution and the police are responsible for the persecutory function, that is, the latter's primary function is the investigation, and the former is responsible for gathering the evidence found during this phase for the filing of criminal proceedings (accusation) with the aim of punishing any perpetrator of a criminal offense. criminal. (Monteiro, 2021). In case of formal accusation by the Public Ministry, the Judge will preside over a process and, in the end, will decide on the guilt or not of the accused.

Such is the nature of the accusatory system: The clear, well defined separation of the powers to investigate (police), to accuse (Public Prosecutor) and to punish (Judiciary), without there ever being an interference—always undue—of one over the other., much less the accumulation of functions. In this system, there will always be a separation between the prosecution and jurisdictional functions (Monteiro, 2021; Capez, 2021).

Brazil's accusatory criminal system is consistent with the fundamental rights and guarantees established in the Federal Constitution of 1988, as well as the general system of checks and balances and separation of powers. The federal Constitution provides not only for limits and checks to entrusted power, but also the most important individual liberties in a State of Law (also Rule of Law?), such as due legal process (article 5, LIV), the presumption of innocence, or reasonable doubt (article 5, LVII), parity of prosecution and defense (both in the investigative phase and in the process) (article 5, caput and I), the natural judge (article 5, XXXVII and LIII), (Capez, 2021).

In respect of the principle of the natural judge, the magistrate must never place himself on the sidelines, keeping himself distant from the prosecution, because otherwise he would no longer act as a judge, since he would accuse together with the competent body for the prosecution and would also judge, thus biasing its performance in the process. (Oliveira, 2022).

In a nutshell, the Brazilian penal and accusatory system, in full harmony with the provisions of the Constitution, carries the corollary principles of inertia and the natural judge, who, in addition to remaining inert and impartial, should not interfere with the function of the Ministry of Justice. Public, which is to accuse, thus characterizing the greatest fullness of a democratic Rule of Law.

REFERENCES

R7.com. (2020, June 18). *É um inquérito do fim do mundo, diz o ministro Marco Aurélio*. Retrieved September 17, 2022, from https://noticias.r7.com/brasil/e-um-inquerito-do-fim-do-mundo-diz-ministro-marco-aurelio-29062022

Abboud, G., & Lunelli, G. (2015, April). Ativismo judicial e instrumentalidade do processo: Diálogos entre discricionariedade e democracia. *Revista De Processo*, *242*, 21–47.

Amado, G. (2022, August 18). *Empresários bolsonaristas espalham fake news contra Dom e Bruno e atacam gays, jornalistas e TV Globo; leia zaps*. Metrópoles. Retrieved September 17, 2022, from https://www.metropoles.com/colunas/guilherme-amado/empresarios-bolsonaristas-espalham-fake-news-contra-dom-e-bruno-e-atacam-gays-jornalistas-e-tv-globo-leia-zaps

Ávila, F. D. (2022, March 4). *Repositório Institucional da Universidade Federal de Sergipe - RI/UFS: O estado de exceção no Supremo Tribunal Federal: análise da atuação do Tribunal como poder soberano nos posicionamentos judiciais e extrajudiciais da última década com enfoque na condução do Inquérito no 4781*. Retrieved September 5, 2022, from https://ri.ufs.br/handle/riufs/15101

Barbosa, O. P. A., & Saracho, A. B. (2018). Estado democrático de direito – Superação do estado liberal e do estado social. *Revista Jurídica Luso-Brasileira*, *4*(5), 1305–1317.

BARROSO. (2011). *Luis Roberto. O Controle de Constitucionalidade no Direito Brasileiro*. Saraiva.

Barroso, L. R. (2009). Judicialização, ativismo judicial e legitimidade democrática. *[Syn]Thesis*, *5*(1), 23–32.

Brazil. (1941). *Criminal Procedure Code. Código de Processo Penal*. DECRETO-LEI Nº 3.689.

Brazil. (1967). *Constituição do Brasil. National Congress, Chamber of Deputies*. Retrieved August 30, 2022 from: https://www2.camara.leg.br/legin/fed/consti/1960-1969/constituicao-1967-24-janeiro-1967-365194-norma-pl.html

Brazil. (1988). *Constitution of the Federative Republic of Brazil* (3rd rev. ed.). The Federal Senate, Undersecretariat of Technical Publications.

Brazil. Federal Supreme Court STF. (2019a). *Ação Penal No. 937/2019*. Retrieved on August 30, 2022, from: https://portal.stf.jus.br/processos/detalhe.asp?incidente=4776682

Brazil. Federal Supreme Court STF. (2019b). *Portaria GP No. 69 de 14 de março de 2019*. Retrieved on August 30, 2022, from: https://www.conjur.com.br/dl/comunicado-supremo-tribunal-federal1.pdf

Brazil. Federal Supreme Court STF. (2020). *Regimento Interno: Atualizado até a Emenda Regimental n. 57/2020*. The Federal Supreme Court, Secretariat of high studies, research, and information. Retrieved August 30, 2022, from: https://www.stf.jus.br/arquivo/cms/legislacaoRegimentoInterno/anexo/RISTF.pdf

Bulos, U. L. (1997). *Mutação constitucional*. Saraiva.

Bulos, U. L. (2011). *Curso de direito constitucional* (6th ed.). Saraiva.

Busato, P. C. (2010). De magistrados, inquisidores, promotores de justiça e samambaias: Um estudo sobre os sujeitos no processo em um sistema acusatório. *Sequência (Florianópolis)*, *31*(60), 133–161.

Capez, F. (2022). *Sistema acusatório e garantias do processo penal*. Conjur. Retrieved September 13, 2022, from https://www.conjur.com.br/2021-out-07/controversias-juridicas-sistema-acusatorio-garantias-processo-penal#_ftnref2

Carnelutti, F. (1994). *Derecho procesal civil y penal*. Editorial Pedagógica Iberoamericana.

Fernandes & Nelson. (2014). O Supremo Tribunal Federal e o ativismo judicial: (re)analisando o dogma do "legislador negativo". *Revista Eletrônica Direito e Política, Programa de Pós-Graduação Stricto Sensu em Ciência Jurídica da UNIVALI, 9*(1).

Garnett, R. W. (2006). Debate: Judicial activism and its critics. *University of Pennsylvania Law Review*, *155*, 112.

Greco, R. (n.d.). *Jurista Rogério Greco alerta para STF 'completamente dominado ideologicamente': 'Não se pode exercitar o seu direito sagrado de opinião.'* Retrieved September 4, 2022, from https://www.folhapolitica.org/2022/09/jurista-rogerio-greco-alerta-para-stf.html

Hamilton, A., Madison, J., & Jay, J. (2012). *The Federalist Papers*. Van Haren Publishing.

Hamilton, S. D. (2001). A ortodoxia do sistema acusatório no processo penal brasileiro: Uma falácia. *Revista Da Faculdade De Direito De Campos*, *2*(2), 247–273.

INQUÉRITO 4781. (n.d.). *Portal Do Supremo Tribunal Federal*. Retrieved August 30, 2022, from https://portal.stf.jus.br/processos/detalhe.asp?incidente=5651823

Koerner, A. (2013, July). Ativismo Judicial? Jurisprudência constitucional e política no STF pós-88. *Novos Estudos CEBRAP*, *96*, 69–85. doi:10.1590/S0101-33002013000200006

Lassalle, F. (1942, January). On the Essence of Constitutions (Speech Delivered in Berlin, April 16, 1862). *Fourth International*, *3*(1), 25–31.

Linz, J. J., & Stepan, A. (1996, August 22). *Problems of Democratic Transition and Consolidation: Southern Europe, South America, and Post-Communist Europe*. The Johns Hopkins University Press.

Locke, J. (2013, May 13). *Second Treatise of Government*. Aziloth Books.

Mello, M. A. (2022). *Dissenting vote. ADPF Nº 572/DF*. Retrieved August 30, 2022, from https://redir.stf.jus.br/paginadorpub/paginador.jsp?docTP=TP&docID=754371407

Montesquieu. (2004). *The spirit of laws*. Barotche Books.

Moraes, A. (2011). *Direito constitucional* (27th ed.). Atlas.

Neto, K. T. (2022). *Inquérito do fim do mundo o apagar das luzes do direito brasileiro*. Editora E.D.A.

Neto, M. J. S. (2009). *Curso de Direito Constitucional* (4th ed.). Lumen Juris.

Nuñez, C. F. A. M., & Quintana, F. (2014, October). Repúblicas em conflito: A separação dos poderes made in America. *Revista de Informacao Legislativa*, *51*(204), 139–161.

O'Donnell, G. (1988). Challenges to Democratization in Brazil. *World Policy Journal*, *5*(2), 281–300. https://www.jstor.org/stable/40209083

Oeste, R. (2022, August 18). *Hang rebate matéria de jornal sobre 'golpe' de empresários: "Fake news."* Revista Oeste. Retrieved August 23, 2022, from https://revistaoeste.com/politica/hang-rebate-materia-de-jornal-sobre-golpe-de-empresarios-bolsonaristas-fake-news/

Oliveira, L. C. (2022, June 2). *O sistema acusatório no processo penal brasileiro e seus desdobramentos*. https://jus.com.br/artigos/98326/o-sistema-acusatorio-no-processo-penal-brasileiro-e-seus-desdobramentos

Prado, G. (2002). *Sistema acusatório: a conformidade constitucional das leis processuais penais*. Editora Lumen Juris.

Reverbel, C. E. D. (2009, April 30). Ativismo judicial e estado de direito. *Revista Eletrônica Do Curso De Direito Da UFSM*, *4*(1). Advance online publication. doi:10.5902/198136947028

Saad, A. (2021, September 9). *O art. 142 da Constituição de 1988: Ensaio sobre a sua interpretação e aplicação* (Portuguese Edition). Independently Published.

Sowell, T. (2011). *Economic Facts and Fallacies* (2nd ed.). Basic Books.

Streck, L. L., Tassinari, C., & Obach Lepper, A. (2015, June 6). O problema do ativismo judicial: Uma análise do caso MS3326. *Revista Brasileira De Políticas Públicas*, *5*(2). Advance online publication. doi:10.5102/rbpp.v5i2.3139

Teixeira, A. V. (2012, January). Ativismo judicial: Nos limites entre racionalidade jurídica e decisão política. *Revista Direito GV*, *8*(1), 37–58. doi:10.1590/S1808-24322012000100002

The Debate Over the Judicial Branch. (n.d.). *Center for the Study of the American Constitution*. Retrieved September 17, 2022, from https://csac.history.wisc.edu/document-collections/constitutional-debates/judiciary/

Tocqueville, A. (2022, September 17). *Da democracia na América*. Principia.

Vasconcelos, P. C. B. (1996). Teoria geral do controlo jurídico do poder público. Lisbon: Cosmos.

Vitória, M. C. (2021, December). A inconstitucionalidade do inquérito n. 4.781 à luz de uma perspectiva democrática. *Revista Jurídica do Ministério Público Catarinense*, *16*(35).

ENDNOTES

1 The idea that the king, or prince, or president or general is implicit, and often very explicit in the Federalist Papers, is the one most likely to usurp power and thus the judiciary and legislative branches are entrusted with checks and balances mechanisms to limit such possibility of abuse. Evidently, the chief of the executive is not the only one to be checked, yet by logic and by history that is where most risk of tyranny stems from. Especially in the Federalist No. 78: "The Executive not only dispenses the honors, but holds the sword of the community. The legislature not only commands the purse but prescribes the rules by which the duties and rights of every citizen are to be regulated. The judiciary, on the contrary, has no influence over either the sword or the purse; no direction either of the strength or of the wealth of the society; and can take no active resolution whatsoever. It may truly be said to have neither FORCE nor WILL, but merely judgment; and must ultimately depend upon the aid of the executive arm even for the efficacy of its judgments".

2 The term was first used by Brazil Supreme Court justice Min. Marco Aurelio Melo, as an expression of concern with the "apocalyptic" gravity of the constitutional infringements performed by his court colleagues. (R7.com, 2020)

3 Lassalle, Ferdinand. (1942). On the Essence of Constitutions (Speech Delivered in Berlin, April 16, 1862). Fourth International, vol.3 No.1, January 1942, pp.25-31. Transcription & mark-up: Einde O'Callaghan for ETOL.

4 Sowell, Thomas. (2011). Economic Facts and Fallacies. Basic Books; 2nd edition. p. 127-128.

5 Barroso, Luis Roberto. (2011). O Controle de Constitucionalidade no Direito Brasileiro. São Paulo: Saraiva. p. 89

6 Garnett, Richard W. (2006). Debate: judicial activism and its critics. University of Pennsylvania Law Review. v 155, p. 112.

7 Fernandes, Rafael Laffitte; Nelson, Rocco Antonio Rangel Rosso. (2014). O Supremo Tribunal Federal e o ativismo judicial: (re)analisando o dogma do "legislador negativo". Revista Eletrônica Direito e Política, Programa de Pós-Graduação Stricto Sensu em Ciência Jurídica da UNIVALI, Itajaí, v.9, n.1.

8 "Não cabe, ao Poder Judiciário, em tema regido pelo postulado constitucional da reserva de lei, atuar na anômala condição de legislador positivo" (RTJ 126/48 – RTJ 143/57 – RTJ 146/461-462 – RTJ 153/765, v.g.), para, em assim agindo, proceder à imposição de seus próprios critérios, afastando, desse modo, os fatores que, no âmbito de nosso sistema constitucional, só podem ser legitimamente definidos pelo Parlamento. É que, se tal fosse possível, o Poder Judiciário – que não dispõe de função legislativa – passaria a desempenhar atribuição que lhe é institucionalmente estranha (a de legislador positivo), usurpando, desse modo, no contexto de um sistema de poderes essencialmente limitados, competência que não lhe pertence, com evidente transgressão ao princípio constitucional da separação de poderes." (STF. Pleno. MS/CE nº 22.690. Rel. Min. Celso de Mello, julgamento em 17-4-97, DJ de 7-12-06, p. 36)

9 FERNANDES, Rafael Laffitte; NELSON, Rocco Antonio Rangel Rosso. O Supremo Tribunal Federal e o ativismo judicial: (re)analisando o dogma do "legislador negativo". Revista Eletrônica Direito e Política, Programa de Pós-Graduação Stricto Sensu em Ciência Jurídica da UNIVALI, Itajaí, v.9, n.1, 1º quadrimestre de 2014. Disponível em: www.univali.br/direitoepolitica - ISSN 1980-7791.

[10]"O PRESIDENTE DO SUPREMO TRIBUNAL FEDERAL, no uso de suas atribuições que lhe confere o Regimento Interno, CONSIDERANDO que velar pela intangibilidade das prerrogativas do Supremo Tribunal Federal e dos seus membros é atribuição regimental do Presidente da Corte (RISTF, art. 13, I); CONSIDERANDO a existência de notícias fraudulentas (fake news), denunciações caluniosas, ameaças e infrações revestidas de animus calumniandi, diffamandi, e injuriandi, que atingem a honorabilidade e a segurança do Supremo Tribunal Federal, de seus membros e familiares; RESOLVE, nos termos do art. 43 e seguintes do Regimento Interno, instaurar inquérito para apuração dos fatos e infrações correspondentes, em toda sua dimensão. Designo para a condução do feito o eminente Ministro Alexandre de Moraes, que poderá requerer à presidência a estrutura e material e de pessoal necessária para a respectiva condução". Retrieved August 30, 2022 from: https://www.conjur.com.br/dl/comunicado-supremo-tribunal-federal1.pdf

[11] "Art. 43. Ocorrendo infração à lei penal na sede ou dependência do Tribunal, o Presidente instaurará inquérito, se envolver autoridade ou pessoa sujeita à sua jurisdição, ou delegará esta atribuição a outro Ministro. § 1o Nos demais casos, o Presidente poderá proceder na forma deste artigo ou requisitar a instauração de inquérito à autoridade competente. §2o OMinistroincumbidodoinquéritodesignaráescrivãodentreosservidores do Tribunal."

[12] Typical fact: hypothesis provided by criminal law of conducts or facts that amount to a crime or felony. Port.: Fato típico. A legal definition of crime.

[13] Allusion to "Sparta's 300".

[14] The Federal Supreme Court, however, took activism to another level. Merely enacting laws on the processes that are brought to them by the thousands annually was no longer enough. They could create laws for the entire Brazilian people, but they could not create laws to defend themselves against the attacks that began to occur with the awareness of the Brazilian people that the highest court was not deciding as one would expect from ministers of unblemished reputation and notorious knowledge legal. This was resolved by 'Process (INQ) 4781'. In it, any real or imaginary fact can suffer (un)due judicial sanction.

Section 3

Innovations in Autonomous Vehicles

Chapter 7
Disruptions From Personal Delivery Devices:
A Survey of the American Regulatory Approach to Sidewalk-Sharing Robotic Delivery Devices

James R. Masterson
https://orcid.org/0000-0002-6620-7202
Morehead State University, USA

ABSTRACT

This chapter examines the use of app-based personal delivery devices (PDDS) in communities and on college campuses in the US. The chapter analyzes the state and local rules and regulations in place to accommodate the usage of such devices. Furthermore, the chapter discusses liability issues arising from incidents and accidents with vehicles and pedestrians and security issues made vulnerable from usage of these devices. Last, the chapter addresses how campuses and surrounding communities can respond to these new delivery systems in an effort to maintain public safety while also allowing consumers access to meals at a time when people faced a greater risk than ever before when visiting grocery stores and restaurants.

INTRODUCTION

For millions of people around the world the global pandemic, brought on by the Covid-19 virus, changed the way people accessed goods and services. Many of these changes were not entirely new, but rather the pandemic led to rapid changes in consumer demand. Increasing numbers of people purchased groceries for pickup at retailers, which had been provided by some stores before the pandemic, while other stores were quick to adapt to the new dynamic environment that directly challenged their ability to survive as a business.

DOI: 10.4018/978-1-6684-6429-8.ch007

On March 11th, 2020, the NBA announced the cancellation of the Jazz vs. Thunder game minutes before the start due to a player's Covid infection. The next day the NBA announced the cancellation of the remainder of the season. By the 13th of March, seven states and the District of Columbia ordered all schools closed to curtail the spread of the virus. More closures would quickly follow suit in the days ahead. On college campuses, changes were abundant and, in many cases, overnight. College courses were quickly moved entirely online facilitated by video conferencing applications, campus dining services shuttered and campus housing facilities scrambled to send students home.

Given the risk of Covid infection associated with person-to-person contact, app-based food delivery services exploded in terms of users. While services such as Uber Eats, Door Dash, and Grubhub all preceded the shutdowns brought on by the pandemic, client usage of these services in the U.S. jumped by nearly 200 percent during the pandemic (Ahuja et al., 2021).

Back on college campuses, there was a new entrant to the meal delivery service market; Starship Technologies. Starship Technologies differed from other app-based meal delivery services by eliminating the human needed for delivery. Starship Technologies uses a multilayered robotic system based on GPS, ground-based sensors, and wireless communications to deliver its meals to customers on nearly a dozen college campuses throughout the US. This system offers the ability to greatly reduce human-to-human contact to deliver much-needed food at a time when cafeterias were shut down and going to the grocery store was thought of as risky.

App-based meal delivery services in general, and Starship Technologies in particular, are disruptive innovations in the food service industry. Disruptive innovations are those that successfully target a segment of a market that is overlooked and gain a foothold in the market by delivering more-suitable functionality, often at a lower cost (Christensen et al., 2015). For decades, in most cities and surrounding areas (save New York and Los Angeles), meal delivery services were largely limited to pizza chains and a small number of sub-sandwich shops. Customers desiring other types of meals, from fast food to typical "dine-in" restaurants, had to resort to the drive-thru or carry-out. Either way, the customer was required to visit the establishment to pick up their order from another human. This opened the door for app-based meal delivery services, such as Grubhub and Door Dash, to focus on customers whose needs were unmet, and those who wanted a different variety of food than was currently available for delivery. These new delivery services catered to a large variety of clients and thus delivered food from a large variety of restaurants, though they shared one thing in common with current delivery provided by pizza and sub-sandwich joints; the need for a human to execute the delivery.

Enter Starship Technologies (Starship hereafter), and other Personal Delivery Devices (PDDs). Starship eliminated the need for a human to execute the delivery and, as such, during the global pandemic, became widely used on college campuses where it was available, and became available on more campuses throughout the U.S. as the pandemic raged on. Starship, in particular, more than tripled its growth in 2021, operating a fleet of more than 1,700 robots daily, according to a March 2022 press release from the company. While previously Starship operated exclusively on college campuses, in 2022 it announced that it would be offering services in Pleasanton, California in partnership with Save Mart, the second community with its delivery service after Modesto.

This chapter examines the use of app-based personal delivery devices (PDDS) in communities and on college campuses in the US. Personal Delivery Devices are ground-based delivery devices that transport cargo autonomously or with the use of a remote operator. The chapter begins with brief literature reviews of disruptive innovations and the government's role in regulating transportation and autonomous vehicles. Then, the chapter discusses PDDs and the opportunities for and threats to local governments considering legalizing the use of such technology within their jurisdiction. The chapter analyzes state and local rules and regulations in place to accommodate the usage of such devices. Furthermore, the chapter discusses liability issues arising from incidents and accidents with vehicles and pedestrians, and security issues made vulnerable by the usage of these devices. Last, the chapter addresses how communities responded to these new delivery systems to maintain public safety while also allowing consumers access to meals at a time when people faced a greater risk than ever before when visiting grocery stores and restaurants.

The findings of this chapter show that PDDs are deployed on dozens of college campuses and are gaining popularity within major cities. Governments at all levels have some impact on PDD operations. The federal government sets a minimum standard for PDDs to operate, to ensure public safety, while state governments take on a vital role by regulating the weight of the devices, the maximum speeds at which they operate, and set the classification of the vehicle. PDDs, for the most part, are considered pedestrians whereas autonomous delivery devices, such as Nuro, are considered vehicles, confined to operating on roads. PDDs, as pedestrians operate on sidewalks, and in some cases, bicycle lanes. Furthermore, the chapter finds that accidents between PDDs and humans are rare. The most common accidents are between PDDs and human-driven automobiles, and here these accidents tend to occur largely at crosswalks as the PDD is crossing the street and the automobile is turning. Accidents between PDDs and humans are quite rare. PDDs' right-of-way issues are a concern among groups with disabilities who may find it difficult to navigate around a PDD blocking access to a crosswalk or parts of a sidewalk. Last, there is growing concern about the cyber-security of PDDs and the potential for a device to be hacked or hijacked for criminal or terroristic purposes.

This chapter contributes to the literature on government responses to disruptive technologies and sheds light specifically on how governments regulate the use of personal delivery device robots in our communities and on our campuses.

LITERATURE REVIEW

A Brief Overview of Disruptive Technology

Government service sectors, specifically those dealing with public safety, face a rapidly changing landscape in dealing with new actors that combine robotic technology and artificial intelligence. Just as businesses must adapt their strategies to their existing revenue model (Markides and Oyon, 2010), governments must adapt to these new business innovations.

At a basic level, local governments (sub-state jurisdictions) provide services to citizens where there is a market failure (for example, mass transit, regulation of air and water pollution, etc.), to provide public goods (for example, roads, bridges, clean water), or to ensure safety and well-being of citizens (licensing, building codes, zoning permits, and traffic safety and regulations). Additionally, local governments facilitate elections on behalf of the state, for federal, state, and local positions. In some cases, technological innovations challenge the way governments provide these services. As this chapter will show, emerging technologies in 5G communication and robots instructed using artificial intelligence have led to an explosion of unmanned delivery vehicles that must comply with local safety, traffic, and pedestrian regulations, as well as public health department codes concerning food storage and handling. These devices put pressure on public entities to maintain safe pedestrian walkways, reduce traffic accidents, and ensure that food is safe for consumers.

The scholarship of disruptive innovations continues to evolve. For Christensen et al., a "disruption" is a process by which market entrants with fewer resources can challenge established firms in a particular industry (Christensen et al., 2015). These new challengers tend to focus on a segment of the market that is underserved by "delivering more suitable functionality." Seen in this light, Starship and other unmanned robotic delivery systems are largely catering to an underserved college student population by providing greater convenience and delivery to anywhere within the delivery zone on a college campus or surrounding community. Traditional delivery services need an address or building and room number, and rarely could the delivery be tracked in real-time. Starship and similar companies can deliver food to the consumer using GPS coordinates while the consumer can track the delivery in real time. Where such devices are available, gone are the days of needing to go to the school cafeteria for a meal. Instead, students can order meals from the classroom, timed for pickup just outside the classroom when the lecture is over.

According to the Organization for Economic Co-operation and Development, a disruptive innovation is one that "has a significant impact on a market and on the economic activity of firms in that market. This concept focuses on the impact of the innovations, as opposed to their novelty" (OECD, 2005, p.58). Seen in this light, personal delivery devices (PDDs)impact the environments within which they operate. Specifically, the rules and regulations within communities and college campuses, along with the infrastructure of these places, by and large, were not created to accommodate unmanned robotic delivery systems. Instead, these systems must operate in the current environment created for manned vehicles and human pedestrians. As such, these systems can significantly impact public safety on these campuses, and the communities in which they operate.

The goal of this chapter is to illustrate the challenges that these PDDs pose to humans in the environment within which they operate. The chapter assesses how state governments have addressed these challenges to minimize risk to others. Last, the chapter concludes with a discussion of the future implications of this technology as it becomes more commonly used in public spaces. But first, it is helpful to understand the role that governments play in the development of technological innovations, particularly in the transportation and delivery sectors, both of which are directly relevant to PDDs, as well as to understand the interests that local governments have in technological innovations such as personal delivery devices.

Brief Overview of The History of Government Responses to Technological Innovations in Transportation

Personal delivery devices sit at the intersection of two developments, relevant to public safety: transportation and artificial intelligence-driven robotics, or autonomous vehicles. Government response to facilitating transportation technologies in the United States dates as far back as 1791 with the legislature of Pennsylvania incorporating private citizens to develop the Schuylkill and Susquehanna Canal. By 1808, the federal government began to take an active interest in the development of roads and canals (Shaw, 1990) and began work on its National Road by 1811. The stage was set for government responses to newer technologies in transportation including steam-powered boats and trains.

In the United Kingdom, the world's first public railway operating a steam engine came about via the Stockton and Darlington Railway Act of 1821 (Kirby, 2002). In the United States, the first business uses of locomotives were to move stones from mines, and were funded with private money.

The American approach to technological innovation, driven by responses to the steam engine within the government was coined in the seminal book by Beard and Beard, *The American Leviathan* (1930).

Meticulously detailing the effects of technological innovation on the U.S. Constitution and each institution delineated therein, Beard and Beard argue that the leviathan that is the American federal government "must now command expertness in all fields of technology and at the same time its work calls for a super-competence able to deal with the interrelations of the various departments (Beard and Beard, 1930, pg. 19)." For them, the problem before the public was how to combine the noblest of political philosophies, as advanced by our Founding Fathers, with the most efficient use of all the instrumentalities of the modern age.

The beginning of U.S. government's response to the new steam locomotive innovation begins with the State of Pennsylvania constructing the Philadelphia & Columbia rail line completed in 1829. This was the first railway work undertaken by a State government (Thompson, 1925). The Baltimore and Ohio Railway was the first common carrier rail system to operate a locomotive in the US, charted by the Commonwealths of Maryland and Virginia in 1826 and 1827, respectively. To accommodate this burgeoning innovation, several responses were needed from various government entities. First, public financing for the project was needed and amounted to nearly $1.5 million by the City of Baltimore (Schley, 2020, pp. 20-21). Second, land was needed to expand tracks throughout the United States. By 1862, the federal government began to provide land grants to railway developers and issued rights of way for track to be laid, creating the conditions for an unprecedented era of mass transcontinental travel, only to be outdone by the subsequent invention of the automobile and the establishment of the national highway system.

The Agricultural Appropriation Act of 1893 established the Office of Road Inquiry, which later became the Office of Public Roads by 1905. This was a federal response to the need to develop paved roads throughout the United States, a need growing shortly after the entrance of a new mode of personal transportation, the automobile (Records of the Bureau of Public Records). The development of Henry Ford's affordable Model T automobile led to a rapid growth in ownership of personal cars. As a result of this growing transportation market, in 1916 the U.S. Congress passed the Good Roads Act. This is the first public funding support for roads, but it would not be until the Federal Aid Highway Act of 1921 that significant improvements were made nationwide with adequate funding (Weingroff, 1996). This latter act provided $75 million to states for road development and led to the creation of the "Pershing Map," a detailed map of roads important to the transportation of military assets for national defense. Though designated as the U.S. Highway System in 1929, this network of roads was built and supported by local and state governments. This web-like network of now-paved roads helped to facilitate the growth of automobile owners in the early 20th century. By 1956, in response to changing U.S. defense and security needs, the Eisenhower administration developed the U.S. Interstate Highway System, making transportation from the east coast to the west coast possible in under 50 hours.

The improvements in the network of roads throughout the U.S. combined with the vastly improved quality of roads led to drivers moving at increasingly greater speeds, particularly on sections of the interstate with few cars. Though the state of Connecticut and New York City both had some form of traffics regulations and speed limits by 1910, many states did not have speed limits. By the 1960s, these states implemented speed limits between 70 and 75 MPH (Farmer, 2019). It would not be until the Nixon administration signed into law the Emergency Highway Energy Conservation Act in 1974, partly in response to the oil embargo, that there would be a national standard for speed limits when traveling on the interstate. According to the National Research Council Committee, this reduction in speed limit resulted in 3,000 to 5,000 fewer traffic fatalities (Transportation Research Board, 1984).

Government intervention in the automobile industry was not limited to how people drove their cars outside, but also how cars protected people inside. National safety standards followed the high-speed limits set among various states to reduce traffic fatalities. In 1966, the National Traffic and Motor Vehicle Safety Act was passed setting the first federal mandatory safety standards by requiring seatbelts in all cars and establishing uniform traffic signs, among other safety-related regulations (Hendrickson, 2003).

In short, as transportation technology developed in the United States, governments at all levels, responded to ensure the public safety of these innovations and help to create a market for the innovations. Without a network of paved roads expanding from sea to shining sea, the full potential of the automobile market would not have been realized. And, just as the automobile market began to take off, the Federal government responded with legislation to mandate safety features in cars and speed limits on the interstate system to protect the personal safety of drivers and their passengers.

Brief Overview of The History of Government Responses to Technological Innovations in Autonomous Vehicles

Relative to transportation regulation in the United States, regulations concerning autonomous vehicles are relatively new, and those are framed into existing laws, rather than having a new body of law specifically to address this new technology. In an effort to get in front of conflicting state regulations about autonomous vehicles, in 2013 the National Highway and Transportation Safety Administration (NHTSA) issued non-binding regulations addressing the deployment of these AI-driven robots on public roads. By 2017, 33 states had issued regulations concerning the use of autonomous vehicles, coupled with a series of follow-up guidelines in January and September of 2016. The NHTSA adopted the Society of Automotive Engineers' six-level schema for automation, to apply to autonomous vehicles, including personal delivery devices. These range from level 0 where the human does all of the driving to level 5 where an automated system can perform all of the driving tasks (SAE, 2021).

The guidelines set forth by NTHSA center on the safety, reliability, and secured utilization of the vehicle as well as the insurance liability of the operator. The framework set forth by these guidelines could be adopted for regulations regarding personal delivery devices (Bascelli, 2018). However, regulation of both PDDs and autonomous vehicles is largely performed within state and local jurisdictions.

Schreurs and Steuwer (2016) presented a regulatory framework for autonomous driving that considered the political, legal, social, and sustainability dimensions of mobility. They argue that rapidly advancing technological developments create a need for a new regulatory environment for autonomous vehicles. Their study highlights competitiveness, innovation, safety, harmonization, and coordination. Concerning PDDs, Bascelli (2018) advances a framework for regulating these devices focusing on device specifications (ie. speed, weight, etc.), safety, insurance, and privacy issues. Similarly, Hoffmann and Prause (2018) discuss the legal challenges presented by personal delivery devices. They discuss two important areas of liability for personal delivery devices: producer liability and tortious liability under traffic law.

Since Florida became the first state jurisdiction to regulate personal delivery devices, more than half of U.S. states have followed suit. By the time of this writing, twenty-six states have adopted such regulations. A review of these is discussed below.

Brief Overview of Technological Innovations in Food Delivery

Rumor has it that in 1889, King Umberto and Queen Margherita received the first delivered pizza while on a visit to Naples (Phillips, 2022). It would take nearly a hundred years for Dominos to institutionalize a new way to get food using an automobile. While food delivery can be traced back to 1920s Los Angeles, it would be pizza that became synonymous with food delivery, with Domino's offering free delivery in under 30 minutes, prompting concerns about unsafe driving during delivery (Sanburn, 2013). By the 2000s, with the advent of the Internet and online ordering, food delivery became mainstream (Jackson, 2021). When food delivery is combined with 4G cellular technology, users can order and track their delivery in real-time. Food delivery app users increased from 66 million in 2015 to 111 million by 2020 in the US. The greatest increase in users was between 2019 and the end of 2020, with a 17% increase, a result of the global pandemic shutting down restaurants for dine-in ordering (Flynn, 2022). And while the pandemic fostered significant growth in app-based food delivery, it also raised awareness of the risks associated with human-to-human contact amidst high transmission rates of the Covid virus, leading to a similar growth in personal delivery device systems deployed.

Starship, one of the first entrants in the robotic food delivery market, unveiled its unmanned delivery services on the campus at George Mason University in 2019. By the end of the year, it would add five more campuses and surpass more than 5 million road crosses. By the end of 2020, it added five more campuses and five more cities to its market. While it would take Starship 20 months to hit 500,000 deliveries from 2019 to 2020, it would only take five months to hit 1,000,000 deliveries by January 2021. The next million would only take Starship 10 months to complete, and just three months to hit its three millionth delivery. Needless to say, as the pandemic raged on across the U.S. and the world at large, more and more markets were incorporating Starship delivery systems and other PDDs, and more and more consumers were using apps to order their next meal (Starship Website).

Whereas natural science and machinery defined an era of technological innovation at the turn of the 19th century, mass communication technology and artificial intelligence define our current era. Shortly after the industrial revolution came to America, government processes were revolutionized by steam, electricity, machinery, and science (Beard and Beard, 1930). Today it has been revolutionized by microchips, batteries, artificial intelligence, and wireless transmission of data. The intersection of these with existing technology comes to form what's broadly called the Internet of Things (IoT). An ever-increasing array of goods are being plugged into the information highway relying on computer processing microchips, cellular or wireless data transmission, smaller and more powerful batteries, and artificial intelligence algorithms to perform more efficient and complex tasks. As Beard and Beard put it in 1930, with prescient relevance today, "new economic facts produce new political facts, evolve novel concepts or variant applications of old ideas" (Beard and Beard, 1930, pg. 4). The role that the public sector plays in harnessing the full potential of these new economic facts is the topic of the next section.

Public Sector Role in Innovation

The supreme task of the American democracy in the new age of technology is not to commit its government to the narrow class of experts who form the spearhead of the penetrating forces of technology in every field, but rather, sensing and appreciating these forces, to shape the operations of its government to the end that the masses of the people who produce the wealth and bear the burdens of civilization may come able progressively to estimate and appropriate the grains of science for the public weal. (Muzzey, 1931, p. 111)

To be certain, the public sector's role in embracing and fostering technology is not new. Through the Constitution's Commerce Clause, the federal government, particularly Congress, has a responsibility to regulate interstate commerce. This has been used to facilitate steamship transport, railroad development, and even the interstate highway system (Beard and Beard, 1930, p. 356). State and local governments are often tasked with providing authority for the intra-state operation of new technologies and passing laws and regulations governing their use.

The impacts of innovative technologies are incremental and often contradictory, and their role has been importantly shaped by the officials of state and local governments in ways that have supported the vitality of the American federal system (Dutton, 1982, p. 109). In the Cold War era, as computer technology gradually developed, governments turned to science to help solve various societal ills (Goldsmith, 1989, pg. 22). This led to a myriad of approaches by thousands of various state and local jurisdictions to craft and implement public policy that relied on the use of technology for the betterment of citizens. To better understand these different tactics, "A Clearinghouse for State and Local Initiatives on Productivity, Technology, and Innovation in the U.S. Department of Commerce was created by the *Omnibus Trade and Competitiveness Act of 1988*. Its primary function was to share information to state and local jurisdictions on who is doing what to stimulate productivity, technology, and innovation to position the American market to be more competitive internationally.

State leaders have three primary roles concerning disruptive technologies: 1) foster innovation in support of its economic development goals, 2) promote innovative technologies from their state, and 3) intervene in the marketplace where conflicts arise with state policies, e.g., public safety, hazardous waste, environmental concerns. (Goldsmith, 1989, p. 22). Goldsmith lays out three such motivations for states to help drive technology within their market. First, in response to the recession of the early 1980s when the duration and level of unemployment were extensive, state policymakers realized that they needed to respond to increased foreign competitive pressures and technological innovations. By some estimates, 70 percent of domestically produced goods were subject to foreign competition by the 1980s, compared to just 20 percent a decade or two earlier (Osborn, 1988). In response to this new reality, restructuring was needed.

Second, Goldsmith argues that policymakers began to realize the need to foster indigenous business development, particularly those that involved a high degree of technological innovation. Policies were crafted by states to incentivize expansion and innovation to provide new job creation, thus supporting a growing state tax base and revenue. As more and more manufacturing facilities moved overseas to take advantage of lower labor costs and a less stringent regulatory policy environment, the battle to attract domestic manufacturing waned throughout the '80s and '90s, as the New Dot Com economy began to emerge. More recently, there has been an effort to revitalize the American manufacturing sector, despite a hollowed-out steel industry (Yudken et al., 2022).

Last, Goldsmith argues that the incentives that once promoted economic development, such as natural resources, cheap labor and land, and low taxes, were not factors of growth for the future. Rather, new firms were looking for assets that would help them attract and retain a highly trained workforce to help the firm innovate (Goldsmith, 1989, pp. 22-23).

As a result of these factors, 1) the need for states to help drive technological developments to better compete with foreign competition, 2) revitalize the American manufacturing sector, and 3) attract a highly trained workforce, policymakers began to position the state to adopt technology-led development strategies as a crucial component to their economic development strategies.

Robotic Delivery Systems as a Source of Economic Development

Concerning autonomous robotic delivery systems, state, and local governments have competing interests they must balance. The upsides of these interests are valuable to communities and their policymakers. These systems offer opportunities for economic expansion. Goldsmith argues that states seek to support their economic developments with technological innovations (Goldsmith, 1989). Unmanned autonomous delivery services foster this through several mechanisms. First, some of these companies are headquartered in the United States, specifically within several cities in California, a state well known for its support of big technology companies such as Apple, Intel, and Tesla, among others. For example, Starship, an Estonian company, has headquarters in San Francisco and engineering plants in Estonia, while Nuro is an American company operating out of Mountain View, CA. Other robotic delivery systems also operate out of California; Toyota's delivery system, Boxbot, operates out of Oakland, CA, Coco operates out of Santa Monica, while Uber's Serve is based out of Santa Monica and Los Angeles, and Marble operates out of San Francisco's Potrero Hill neighborhood. Last, Refraction AI operates in Ann Arbor, MI and Austin, TX. States that can attract these startups are positioned to capitalize on this new and rapidly expanding employment market, where high-paid engineers and software developers are employed.

Second, state and local governments, by adopting these technology services into their jurisdiction, can support local businesses. As consumer demand for more types of goods delivered in real-time increases, the need to support local businesses to meet these newly emerging challenges increases. Two-day delivery was made famous by Amazon, which has now moved to same-day delivery in major cities. However, many of the goods provided by Amazon are not locally sourced. As such, communities supporting autonomous delivery systems support local businesses that take advantage of this technology. These robots offer the potential to boost local business sales and provide consumers with delivered goods in a moment. Restaurants and grocery stores alike can plug into this service, as well as convenience stores among others (such as pharmacies). Demand from both the consumer side and the supplier side has increased dramatically over the course of the pandemic. "We saw demand for robot usage just go through the ceiling," said Alastair Westgarth, the CEO of Starship Technologies. "I think demand was always there, but it was brought forward by the pandemic effect" (Durbin, 2021).

Third, delivery bots help managers in the food service and hospitality sectors meet labor challenges they faced as a result of the pandemic. Delivery bots can supplement employment needs in food delivery allowing businesses to redirect their human workforce to tasks that are not able to be completed by robots (Dumas, 2022). As mentioned earlier, the pandemic resulted in nearly all "non-essential" businesses shutting down. During the shutdown, millions of workers lost their jobs. A total of 8.3 million jobs were lost when hotels shuttered in March and April of 2020 (Singh and Wang, 2021). In the restaurant industry, 3.1 million jobs were lost (Lippert et al., 2021). When restrictions were lifted, businesses found it difficult to bring their full workforce back to work (Kreitzberg and Lucas, 2022). During the pandemic, work-from-home was an option for many employees whose tasks could be moved from a company office and computer to a home office and computer, but for those working in many service sectors, such as the food services and hospitality sectors, working from home was simply not an option. And, when restrictions were gradually lifted throughout the US, threats from the Covid virus were not eliminated. As a result, many workers were reluctant to enter back into the workforce in places where these jobs required close contact with many different people, and in particular, where, as in the service and hospitality sectors, wages were the lowest across all industries (BEA, 2021). Robots are now being used in the food services and hospitality sectors where they can complete tasks traditionally completed by humans.

Last, these delivery services can help recruit and retain high-skilled labor in communities employing delivery bots. A survey of employment opportunities from Kiwi, Coco, Nuro, and Starship shows that among all of these firms' software and engineering jobs are available both in the city in which the headquarters are located as well as in the cities in which the device operates[1]. High-skilled labor is not only useful where the device is designed and manufactured, but also in the communities in which the bots operated. For example, Starship hires Fleet Attendants to service the robots, and Kiwibot hires Operations Coordinators to set up new routes and service fleets of devices, seeking those with a technical background in electronics or electrical engineering. Coco hires human drivers for their bots and software engineers at their headquarters. Local communities benefit from the deployment of PDDs through the increasing high-skilled high-paid labor force, whose occupational and income taxes support local and state governments.

Delivery Systems as a Source of Public Consternation

Local communities also face downsides by allowing these robotic delivery vehicles to operate in their jurisdictions. Two significant issues are of paramount concern for locally elected officials. First, public safety and right-of-way issues are the most visible and pressing concerns. Innovative technology, when introduced to the public, brings not only uncertainty from the public about how the device will function and what to expect from it but also about how the public will respond to the device and interact with it. LA Department of Transportation Colin Sweeny remarked that LADOT wants to incentivize the use of these devices without compromising safety and public right of way. For Starship CEO Alastair Westgarth, "Safety is always our number one priority, so if a robot encounters something unusual, it will stop and send an alert to our remote operators (Marr, 2021).

The second issue of concern is the displacement of local jobs, in essence, undermining the economic incentives they hoped to have created. The loss-labor argument suggests that as more delivery robots are deployed, fewer human deliverers are needed. Uber's Serve, in September 2022, ran a trial program in Vancouver Canada to deliver Pizza Hut pizzas, reducing the number of human delivery personnel needed (Short, 2022). A 2016 report under the Obama administration estimates that between 2.1 and 3.3 million full- and part-time jobs are threatened by autonomous vehicle technology (The White House, 2016). Human workers pay occupational taxes in the cities and counties in which they work, while delivery bots do not.

The loss-labor argument does not address labor increases as a result of automation, as mentioned above. AI-driven autonomous vehicles are likely to create jobs, as well as displace others. The White House report identifies four types of job growth associated with AI growth: jobs where humans engage with AI devices (e.g. packing goods to be delivered), jobs where humans develop AI devices (e.g. software development, engineering, and data analysis), jobs where humans supervise AI devices (e.g. operational managers, repair, and maintenance), and jobs to prepare for more AI devices (e.g. urban planners and designers) (The White House, 2016, pp. 18-19).

To offset this loss in local occupational taxes, a sizable expansion in sales taxes from local economic activity driven by these autonomous vehicles would have to occur. Alternatively, an automation tax could be implemented using reverse depreciation/capital allowances as a mechanism. Local government policies can implement an automation tax by building on the existing depreciation/capital allowances framework where necessary (Ooi and Goh, 2022, p. 303). Recently, Bill Gates and former NYC Mayor

Bill DeBlasio have called for an automation tax. In essence, this tax would slow down the rate of automation implementation and help offset losses to payroll taxes (Kharpal, 2017; Brown, 2019).

However, recent research contradicts the loss-labor argument. In their examination of automation-adopting Canadian firms, Dixon et al. find that these firms experience an increase in employment and that this increase is largely among low-skilled workers (Dixon et al., 2021). Similarly, Acemoglu et al., find that among French firms that adopted automation, jobs were added (Acemoglu et al., 2020). Seamans argues taxing robots is not only a bad idea because firms adopting them tend to increase employment, but that identifying exactly what a robot is represents a very difficult process that makes taxing them nearly impossible (Seamans, 2021). He argues that settling on an appropriate definition that does not disproportionally favor one type of automation-adopting firm over another will not be easy. Rather, policymakers should consider other policy changes to help workers, possibly changing how capital and labor are taxed, but also focusing more broadly on labor market reform (Seamans, 2021).

Near the onset of the deployment of PDDs, not all communities were excited about the potential opportunities that came with them and were focused on the downsides. In 2017, San Francisco's City Council voted to strictly limit personal delivery devices. The new rules were among the strictest in the nation at the time. "Sidewalks, I believe, are not playgrounds for the new remote-controlled toys of the clever to make money and eliminate jobs," stated one community activist who represented seniors and those with disabilities (Nichols, 2007). Tortoise, a mobile bakery bringing cookies to customers in the San Francisco area, found a work-a-around to the strict Bay area regulations. Being a "mobile smart store" frees them of the restrictions that delivery bots such as Starship, Kiwibot, and Coco faced. Despite the initial concern expressed by some communities, more communities eased restrictions and permitted delivery bots (Wells, 2022). It is only a matter of time before the line separating delivery bots and mobile smart stores becomes blurred beyond reasonable distinction.

METHODOLOGY

This chapter relies on a narrative methodology based on a survey of state laws, news reports, and social media posts drawing attention to unmanned delivery systems, or PDDs. What narrative researchers hold in common is the study of stories, narratives, or descriptions of a series of events (Clandinin, 2007). State and local regulations are surveyed to highlight how different jurisdictions treat these devices. News articles and reports that address incidents and accidents with these devices are included as well. News aggregators such as Google News, Apple News, and Lexis Nexis were used to gather comprehensive data on the uses, benefits, costs, and accidents involving autonomous robotic delivery systems, from 2015 through 2022. The goal of this chapter is not to ascertain statistical probabilities of incidents and responses to these incidents as a quantitative analysis may do, but rather to illustrate through narrative media text and video the types of uses of PDDs, the locations where these devices are used, the types of problems experienced between humans and PDDs, and how governments have responded to the use of PDDs in their respective jurisdiction, through the analysis of state laws.

Narrative methodology is an appropriate methodology to study the nascent deployment of PDDs. These devices have not been in widespread use among the population for very long, though they are gaining increased popularity. As these are new and emerging technologies, there is not a wide body of evidence from which to draw broad conclusions. There do not exist data sets of information about accidents and uses of PDDs. Therefore, using narrative methodology to illustrate the beginning introduction of PDDs into certain jurisdictions will allow the following questions to be answered: Where are PDDs commonly deployed? How have state and local governments responded to PDDs within their jurisdictions? What issues arise from the public use of PDDs interacting among pedestrians and human-driven automobiles, where these devices are connected to the internet to function? How have state governments regulated the use of PDDs? How can local communities prepare for PDD where they are not already deployed? The answers to these questions are addressed in the remaining sections of this chapter.

Robotic Delivery Systems in Use

As mentioned above, the impact of the Covid pandemic had far-reaching consequences across various markets both within and outside of the U.S. economy, particularly in the food delivery market. Simply put, more people began to utilize personal delivery devices to get food and other goods. This led to a surge in the deployment of these devices around the United States. Demand for delivery robots is expected to surpass the one-billion-dollar marker by 2030, a 30% compounded growth rate (Market Research Future, 2022). By and large, robotic delivery systems are operating on college campuses and major cities. College campuses offer providers an ideal consumer market and an easier technical terrain for the devices to maneuver compared to the hectic cities, and as such, are the most common environment in which robotic delivery systems are found. Starship, for example, operates on 18 campuses, whereas Kiwibot has partnered with Sudexo on 26 campuses that it serves.[2]

Table 1 shows the most commonly deployed PDDs in the U.S. While college campuses are the most common service area for PDDs, densely populated communities and major cities are attracting more PDD firms given their larger consumer base. In some cases, PDD firms are partnering directly with restaurants or grocery stores, generally working with multiple suppliers in the community, expanding their customer base. Where Starship has an app for consumers to order through, Coco works directly with the businesses that utilize the Doordash app for delivery within 1.5 miles. Similarly, Serve works with Uber Eats customers to have food delivered by their bots.

Table 1. PDD deployed in US

PDD Firm	Colleges Served	Cities Served	Cities	Partnerships
Starship	22	2	Modesto, CA; Pleasanton; CA	Save a Lot
Kiwibot	26			Sodexo
Coco		2	Los Angeles, CA; Austin, TX	Doordash
Serve		1	Los Angeles, CA	UberEats
Refraction AI		1	Austin, TX	Chick-Fil-A

With the surge in demand for the use of robotic delivery vehicles comes a growing need for them to be regulated. In the U.S., these vehicles must first meet Department of Transportation safety standards (DoT, 2021, pg. 10). At the state and local levels, these vehicles must be approved for use on the streets and defined as to what type of vehicle they are considered. For example, the Florida Legislature, as well as nearly all other states, classify personal delivery devices as pedestrians. However, they must yield the right of way to actual humans and bicyclists (Deto, 2020).

Research on legislation and policy invention suggests that states learn from other states when reviewing legislation, making adjustments that are increasingly comprehensive and converge upon an optimal policy (Glick & Hays, 1991; Mooney & Lee, 1995). However, other scholars find some states simply adopt legislation and policy language that is essentially copied from other state legislatures (Hertel-Fernandez, 2014; Hertel-Fernandez and Kashin, 2015; Kroeger, 2015). This latter body of scholarship is certainly true of legislation addressing personal delivery devices among U.S. states.

On July 1st, 2017, House Bill 1027 was passed, making Florida the first state to pass regulations for personal delivery devices, detailing where they can operate, at what speed, and under what conditions. Furthermore, Florida law requires specific operator information to be posted on each device and requires operators to insure the device for damages up to $100,000, a legal framework flowing from the Bascelli recommendations discussed above. Twenty-six states and the District of Columbia have enacted legislation regulating personal delivery devices, by the time of this writing. Two states, Massachusetts and Illinois have not yet passed drafted bills into law, and the Governor of Kansas vetoed legislation passed by both of its law-making chambers because enforcement and liability clauses were too vague. Other states, such as California, revised DMV regulations to allow for autonomous delivery vehicles, whereby municipalities further regulate these devices. Table 2 displays state regulations for the twenty-six states and the District of Columbia that have passed regulations for personal delivery devices. Three additional states are included in the table, though legislation has not been signed into law.

Among the twenty-seven state/D.C. laws enacted, many of the specific codes are identical between them, with some codes copied and pasted into later adopters' legislation, consistent with the findings of Jansa, Hensen, and Gray (2018). For example, Jansa et al. argue that late adopters of legislation often resort to cut-and-paste policy, rather than reinvention (Jansa et al. 2018). For example, Indiana introduced a bill addressing personal delivery devices on January 4th, 2021 containing this language limiting political subdivision regulation or PDDs:

1. Except as otherwise provided by law, a political subdivision shall not enact or enforce an ordinance or resolution relating to any of the following:

    a. The design, manufacture, maintenance, certification, licensing, registration, taxation, assessment or insurance requirements of a personal delivery device; and
    b. The types of property that may be transported by a personal delivery device.

Oklahoma introduced a bill on January 21st addressing PDDs and adopted this exact language in its bill's text. Regardless of the level of exact duplication between these bills, there is significant overlap in specific codes adopted.

Florida law requires the operator to maintain $100,000 in liability insurance. Twenty states followed suit writing this into their regulations. One state, Idaho, requires $90,000 in liability insurance, while Georgia and Kansas require $250,000 and $500,000 respectively. Similarly, Florida law sets the maximum speed at 10 miles per hour, a speed adopted by 7 states that passed legislation in 2017 and 2018. Nine more states would adopt this rate between 2019 and 2022. Pennsylvania set the rate at four mph, the slowest, while New Hampshire set the rate at 12.5 mph, the fastest. The maximum weight limit adopted by states has a greater amount of variation than the speed limits set. For instance, Florida set a limit of 80 pounds, without cargo, while Michigan set a limit of one thousand pounds. The most frequent weight limit written into state codes is five hundred pounds, adopted by seven states. Furthermore, six states adopted a limit of five hundred fifty pounds. All but four of the state laws in effect at the time of writing required the operator to clearly display contact information on the device. Last, sixteen of the state laws explicitly permit local communities to further regulate PDDs so long as it is done consistent with the state regulation. North Carolina set a two-year time limit for local jurisdictions to pass ordinances regulating PDDs.

At the time of writing, twenty-four states have yet to pass state-wide legislation regulating the use of personal delivery devices on their sidewalks and streets. To be sure, some states may alter current policies allowing for PDD use with the caveat that local jurisdictions provide specific regulations, as in the California approach.

Table 2. State (and DC) regulations on PDDs

State	Year	Maximum Weight	Max Speed Sidewalks	Max Speed Roads	Operator Contact Information	Insurance	Local Control (consistent with State law)	Legislation
Florida	2017	80	10	Unspecified	Required	$100k/device	Allowed	HB 1027
Idaho	2017	550	10	Unspecified	Unspecified	$90k/device	Allowed	HB 204, HB 528
Ohio	2017	90	10	Unspecified	Unspecified	Unspecified	Allowed	HB 49
Virginia	2017	500	10	25	Unspecified	Unspecified	Allowed	SB 1207, SB 758
Wisconsin	2017	80	10	Unspecified	Required	Unspecified	Local restriction not permitted	SB 148
Arizona	2018	120	10	Unspecified	Required	$100k/device		HB 2422
Washington, D.C.	2018	90	10	Unspecified	Unspecified	Unspecified		DC 22-137
Utah	2018	150	10	20	Required	$100k/device		HB 217, HB 277
Texas	2019	110	10	20	Required	$100k/device	Allowed	SB 969
Washington	2019	120	6	Unspecified	Required	$100k/device	Allowed	HB 1325
Colorado	2020	Unspecified	12	20	Required	$100k/device	Allowed	SB 20-092
Michigan	2020	1000	10	20	Required	$100k/device	Allowed	SB 892
Minnesota	2020	500	12	20	Required	$100k/device	Allowed	HF 270
Missouri	2021	200	10	Unspecified	Required	$100k/device	Allowed	SB 176
North Carolina	2020	500	12	20	Required	$100k/device	Allowed (until 2022)	SB 739
Pennsylvania	2020	550	12	25	Required	$100k/device	Limited by bill	SB 1199
Tennessee	2020	Unspecified	10	Unspecified	Required	$100k/device		SB 2836
Arkansas	2021	500	10	20	Required	$100k/device		Act 926
Indiana	2021	500	10	Unspecified	Required	$100k/device	Allowed	HB 1072
Iowa	2021	550	12	20	Required	$100k/device	Allowed	HB 25
Louisiana	2021	500	12	20	Required	$100k/device	Allowed	Act 214
Maryland	2021	550	7	Unspecified	Required	$100k/device		SB 726
Massachusetts		550	10	Unspecified	Required	$100k/device	Allowed	H 3331*
New Hampshire	2021	80	12.5	Not allowed	Required	$100k/device		HB 116
Oklahoma	2021	550	10	Unspecified	Required	$100k/device		HB 706
Georgia	2022	500	4	20	Required	$250k/device	Allowed	HB 1009
Illinois		300	10	Unspecified	Required	$500k/device	Allowed	HB 2902*
Kansas	2022	550	10	Unspecified	Required	$500k/device	Allowed	SB 161**
Mississippi	2022	Unspecified	10	Unspecified	Required	$100k/device	Allowed	HB 961
Wyoming	2022	550	12	20	Required	$100k/device	Allowed	SF 0017

*Bill introduced but not passed
**Bill vetoed

Some states, like Missouri, allow municipal governments to enact further legislation to set parameters and restrictions for personal delivery devices, while other states, such as Pennsylvania, passed legislation that does not allow local governments to further restrict PDDs. Stringent regulations are seen as a restriction to delivery robot deployment (Markets and Markets, 2021). However, regulations are necessary to address public safety and right-of-way issues, as well as food safety and data security concerns.

Tort Issues and Public Safety

The use of personal delivery devices (as in Starship, Coco, etc.) and remote delivery systems (as in the human remote-controlled Kiwibot) give rise to tort issues as they operate within public spaces. A tort is an "act or omission that gives rise to injury or harm to another and amounts to a civil wrong for which courts impose liability (Legal Information Institute)." Pedestrians must share public sidewalks and crosswalk space with these robots, increasing risks of potential bodily harm should they collide with a robot. To reduce the impact and harm of such a collision, Nuro, an automated delivery device not considered a pedestrian, added airbags to the outside of their robotic delivery vehicle (Knapp, 2022). Additionally, there are concerns that bicyclists and automobile drivers may run into a delivery robot.

These concerns are not merely academic. While they occur infrequently, there have been documented cases. Accidents between PDDs and humans are uncommon, but there is a documented case in the UK where a Starship robot ran into a toddler at a grocery store. According to the report, the robot ran into the toddler and never stopped, continuing along its journey (Nott, 2022).

More common are collisions between vehicles on the road and a PDD. For example, four such documented collisions have occurred between a Starship delivery bot and an automobile. One of the first documented cases of a vehicle running into a Starship delivery bot occurred in June 2020. A Frisco, TX resident alleges that a Starship robot struck the left side of her car as she was waiting at a red light, just as it turned green (Parks & Zoga, 2020). Aside from the damage that occurred, reportedly in the amount of $2,600, the delivery robot contained no contact information on it in the event of an accident. When the incident was reported to the driver's insurance company, the insurer was, reportedly, initially reluctant to cover the accident as a collision, and required the driver to pay the deductible, as though she was at fault. Furthermore, the police officer to whom she reported the incident to was unsure if the delivery robot was classified as a vehicle or not.

Her story is not unique. Near the campus of Bowling Green State University, a semi-truck driver ran over a Starship bot that "[d]arted out into the street (Terbrack, 2021)." After having contacted the Bowling Green Police Department, he claims, "the officer was also unsure of what to do, an issue created from the robots not having any contact information on them (Terbrack, 2021)." And, according to a BG Media public records request, this was one of two accidents with Starship bots in the jurisdiction reported since their debut in March of 2021.

Despite an orange flag waving three to four feet in the air to promote the visibility of the robot on the ground, they have been known to get run over by higher-sitting SUVs and trucks. In a viral Tik Tok video posted on August 31st, 2021, a Starship robot can be seen crossing the street on the edge of the University of Kentucky campus. As the PDD slows onto the crosswalk to make the risky traverse across the street, a Ford Escape can be seen taking a right turn, clipping the driverless device with its rear passenger wheel (Rach.ipsa, 2021). Similarly, on the edge of the Purdue University campus, a Starship bot met its demise under the weight of a garbage collection truck. As reported by the truck driver, he did not see the delivery vehicle crossing the street until after he ran over it (Staff Reporter, 2021). In cases where the bot is crossing the street whilst the crosswalk permits walking, the bots are treated like pedestrians and have the right of way. These drivers could be on the hook for the $5,500 cost to manufacture these robots (Palmer, 2019).

Right of Way Issues

In addition to risks when the personal delivery devices move across the street, there are also concerns when the bots block public access to the right of way or when the bots themselves violate a right of way when entering prohibited public spaces. When not moving, these autonomous devices may inadvertently block ingress and egress to and from a crosswalk. When moving, these devices may inadvertently enter public spaces in which they are prohibited from entering, such as a crime scene or road/sidewalk closure.

The University of Pittsburgh temporarily suspended Starship operations on its campus when a Ph. D. student, who uses a wheelchair, was allegedly blocked ingress onto the sidewalk from the crosswalk by a Starship robot. Fortunately, the student was able to maneuver around the bot. Additionally, a disability advocate in the same city also reported sighting a Starship blocking the sidewalk exit in a different incident. It is worth noting that in a review of video footage of the Ph.D. student that was allegedly blocked from entering the sidewalk, Starship states that they dispute that she was blocked (Davis, 2019).

Off campus, in the streets of Hollywood, a Serve robot was caught on camera during a delivery passing through a crime scene. The event occurred when, during a hoax school shooting call, Hollywood PD responded to the event and cordoned off a crime scene area with yellow tape. En route to its delivery, Connie, the delivery bot, stopped at the yellow tape at the edge of the sidewalk, just before the crosswalk. A cameraperson covering the event saw the PDD move to the crime scene tape and stop. After a few seconds, the cameraperson lifted the tape and the bot proceeded under it, crossed the street, and moved past a half dozen or so Hollywood police officers (Collins, 2022). While human error led to the events whereby the PDD passed through the crime scene, the robot's human owners could be held liable for contaminating a crime scene or obstructing justice, which, according to *Black's Law Dictionary* is interference with the orderly administration of law and justice.

Cybersecurity/Data Breaches

As more devices become internet-based and require sensitive user information, such as name, address, and banking information, the concern for resilient cybersecurity measures grows. Cybercrime in the U.S. is pervasive. Nearly 59% of Americans report that they have experienced cybercrime in some fashion. In 2021 alone, there has been an estimated $6 trillion economic loss caused globally by cybercrime (Morgan, 2021). In general, universities have been a prime target for data breaches. For example, in 2021, there were a reported 1,241 cyber incidents involving educational institutions (Braue, 2022).

For Starship Technologies, users must download the app first, then select a participating restaurant from which to order food. Starship also works in conjunction with DoorDash, as does Coco. Users unlock the bot with a link sent to a Smartphone. In August of 2022, DoorDash confirmed that it suffered a data breach where the names, email and delivery addresses, and phone numbers of customers were stolen (Page, 2022). Similarly, Yandex, a Russian-owned company that operated delivery bots on the Ohio State University campus, suffered a data breach in 2022, during the onset of the Russian-Ukraine war. Hacker group Anonymous claimed to have stolen account information from 150 thousand users and posted the information publicly online (Greenwood, 2022). Like any other e-commerce business, food delivery services must keep sensitive user data safe from cyber criminals. However, unlike many other e-commerce businesses, autonomous delivery services must also ensure the security of the delivery robot and, under some state and local regulations, must also be able to be remotely controlled by a human when in need of overriding the AI system (e.g., the state of Washington requires this).

There are several features that a Starship robot, and other similar PDDs, employ to deter theft and to recover the robot if deterrence fails. First, most PDDs have cameras that actively record their surroundings, 360 degrees around the device. These cameras, in some cases, are monitored in real-time by a human. Second, Starship employs a loud "screech" should it be picked up, drawing attention to it from bystanders. Last, if deterrence fails, the PDDs have live GPS that feeds back to their human controllers to assist in device recovery. In one such case, a Kiwibot was stolen near the Berkley campus of the University of California. In this case, police were able to use the GPS tracking capabilities built into the Kiwibot to locate it, and initiate movement commands to the robot to determine which car trunk the stolen bot was in (Reguso, 2019).

While the Starships, Kiwibots, and others have protections against theft, it is less known whether they have protections against being hijacked. Hijacked bots could be used to deliver "trojan-horse" like explosive devices. Recalling the Starship bot that breached a crime scene, had the PDD been hijacked, it could have delivered an explosive device to nearly a dozen police officers standing within a twenty feet radius of it as it moved along the sidewalk. In fact, in the video of this event, one of the camerapersons nearby, half-jokingly made such a comment, stating he hoped it didn't have a bomb in it.[3] In 2018, while traversing the University of California campus, a KiwiBot burst into flames (Brown-Heidenreich, 2018). While no one was reportedly injured, the cause of the fire is unknown.

In addition to physical security, if PDDs can be hijacked, food safety procedures, which rely on a known chain of custody (from restaurant to bot to consumer), may become compromised. Foods must be stored at specific temperatures depending on the item and must be contained to eliminate cross-contamination, in accordance with local public health department standards. Hijacked food delivery devices can no longer guarantee this basic food safety requirement. Last, hijacked bots could simply be deployed off their dedicated path and into high-speed high-traffic areas causing traffic accidents.

Few studies have shed light on this issue and potential solutions. In a recently published article, Ferrer et al. called attention to this issue and demonstrate how the use of blockchain technology can be used to raise the costs of malicious autonomous device activity and reduce its occurrence (Ferrer et al, 2022). Their experiment shows that employing blockchain technology in autonomous vehicles can impose costs on vehicles that attempt to deviate from their benign instructions. Their results provide a path toward the deployment of Byzantine-resistant (malicious or erroneous instructions) real-world multi-robot systems, ensuring a situation guarantying convergence of intended algorithmic instructions dictating autonomous vehicle behavior for its intended purpose, despite alternative instructions (Byzantine instructions/algorithms) that may arise to alter the vehicle's behavior for malicious intent. These alternative instructions may arise through a hack into the device's communication hardware or a malfunction within the device.

CONCLUSION

Personal delivery devices are here to stay, and there will be a greater reliance on them for last-mile delivery as well as first-mile delivery in densely populated urban areas. They are being deployed on college campuses throughout the U.S. and U.K. where there is an underserved student population in need of on-demand food delivery suitable to their often-hectic studying habits that require their time and attention at various locations across campuses, where food services may not be readily accessible. Food delivery can save students time when studying, eliminating the time spent traveling to food service facilities or off-campus restaurants, and eliminating the time standing in line or being waited on. In short, robotic delivery can save students precious time when they need it most. Also, PDDs are increasingly being deployed in cities where consumers are located near restaurants, delivery routes that human deliverers often find inefficient in terms of compensation.

This chapter surveyed local regulations regarding PDDs which attempt to provide a safe environment for these devices to operate while also trying to facilitate the deployment of this technology for economic development opportunities within the community. This chapter shows that while problems with PDDs and the public are rare, some types of incidents are more common than others. Specifically, this research finds that incidents of delivery robots colliding with vehicles are the most common type of accident reported. Incidents between robots and pedestrians are less common, and in fact, are rare. When these delivery robots encountered a moving vehicle it was almost always as the robot was crossing the road. This raises issues about the visibility of the robotic device. In some of the reported accidents, the human drivers of the vehicles, reportedly, did not see the robot while making a turn as the robot was crossing the street.

Issues involving pedestrians were of two types: the PDDs blocking access to a sidewalk/crosswalk and PDDs colliding with a toddler. Though these devices have nearly 360-degree camera capabilities, they do not seem to be able to detect when objects near them are trying to get around them. They also may not be well-calibrated to detect small objects, such as children or animals. Some devices, such as Nuro (an ADD, not PDD), are preparing for such collision by deploying external airbags. These automated delivery devices are different from personal delivery devices in that they operate on roads, rather than sidewalks. However, the sidewalk trolling devices would be well-served by deploying some similar protection just before impact with an object.

Last, while the most common PDDs have integrated anti-theft protections, it is less known the degree to which they are hack- and hijack-proof. Commandeering such a device raises significant questions about physical security threats. For example, what protections exist that defend against an unauthorized user from hacking the bot and opening it up? What protections exist to keep a user from putting an explosive device in a bot after opening it legitimately as a customer? Last, what protections exist to keep the GPS tracking from being tampered with or jammed, perhaps with electromagnetic interference, leading the autonomous vehicle to veer into fast-moving traffic?

PDDs are here to stay. Current trends indicate that they will be in a neighborhood near you if they are not already. Local governments would be well served to begin to understand the implications of these delivery bots operating within their community. To that end, creating strategic plans addressing this current trend would be wise. Communities should assess how their local businesses would be affected by the use of these devices and what new businesses may be created as a result of having these in the community. Furthermore, the implications that PDDs have on public safety and access to public rights of way should be well studied within the community. Community leaders will need to assess the conditions of the infrastructure needed to support PDDs and what changes need to be made to ensure a safe and efficient functioning environment for PDD operation. For example, broken sidewalks may make it difficult or impossible for delivery bots to traverse, while push-button crosswalks may delay the delivery of goods as it waits for a human to press the button. However, it is useful to note here that many state laws do not require local jurisdictions to provide improvements to infrastructure to accommodate PDDs.

Community business leaders and residents should discuss the impact of these devices to inform decision-makers about how to make the best preparations for them. Such preparations are not novel and are the backbone of good public policy. The City of Frisco has developed such a plan. Here, city managers created a plan to pilot test driverless vehicles in their city (City of Frisco Website). This pilot plan includes testing robotic delivery food services Starship and Serve, as well as package delivery through Fed Ex's Delivery Robot. Included in the Frisco plan is pilot testing UAV drone delivery by Google's Wing device. Reports about each of the test-piloted programs are listed on the city's website to inform residents.

While more and more PDDs will be roaming our streets and sidewalks serving our communities, more and more aerial vehicles will be deployed to deliver goods as well. Autonomous vehicle delivery is a growing trend. With the adoption of any new technology, citizen engagement with community leaders about the implications of deploying such technology will help to ensure a smoother transition to the changes that these robotic creatures bring to our communities.

REFERENCES

Acemoglu, D., Lelarge, C., & Restrepo, P. (2020, May). Competing with robots: Firm-level evidence from France. *AEA Papers and Proceedings. American Economic Association*, *110*, 383–388. doi:10.1257/pandp.20201003

Ahuja, K., Chandra, V., Lord, V., & Peens, C. (2022, June 21). *Ordering in: The rapid evolution of Food Delivery.* Retrieved February 15, 2023, from https://www.mckinsey.com/industries/technology-media-and-telecommunications/our-insights/ordering-in-the-rapid-evolution-of-food-delivery

BEA. (2021, September 30). *Wage and salary of U.S. employees, by industry 2020.* bea.gov. Retrieved October 7, 2022, from https://apps.bea.gov/iTable/iTable.cfm?reqid=19&step=2#reqid=19&step=2&isuri=1&1921=survey

Beard, C., & Beard, W. (1930). *The American Leviathan.* McMillian.

Bischoff, L. A. (2022, March 2). Ukraine conflict: Ohio politicos push Russian divestment, OSU cancels food delivery robots. *The Columbus Dispatch.* Retrieved August 9, 2022, from https://www.dispatch.com/story/news/2022/03/03/ohio-state-russia-ukraine-divest-pensions-robots-meal-delivery-yandex/9350250002/

Braue, D. (2022, July 14). Will schools ever learn cybersecurity? *Cybercrime Magazine.* Retrieved October 4, 2022, from https://cybersecurityventures.com/will-schools-ever-learn-cybersecurity/

Brown, S. (2019, September 9). *Presidential hopeful Bill de Blasio wants a tax on robots.* CNET. Retrieved September 28, 2022, from https://www.cnet.com/science/presidential-hopeful-bill-de-blasio-new-york-mayor-wants-tax-on-robots/

Brown-Heidenreich, S. (2018, December 14). Kiwibot catches fire outside Martin Luther King Jr. student union. *The Daily Californian.* Retrieved October 6, 2022, from https://www.dailycal.org/2018/12/14/kiwibot-catches-fire-outside-mlk-student-union

Christensen, C., Raynor, M., & McDonald, R. (2015). What is Disruptive Innovation? *Harvard Business Review*, (December Issue), 44–53. PMID:17183796

Clandinin, D. J. (2007). *Handbook of narrative inquiry: Mapping a methodology*. SAGE Publications, Inc., doi:10.4135/9781452226552

Collins, J. (2022, September 27). *Food delivery robot barges through crime scene, has no time for our human nonsense*. LADbible. Retrieved October 4, 2022, from https://www.ladbible.com/news/food-delivery-robot-barges-through-crime-scene-20220916

Davis, K. (2019, October 23). *Food Delivery Robots pulled from Pitt campus after backlash about mobility*. 90.5 WESA. Retrieved October 4, 2022, from https://www.wesa.fm/identity-justice/2019-10-22/food-delivery-robots-pulled-from-pitt-campus-after-backlash-about-mobility

Deto, R. (2020, December 2). *Pennsylvania legalizes autonomous delivery robots, classifies them as pedestrians*. Pittsburgh City Paper. Retrieved October 1, 2022, from https://www.pghcitypaper.com/pittsburgh/pennsylvania-legalizes-autonomous-delivery-robots-classifies-them-as-pedestrians/Content?oid=18482040

Dixon, J., Hong, B., & Wu, L. (2021). The Robot Revolution: Managerial and employment consequences for firms. *Management Science*, *67*(9), 5586–5605. doi:10.1287/mnsc.2020.3812

Dumas, B. (2022, September 22). *Delivery Robots emerge to bridge gaps as labor shortage drags on*. Fox Business. Retrieved October 7, 2022, from https://www.foxbusiness.com/technology/delivery-robots-emerge-bridge-gaps-labor-shortage-drags

Durbin, D. (2021, November 2). *Robots hit the streets as demand for food delivery grows*. AP News. Retrieved September 26, 2022, from https://apnews.com/article/coronavirus-robot-food-delivery-technology-business-5b7dbb782f89ef3d586eefb9a33d5e1c

Dutton, W. (1982). Technology and the Federal System. *Proceedings of the Academy of Political Science*, *34*(4), 109–120. doi:10.2307/3700973

Ferrer, E., Jiménez, E., Lopez-Presa, J. L., & Martín-Rueda, J. (2022, April). Following Leaders in Byzantine Multirobot Systems by Using Blockchain Technology. *IEEE Transactions on Robotics*, *38*(2), 1101–1117. doi:10.1109/TRO.2021.3104243

Flynn, J. (2022). *18+ Food Delivery Statistics [2022]: Online Ordering Industry Numbers You Need To Know*. Zippia.com. Feb. 2, 2022, https://www.zippia.com/advice/food-delivery-industry-statistics/

Goldsmith, M. (1989). State Policies and Programs for Encouraging Innovation and Technology Development. In *State and Local Initiatives on Productivity, Technology, and Innovation: Enhancing a National Resource for International Competitiveness*. Advisory Commission of Intergovernmental Relations.

Greenwood, J. (2022, March 7). *Anonymous claims Yandex data breach affecting 150K Users*. The Cybersecurity Times. Retrieved October 4, 2022, from https://www.thecybersecuritytimes.com/anonymous-claims-yandex-data-breach-affecting-150k-users/

Hendrickson, K. A. (2003). National Traffic and Motor Vehicle Safety Act. In S. I. Kutler (Ed.), *Dictionary of American History* (3rd ed., Vol. 5, pp. 561–562). Charles Scribner's Sons.

Hertel-Fernandez, A. (2014). Who passes business's "model bills"? policy capacity and corporate influence in U.S. state politics. *Perspectives on Politics*, *12*(3), 582–602. doi:10.1017/S1537592714001601

Hertel-Fernandez, A., & Kashin, K. (2015). Capturing business power across the states with text reuse. Academic Press.

International, S. A. E. (2021, April 30). *J3016_202104: Taxonomy and definitions for terms related to driving automation systems for on-road motor vehicles*. SAE International. Retrieved February 15, 2023, from https://www.sae.org/standards/content/j3016_202104/

Jackson, E. (2021, November 18). *History of food delivery and how it's changed*. Thistle. Retrieved September 10, 2022, from https://www.thistle.co/learn/thistle-thoughts/history-of-food-delivery-and-how-its-changed

Jansa, J. M., Hansen, E. R., & Gray, V. H. (2018). Copy and paste lawmaking: Legislative professionalism and policy reinvention in the States. *American Politics Research*, *47*(4), 739–767. doi:10.1177/1532673X18776628

Kharpal, A. (2017, June 2). *Bill Gates wants to tax robots, but the EU says, 'no way, no way'*. CNBC. Retrieved September 28, 2022, from https://www.cnbc.com/2017/06/02/bill-gates-robot-tax-eu.html#:~:text=Earlier%20this%20year%2C%20billionaire%20Gates,robot%20at%20a%20similar%20level.%E2%80%9D

Kirby, M. (2002). *The Origins of Railway Enterprise: The Stockton and Darlington Railway 1821–1863*. Cambridge University Press.

Knapp, M. (2022, January 12). *Nuro adds airbags (for people) to self-driving delivery vehicles*. PCMAG. Retrieved October 3, 2022, from https://www.pcmag.com/news/nuro-add-airbags-for-people-to-self-driving-delivery-vehicles

Krietzberg, I., & Lucas, A. (2022, July 17). *Restaurants are short-staffed, and that's taking a big toll on customers and workers alike*. CNBC. Retrieved October 7, 2022, from https://www.cnbc.com/2022/07/17/customer-service-suffers-at-short-staffed-restaurants-as-covid-takes-toll.html

Kroeger, M. (2015, May 28-30). *Plagiarizing policy: Model legislation in state legislatures*. Academic Press.

Legal Information Institute. (n.d.). *Tort*. Legal Information Institute. Retrieved October 2, 2022, from https://www.law.cornell.edu/wex/tort

Lippert, J. F., Furnari, M. B., & Kriebel, C. W. (2021). The Impact of the COVID-19 Pandemic on Occupational Stress in Restaurant Work: A Qualitative Study. *International Journal of Environmental Research and Public Health, 18*(19), 10378. doi:10.3390/ijerph181910378 PMID:34639678

Market Research Future. (2022, May 19). *Delivery robots market projected to reach valuation of USD 971.3 million by 2030, growing at a 31.3% CAGR - Report by Market Research Future (MRFR)*. GlobeNewswire News Room. Retrieved September 28, 2022, from https://www.globenewswire.com/en/news-release/2022/05/19/2446711/0/en/Delivery-Robots-Market-Projected-to-Reach-Valuation-of-USD-971-3-Million-by-2030-Growing-at-a-31-3-CAGR-Report-by-Market-Research-Future-MRFR.html

Markets and Markets. (2021, April). *Delivery robots market size, share: Industry report (2021-2026)*. Markets and Markets. Retrieved October 2, 2022, from https://www.marketsandmarkets.com/Market-Reports/delivery-robot-market-263997316.html

Markides, C., & Oyon, D. (2010). What to do against disruptive business models (when and how to play two games at once). *MIT Sloan Management Review, 51*(4), 25–32.

Marr, B. (2021, December 10). *The Future of Delivery Robots*. Forbes. Retrieved September 26, 2022, from https://www.forbes.com/sites/bernardmarr/2021/11/05/the-future-of-delivery-robots/?sh=3bfc300d7337

Morgan, S. (2021, April 27). Cybercrime to cost the world $10.5 trillion annually by 2025. *Cybercrime Magazine*. Retrieved October 4, 2022, from https://cybersecurityventures.com/cybercrime-damages-6-trillion-by-2021/

Muzzey, D., Beard, C. A., & Beard, W. (1931). Review: The American Leviathan. *Political Science Quarterly, 46*(1), 109–111. doi:10.2307/2143112

Nichols, G. (2017, December 12). *San Francisco bans delivery robots in most of the city*. ZDNET. Retrieved October 8, 2022, from https://www.zdnet.com/article/san-francisco-bans-delivery-robots-in-most-of-the-city/

Nott, G. (2022, April 12). *Robot Grocery Delivery Company starship investigating toddler 'hit and push' incident*. The Grocer. Retrieved October 5, 2022, from https://www.thegrocer.co.uk/technology-and-supply-chain/robot-grocery-delivery-company-starship-investigating-toddler-hit-and-push-incident/666529.article

Office of the Bureau of Public Roads. (n.d.). https://www.archives.gov/research/guide-fed-records/groups/030.html#30.1

Ooi, V., & Goh, G., (2022). Taxation of automation and artificial intelligence as a tool of labour policy. *eJournal of Tax Research, 19*(2), 273-303.

Osborne, D. (1988). *Laboratories of Democracy*. Harvard Business School Press.

Page, C. (2022, August 25). *Doordash hit by Data Breach linked to Twilio hackers*. TechCrunch. Retrieved October 4, 2022, from https://techcrunch.com/2022/08/25/doordash-customer-data-breach-twilio/

Palmer, M. (2019, September 4). *Europe's 6-wheeled delivery robots begin invasion of US campuses*. Sifted. Retrieved October 3, 2022, from https://sifted.eu/articles/starship-robot-delivery/

Parks, E., & Zoga, D. (2020, September 11). *NBC 5 responds: What happens if you're in a crash with a robot?* NBC 5 Dallas-Fort Worth. Retrieved October 3, 2022, from https://www.nbcdfw.com/news/nbc-5-responds/nbc-5-responds-what-happens-if-youre-in-a-crash-with-a-robot/2442345/

Phillips, B. (2022, April 14). Pizza Margherita may be fit for a queen, but was it really named after one? *National Geographic*.

Rach.ipsa. [@rach.ipsa]. (2021). *rach.ipsa* [TikTok profile]. TikTok. Retrieved September 22, 2022, from https://www.tiktok.com/@rach.ipsa/video/7002708616255950085

Raguso, E. (2019, April 26). *Police: Berkeley Man, fed up with Kiwibot delivery robots, steals one*. Berkeleyside. Retrieved October 6, 2022, from https://www.berkeleyside.org/2019/04/26/police-berkeley-man-fed-up-with-kiwibot-delivery-robots-steals-one

Sanburn, J. (2013, May 18). Slices of History: Great Moments in Pizza Innovation. *Time: Business*. Retrieved September 10, 2022, from https://business.time.com/2013/05/18/slices-of-history-great-moments-in-pizza-innovation/

Schley, D. (2020). *Steam City: Railroads, Urban Space, and Corporate Capitalism in Nineteenth-Century Baltimore*. University of Chicago Press. doi:10.7208/chicago/9780226720395.001.0001

Schreurs, M. A., & Steuwer, S. D. (2016). *Autonomous Driving—Political*. Legal, Social, and Sustainability Dimensions.

Seamans, R. (2022, March 9). *Tax not the Robots*. Brookings. Retrieved September 28, 2022, from https://www.brookings.edu/research/tax-not-the-robots/#footnote-1

Shaw, R. (1990). *Pioneer Canals and Republican Improvements. In Canals For A Nation: The Canal Era in the United States, 1790-1860*. University Press of Kentucky. https://www.jstor.org/stable/j.ctt5vkjxc.5

Short, H. (2022, September 26). *Pizza Hut Canada launches door-to-door robot delivery service: Dished*. Retrieved September 26, 2022, from https://dailyhive.com/vancouver/pizza-hut-canada-robot-delivery-service

Singh, A. J., & Wang, B. (2021, May 31). *Impact of covid-19 on the hospitality industry and implication for operations and Asset Management*. Boston Hospitality Review. Retrieved October 7, 2022, from https://www.bu.edu/bhr/2021/05/31/impact-of-covid-19-on-the-hospitality-industry-and-implication-for-operations-and-asset-management/

Starship. (2022, March 1). *$100m raised in last 30 days by Starship Technologies as demand for its autonomous delivery service triples in 2021* [Press release]. Retrieved February 15, 2023, from https://www.starship.xyz/press_releases/100m-raised-in-last-30-days

Terbrack, J. (2021, March 31). *Starship vs. whips: Robots pose potential risk on streets*. BG Falcon Media. Retrieved October 3, 2022, from https://www.bgfalconmedia.com/news/starship-vs-whips-robots-pose-potential-risk-on-streets/article_f31331b2-91af-11eb-af
2b-d7feb69985a3.html

The White House. (2016, December). *Preparing for the future artificial intelligence, of artificial*. The White House of President Barak Obama. Retrieved September 27, 2022, from https://obamawhitehouse.archives.gov/sites/whitehouse.gov/files/documents/Artificial-Intelligence-Automation-Economy.PDF

Thompson, S. (1925). *A short history of American railways, covering ten decades*. Appleton and Company.

US DoT. (2021, January). *Automated Vehicles: A comprehensive plan*. U.S. Department of Transportation. Retrieved October 1, 2022, from https://www.transportation.gov/sites/dot.gov/files/2021-01/USDOT_AVCP.pdf

Weingroff, R. (1996). From 1916 to 1939: The Federal-State Partnership at Work. *Public Roads, 60*(1).

Wells, M. (2022, March 4). *Trendy bakery unleashes robots in San Francisco parks*. SFGATE. Retrieved October 8, 2022, from https://www.sfgate.com/food/article/treat-vending-robots-at-SF-parks-16975139.php

Yudken, J. S., Croft, T., & Stettner, A. (2022, April 27). *Revitalizing America's manufacturing communities*. The Century Foundation. Retrieved September 16, 2022, from https://tcf.org/content/report/revitalizing-americas-manufacturing-communities/?agreed=1

ENDNOTES

1. Survey of employment opportunities from all mentioned PDD providers via company websites conducted by the author on October 5 through October 7, 2022.
2. Data taken from each company's website accessed October 9th, 2022; https://www.kiwibot.com/college-campuses and https://starshipdeliveries.com/campus/.
3. Today Show video found here shows incident with audio: https://www.today.com/food/news/food-delivery-robot-rolls-crime-scene-viral-video-rcna47810.

Chapter 8
Government Response and Perspective on Autonomous Vehicles

Victor Kilanko
Claremont Graduate University, USA

ABSTRACT

Autonomous vehicles are artificial intelligence-based mobility devices. They are not a new technological development as the concept has been a work-in-progress for centuries. However, this century has the unique opportunity to witness the emergence of self-driving cars on roads all over the world. This, thus, makes it imperative to quickly address how the world would change with robotaxis on our roads and how we ought to prepare for them. The advantages and disadvantages of autonomous driving machines abound; however, the onus is on governments to minimize the threats and harness the benefits. This chapter, therefore, is about how governments can effectively leverage autonomous vehicles to promote humans' quality of life and the necessary considerations on how to minimize the risks associated with self-driving vehicles.

INTRODUCTION

Autonomous vehicles (AVs) are like every other technological application built on the agency of artificial intelligence. These vehicles aim to ease transportation, minimize road accidents, and increase human life quality. Cars have had automated technology for decades (Schreurs & Steuwer, 2015). This technology had been presented in the form of features such as anti-lock brakes, cruise control, lane departure warning signs, and many more. However, the focus of this paper is on fully automated vehicles. Like every technology with good intentions, there may be unintended harmful tendencies. Therefore, this chapter will dive deep into this disruptive technology by analyzing the impacts of autonomous vehicle technology, governments' policies and regulations, and limitations to AVs regulations.

DOI: 10.4018/978-1-6684-6429-8.ch008

THE TECHNOLOGY

In the 1500s, Leonardo da Vinci designed an autonomous vehicle (AV) - a self-driving cart that would move along preset paths (Stanchev & Geske, 2016). This is believed to be the first AV design. Several centuries later, the city of Manhattan was home to the demonstration of a radio-controlled car in 1925. However, this demonstration was discontinued when the car crashed into other vehicles. Further in the 20th century, General Motors created a self-driving car controlled by an electromagnetic field on a road with implanted magnets (Bimbraw, 2015). The road was modified to aid the car using sensors mounted on the car. By the 1960s, cameras were attached to AVs to aid their navigation. Japan improved on this idea to gather enough data with the cameras, enabling the cars to go as fast as twenty miles per hour. By 1995, Carnegie Mellon University had made a self-driving car that could travel from Pittsburgh to San Diego through advanced image processing using neural networks. Advanced image processing helped the car to steer, but the speed and braking were human-controlled (Stanchev & Geske, 2016). In the 2000s, the United States Department of Defense became involved in AV challenges and testing. In the 2010s, ridesharing companies tried to push the current research limit of AVs, seeing as AVs are critical to their profitability and bottom line. AV research has come a long way, even though commercial AVs have proved elusive thus far.

To further understand AV research and appreciate how AVs should be regulated, let's explore the levels of vehicle automation as defined by the International Society of Automotive Engineers (SAE) (Rivard, 2018). At level zero, we have no automation. Humans, at the wheel, make navigation, brakes, and speed decisions. At level one, we have the driver assistance phase where the car can control either its speed or steering, but not the two simultaneously. At this phase, a driver is at the wheel to take total responsibility for the car. Level 2 has partial automation in which the car can accelerate, brake, and steer itself in some circumstances. Tesla's autopilot is an excellent example of level 2 vehicles in which drivers must remain at the wheel to respond to traffic signals and hazards. Conditional automation is achieved at level 3. Here, the car can drive itself while also scanning its environment. Nevertheless, a driver must be at the wheel if the car needs help navigating unfamiliar scenarios. Level 4 is a high automation level where mere human oversight is needed. The car still has a steering wheel and pedals, but can essentially drive itself. Level 5 is the peak of autonomous vehicles. Every human in the car is a passenger as there are no pedals or steering wheel (Stanchev & Geske, 2016). Thus, AVs at level 5 are considered the future of mobility and the holy grail of automation, the standard on which this chapter is based.

There are different stakeholders involved in delivering these vehicles of the future (Schreurs & Steuwer, 2015). There are auto manufacturers who have mastered the art and science of auto assembling; technology companies who have often been associated with software; ride-hailing companies with platforms for shared mobility; politicians and legislators; activists; city planners; researchers who could be artificial intelligence scientists or ethicists; and investors. The interaction of these different players has disrupted the transportation industry from the emergence of ride-hailing platforms to the creation of car-subscription services to electric vehicles, and now, self-driving cars (Gomez-Ibanez, n.d.). All disruptions seem to be connected and culminating in the technological climax of connected vehicles that can interact with one another, road infrastructure, and humans via a phone. Hence, self-driving cars are in the near future. While it may take a while to reach full-scale commercialization and adoption of AVs, the future road might be laced with fleets of robotaxis ordered via phones. And of course, the interaction among these stakeholders would continue to facilitate the dream of self-driving cars.

Many big-time players in the autonomous vehicle world have developed valuable partnerships and invested substantially towards the realization of driverless vehicles. In June 2020, Amazon acquired Zoox, a company manufacturing fully autonomous all-electric cars, for over $1 billion (Porter, 2020). Zoox's cars are bi-directional with no pedals and steering wheels pursuing a level 5 automation (Gale, 2022). They have also been described as transportation devices purposely built to offer mobility-as-a-service (Gale, 2022). These robotaxis have been widely assessed on roads and highways and could be on the market sooner than imagined. Yet, Amazon still has other investments in AV companies such as Aurora Innovation (Porter, 2020). Google's parent company, Alphabet, also has an AV arm called Waymo. Waymo operates a self-driving ride-hailing service called Waymo One in Phoenix, Arizona (Elias, 2020). Waymo also has a partnership with Geely (owners of Volvo in China) to take all-electric AVs to the market (Fannin, 2022). GM Cruise, owned by General Motors, currently allows public members to order its driverless robotaxis (Korosec, 2022). General Motors announced an investment of $2 billion in GM Cruise in 2022 and would commence the mass production of a fully autonomous car in 2023 (Bellan, 2022; Korosec, 2022). General Motors expects Cruise to fetch the company about $50 billion in profit by 2030, though the subsidiary has not yet made a profit (Bellan, 2022). Cruise also formed an alliance with Honda to commercialize all-electric AVs (Fannin, 2022). There is also a partnership between GM Cruise and Walmart to robo-deliver groceries in Phoenix, Arizona (Fannin, 2022). Ford Motors and Volkswagen-backed Argo AI began robotaxi operations on the busy streets of Miami and Austin (Sensiba, 2022). Meanwhile, Argo AI has also partnered with ride-hailing services through its application programming interface (API) called Argo Connect to offer AV options to riders (Sensiba, 2022). Many partnerships have been formed on this mission to put self-driving cars on roads. Motional was created in 2021 between Hyundai and Lyft. Baidu also partnered with Geely to mass produce AVs in 2023. Furthermore, Baidu has developed another partnership with Pony.ai to start a robotaxi service in China in 2021. Resultantly, over $12 billion of equity funding was invested in AV technology companies in 2021 (Fannin, 2022), and many partnerships continue to be formed.

The trucking sector of the transportation industry has also seen many developments. Embark Truck recently completed road testing of its trucks in snowy weather conditions in Montana, USA (Edwards, 2022). Embark gathered and analyzed multiple weather data points which were then programmed into its trucks' Vision Map Fusion (VMF) technology to ensure they can navigate extreme weather conditions. Aurora Innovation likewise recently announced the release of road-ready trucks by 2024 (Magill, 2022). In the next few years, AVs will take over the trucking industry, and routine long-distance truck delivery will become automated.

This account of the developments of autonomous vehicle would not be complete without mentioning that some accidents have occurred involving autonomous vehicles. One of Pony.ai's cars hit a lane divider in Fremont, California on October 28, 2021 while changing lanes in autonomous mode (Liao, 2021). There was also a fatal crash in 2018 involving an Uber AV which killed an Arizona woman, even though there was a test driver in the vehicle at the time of the crash (Levin, 2018). However, more than ever, the dream of driverless cars on our roads is becoming more evident. The accidents have not deterred self-driving vehicles' innovation. Instead, they have presented opportunities to all stakeholders to understand the risks of self-driving cars and become prepared to mitigate them. Hence, legislators and ethicists must work with industry players to minimize fatalities and stop this dream from becoming a nightmare.

IMPACT

The emergence of cars expanded urban development, created suburbs, and influenced how humans live. People live further away from families, work, and entertainment because of the luxury of faster transportation. In the same way, the advent of AVs will further disrupt life on earth as we know it and affect decisions on work, entertainment, commuting, home buying, education, and much more. Self-driving automobiles will change cities. Legislators need to be aware of these impacts and craft policies accordingly.

The former US Secretary of Labor, Elaine Chao, publicly expressed her concern for the impact of AVs on driving-related jobs ("Autonomous Vehicles Confront a Political Viability Crisis - Baron Public Affairs," n.d.). Progressives have claimed that AVs may further economic oppression among minority groups as members of these groups are strongly represented in the driving occupation. About 2.9% of the US labor force is in the driving occupation (Klaver, 2020). If autonomous vehicles replace these people, about $180 billion of income will be lost yearly (Klaver, 2020). Hence, the adoption of AVs may cause an upward trend in unemployment rate. On the other hand, there is an imminent shortage of professional truck drivers (Talbott, 2020). So, the commercialization of AVs might support truckers' shortages and cause a positive shift in the trucking industry. Thus, as AVs create unemployment in the transportation industry, they also meet the increasing demand for truckers.

AVs may disrupt public transit, commuting, and private vehicle ownership as well (Gomez-Ibanez, n.d.). With AVs, cars may change from a product to a service with several companies offering riding services (Klaver, 2020). If people can order rides efficiently, the attraction of car ownership may wane. The reduction in private car ownership may negatively affect the auto industry (Hayes, 2021). By 2030, the US could move from 247 million vehicles to 44 million vehicles (Editors, 2017). Nevertheless, commuting may increase as it becomes easier to travel due to no driving stress or burden. Additionally, people are willing to commute more because commuting time may be spent doing other productive activities (Polly, 2020). A multidisciplinary research team from eight universities found that AVs could eliminate stop-and-go traffic (Editors, 2017). An average US driver losses about $1,348 per year to congestion (Klaver, 2020). The cost of congestion in freighting is about $74 billion a year (Klaver, 2020). Thus, we can expect more people to live farther from work as they are more willing to commute with AVs. If AVs are electric, people are also incentivized to travel more due to no fuel cost and low variable costs. Transportation cost will reduce. An average US family will save about $5600 yearly from reduced transportation cost (Editors, 2017). Vehicles may be shared more as well, which increases the productivity of vehicles. Putting vehicles to more efficient use will unavoidably free parking spaces. Moreover, the mobility of disabled and senior citizens may be increased (Hayes, 2021). Consequently, the widespread adoption of AVs will put transportation resources to more efficient use and substitute driving time with more productive tasks.

About 90% of traffic fatalities are due to the driver's error (Gomez-Ibanez, n.d.; Klaver, 2020). More people die from road accidents than AIDS (Klaver, 2020). Removing drivers means ending these fatalities. The economic cost of vehicle accident in the US was $836 billion in 2010 when human cost, economic damage and productivity loss are added (Klaver, 2020). In the Netherlands, the economic cost of vehicle accident was 17 billion euros in 2018 (Klaver, 2020). If the US could reduce road fatalities by 1%, it would save about $8 billion annually (Klaver, 2020). As a result, the auto insurance industry may be affected if AVs are sold to be safer and less prone to accidents (Hayes, 2021). The US vehicle insurance sector is about $300 billion in market size (Klaver, 2020). Reduced accidents will lead to reduced premium and reduced income for insurers. Reduced road accidents and efficient use of travel time is worth about $750 billion to the US in added societal benefits (Klaver, 2020). As level 5 automation is achieved, road fatalities will be minimized.

Self-driving vehicles will also impact the government's revenue from the transportation industry. The Texas Department of Motor Vehicles makes about $4.5 billion annually from car registration, fees, and licensing (Talbott, 2020). This money goes towards highways and road maintenance and development. The dawn of self-driving cars may wipe out the need to get a driving license and therefore reduce vehicle licensing and fee revenue. Illinois makes about $60 million per year on parking meters and parking garage taxes (Talbott, 2020). If AVs use less parking, this funding is gone. Of course, this means that parking garages can be put to more productive economic usages like parks, schools, malls etc., and the government can continue to get revenue from taxes. Washington DC makes about $300 million from traffic violation fines and parking infractions annually (Talbott, 2020). If AVs obey all traffic laws, where would cities make such money to augment their budget? Also, it should be noted that vehicle accidents generate about $71 billion in repair cost yearly (Klaver, 2020). Autonomous vehicles may reduce repair cost if they seldomly break down and resultantly reduce sales taxes and tax revenue from auto repair shops. As the government may benefit from the massive use of driverless vehicles, traditional transport-related revenue will decline.

AVs may also affect urban development and cities' layout. Cities may be built to give more space to pedestrian traffic and cyclists (Polly, 2020) as autonomous vehicles are very effective in communicating with one another and road infrastructure. As established above, with reduced transportation cost (Klaver, 2020), people may be willing to live farther from work and entertainment. This totally changes the outlook of cities. Cities may become more spread out because real estate development may become more dispersed. This could lead to declining population in some cities. Therefore, taxes per capita may increase because there are less people to contribute to road and infrastructure maintenance and development. Moreover, because people are willing to live outside major cities, suburbs may experience faster population growth and increased development. However, many environmental activists have argued against AVs because of the possibility to raise real estate development. The impact of AVs on humans' quality of life is enormous, however, the onus lies on all stakeholders to minimize the negatives while the positives are amplified.

GOVERNMENT'S RESPONSE

Due to the advancement of autonomous technology and potential reality of self-driving vehicles, the regulatory environment has seen many developments lately. Many of which will be discussed in this section. However, seeing as autonomous vehicle technology is more drastic, a more robust regulatory intervention is needed. Questions about data protection, safety, and privacy need to be answered. Stakeholders must work together to facilitate a regulatory and technological environment to realize these future cars while protecting human lives, self-agency, and dignity. From testing to commercialization to parking to city planning, regulations must be made to protect people and protect the technology from criminals.

Historically, governments have responded to innovative technologies with skepticism, cynicism, and neglect. This, however, changed during World War II as the United States experienced unprecedented partnerships between the government, industry, and universities. Scientific and technological innovations were perceived as active contributors to the US victory. Nevertheless, the government still struggled with peacetime applications of innovations, building science and education infrastructures, and fundings.

The establishment of the National Science Foundation (NSF) took years before it could pass through Congress. Even at this, while the defense budget was about $50 billion in 1952 with $1 billion allocated to the Department of Defense Research projects, NSF only got $3.5 million.

Fast forward to the 2000s, the government relied on high-tech innovation in internet technology, green energy, software, and artificial intelligence to increase the US economic competitiveness in the global comity. As much as military applications are still important, maintaining a balance of trade surplus is just as crucial. Hence, the federal government emphasized an innovation technology that could replace old technology while providing abundant jobs and wealth for Americans and the United States of America (Block & Keller, 2015). Thus, from the American Recovery and Reinvestment Act to the Infrastructure Bill (Kilanko, 2021), the headline is building infrastructure to promote innovation and American competitiveness.

Therefore, political actors can hasten autonomous vehicles' development through quicker and broader licensing, funding, infrastructural support, and reasonable risk management in the use of the technology. Fiscally, government may direct funds towards the AV network and ecosystem with the aim to promote demand and boost public confidence in the technology. To raise public confidence in the technology and increase demand for AVs, government may also procure AVs as government service vehicles and create positive awareness for the technology. Supply can be increased through research and development fundings. Monetarily, tax credit may be extended towards buyers of autonomous cars, while tax breaks are given to manufacturers and vendors. Likewise, the government must support research geared towards the ethics of artificial intelligence. Public innovation policies must be promoted to facilitate safe adoption of AVs.

How a technology is framed and promoted determines the political support and public opinion it receives. If AV is framed as a technological advancement to promote national competitiveness by creating a new industry that can offer an avalanche of jobs for citizens while ensuring senior citizens and the disabled are safely transported, the public will accept it and politicians will push for it. If framed as a loss of individual control and the resignation of individual freedom to big corporation and government, the public's goodwill is lost, and political actors are forced to stop the technology. Hence, a significant part of promoting AVs is selling the benefits to the public. Considering all the gains inherent in the development and commercialization of AVs, this could be a task undertaken by the government. State leadership should be eager to sell the vision of the future of humanity with self-driving cars to its citizens. This is not a time for cynicism or partisan politics but for visionary leadership. We all stand to benefit from the adoption of AVs when the risks are effectively mitigated.

The European Union is in support of autonomous vehicles. Project AdaptIVe was funded by the European Union to demonstrate the possibilities of automated driving in complex traffic situations (Schreurs & Steuwer, 2015). There are also multiple European-funded projects directed towards AV technology to make Europe more competitive. Project HAVE-it was created to demonstrate high level automation is possible in vehicles (Schreurs & Steuwer, 2015). Project SMART investigated the possibilities and benefits of connected cars (Schreurs & Steuwer, 2015). It submitted that connected cars promote efficient use of infrastructure and fuel and are safer. Project Citymobil was commissioned to bring to fruition the promise of automated public transport (Schreurs & Steuwer, 2015). Internationally, many governments have realized that their long-term competitiveness and economic sustainability may be judged by how artificial intelligence-based applications like AVs are funded, evaluated, and adopted. Europe has been making all necessary changes to be technologically friendly.

The United Nations 1968 Convention on Road Traffic (the Vienna Convention) Article 8, paragraph 5 states, "Every driver shall at all times be able to control his vehicle or to guide his animals" (Nations, 1977). In May 2014, Austria, Belgium, France, Germany, and Italy proposed an amendment to this convention that would allow self-driving cars on roads provided they can be overridden by the driver (Schreurs & Steuwer, 2015). The amendment was accepted by the UN Working Party on Road Traffic Safety. Consequently, AVs can be tested on roads in Europe.

Other governments have also been in support of AVs. In 2013, Nissan was approved by the Japanese government to test self-driving cars (Schreurs & Steuwer, 2015). After that, many Japanese politicians were seen in self-driving cars to create awareness and boost the public's confidence in them. The Swedish government has been highly pro-self-driving cars. Volvo was commissioned in 2014 to provide self-driving cars for every Swedish citizen (Schreurs & Steuwer, 2015). The Swedish government recognizes the benefit of AVs and seeks to create a positive legal environment to promote it. In Germany, AVs are well-funded and well-tested as well. There is much research into autonomous vehicles and intelligent logistics. The German Federal Highway Research Institute has stressed the importance of a national nomenclature and standard for automation (Schreurs & Steuwer, 2015). China has also been highly supportive of AV development and testing with dedicated testing routes and road sections (Fannin, 2022). Many countries around the world are adequately preparing for the future of transportation.

In the United States, states like California, Nevada, and Florida have passed legislations on AVs. Many other states are considering similar legislation. This legislation focuses on the definition of AV, testing, operation permission/licensing, and establishment of legal liability (Schreurs & Steuwer, 2015). Many technology companies have further pushed for legislative clarifications, demystification, and alignment. It would be burdensome for companies to have to deal with different legal standards across states. Hence, these laws must be standardized. In May 2013, the US National Highway Traffic Safety Administration (NHTSA) instituted vehicle classification ranging from level 0 to level 4, rising from no automation to full automation (Schreurs & Steuwer, 2015). NHTSA also works with state legislative bodies to synchronize AV laws across states (Schreurs & Steuwer, 2015). Many states envision autonomous vehicles as the future and are willing to enact favorable laws to attract AV benefits and innovation to their state economies.

However, there have been some hostile public policy environments for autonomous vehicles due to some skepticisms towards the technology. A review of more than 200 political publications focused on AVs revealed that more than a few politicians across the aisle view AVs as too dangerous for centralized control. They also look unfavorably on the possibility of AVs to displace low and middle-income jobs. Some progressives and left-leaning politicians mistrust corporations and are unwilling to leave the control of fleets of AVs to them. They believe that the government must serve as a check. There have also been calls for cities to disavow and disallow AVs on their roads. Many conservatives think that the possibilities of AVs equal the potential for increased state and government control. They suppose that the government may hack it or legislate it to limit their freedom and rights. Should there be another pandemic or a national emergency, they think that the government can easily limit their freedom of movement by shutting down AVs and the transportation network (News, 2020). Businesses could easily become unreachable by cars, trains, or trucks. Some have also argued that yielding vehicle steering control is releasing our human decision-making capacity to algorithms and technocratic elites (News, 2020). Our pattern of movement and commute would be known by algorithms and companies intimately. This is interesting because our cellphones, mobile applications, and telecommunication service providers already have more than enough information on our commute and travel patterns. Though, it is understandable to see people refuse to yield the control of another device to technology companies, allowing them to intrude their privacy. As long as it is unclear how passenger's rights are protected under the AV era and how secure AVs are from villainous attacks or central control, many would remain suspicious of the technology.

The US Government's Response to China's Competition

It should be noted that the utopian dream of robotaxis fleeting our roads will not be achieved if the access and distribution of key components are restricted by government policies due to trade war or fear of competition. One of the most important components of artificial intelligence and autonomous vehicles is high-performance semiconductor chip technology. As of now, there will be no AVs without advanced microchips. However, on October 7, 2022, the United States government broadly banned China from manufacturing and importing advanced semiconductor chips (Woods, 2022). The United States Commerce Department entirely prohibits semiconductor chips and chip manufacturing equipment export to China (Bluhm, 2022). This cuts off China from the world semiconductor supply chain. Countries that use controlled US technology are compelled to abide by these export restrictions (Sheehan, 2022). Leveraging its global power as a semiconductor powerhouse, these countries are compelled to forego huge revenue from exporting semiconductors to the biggest semiconductor market in the world, China (He, 2022). In 2021, China imported about $414 billion worth of semiconductors (He, 2022). Without access to chips and advanced chip making equipment manufactured with US technology, China will lag behind in the AVs race and its future in the AV space is greatly jeopardized (Sheehan, 2022).

Furthermore, US citizens and green card holders cannot help with manufacturing of chips at any Chinese facilities (He, 2022). This has a huge threat to annihilate China's current semiconductor industry as American workers are forced to quit. Resigning executives, research workers and engineers will surely deplete the Chinese chips talent pool. One of China's biggest semiconductor equipment manufacturers, Advanced Micro-Fabrication Equipment China (AMEC) has a founder and chairman with US citizenship (He, 2022). Many other companies also have executives with US citizenship or green cards. These people would be forced to resign per the Biden administration's policy, leaving China deprived of chip manufacturing talents, leadership, and knowledge. This policy has the potential to set China back in its AVs pursuit and limit its technology ambition and innovation. The most important question now is how much of China's contribution to the autonomous vehicle space would the world miss?

The above policy is a perfect case of how a seemingly AV-neutral policy can disrupt industry and market. Some players are left in the dust, while some soar. The Chinese market which has been excelling in autonomous vehicle technology and innovation will probably slow down as soon as China runs out of chips. This will consequently delay the wide adoption of self-driving cars in China. We can trust China to retaliate against these policies by restricting the importation of autonomous cars to China. Currently, the US looks to benefit from this Biden era policy as its big AV competitor is constrained. However, we have yet to see how China may strike back and impede US AV domination.

The Ethical Dilemma of Autonomous Vehicles

It is imperative to consider how these vehicles are programmed to handle accidents because unlike humans, autonomous vehicles do not make split-seconds decisions. Whatever they do is predesigned, hence, it is critical to be certain that these vehicles are not committing premeditated murder. In the case of an accident, does the AV self-sacrifice the car with the passenger in it or crash into pedestrians or buildings

(Klaver, 2020)? How does it assess risks and consequences? Are these life and death decisions left to some computer programmers or handled in some other way? In the case of an inevitable fatality, does the AV crash into a cyclist with a helmet who has a better chance of survival and avoid the one without a helmet? Wouldn't the safety-conscious cyclist be punished for doing the right thing? Or, does it crash into an elderly person who might have a low quality of life? Who decides this? Should any human or company or government be given this kind of authority? This is a hornet's nest.

Literatures agree that AVs cannot completely avoid difficult traffic situations. Hence, crash optimization algorithms must pay attention to ethics (Martinho, Herber, Kroesen, & Chorus, 2021). There will always be moral dilemmas; circumstances where the AV needs to undertake a challenging moral choice. What ethical principles should guide machine behavior (Frank, Chrysochou, Mitkidis, & Ariely, 2019)? As ethics are perceived as ambiguous, Martinho, et. al. (2021) compiled a list of artificial intelligence's ethical issues to guide the ethical discussions of AVs. The list of ethical issues includes: safety and cybersecurity; common good and sustainability; human oversight, control and auditing; science-policy link; solidarity, inclusion, and social cohesion; diversity; public awareness and education; privacy; accountability; transparency; ethical design; legislative framework; future of employment; protection of whistleblowers; explanability; dual use problem and military; certification; human autonomy; responsible research funding; fairness, non-discrimination, and justice; and the absence of hidden costs (Martinho et al., 2021). This list offers a deontological approach to managing AVs' ethical issues through principles only. Designing a universal moral code for AVs should be based on principles and value considerations (Martinho et al., 2021), though this is complicated by human moral biases (Frank et al., 2019).

Of the above 21 ethical issues, the most commonly discussed and considered in AV companies' technical and business reports is safety and cybersecurity in the case of extreme traffic situations (Martinho et al., 2021). This consideration is often known as the trolley problem made popular by Philippa Foot in 1967. The trolley problem is a thought experiment where an agent needed to decide how to handle a runaway trolley (Foot, 1967). Should the trolley proceed its course and kill five trackworkers or should it be diverted off course to kill one worker (Foot, 1967)? This is a moral choice of diverse combinations of lives saved and sacrificed, and it bears similarities to the decisions AVs must be pre-programmed to do in the case of inevitable fatality (Frank et al., 2019; Martinho et al., 2021). Though researchers have pointed out that the extreme trolley case described above would be, at best, rare occurrences. Weak trolley cases like AVs deciding whether to hit a bus that swerved into its lane or to react and swerve into another lane thus hitting a curb may be more common (Goodall, 2020). This, however, does not mean that extreme cases are impossible. Hence, the best point of action is to figure out how to deal with diverse cases of moral dilemmas.

A solution was provided by the Moral Machine Experiment. This experiment gathered datapoints on different humans' moral decisions. These collected datapoints are used to train AV algorithms (Frank et al., 2019). The resulting algorithms formed the moral preferences and societal expectations of the ethics of AVs (Martinho et al., 2021). However, it has been discovered that algorithms based on humans' moral decisions can also be clouded by their biases and predispositions, and limited by their beliefs and cognitions (Frank et al., 2019). Moreover, it is also argued that the point of ethics is that personal preferences and features should not count because of the ethical guidelines stated above, such as transparency and accountability (Martinho et al., 2021). However, everyone would agree that kids should be afforded preferential treatment (Martinho et al., 2021). In the thought experiment, people generally prefer a utilitarian approach to AVs' moral dilemma- minimize casualties for the greater societal good (Bonnefon, Shariff, & Rahwan, 2016; Frank et al., 2019). The utilitarian preference changes, however, when the interviewees are an active party in the casualty. Then, they prefer utilitarian regulations to be suspended and the lives of their families and themselves be protected at all costs (Morita & Managi, 2020). Therefore, there are ongoing debates about what ethical frameworks to use. Utilitarianism or deontology?

Many AVs companies have dealt with this extreme trolley case by either ignoring it or accepting that there may be some edge cases and offering more technical solutions. Mercedes-Benz proposed the use of Object and Event Detection and Response to help AVs use sensors, computation, and actuators to deal with traffic situations (Martinho et al., 2021). This is based on the understanding that non-routine traffic situations are composed of many blind spots. The use of radars and speed limitations may help minimize crashes in extreme situations. Some researchers have considered the possibility of humans taking control of the vehicle in the case of extreme situations. Meaningful human control and oversight should be explored. AVs can be adjustable by human owners and operators (de Sio & van den Hoven, 2018). Or, complete human control may be necessary to take charge of the AV and take responsibility for ethical dilemma situations in extreme traffic situations (Nyholm & Smids, 2016). In practice, it is not clear how this control process would work in a level 5 automation.

On the other hand, Contissa, Lagioia and Sartor (2017) called for offering AVs' owners an ethical knob to customize their AVs ethical approach, while manufacturers must only ensure owners can choose and the AV would always implement their choice. AVs, thus, have turned to moral proxies where owners become responsible for their choices (Contissa, Lagioia, & Sartor, 2017; Millar, 2015). AVs seem comfortable with deontological or utilitarian ethical approaches (Gogoll & Müller, 2017). In a society that never agrees on ethical standards, Personal Ethics Setting (PES) could be considered. PES democratizes ethical decisions in moral dilemmas with the burden of judgment and consequences shifted to owners (Gogoll & Müller, 2017). Rather than following universal rules, individual moral agents are allowed to choose their AVs moral actions in extreme traffic situations and face the consequences of their actions. Promoting individual moral agency is seen as respectful to citizens as moral agents capable of making their own decisions. These human decisions can be accomplished by the AVs predesigned with a selection of ethic settings describing diverse ethical theories from which users can select (Gogoll & Müller, 2017).

Many scholars have argued against PES, seeing as discrimination can be programmed into it (Gogoll & Müller, 2017). What if races, groups of people, or neighborhoods are considered dispensable? Would the car of a middle-class citizen be programmed to protect him at all costs should a crash occur in inner cities and instructed to react differently when driving through expensive neighborhoods uptown? We may object that manufacturers can easily limit selections by offering only 3 options, namely: altruistic mode, impartial mode and egoistic mode (Contissa et al., 2017). Nevertheless, it is possible that manufacturers may offer more customizable options and these features can even become unique selling points. However, these options may become too burdensome for some types of owners.

Furthermore, PES could lead to a prisoner's dilemma situation and crowd out morality. The trolley problem does not sufficiently mirror real-life AV situations as not all variables are represented in the thought experiment (Gogoll & Müller, 2017). For example, the trolley problem involves only one moral agent whose actions are not influenced by the actions of others, and only his actions determine the moral consequences. In the real world, there are interactions between moral agents. If PES is the standard and everyone selects a PES option to protect themselves with willingness to face the legal consequences, why would anyone be motivated to be altruistic? Additionally, when we are not sure of other people's PES choice, we choose the option that best protects us and our families. If moral agents are not sure what to expect, would they still be ethical? Each moral agent not willing to be a loser might rescind altruism which then leads to more fatal societal consequences. In game theory, when every agent defects and tries to self-preserve, the overall casualty increases.

Crowded-out morality seems like a serious problem and there have been calls for government intervention. PES may give way to a mandatory ethical setting (MES). Government needs to mandate an ethic setting that maximizes traffic safety and minimizes harm to all agents involved (Gogoll & Müller, 2017). It has been established that people are generally comfortable with utilitarian ethical principles (Bonnefon et al., 2016); lawmakers should code this into law for auto manufacturers to follow. However, if the government intervenes and imposes rules, would single drivers not be at a disadvantage? Mandatory settings may mean that automakers can face criminal or civil litigations based on the actions of their vehicles (Martinho et al., 2021). This burden of criminal liability may impede the development and deployment of AVs. Thus far, most governments have implemented lenient rules to promote AVs (Taeihagh & Lim, 2018). Additionally, the introduction of legal liability into designing crash collision algorithms means programmers have to consider the cost of punitive and compensatory damages in each combination of coded fatality reaction responses (Martinho et al., 2021). Hence, the AVs might be programmed to minimize damages paid even if more lives were sacrificed. Thus, we have successfully introduced a legal dilemma to the ethical dilemma issue. AV manufacturers would, of course, lean towards legal protection and give little regard to moral perspectives. So, as governments consider their actions, they need to be aware of their counteractions and consequences.

There have been arguments for the use of PES, seeing as legal liability can be clearly handled as it is currently handled with human-driven cars (Contissa et al., 2017). In a human-driven car, a pedestrian death caused by a crash on the course of the car due to the pedestrian fault may be legally justified. The number of pedestrians may not matter as long as the driver didn't cause the accident and had to save his own life. A necessity argument can be made in court. However, if an AV is preprogrammed to drive through a group of pedestrians when it could have swerved and killed one pedestrian in the case of an inevitable accident, the manufacturers may find it challenging to defend the premeditated mass murder (Contissa et al., 2017). The necessity argument cannot be used if the passenger is not in immediate danger. Hence, there seems to be no clear-cut choice between PES and MES.

It has been highlighted above how the government may respond to the ethical and legal dilemmas created by autonomous vehicles. While all these perspectives are still under consideration, governments must do the utmost to protect the technology and people as well. The 21 ethical issues raised earlier must be the guiding principles to arrive at a utility-maximizing solution for autonomous vehicles' safety regulations.

AVs, Crime, and Policing

An artificial intelligence' threat assessment event called AI and Future Crimes was held at the University College London on February 2019 (Caldwell, Andrews, Tanay, & Griffin, 2020). This workshop was attended by artificial intelligence and crime stakeholders from industry, academia, and government. At this workshop, a potential AI-powered crime list was created, and the criminal items ranked based on the social impact of the crime, the achievability of the crime, criminal profit involved, and how easily the crime can be defeated. The sixth-ranked item on the list was autonomous vehicle facilitated terrorist attacks (Caldwell et al., 2020). When AVs become widely available and accessible to everyone, criminals and crime organizations will also have access to the vehicles. The threat of cyberattacks on AVs is highly possible and has been highlighted earlier. However, it should be noted that AVs can simply be co-opted by criminal elements to undertake criminal activities. AVs may be used in vehicular terrorism by driving into gatherings of people (Caldwell et al., 2020; Cowper, 2018). There are terrorist organizations with the wealth and technical knowledge to execute large scale vehicular terrorist attacks. Moreover, drugs and weapons trafficking may be boosted by self-driving trucks (Bulrush, 2022). Drug cartels will find this as an interesting avenue to distribute drugs. Sex traffickers and kidnappers as well will explore AVs for their criminal needs (Cowper, 2018). In other words, there could be a network of autonomous vehicles trafficking drugs, weapons, and humans all over the world. Assuming these vehicles obey all traffic rules, lack drivers with criminal history, and have never been flagged, these vehicles would operate freely without resistance from law enforcement. Therefore, the police will face the challenges of stopping crime, ensuring public safety, and ensuring cybersecurity in a reality driven by AVs (Felitto, 2021). Therefore, the government response to autonomous vehicles must include proactive measures to police autonomous vehicles.

It has been determined that traffic control and enforcement may be an irrelevant police activity with level 5 AV automation (Cowper, 2018), as AVs automatically obey traffic rules and ensure smooth traffic. This means that fewer officers are needed to police traffic (Winston, 2020) and the size of police department's traffic division shrinks (Hurtado, n.d.). About 20 million motorists are pulled over in the US per year for traffic violations and the paperwork involved can be significant (Winston, 2020). AVs ensure that police resources aren't necessary in enforcing traffic laws. However, it should be noted that traffic stops represent the most significant pretext by which officers investigate crimes (Hurtado, n.d.). The continuation of these investigative stops will be threatened by AVs. The police, therefore, need to figure out how to undertake proactive policing in the AV era. This may involve understanding crime patterns.

Police departments can also use autonomous vehicles to patrol, monitor, and investigate suspected criminal activities (Cowper, 2018). Many US police departments have implemented predictive policing technology to match police officers and patrol vehicles with high crime areas to proactively prevent crimes (Hurtado, n.d.). In the AV era, predictive policing technology can automatically direct patrol cars and help to intelligently patrol neighborhoods. With the cameras installed on AVs, patrol cars can be used for surveillance, crime detection, and monitoring (Cowper, 2018). Many scholars have even stated that policing is embedded into the AV infrastructure as the cameras and sensors can be used by law enforcement to monitor and prevent crimes (Joh, 2019). AVs with all the cameras and sensors are potential crime witnesses in criminal investigations (Muller, 2022). They can record the inside of the vehicle as well as their surroundings and can flag suspicious activities. In addition, police resources can be diverted from traffic enforcement to surveillance, crime monitoring, and data gathering (Bulrush, 2022). Data gathered from cameras and sensors may be used to predict future crimes (Joh, 2019). Hence, policing will focus on crime prevention and not making arrests. However, we need to be careful with predictive policing so as not to become a cesspool of discriminatory policing practices. We don't want to promote the narrative of a group of people being overpoliced and excessively surveyed.

Many scholars have also raised the possibility of the use of facial recognition software (Joh, 2019). AVs may be equipped with facial recognition technology to fight crimes (Bulrush, 2022). In Shenzen China, jaywalkers are captured by facial recognition technology and fined via text messages (Tao, 2018). Many police departments in the US already implement facial recognition software to track criminals. Additionally, the technology is widely used at airports, borders, and stadiums. However, this technology can turn a country into an oppressive surveillance state and be used discriminatorily against minorities. In China, this technology is used to monitor a Muslim minority group called Uighurs (Conger, Fausset, & Kovaleski, 2019). In 2018, the landmark project "Gender Shades" discovered that facial recognition algorithms from IBM and Microsoft are highly inaccurate on darker-skinned females (Buolamwini & Gebru, 2018). This technology can also lead to the disproportionate arrest of minorities as they may be more likely to be over-policed. Considering the current racial disparity in the US arrest rates, facial recognition technology may only amplify these disparities. Hence, it was no surprise that San Francisco and many other US cities shut down the use of facial recognition technology by their police department (Conger et al., 2019).

There have also been calls for the police to have remote access to AVs or the artificial intelligent system controlling the vehicles (Felitto, 2021). The logistics of such haven't been worked out yet, but the police may want to maintain an oversight on these vehicles. This may involve a backdoor access to be able to remotely control speed and bring criminal elements to a halt (Bulrush, 2022). AVs must be publicly accountable; hence, law enforcement may want as much control as possible. Waymo has considered implementing a kill-switch in its AV to allow law enforcements to shut down vehicles (Bulrush, 2022). The need for this type of oversight must be weighed with rights to privacy and self-determination.

CONCLUSION

Undoubtedly, autonomous vehicles are the transportation device of the future. They represent the dreams and hard work of many stakeholders. A lot has been said on the technological benefits of the realization of this dream. In addition, the economic gains and accompanying societal benefits are immense. Self-driving cars embody a significant leap in the quality of life of humans on earth. However, AVs also characterize the nightmares and fears of many. What happens if the self-driving transportation network of a place like New York City is attacked and hijacked by terrorists? What about privacy? And how do we stop driverless cars from being used to traffic kids, weapons, and drugs? These are all important questions that have yet to be answered comprehensively. Hence, the hesitations of many towards autonomous vehicles seem understandable. However, like with every other technology, there are pros and cons. The possibilities of undertaking drug trades with phones did not stop the development of handheld communication devices. The existence of the dark web has not hindered the use of internet. In fact, the recognized dangers of autonomous vehicles present a perfect opportunity for governments to intervene and regulate this technology. Now is the time for lawmakers and politicians to understand this technology and engage manufacturers to ensure that the fears associated with AVs are never realized. States need to act on behalf of the average citizen to ensure a safe future filled with advanced technology. We are watching!

REFERENCES

Autonomous Vehicles Confront a Political Viability Crisis. (n.d.). *Baron Public Affairs*. Retrieved July 13, 2022, from https://baronpa.com/library/autonomous-vehicles-confront-a-political-viability-crisis/

Bellan, R. (2022, April 26). *GM expects to spend $2B on Cruise in 2022*. Retrieved May 25, 2022, from https://techcrunch.com/2022/04/26/gm-expects-to-spend-2b-on-cruise-in-2022/

Block, F. L., & Keller, M. R. (2015). *State of innovation: the US government's role in technology development*. Routledge. doi:10.4324/9781315631905

Bluhm, M. (2022, November 5). *Biden's hugely consequential high-tech export ban on China, explained by an expert.* Retrieved December 17, 2022, from https://www.vox.com/world/2022/11/5/23440525/biden-administration-semiconductor-export-ban-china

Bonnefon, J.-F., Shariff, A., & Rahwan, I. (2016). The social dilemma of autonomous vehicles. *Science*, *352*(6293), 1573–1576. doi:10.1126cience.aaf2654 PMID:27339987

Bulrush, J. (2022, July 7). *Autonomous vehicles are a dream for drug smugglers — and the law must prepare.* Retrieved August 19, 2022, from https://thenextweb.com/news/law-enforcement-arent-ready-for-autonomous-vehicle-faciliated-crime-syndication

Buolamwini, J., & Gebru, T. (2018). Gender shades: Intersectional accuracy disparities in commercial gender classification. *Conference on fairness, accountability and transparency*, 77.

Caldwell, M., Andrews, J. T. A., Tanay, T., & Griffin, L. D. (2020). AI-enabled future crime. *Crime Science*, *9*(1), 14. doi:10.118640163-020-00123-8

Conger, K., Fausset, R., & Kovaleski, S. F. (2019). San Francisco bans facial recognition technology. *The New York Times, 14.*

Contissa, G., Lagioia, F., & Sartor, G. (2017). The Ethical Knob: Ethically-customisable automated vehicles and the law. *Artificial Intelligence and Law*, *25*(3), 365–378. doi:10.100710506-017-9211-z

Cowper, T. J. (2018, February 13). *Autonomous Vehicles- How Will They Challenge Law Enforcement?* Retrieved August 19, 2022, from https://leb.fbi.gov/articles/featured-articles/autonomous-vehicles-how-will-they-challenge-law-enforcement

de Sio, F. S., & van den Hoven, J. (2018). Meaningful Human Control over Autonomous Systems: A Philosophical Account. *Frontiers in Robotics and AI*, *5*, 15. doi:10.3389/frobt.2018.00015 PMID:33500902

Editors, A. (2017). *Potential impacts of self-driving cars*. Access Science.

Edwards, D. (2022, July 26). *Embark Trucks completes autonomous vehicle testing in snowy conditions.* Retrieved September 10, 2022, from https://roboticsandautomationnews.com/2022/07/26/embark-trucks-completes-autonomous-testing-in-snowy-conditions/53461/

Fannin, R. (2022, May 21). *Where the billions spent on autonomous vehicles by U.S. and Chinese giants is heading*. Retrieved May 25, 2022, from https://www.cnbc.com/2022/05/21/why-the-first-autonomous-vehicles-winners-wont-be-in-your-driveway.html#aoh=165336535384
39&csi=0&referrer=https%3A%2F%2Fwww.google.com&_tf=From%20%2
51%24s&ampshare=https%3A%2F%2Fwww.cnbc.com%2F2022%2F05%2F21%
2Fwhy-the-first-autonomous-vehicles-winners-wont-be-in-your-
driveway.html

Felitto, G. (2021, June 7). *Is Law Enforcement Ready for Self-Driving Cars?* Retrieved August 21, 2022, from https://thecrimereport.org/2021/06/07/1333055/

Foot, P. (1967). The problem of abortion and the doctrine of the double effect. *Oxford Review, 5.*

Frank, D.-A., Chrysochou, P., Mitkidis, P., & Ariely, D. (2019). Human decision-making biases in the moral dilemmas of autonomous vehicles. *Scientific Reports*, *9*(1), 13080. doi:10.103841598-019-49411-7 PMID:31511560

Gale, M. (2022, April 12). *Meet Zoox, The 100% EV And Autonomous Human Transportation Company From Amazon*. Retrieved May 25, 2022, from https://www.forbes.com/sites/michaelgale/2022/04/12/meet-zoox-the-100-ev-and-autonomous-human-transportation-company-from-amazon/?sh=2ce3af5a3cd0

Gogoll, J., & Müller, J. F. (2017). Autonomous cars: In favor of a mandatory ethics setting. *Science and Engineering Ethics*, *23*(3), 681–700. doi:10.100711948-016-9806-x PMID:27417644

Gomez-Ibanez, T. (n.d.). *Implications of Self Driving Vehicles on Public Policy with Tony Gómez-Ibáñez.* Retrieved July 13, 2022, from https://www.hks.harvard.edu/faculty-research/policy-topics/cities-communities/implications-self-driving-vehicles-public-policy

GoodallN. J. (2020). More than trolleys: plausible, ethically ambiguous scenarios likely to be encountered by automated vehicles. doi:10.31219/osf.io/bzkas

Hayes, A. (2021, August 31). *The Unintended Consequences of Self-Driving Cars*. Retrieved July 13, 2022, from https://www.investopedia.com/articles/investing/090215/unintended-consequences-selfdriving-cars.asp

He, L. (2022, October 31). *US curbs on microchips could throttle China's ambitions and escalate the tech war.* Retrieved December 10, 2022, from https://www.cnn.com/2022/10/31/tech/us-sanctions-chips-china-xi-tech-ambitions-intl-hnk

Hurtado, P. (n.d.). *Implications of Self-Driving Vehicles*. Retrieved August 21, 2022, from https://www.policechiefmagazine.org/implications-of-self-driving-vehicles/

Joh, E. E. (2019). Policing the smart city. *International Journal of Law in Context*, *15*(2), 177–182. doi:10.1017/S1744552319000107

Kilanko, V. (2021). The potential effects of Biden's infrastructure bill on the american economy. *International Journal Of Scientific Advances*, *2*(5). Advance online publication. doi:10.51542/ijscia.v2i5.1

Klaver, F. (2020, April). *The Economic and Social Impacts of Fully Autonomous Vehicles*. Retrieved July 13, 2022, from https://www.compact.nl/en/articles/the-economic-and-social-impacts-of-fully-autonomous-vehicles/

Levin, S. (2018, March 19). *Self-driving Uber kills Arizona woman in first fatal crash involving pedestrian.* Retrieved September 10, 2022, from https://www.theguardian.com/technology/2018/mar/19/uber-self-driving-car-kills-woman-arizona-tempe

Liao, R. (2021, December 14). *California suspends Pony.ai driverless test permit after crash.* Retrieved September 10, 2022, from https://techcrunch.com/2021/12/14/pony-ai-suspension-driverless-pilot-california/

Magill, K. (2022, August 12). *Aurora delays autonomous truck launch following supplier constraints.* Retrieved September 10, 2022, from https://www.supplychaindive.com/news/aurora-paccar-autonomous-truck-launch-delay-supply-constraints/629546/

Martinho, A., Herber, N., Kroesen, M., & Chorus, C. (2021). Ethical issues in focus by the autonomous vehicles industry. *Transport Reviews*, *41*(5), 1–22. doi:10.1080/01441647.2020.1862355

Millar, J. (2015). Technology as moral proxy: Autonomy and paternalism by design. *IEEE Technology and Society Magazine*, *34*(2), 47–55. doi:10.1109/MTS.2015.2425612

Morita, T., & Managi, S. (2020). Autonomous vehicles: Willingness to pay and the social dilemma. *Transportation Research Part C, Emerging Technologies*, *119*, 102748. doi:10.1016/j.trc.2020.102748

Muller, J. (2022, May 13). *Self-driving cars could be potential crime witnesses.* Retrieved August 19, 2022, from https://www.axios.com/2022/05/13/self-driving-cars-video-police-witness

Nations, U. Convention on Road Traffic, 1042 § (1977). Retrieved from https://treaties.un.org/Pages/ViewDetailsIII.aspx?chapter=11&mtdsg_no=XI-B-19&src=TREATY

News. (2020, July 21). *The Political Case Against Self-Driving Cars.* Retrieved July 13, 2022, from https://mindmatters.ai/2020/07/the-political-case-against-self-driving-cars/#:~:text=in%20political%20philosophy%2C%20raises%3A,less%20harm%20to%20the%20environment

Nyholm, S., & Smids, J. (2016). The Ethics of Accident-Algorithms for Self-Driving Cars: An Applied Trolley Problem? *Ethical Theory and Moral Practice*, *19*(5), 1275–1289. doi:10.100710677-016-9745-2

Polly. (2020, June 22). *6 Ways Autonomous Vehicles Will Impact Life As We Know It.* Retrieved July 13, 2022, from https://roboticsandautomationnews.com/2020/06/22/6-ways-autonomous-vehicles-will-impact-life-as-we-know-it/33326/

Porter, J. (2020, June 26). *Amazon is acquiring self-driving car startup Zoox.* Retrieved May 25, 2022, from https://www.theverge.com/2020/6/26/21304111/amazon-zoox-acquisition-self-driving-car-autonomous-vehicles-startup-robotaxi-bi-directional-vehicle

Schreurs, M. A., & Steuwer, S. D. (2015). Autonomous Driving - Political, Legal, Social, and Sustainability Dimensions. In M. Maurer, J. C. Gerdes, B. Lenz, & H. Winner (Eds.), *Autonomes Fahren* (pp. 151–173). Springer Berlin Heidelberg. doi:10.1007/978-3-662-45854-9_8

Sensiba, J. (2022, May 21). *Ford-Backed Argo AI Begins Driverless Taxi Operations, & Why This Is Big News.* Retrieved May 25, 2022, from https://cleantechnica.com/2022/05/20/ford-backed-argo-ai-begins-driverless-taxi-operations-why-this-is-big-news/amp/

Sheehan, M. (2022, October 27). *Biden's Unprecedented Semiconductor Bet*. Retrieved December 17, 2022, from https://carnegieendowment.org/2022/10/27/biden-s-unprecedented-semiconductor-bet-pub-88270

Taeihagh, A., & Lim, H. S. M. (2018). Governing autonomous vehicles: Emerging responses for safety, liability, privacy, cybersecurity, and industry risks. *Transport Reviews*, *39*(1), 1–26. doi:10.1080/01441647.2018.1494640

Talbott, S. (2020, June 23). *The Political Economy Of Autonomous Vehicles*. Retrieved July 13, 2022, from https://www.forbes.com/sites/selikajosiahtalbott/2020/06/23/the-political-economy-of-autonomous-vehicles/?sh=74f629d4155
5

Tao, L. (2018). Jaywalkers under surveillance in Shenzhen soon to be punished via text messages. *South China Morning Post, 27.*

Winston, C. (2020, August 25). *Autonomous vehicles could improve policing, public safety, and much more*. Retrieved August 21, 2022, from https://www.brookings.edu/blog/techtank/2020/08/25/autonomous-vehicles-could-improve-policing-public-safety-and-much-mor
e/

Woods, D. (2022, October 19). *Why the US cut China off from advanced chips*. Retrieved November 17, 2022, from https://www.npr.org/2022/10/19/1129990818/why-the-us-cut-china-off-from-advanced-chips#:~:text=On%20October%207%2C%20tho
ugh%2C%20just,or%20buy%20advanced%20semiconductor%20chips

Section 4

Innovations in Blockchain Technology

Chapter 9
Cryptocurrencies and Contentious Politics

Benjamin James Alexander Leo
University of Kentucky, USA

ABSTRACT

Cryptocurrencies, first introduced in 2009, have rapidly grown in value, volume, and use. The sentiment that they are no more than a passing fad and need not be dealt with by scholars or policymakers remains but is no longer legitimate. Hostile actors such as terrorist groups and sanctioned states use cryptocurrencies to finance activities ranging from deadly attacks to rogue nuclear programs. The use of cryptocurrencies by legitimate actors brings far less dramatic externalities but still presents serious issues, such as the vulnerability of citizens to cryptocurrency-related scams, lost tax revenue, and the stress cryptocurrencies place on countries' electrical grids. This chapter provides readers with an understanding of exactly how cryptocurrencies function, how hostile actors use them, and how governments respond so that they may better understand the challenges posed by cryptocurrencies and how they might be addressed.

INTRODUCTION

Cryptocurrencies are a new and exciting phenomenon. Their decentralized nature and the anonymity they provide stand in stark contrast to traditional, government-controlled fiat currencies. Furthermore, their use, value, and effects are subject to a great deal of interest from the public, media outlets, and governments worldwide. But is this attention merited? Do cryptocurrencies matter to political outcomes of interest, and if so, how? This chapter presents case evidence of cryptocurrencies as a meaningfully disruptive global economic force.

As discussed more thoroughly in the following sections, cryptocurrencies rely on blockchain technology and one-way hashing functions to create a digital currency independent of privileged third parties. Using a system of related public and private cryptographic keys, users can transmit funds directly to one another pseudo-anonymously. Transactions are verified and recorded on the blockchain, a publicly available ledger of every transaction that has ever occurred for each type of cryptocurrency and each user's

DOI: 10.4018/978-1-6684-6429-8.ch009

account balance. However, these cryptographic keys allow for anonymity and security with transactions despite using a public ledger system, as transactions occur without identifying information.

Since their introduction, cryptocurrencies have rapidly increased in value, volume, and use. As of May 28, 2022, a single bitcoin is worth roughly $29,000, and bitcoin alone has a market cap of $550 billion, approximately the same size as Facebook's parent company, Meta. Bitcoin is still the most valuable cryptocurrency, but there are now over 10,000 alternative cryptocurrencies, or "altcoins," that consumers can purchase. No altcoin is close to bitcoin in terms of value, but there are now fifty altcoins with a market cap of over $1 billion, and the market cap of cryptocurrencies is now at $1.5 trillion (*Statista Report: Cryptocurrencies*, 2022). With this rapid increase in volume and value, the mainstream acceptance of cryptocurrencies is also increasing. Consumers can use cryptocurrencies to purchase goods or services at companies such as Tesla, Microsoft, PayPal, Newegg, Home Depot, Shopify, AT&T, Dish, Bloomberg, Expedia, and Wholefoods. Over the previous two years, the twenty-four-hour trade volume of cryptocurrencies has grown from $500 million to highs of over $500 billion (*Statista Report: Cryptocurrencies*, 2022). While the cryptocurrency market remains volatile, its growth since its inception has been exponential.

Cryptocurrencies' popularity with law-abiding citizens is growing, but its earliest and most loyal adopters were those operating outside the bounds of domestic and international laws. For example, shortly after the creation of cryptocurrencies, an online marketplace named "The Silk Road" appeared on the dark web, offering illicit goods and services such as firearms, narcotics, assassinations, and erotica, but only in exchange for Bitcoin. In its two years of operation, the Silk Road processed an estimated one billion US dollars (USD) worth of transactions (Norry, 2017; Kethineni & Cao, 2020). Despite the Silk Road's demise following its owner's arrest on pornography charges, illicit trade on the dark web with cryptocurrencies has only increased, with numerous copycat websites emerging. As a result, the cryptocurrency-based black market has grown exponentially, with tens of billions of USD in illegal cryptocurrency transactions occurring yearly (Foley et al., 2019; Grauer et al., 2022).

Among the growing number of actors using cryptocurrencies for illicit means are anti-state and terrorist groups. When these groups attempt to solicit donations from supporters or distribute funds to their operators using traditional financial services such as Western Union or PayPal, their transactions are often flagged, and their operators are identified and captured. Cryptocurrencies are easily accessible alternatives to fiat currencies and a potential solution to anti-state organizations' challenge from increasingly sophisticated counter-terrorism financing (CTF) strategies. Hamas, Al-Qaeda, ISIS, Boko Haram, and many other organizations already use cryptocurrencies to avoid CTF efforts. Still, they are not the only hostile actors making use of crypto. States facing heavy sanctions from the international community, such as Russia, North Korea, and Iran, use cryptocurrency-based strategies to mitigate damage from Western economic sanctions. The most popular of these strategies include generating profits through large-scale mining efforts and criminal operations, as well as the development of state-backed cryptocurrencies designed to serve as an alternative to the Society for Worldwide Interbank Financial Telecom (SWIFT).

On an international scale, states must address hostile actors' use of cryptocurrencies. However, they must also determine how to regulate cryptocurrency domestically to limit lost tax revenue, consumer harm, and electrical grid failures (Turak, 2021; Thiemann, 2021). In 2020, for example, European Union (E.U.) states could have generated an estimated one billion USD in additional revenue by applying capital gains taxes to cryptocurrencies (Thiemann, 2021). Cryptocurrencies are also prone to cyber theft and scams, with consumers losing billions to cryptocurrency scams. Furthermore, if left unchecked, cryptocurrency mining centers can overwhelm electrical grids (Turak, 2021). There is an extreme variance among states' chosen regulatory measures for cryptocurrencies. They range from China's absolute ban on cryptocurrencies to El Salvador's recognition of Bitcoin as one of their national currencies (Huo, 2022; Riley, 2021). Most states lie in the middle of these two extremes by allowing cryptocurrencies to exist as a commodity but with some form of regulations present (Library of Congress, 2021).

Cryptocurrencies are new and exciting assets that many speculate could be the next evolution of currency. They offer many potential benefits, such as greater access to the financial system, protection from inflation, and fee-free transactions. However, to limit the negative externalities of cryptocurrencies, individuals, scholars, and policymakers must be aware of what cryptocurrencies are, how they work, and what these negative externalities are.

The Blockchain

Before discussing the mechanisms of a cryptocurrency transaction, it is essential to understand the blockchain. Attempting to do otherwise would be akin to learning about Facebook without any knowledge of the Internet.

When using a traditional fiat currency, banks are used to manage and protect ledgers.

Every time a transaction is made with a debit card, for example, the bank verifies the transaction as legitimate, transfers money from an account to the store's account, and records the transaction. While one can view their side of any transaction on their monthly statements, they cannot view the ledger in its entirety. The bank retains exclusive access to the complete ledger, and in exchange for maintaining it, a small fee is paid with every transaction. In contrast, cryptocurrencies decentralize both their ledger and the verification process. The process of validating transactions with cryptocurrencies is called "mining." The following section will explain it in greater detail, but the focus is now on the blockchain.

One of the challenges cryptocurrencies face in decentralizing their ledger is the need to ensure that some deviant actors do not inaccurately alter the information on it. The solution is the blockchain. A blockchain is a digital database where each "block" of data, a fixed amount of data such as a gigabyte, is chained to the previous block with a unique hash. [1] The unique hash for each block results from the block's content and the hash of every previous block being entered into a one-way hashing algorithm (Crosby et al., 2016; Nofer et al., 2017; Pierro, 2017). To clarify, the hash is the output, and the hashing function is the mathematical equation that creates the hash.

A simple example of how this process works is provided in Figure 1. The initial hash on a blockchain is simply the output of its contents being entered into a hashing function, such as SHAH-256. This hash is then entered into the same hashing function along with the contents of the second block to create the hash for the second block. This process is repeated in perpetuity for each block, and it is in this way that blocks of data are "chained" together on the blockchain. Any alteration in the data on some previous block would alter that block's hash and, by extension, the hash of every proceeding block.

Figure 1. The blockchain

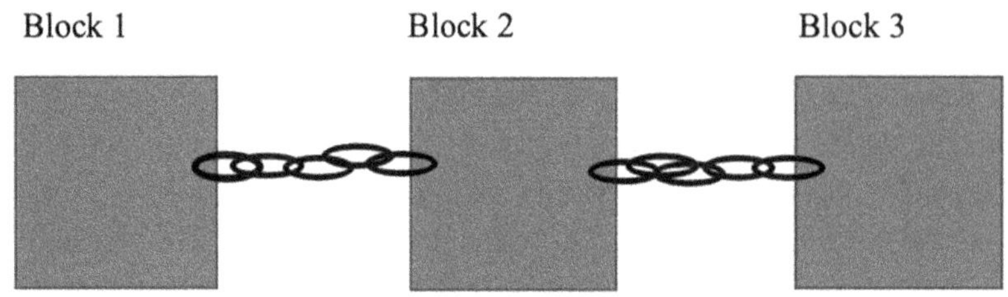

Previous Hash:	000	Previous Hash:	8aefb06c42 6e07a0a671 a1e2488b48 58d694a730	Previous Hash:	3cc91b175c 9e878e6b02 9cfd42d08e0 6e516e99e
Block 1 Hash:	8aefb06c42 6e07a0a671 a1e2488b48 58d694a730	Block 2 Hash:	3cc91b175c 9e878e6b02 9cfd42d08e0 6e516e99e	Block 3 Hash:	b15bbdf49 adb764c82 f6627c7243 9e33d5098355

Chaining blocks together with hashing functions creates a highly robust level of security, despite the decentralized nature of most blockchains. The hash of any block is the product of its content and that of every preceding block. As such, even the slightest change to the content of a block would radically change its hash and that of every proceeding block. The decentralized nature of most cryptocurrencies also means that if a deviant actor attempts to pass a fraudulent version of the blockchain as valid, all other users would be able instantly to recognize it.

Unlike Figure 1, which was meant only to help readers visualize how a blockchain works, Figure 2 demonstrates how a relatively minute change in a string of data entered into a hashing function can produce radically different outputs when using the SHAH-256 algorithm. For example, changing the input from "RedBlue1234" to "RedBlue12345" creates a much different output. This is because any block on a blockchain contains data on hundreds of transactions, such as the sender, the recipient, the amount sent, and the time it occurred. The slightest change to any part of that information for even a single transaction would produce a radically different hash for that block of data and every following block.

Figure 2. Changes in hash input and output (SHAH-256)

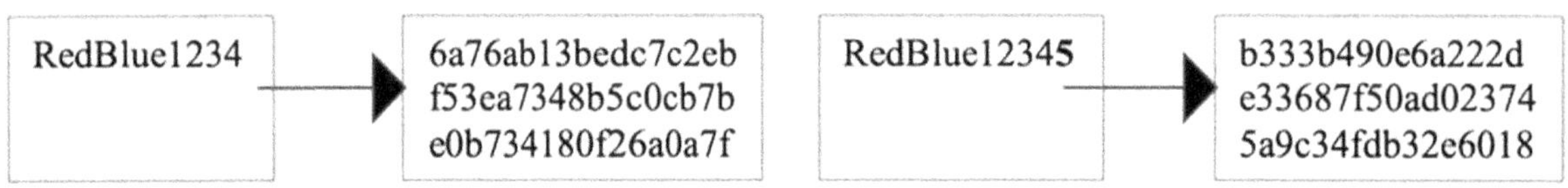

Blockchain technology is the facilitating technology behind every cryptocurrency transaction. User account balances and transactions are stored on it, and miners use it in the verification process. Cryptocurrency blockchains are typically decentralized in that anyone can access them, but because each block is "chained" to the previous block with a one-way hashing function, passing a fraudulent version of the blockchain becomes impossible. In this way, blockchains allow transactions to happen securely and without a privileged third party.

Other Applications of Blockchain Technology

Blockchain technology is best known as the underlying technology for digital assets such as cryptocurrencies, but its potential application expands far beyond that of a simple ledger. Blockchain technology can securely store and transmit any form of information, such as votes, healthcare records, messages, etc. Encrypted messaging services such as Dust and Telegram, as well as the dark web browser the Onion Routing Project, or TOR, are also built on blockchain technology. Microsoft's Outlook email service is already integrated with blockchain technology, allowing users to encrypt their emails. Furthermore, private companies have held limited trial runs for blockchain voting systems in the U.S. More notably, the United States postal service has filed a patent for a blockchain-based voting system and the use of third-party blockchain voting systems on a local level (United States Postal Service, 2020). Countries such as Thailand, Brazil, Japan, Russia, and South Korea are also experimenting with or have already implemented a blockchain voting system on a limited scale (Couger, 2021). However, the CCP (Chinese Communist Party) is the most bullish party regarding the blockchain. They see blockchain technology not only as a tool for recording data or sending messages but as the future of the global economy and believe it will become a vital tool for international trade (Riley, 2021).

Cryptocurrencies: How They Work and Who Is Using Them

By its simplest definition, cryptocurrencies are a tool for peer-to-peer digital transactions. They enable financial transactions without the involvement of some privileged third party, such as a bank or credit card company. Doing so offers benefits such as heightened privacy from law enforcement, greater security, lower or non-existent transaction fees, and protection from inflation. However, cryptocurrencies must also contend with the negative ramifications of removing these privileged third parties. For example, these organizations are typically responsible for confirming transactions and preventing double-spending. A simplified version of how cryptocurrency transactions differ from those made with fiat currencies, such as the U.S. dollar, is presented in Figure 3.

Figure 3. Simplified comparison of transaction processes

Debit/Credit Card Transactions:	1) Transaction information sent to a bank	2) The bank verifies the transaction	3) Funds are moved to the new account
Crypto Transaction:	1) Transaction information is broadcast publicly to nodes (miners)	2) A pool of users collectively verifies the transaction	3) Funds are moved to the new account

As demonstrated in Figure 3, when someone uses a debit or credit card to make a transaction, the information from it is first sent to the banks of the buyer and the seller. Then, using the ledger of users' balances that only they control, the buyer's bank confirms that they possess the currency they are trying to spend and are not trying to spend it in multiple places. If verified, the buyer's bank will transfer the funds to the seller's bank account from the buyer's account. By holding the ledger private and using a massive infrastructure to verify transactions, everyday purchases people make—such as using one's VISA debit card at the grocery store—happen within seconds; the buyer and seller rarely give a second thought to the validity or privacy of the transaction.

Because there is no privileged third party to verify transactions and maintain transactional privacy, cryptocurrencies must provide alternative solutions. The solutions used by cryptocurrencies, summarized in Table 1, are the decentralization of their ledger, decentralization of the transaction verification process, and the use of a public and private key system. Like with fiat currencies, there are still three main actors in a transaction. The first two are still the initiator (buyer) and the receiver (seller)[2]. The difference lies in who the third actor is. Instead of using a privileged third party such as a bank to verify transactions, cryptocurrencies rely on 'miners,' individuals who voluntarily participate in the verification process (also called 'nodes'). The next section will detail each step of a cryptocurrency transaction from Table 1.

Table 1. Detailed example of a cryptocurrency transaction

1) User A creates a transaction, specifying the amount of cryptocurrency being sent and the public address of the user to which it is being directed.
2) Using the transaction information and the private key associated with the cryptocurrency being sent, User A creates a unique digital signature for the transaction.
3) User A broadcasts this transaction to their cryptocurrency's network and waits for a miner to select it.
4) A miner selects the transaction and verifies its validity against the blockchain (that cryptocurrencies' ledger).
5) Once the miner has enough other transactions to fill a block, they try to solve the Proof of Work (POW) algorithm.
6) Upon successful completion of the POW puzzle, the mining node broadcasts its solution and the contents of its block to the other nodes on the network.
7) If a majority of active nodes confirm the validity of the transactions within the submitted block and the validity of the solution to the POW puzzle, the block is added to the blockchain.
8) User B now controls the cryptocurrencies previously associated with User A, and the Miner verifying the transaction receives their cryptocurrency reward.

Initiating the Transaction

The first three steps of a cryptocurrency transaction, outlined in Table 1, are much like what is seen in everyday fiat transactions. In step 1, the initiator chooses some amount of cryptocurrency and who to send them to. This is done using a system of public and private keys and a unique digital signature for each transaction. With cryptocurrencies, the term "keys" refer to complex alphanumeric strings. Each cryptocurrency account possesses at least one public and private key. Public keys are used to lock and receive cryptocurrencies, and private keys are used to unlock and send cryptocurrencies. The relationship between public and private keys is like that between email addresses and the corresponding account password. Knowing another's email address enables another user to send them an email, not access their account and information. Similarly, with cryptocurrencies, knowing a user's public key only allows one to send them funds and does not enable them to access that user's account. However, because the public key is the output of the private key entered into a one-way function, the relationship between keys can be verified. Figure 4 provides an example of public and private keys and their primary attributes.

Figure 4. Public and private keys example (bitcoin)

Private Key	Public Key
12eDyw84U9Wmk9g7hNU8RWySdkQ BpvQYqP	5Jg53xSeGAHt7pFymaaeav5ScPhpt E7icq65vyXe6QHwTNishdb
• For unlocking and sending cryptocurrencies. • Like an account password. • Irretrievable if lost.	• For locking and receiving cryptocurrencies. • Like an account username. • Can be generated with the private key if lost.

Referring to Table 1, the second step of the cryptocurrency transaction process is for the initiator to create a unique digital signature. Digital signatures are somewhat like the unique purchase number banks create for each transaction we make daily (often the first column in one's monthly bank statement). Recall that there is no bank to verify that the initiator has sufficient funds. As such, cryptocurrencies use digital signatures to allow receivers and miners to confirm if the initiator possesses the private key associated with the cryptocurrencies they are attempting to spend without anyone sharing their private key. Digital signatures are needed because using one's private key to verify ownership would compromise account security (in the same way that one does not provide the grocery store with a username and password to verify account information in the checkout line). The digital signature and the user's public key are then used to confirm transaction validity using the formula used to create the digital signature (Kaur & Kaur, 2012). As demonstrated in Figure 5, the procedure used to create the digital signature is the product of a one-way function that takes the buyer's private key, combines it with the transaction amount, and produces a complex digital signature for each transaction.

Figure 5. Simplified example of digital signature calculation

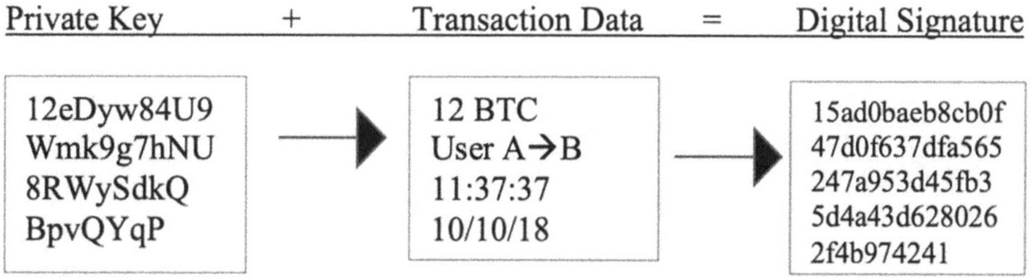

The third step is for the initiator to broadcast the transaction information—including the amount of cryptocurrency to be sent, the public key of the receiver, and the digital signature from the transaction—to their cryptocurrency's network. Though seemingly complex, steps 1-3 happen within a matter of seconds (commonly two clicks on a smartphone) once the initiator has chosen who to send their cryptocurrency to and how much to send. It is important to note that the public and private keys system here is critical for ensuring anonymity throughout the transaction process. The key system allows the initiator to publicly broadcast their transaction information without anonymizing themselves or the user they are sending cryptocurrency. This is possible because public keys, being nothing more than alphanumeric strings, do not provide any information about the identity of an account owner. Figure 6 provides an example of the information available on the blockchain and evidences the lack of identifying information available there.

Figure 6. Example account on the bitcoin blockchain

Address:	bc1qdwdnmp3dul64wzq5pzqj6dnqwrhhfy2dr0anwp
Balance:	0.00000000 BTC
Total Received:	0.00111670 BTC
Total Spent:	0.00111670 BTC

Mining Transactions: The Verification Process

Referring to Step 4 in Table 1, once the initiator has broadcast their transaction to their cryptocurrency's network, the verification process occurs. This process is like how a transaction with a VISA card is sent to your bank for confirmation, but here the transaction is validated by any individual that has chosen to participate in this process (which is what is meant by the process being "decentralized"). Instead of a bank, now "miners" enter the process. In the mining process, a miner (or "node") is an active computer, or pool of computers, running on a network that downloads and runs the requisite software to verify transactions. In this example, a given node would select the proposed transaction and verify it using the cryptocurrency's blockchain. They would then add it to a "block" of transactions. For example, on the Bitcoin blockchain, a block is 1 megabyte large and contains around 500 transactions.

Proof of Work

Once a node has filled its "block" of data, it would then progress to Step 5 in Table 1, solving the proof of work algorithm. One challenge associated with decentralizing the verification process is the possibility that some bad actor(s) might attempt to submit fraudulent blockchains. The solution many cryptocurrencies use is Proof of Work (POW). POW requires nodes to solve a complex algorithmic puzzle, with nodes entering potential answers until the correct solution is found. As a grossly simplified example of this algorithmic puzzle, imagine trying to write a computer program that attempts to open a 1000-digit combination lock by randomly choosing numbers. Solving this puzzle would be very hard to do via your random chance program, but it would be simple to validate once a solution is found; others could just enter the combination to see if the lock opens. Hundreds of thousands of active nodes are verifying transactions at any given time. However, only the node that first completes this puzzle will have its block of transactions added to the blockchain. Solving a POW puzzle is a brute-force process where nodes continuously submit possible solutions until the correct answer has been found. This is a game of chance where nodes can submit more solutions with more computing power and increase their chances of verifying their block. The difficulty of this process entails considerable cost, which dissuades any such bad actors. For context, a single, entry-level bitcoin mining device, the Antminer, costs roughly 4,000 USD. Some large mining farms will contain hundreds, if not thousands, of these devices and consume massive amounts of electricity. Furthermore, in 2022 verifying a single bitcoin transaction would require 1,449 kWh, equating to an American household's average power usage over fifty days (de Vries et al., 2022).

Once the node verified their block of data and successfully solved the POW puzzle, they would move to Step 6 in Table 1 by submitting their solution to all other active nodes on their cryptocurrency's network. The time it takes to produce a block or a cryptocurrency's hash rate varies significantly. For example, the hash rate for a Bitcoin block is ten minutes, while some cryptocurrencies, such as Z-Cash, can be hashed in as little as 10 seconds. Next, in step "7;" POW solutions are confirmed by the other active nodes. Cryptocurrencies not only dissuade bad actors in a POW system but also operate on a majority rules basis where each transaction and POW solution must be confirmed by more than half of all active nodes on the network. As such, for any bad actor to seize control of a cryptocurrency, they would need to possess 51% of that cryptocurrency network's computing abilities. Therefore, such attacks are a remote possibility, as they are incredibly costly, and actors with such mining power would generate more profit by participating in the legitimate transaction verification process. Furthermore, the value of any coin on a compromised blockchain would plummet, depriving the deviant actor(s) of any benefit.

Once fifty-one percent of all active nodes verified the transaction within the nodes block and their solution to the POW puzzle, step 8, the confirmation process, follows. The cryptocurrency the initiator sent would now be associated with the receiver's private key. Note that the cryptocurrencies are not actually moving here from the sender to the receiver. What has changed is that the receiver of the cryptocurrency can now unlock and send them with their private key. This is not unlike how banks work, as they do not move dollars around from account to account. Instead, they simply note the change in the account with which the transacted dollar amount is associated. Also, in this final step, the mining node that has successfully added a block to the blockchain receives a set amount of the cryptocurrency they are mining as a reward. For example, when a node mines a block on the bitcoin blockchain, its reward would be 6.25 bitcoins. Through this reward process, "good" users are incentivized to invest and participate in the cryptocurrency verification process.

Additional Information on the Cryptocurrency Transactions

Cryptocurrency Exchanges

Most users will use their cryptocurrencies through some exchange service, such as Coinbase or Venmo, but this does not necessarily mean a privileged third party becomes a part of the transaction process. For example, many exchanges, such as Binance or Bybit, facilitate peer-to-peer transactions. For these transactions, the only purpose of the exchange is to help interested buyers and sellers find one another. These buyers then conduct transactions using the key pairings only they possess. However, many users opt for cryptocurrency transactions using exchanges that function as a third party, otherwise known as over-the-counter exchanges. These exchanges hold users' public and private keys on their servers and initiate transactions on behalf of their users. Such exchanges lack the benefits of traditional cryptocurrency use but are more easily accessible.

Transaction Anonymity

Decentralized cryptocurrencies like Bitcoin do not provide any identifying information on their blockchain, but in some instances, it is possible to deanonymize cryptocurrency users (Turner & Irwin, 2018; Yin et al., 2019). Therefore, it is most accurate to categorize cryptocurrencies as pseudo-anonymous rather than fully anonymous. However, the security and anonymity offered by cryptocurrencies are still considerable and require several criteria to be met before it is broken. For example, users must use the same public key for repeated transactions, not obscure their I.P. address with a VPN, and not utilize other tools to obfuscate their identity further. In addition, individuals can also now choose from a growing number of stealth cryptocurrencies, such as Monero, that take additional steps like hiding their blockchain, pooling transactions together, and generating new keys for each transaction to ensure user anonymity. Nonetheless, the demonstrated ability of law enforcement agents and cybersecurity experts to discern the identities of cryptocurrency users suggests that there are vulnerabilities in the security of some cryptocurrencies.

Proof of Work Alternatives

POW is not only a computationally intensive process but also requires investment in expensive, high-tech mining equipment and enough electricity to strain some nations' power grids seriously. For example, high power consumption levels from cryptocurrency mining have contributed to intermittent failures on the Iranian electrical grid and contributed to China's decision to outlaw all cryptocurrency mining operations (Chudinovskikh & Sevryugin, 2019; Natasha Turak, 2021). To combat the high levels of cost and energy consumption required by the POW process, newer cryptocurrencies are replacing POW with a Proof of Stake (POS) system where the nodes must prove ownership of a coin to participate in the verification process. With POS, a node's influence and reward in the mining process are relative to the amount of that cryptocurrency they own.

Cryptocurrency Use

Cryptocurrency use has grown exponentially since its introduction, with an average of USD 100 billion in cryptocurrency traded every 24 hours and a market capitalization that has previously surpassed 3 trillion USD (*Industries & Markets: Cryptocurrencies*, 2021). However, as demonstrated in Figure 4, cryptocurrency use remains relatively rare when measured against fiat currencies such as the U.S. dollar. For example, a Singapore-based cryptocurrency firm, TripleA, estimated that there were around forty-six million individual cryptocurrency users in the United States, or approximately 14% of the population (TripleA, 2022). Similarly, a survey from Pew Research Center found that 16% of Americans have purchased cryptocurrencies in some capacity (Perrin, 2021). Comparatively, an estimated 95% of Americans maintain an account at a banking or credit company, demonstrating the continued dominance of fiat currencies despite cryptocurrency growth (FDIC, 2021).

Figure 7. Top 10 countries for crypto use (2021)

Country	Number of Users	Percentage of Population
Vietnam	20,210,834	20.27%
Ukraine	6,516,114	15.72%
United States	46,020,521	13.74%
South Africa	7,712,116	12.45%
Kenya	6,101,599	11.60%
Pakistan	26,457,317	11.50%
Nigeria	22,332,791	10.34%
Russia	14,647,694	10.10%
Venezuela	3,015,636	9.02%
Brazil	16,652,150	7.75%

(TripleA, 2022)

Why and What Are Cryptocurrencies Used For?

Presently there are more than 10,000 different cryptocurrencies available. Some have unique and specialized functions, while others serve simply as a digital form of money. For example, some cryptocurrencies provide heightened security, others allow for convenience with quick verification speeds, and others have niche purposes. The traits of a cryptocurrency dictate the motivations for its use. Still, cryptocurrency use is generally motivated by its perceived accessibility and ease of use, internet access, and the expected benefit it can provide (Arias-Oliva et al., 2019; Nadeem et al., 2021). Cryptocurrency use is most common among young, educated men and those whose families use cryptocurrencies (Alvarez et al., 2022).

Bitcoin remains the most valuable and popular cryptocurrency, but many new cryptocurrencies provide significant technological upgrades (Baur & Dimpfl, 2021). Cryptocurrencies such as Litecoin, which has a faster hash rate, and stablecoins such as Tether or USD Coin, which are tied to the U.S. dollar, are better suited to replace fiat currencies. Bitcoin is also popular with illicit actors, such as those dealing in money laundering, facing financial sanctions, or wanting to evade taxes (Dupuis & Gleason, 2021; Foley et al., 2019; Marian, 2013). However, new coins such as Monero and Z-Cash offer higher levels of security and anonymity and will likely take Bitcoin's place as the default cryptocurrency for criminal actors.

How Hostile Actors Use Cryptocurrencies

Cryptocurrencies enable financial transactions beyond the purview of state regulation, and while this offers numerous benefits to the average consumer, hostile actors have much to gain from using them. Whether it is an anti-state organization in need of donations or a sanctioned state seeking to transform its frozen oil assets into currency, there are numerous avenues of cryptocurrency use through which hostile actors are already benefiting.

Anti-State Organizations

Anti-state organizations, such as rebel, terrorist, or separatist groups, all require significant funding levels and the capacity to retain and spend funds to achieve their goals. Without financial capabilities, these organizations will likely fail in whatever operations they undertake and eventually dissolve (Keatinge & Danner, 2021). Funds are typically acquired via personal donations, state donations, or business ventures of varying legality ranging from automobile dealerships to the drug trade (Clarke, 2016). Recent advances in counter-terrorism funding (CTF) operations post-9/11 have made it increasingly difficult for these groups to develop and maintain their revenue streams with traditional financial services and fiat currencies (Dion-Schwarz et al., 2019). ISIS, Hamas, Al-Qaeda, Boko Haram, etc., have had funds seized that were held in fiat currencies or had donors arrested after attempts to transmit funds with traditional financial services such as PayPal, Venmo, and Western Union. In response, many of these groups are transitioning or, at the very least, experimenting with cryptocurrency use.

Since cryptocurrencies are not managed by any privileged third party, require no incriminating information, and do not allow transactions to be reversed, they are an attractive financial tool to any anti-state group. Furthermore, they are a highly accessible technology for these groups and their supporters. For example, a simple paper wallet, a public and private key pair printed off on paper, can be generated instantly and for free using software available on websites such as "walletgenerator.net" (Figure 8). Similarly, the mediating technology needed to ensure the anonymity and security of transactions, such as VPNs, hard wallets, and stealth exchanges, are all easy and free to use.

Figure 8. Example paper wallet

Other criminal organizations have long since discovered the usefulness of cryptocurrencies in facilitating illicit transactions. Bitcoin is the default currency for illegal marketplaces on the dark web, where illicit goods such as narcotics, weapons, organs, and passports, among other items, are relatively easy to obtain (Finklea, 2017). It is estimated that Bitcoin alone is used to process billions in USD of illicit transactions annually, some of which can be attributed to these anti-state organizations (Foley et al., 2019; Kim Grauer et al., 2022; Zahirah & Ridwan, 2018). One estimate found that terrorist organizations were responsible for over 40% of illicit transactions made with cryptocurrencies in the E.U. (Zahirah & Ridwan, 2018). These groups use cryptocurrencies to collect donations and ransom payments, purchase firearms and other weapons, fund attacks abroad, and facilitate the general cross-border movement of funds (Malik, 2018; Teichmann, 2018; Zahirah & Ridwan, 2018). Cryptocurrencies are a more efficient and effective means for completing these tasks and provide a high level of security when used effectively.

Among anti-state organizations, Islamic extremist groups were the earliest and most proactive in using cryptocurrencies (Hasbi & Mahzam, 2018; Weimann, 2016; Whyte, 2018; Zahirah & Ridwan, 2018). Still, non-Islamic groups such as the Proud Boys in the United States, other far-right groups in Europe, and several Pro-Russia militant groups in Ukraine also use cryptocurrencies (Chainalysis Team, 2022; Dion-Schwarz et al., 2019; Zahirah & Ridwan, 2018). The earliest available evidence of terrorist groups using cryptocurrencies comes from a 2014 article in the ISIS-published magazine, Al-Haqiqa, which urged their followers to begin sending donations in Bitcoin. This initial foray into crypto was strongly motivated by advances in CTF and physical strikes on the buildings housing their cash stockpiles (Keatinge & Danner, 2021). Since then, numerous other groups, such as Hamas, Al-Qaeda, Boko Haram, Merciful Hands, Al Ikhwan, The Proud Boys, etc., have followed suit (Gauer et al., 2022). Furthermore, these groups have deployed increasingly sophisticated technology to acquire and protect their crypto assets (Figure 9). Rather than tweeting links to their Bitcoin wallets, these groups now almost exclusively use encrypted communication services such as the dark web and the Telegram messenger service to communicate with supporters and facilitate donations (Weimann, 2016).

Figure 9. Advances in crypto fundraising

ISIS and Al-Qaeda have constructed significant infrastructure for procuring and managing their cryptocurrency assets. For example, a sub-division of ISIS called the Electronic Horizons Foundation manages its cryptocurrency portfolio and operates a help desk for potential donors (Johnson, 2021; Irwin & Milad, 2016; Frost, 2020). Similarly, an Al-Qaeda affiliate group, Hayat Tahrir al-Sham, operates several brick-and-mortar cryptocurrency stores called "Bitcoin Transfer" in northwest Syria. These stores function as cryptocurrency trading centers for locals but also as hubs for collecting and investing donation funds (O'Leary, 2021). Instead of only soliciting contributions, they also teach their global network of supporters how to invest in cryptocurrencies, creating a diverse network of crypto portfolios and donors.

Cryptocurrencies are an effective tool for anti-state organizations for donation collection and investments. However, their prevalence on the dark web makes them the ideal currency for illicit transactions such groups may need to conduct. Anti-state organizations typically fund themselves with criminal activities. These may include the sale of narcotics, hostage-taking, and human trafficking (Clarke, 2016; Finklea, 2017). For example, in 2014, ISIS earned an estimated 20 million USD from hostage ransoms, and in

2017 they entered the drug trade. Hamas and Al-Qaeda are also involved in the drug trade and maintain relationships with narcotics organizations in the Americas (Clarke, 2016; Clarke, 2017). So long as the proper mediating technology is employed, anti-state groups can use cryptocurrencies to manage these revenue streams more effectively. They can more easily expand their narcotic sales to broader markets and collect untraceable ransoms. Once these profits are collected, they can instantly transfer them back to their organizations without fees.

Anti-state organizations also use cryptocurrencies to conduct operations outside their home territory. They are used for paying operators and purchasing the requisite supplies off dark web marketplaces. For example, the 2016 Charlie Hebdo attacks in Paris, France, which left 12 dead and 11 injured, were financed by a 3 million USD payment in bitcoin from Al-Qaeda. Similarly, the weapons used in the Bataclan and Stade de France attacks in 2015 that left 130 dead were purchased on the darknet using bitcoin from an arms dealer operating under the pseudonym "D.W. Guns" (Hasbi & Mahzam, 2018). Other valuable equipment for these operations, such as explosive devices and fake passports, are also available on the dark web and could be used to conduct future attacks.

Cryptocurrencies have several practical applications for anti-state groups, but they also possess several shortcomings that prevent these organizations from abandoning their use of fiat currencies entirely. First, the value of most cryptocurrencies is incredibly volatile. For example, from November 2021 to July 2022, the price of a single Bitcoin plummeted by 40,000 USD, and the total value of the cryptocurrency market decreased by USD 2 trillion. However, market volatility alone does not entirely negate crypto's usefulness to these groups, as they are easily and instantly interchangeable with fiat currencies. This allows groups to use cryptocurrencies for whatever purpose is needed and then exchange them back into a more stable fiat currency to avoid any loss of value. Furthermore, while there is a substantial risk of losing value with cryptocurrencies, they can also be highly profitable investments depending on the state of the market.

Cryptocurrencies are increasingly popular and gaining mainstream acceptance, but their use is still relatively niche. As such, they are not a viable means for paying wages or purchasing legal goods such as food or clothes. Finally, while cryptocurrencies are superior to fiat currencies in terms of anonymity and protection from authorities that they can provide, the security they provide is not absolute. For example, the United States government has seized millions of dollars in cryptocurrency assets from Hamas, ISIS, and Al-Qaeda, showcasing the vulnerabilities of cryptocurrencies (United States Department of Justice, 2020).

Cryptocurrencies are not yet able to completely replace fiat currencies for anti-state groups, but they still offer numerous valuable applications. Although Crypto use is not yet rampant among anti-state organizations, if the perception that they can provide such benefits remains, their usage will continue to proliferate, as terrorist groups are prone to imitating successful innovations by their peers (Keatinge & Danner, 2021). Such a trend represents a serious challenge to law enforcement officials, but not one that is insurmountable.

Hostile States

Rogue and hostile states, or those in opposition to Western powers such as the U.S., are often targeted with economic sanctions. Many of these states are developing or are already using cryptocurrencies as a sanction-busting tool. Sanctions are meant to stifle their economy by limiting their access to international markets through import and export limits, heightened tariffs, trade embargos, etc. The damage from these policies is designed to enforce compliance and deter similar actions by other states by signaling the repercussions violators will face (Gould-Davies, 2020; Wronka, 2021). America and the western world have been able to effectively sanction states since the end of the second World War primarily due to the dominance of the American dollar in international trade (Dudely, 2021). Not only can they freeze the assets of any state held in U.S. dollars, but the dominance of the U.S. dollar also gives America a great deal of influence over countries' access to SWIFT (Society for Worldwide Interbank Financial Telecommunication). SWIFT is the primary means by which banks complete international transactions and losing access to it locks a country out of the global economy (Walsh, 2017; Maharrey, 2018). While SWIFT is headquartered in Brussels, its reliance on the U.S. dollar forces it to comply with U.S. sanctions.

Cryptocurrencies, both decentralized and those controlled by states, present severe challenges to the dominance of the U.S. dollar as the currency of international trade and the American-led economic regime. States such as Russia, Iran, North Korea, and Venezuela, among others, recognize this and are using several different cryptocurrency strategies to sanction bust. It is important to note that cryptocurrencies are not yet capable of undoing the effect of sanctions entirely. Their market cap and the profit that can be gained with them are not large enough. Still, they allow states to generate wealth and conduct international business despite the most severe sanctions. For example, oil-rich states have found that they can use embargoed resources to generate wealth through the mining process, others have capitalized on the potential for cryptocurrency-related crime, and many are developing central bank coins (CBC), also referred to as state-controlled cryptocurrencies. Each of these strategies presents a severe and unique challenge to the western world's ability to enforce sanctions on hostile states. The section below summarizes how three major targets of western sanctions, Iran, Russia, and North Korea, use cryptocurrencies to sanction bust.

Iran

As a result of their support for terrorist organizations and their upstart nuclear program, Iran faces heavy sanctions from the western world. As a result, their banks have lost access to SWIFT, state and privately-owned foreign assets have been frozen or seized, their primary commodity export, oil, is heavily sanctioned, and the import and export of arms have been severely limited (Katzman, 2022). As a result, they have been forced to explore various pathways to sanction busting, such as exporting oil into neighboring Afghanistan in exchange for literal truckloads of cash to keep their economy afloat (Dagres & Slavin, 2018). More notably, they are at the forefront of cryptocurrency-based sanction busting by working to develop a sanction-proof domestic cryptocurrency, another cryptocurrency to support non-dollar-based international trade, and their use of frozen oil assets to power bitcoin mining on a national scale at unprecedented levels (Konowicz, 2018; Kim Grauer et al., 2022; Mahdavieh, 2019).

Like most other states, Iran's reliance on the dollar has made it vulnerable to sanctions, a weakness they mean to address with a state-backed cryptocurrency. The first attempt at this was led by four private banks, Mellat, Melli, Pasargad, and Parsian, who together issued the cryptocurrency Paymon (Clautice, 2019). However, Paymon has been a complete failure. A single Paymon is only worth USD 0.000018, and with 611.23 million in circulation, it has a meager market cap of USD 11,002.14 (Tassev, 2022a). Despite this failure, the Iranian central bank is moving forward with a CBC, the crypto-rial, a centrally controlled crypto version of their fiat currency, the rial (Tassev, 2022b). The crypto-rial will be exclusively distributed and managed by the central bank, in stark contrast to the decentralized nature typically associated with cryptocurrencies. It is set to be released in the fall of 2022, but Iran's crypto-based aspirations do not stop here. At the 2018 Chain 18 conference in Armenia, Iran signed a trilateral agreement with Russia and Armenia meant to facilitate the dispersion of expertise in blockchain technology among each other with the primary goal of developing a SWIFT cryptocurrency alternative (Motamedi, 2018).

The centerpiece of Iran's cryptocurrency-based sanction-busting strategy is a multi-billion-dollar national cryptocurrency mining effort. The government licenses cryptocurrency miners, who can use relatively low electric prices to mine so long as they sell the cryptocurrencies, they earn to the Iranian central bank. Electricity is one of the most significant expenses when mining cryptocurrencies, making Iran an attractive location for domestic actors and foreign miners. Chinese investment firms have established a sizeable presence following the recent ban on cryptocurrency mining in China. With this, Iran directly converts frozen oil reserves into currency. In 2020 they were able to turn roughly 10 million barrels of crude oil, 4% of their 2020 export total, into what is estimated to be worth well over 1 billion USD in bitcoin (Robinson, 2021).

Iran has been unable to completely mitigate the negative effect of the sanctions they are under, but its limited success is inspiring copycat behavior from states in similar positions. For example, the following sections on Russia and North Korea show how they are already copying their crypto-mining strategy (Grauer et al., 2022; Sigalos, 2022). This strategy should be of extreme interest to policymakers, as it allows any sanctioned state, especially those with oil resources, to directly exchange frozen assets into cryptocurrencies. Presently, this is an insufficient strategy, even for oil-rich nations, to completely alleviate the damage from international sanctions. Still, as the cryptocurrency market continues to develop and grow in value, cryptocurrency mining will become an increasingly effective tool for sanctioned states to generate wealth.

Russia

The annexation of Crimea in 2014, followed by a full-scale invasion of Ukraine in early 2022, brought on a deluge of sanctions against Russia. Their banks are no longer allowed to use SWIFT, 600 billion USD worth of assets held abroad by the central bank are frozen, and oil exports from Russia are banned by the U.S. and limited by most European countries (BBC Staff, 2022; Makhlouf & Selmi, 2022). While these sanctions have been unable to force compliance, they have inflicted considerable economic damage. Before the latest wave of sanctions, the Russian ruble had depreciated 50% against the U.S. dollar (Gould-Davies, 2020; Mahdavieh, 2019). Like Iran and many other States, Russia is implementing crypto-based strategies to lessen the damage these sanctions cause.

Initially, Russia was wary of domestic cryptocurrency use; both Putin and representatives of the Central bank expressed concern that they might be used for money laundering or terrorist financing. However, they are now aggressively proactive in their use of cryptocurrencies to combat western sanctions (Dorokhova et al., 2021). Several Russian oil companies are building crypto-mining centers, and the Russian central bank is working to develop a sanction-proof national cryptocurrency and a crypto alternative to SWIFT (Konowicz, 2018; Mahdavieh, 2019; Amick, 2022). Of note is that their BRICs allies and Iran back their cryptocurrency alternative to SWIFT.

In 2017 Russia announced it would be creating its own CBC, the crypto-ruble, a digital counterpart to the ruble meant to be sanction-proof (Clautice, 2019). The crypto-ruble's expected launch date is late 2023 or early 2024, and trial runs are already underway. However, it is unclear if the crypto ruble is also meant to facilitate international trade or if it is solely for domestic transactions. Like Iran, Russia is also involved in efforts with several friendly states to create a crypto alternative to SWIFT. This may include integration with the planned Chinese CBC. However, the more likely outcome is a shared cryptocurrency and cryptocurrency trading platform for the BRICS member states of Brazil, Russia, India, China, and South Africa, which represent an increasingly large portion of the global economy (Clautice, 2019; Mahdavieh, 2019).

Russian companies and citizens alike are using cryptocurrencies at an increasing rate in response to these heightened international sanctions. Like Iran, Russia is an oil-rich state; as such, oil exports have been one of the primary targets of these sanctions. The state-backed oil company, Gazpromneft, announced at the 2022 St. Petersburg International forum that they would partner with Bitriver, a bitcoin mining firm based out of Sweden, to build and power bitcoin mining centers in Russia (Mahdavieh, 2019; Amick, 2022). On a smaller scale, cryptocurrencies are increasingly popular among Russia's citizenry, with a recent uptick in trading volume placing them 18th in the Global Crypto Adoption index from Chainalysis (Grauer et al., 2022).

Criminal activities related to cryptocurrencies have also increased dramatically in Russia and, to an uncertain extent, are supported by the state. Chainalysis, a cryptocurrency analysis firm, found that in 2021 they were first in revenue generated from crypto-ransomware attacks, with an estimated 400 million USD in revenue. Much of this was acquired by the cyber-criminal organization Evil Corp, whose leadership has ties to the Russian government (Grauer et al., 2022). Beyond ransomware, Moscow has established itself as the capital of cryptocurrency laundering, washing over 2 billion USD in cryptocurrencies between 2018 and 2021 (Grauer et al., 2022). Unlike the ransomware industry, money laundering operations in Russia operate publicly and are generally accepted by authorities, with roughly a dozen laundering companies freely inhabiting Vostok, the largest skyscraper in Moscow (Kartikay Mehrotra & Olga Kharif, 2021).

As with Iran, cryptocurrencies are an insufficient tool for Russia to recoup the damages lost to sanctions. However, their continued and escalating use of cryptocurrencies is particularly notable, considering their recent belligerent actions in Ukraine. Moreover, that they have copied Iran's strategy for turning frozen oil assets into currency through the mining process shows that these sanctioned states are learning from each other as they continue innovating new methods for sanction-busting with cryptocurrencies.

North Korea

Like Iran and Russia, North Korea is heavily sanctioned, primarily due to its rogue nuclear program and support for terrorist organizations. As such, it is prohibited from importing or exporting most goods (Clautice, 2019). Although they receive support from China through aid and trade, sanctions have shrunk their GDP to USD 18.5 billion in 2021, which for comparison, is roughly 3% of the state of Ohio's GDP in that same year. Unlike those Iran and Russia, North Korea lacks oil resources to draw on for large-scale mining efforts. However, with their preexisting proficiency in cyberspace, they have been able to alleviate some of the damage from these sanctions by mining and stealing cryptocurrencies. While their efforts are not as profitable as those of other states, they should still be of great interest to scholars and policymakers alike, as the revenue generated from these efforts is typically used to fund their nuclear program (Macfarlane, 2021).

Because North Korea lacks any considerable oil deposits, its primary means of generating revenue through cryptocurrencies is cybertheft, typically by targeting vulnerable cryptocurrency exchanges with malware and seizing users' funds (Wertz, 2020). Despite their impoverished state, North Korea maintains robust offensive cyber capabilities, most notably with the Lazarus group. This division of North Korea's military is made up of 6,000 operators dispersed throughout the world who conduct offensive cyber operations to generate revenue. They are best known for their 2014 attacks on Sony Pictures in response to a movie parodying Kim Jong-un's death. Operations by the Lazarus group have generated roughly 2 billion USD in cryptocurrencies, with several individual attacks that have generated hundreds of millions in profits alone (Park, 2021). A smaller subset of the Lazarus group involves foreign, undercover North Korean operatives infiltrating foreign technology firms, stealing technology, and sending their earnings home to the regime (Lyngaas, 2022).

Despite its limited financial and natural resources, North Korea also mines cryptocurrencies. Due to their limited resources, they will unlikely be able to replicate Iran's success with cryptocurrency mining. However, it is noteworthy that North Korea exclusively mines Monero, a stealth coin that offers heightened security compared to Bitcoin (Park, 2021). This transition to stealth coins, which is also observed in non-state groups, is a result of the increased ability of western intelligence agencies to seize cryptocurrencies, which has led to the seizure of North Korean crypto assets as well (Kim, 2021).

North Korea's cryptocurrency strategies demonstrate that even impoverished states lacking significant financial or natural resources can leverage cryptocurrencies to generate wealth. Powering cryptocurrency mining centers with natural resources that would otherwise be frozen on the international market is the easiest way for sanctioned states to generate wealth with cryptocurrencies. However, the general lack of understanding of cryptocurrencies means countries with limited resources can still generate revenue with cyber-attacks.

How States Are Responding

In general, the international community's response to the threats posed by the use of cryptocurrencies by hostile actors has been relatively limited, but some meaningful success has been observed. While there is no centralized effort, individual state actors have seized substantial amounts of cryptocurrency from anti-state actors on numerous occasions (U.S. Department of Justice, 2020). While this news is encouraging, there are a few key points to consider. As discussed earlier in this chapter, not all cryptocurrencies offer their users the same level of privacy and security. Anti-state groups initially only used Bitcoin, but in the face of these developments, many are transitioning to stealth coins, such as Monero (Bridget Johnson, 2021). Furthermore, cryptocurrency assets can only be seized if the private key is acquired in the physical world or if those assets are being held on a centralized exchange that cooperates with law enforcement officials. As such, while states have achieved some success in seizing cryptocurrencies held by hostile groups, it is unclear if they will achieve the same success as these groups transition to only using stealth coins.

The international community's response to the use of cryptocurrencies by hostile states has been even more limited and less successful than their response to anti-state groups' use of crypto. For example, the United States has forbidden cryptocurrency exchanges to do business with sanctioned states and pursued legal action against those that have (Cox, 2022; Iddon, 2022). However, these efforts have been unable to achieve even limited success. Iran, for example, has been able to process nearly eight billion USD in cryptocurrency transactions since 2018 on the Coinbase exchange alone (Iddon, 2022). Of note is that Coinbase was founded and previously headquartered in the United States.

Cryptocurrency use by hostile groups and states presents a new and serious challenge to the international community. While thus far, the international community has achieved some success in preventing anti-state groups from acquiring cryptocurrencies, the same cannot be said for efforts to prevent hostile states from using cryptocurrencies. However, in the case of cryptocurrency use by anti-state groups, it remains to be seen if states will still observe success in seizing their assets as they transition into using only stealth coins.

Domestic Regulation of Cryptocurrencies

Cryptocurrency use by hostile states and groups is one issue states must contend with, but domestic use of cryptocurrencies by the public also presents various issues. It challenges their ability to control and implement fiscal policies and poses a serious challenge to counter-terrorism funding, anti-money laundering, and general illicit transactions (Huo, 2022). Furthermore, with the growing value of the cryptocurrency market, taxes related to cryptocurrencies are increasingly valuable. Estimates suggest that E.U. states could have increased their total tax revenue by 850 million euros in 2020 (Thiemann, 2021). Cryptocurrency mining can also put a tremendous strain on a nation's electrical grid when left unchecked. Iran, for example, has been able to generate a great deal of revenue with its state-sponsored cryptocurrency mining program. However, they also suffer from sporadic power grid failures due to the massive amount of electricity this endeavor requires. (Turak, 2021).

State policies towards cryptocurrencies vary throughout the world. On one end, El Salvador has embraced crypto and uses it as an officially recognized state currency. On the other end, states such as China, Algeria, Bolivia, Morocco, Nepal, Pakistan, and Vietnam have issued a complete ban on all cryptocurrency-related activities (Riley, 2021). Most states lie between these two poles, either implementing limited cryptocurrency regulations or simply extending existing financial regulations to cryptocurrencies. The purpose of this section is to show the various policy approaches states are taking to cryptocurrencies and the challenges they are facing.

China

Once home to the largest market for cryptocurrency transactions and mining activity, China gradually tightened restrictions on crypto use until it implemented an outright ban on all cryptocurrency activity in September 2022 (Xi, 2022; Riley, 2021). Official statements from the government claim that cryptocurrencies were banned to protect the citizenry from volatile investments and fraudulent coin offerings. However, reports suggest the ban was also motivated by concerns about mass capital outflow via cryptocurrencies. Estimates suggest tens of billions of dollars in cryptocurrency leave China yearly (Xi, 2022). The ban also removes any competition for the new Chinese CBC, the e-CNY, or the digital yuan. The e-CNY has been in development since 2014 and is currently in a trial phase active in 4 cities, with 140 million personal accounts and 10 million corporate accounts (Xi, 2022).

China and other authoritarian regimes, like Russia and Iran, are implementing state-controlled cryptocurrencies not only to sanction proof but because using a blockchain-based digital currency will provide heightened control over their population's financial activity. With them, a government has direct control over who can access their national currency and has an immutable record of every transaction that has occurred with it, exactly who it was between when it happened, and how much was exchanged. Of course, such a strategy is only effective if there are no cryptocurrency alternatives, which is why many states implementing central bank cryptocurrencies have banned all other alternatives. However, evidence suggests these bans have been ineffective.

The issue any cryptocurrency ban will encounter is that many cryptocurrencies are designed to resist government censorship, and the tools needed to access them, even in the face of China's cyber expertise, are readily accessible (Chang et al., 2021). Virtual private networks (VPNs) are computer applications that enhance privacy and allow one to circumvent firewalls and access banned content. Using VPNs and encrypted messaging services like Telegram, consumers in China can still access cryptocurrency exchanges themselves or purchase them through third-party intermediaries in less-regulated countries, namely Japan and South Korea (Parker, 2018; Nabilou, 2019). China's ban on cryptocurrency mining has been equally ineffective. Despite an initial decrease in China's national hash rate, Mining activity in China surpassed pre-cryptocurrency ban levels in less than six months and now tails only the U.S. in terms of mining activity (*Bitcoin Mining – an (Un)Surprising Resurgence?*, 2022). Some Chinese mining operations have immigrated to nations with friendlier crypto-mining policies, such as Iran or South Korea. However, that covert mining activity in China has already surpassed pre-mining ban levels showcases the inability of states to ban cryptocurrency activity.

Cryptocurrency regulation in China demonstrates the limited ability of governments to implement an outright ban on cryptocurrency activities and is an example of the popularity of central bank currencies among authoritarian states. Despite possessing the most sophisticated cyber capabilities, China has been unable to limit cryptocurrency activity within its borders, much less ban it. The tools needed to circumvent any limitations or firewalls on cryptocurrencies are simply too easily accessible. Despite this failure in regulating cryptocurrencies, China thus far has had the most success of any state in implementing a central bank digital currency and is on track to have the first widely used CBC.

El Salvador

In sharp contrast to China's absolute ban on cryptocurrency activity, the government of El Salvador made Bitcoin a national currency with the "Bitcoin Law" on September 7, 2021. This radical approach to regulating has, similarly to China's crypto ban, failed to achieve any success. Despite an aggressive push by the government to make Bitcoin use mainstream, it is still exceedingly rare, and the recent downturn in the cryptocurrency market has seen the government's Bitcoin portfolio depreciate by roughly 60%. China's outright ban demonstrates the inability of states to ban cryptocurrencies, but El Salvador's failed Bitcoin gambit shows the dangers of adopting a decentralized cryptocurrency.

El Salvador would seem the ideal setting if a decentralized cryptocurrency were to succeed as a national currency. Bitcoin enables instantaneous and fee-free international transactions and can be easily accessed by the unbanked. Such traits would be expected to attract users in a nation where 70% of the population is unbanked and 25% of the national GDP comes from remittances. The government also supported a widespread incentive program that included $30 in Bitcoin for downloading El Salvador's bitcoin wallet, Chivo, and lower gas prices (Alvarez et al., 2022). Despite all this, only 12% of the population regularly use Bitcoin, and even fewer use it for applications such as receiving remittances or becoming banked (Alvarez et al., 2022).

Countries cannot ban cryptocurrency use, but the case of El Salvador shows that they cannot, and should not, try to force the use of a decentralized cryptocurrency on their citizens. Decentralized cryptocurrencies are highly volatile assets, making them risky for poorer states like El Salvador. Of note is that El Salvador has continued to double down on their cryptocurrency gamble and recently purchased an additional USD 1.5 million in bitcoin. However, bitcoin use remains low in El Salvador, and this endeavor is widely condemned as a failure.

United States

In between these two poles of extreme cryptocurrency policies are states such as the U.S., which have taken a more measured approach to regulation. In the U.S., citizens are free to use cryptocurrencies, but the exchanges on which most citizens use them are heavily regulated on a federal level. While the goals of the U.S. government's cryptocurrency regulatory policies are much more limited in scope than El Salvador or China, their policy of exchange-targeted regulation appears to be more successful in terms of achieving their stated goals.

The 2021 Infrastructure Investment and Jobs act requires all cryptocurrency brokers to report any transaction made over USD 10,000 and to issue reporting forms to their customers for any transactions made (Library of Congress, 2021). They are also required to collect identifying information from their customers. This policy effectively deanonymizes all cryptocurrency transactions on over-the-counter exchanges in the U.S. and incurs minimal effort and costs to the state. Of course, not all cryptocurrency transactions are legal, but this anonymizes most cryptocurrency transactions and users while making them taxable. This approach has already produced notable results. Working with several cryptocurrency exchanges, the U.S. government captured 3.6 billion USD worth of stolen Bitcoin in February of 2022 (Meshulam et al., 2022). This softer approach has also yielded success in regulating the cryptocurrency mining industry. Texas, which has become one of the world's largest cryptocurrency mining hubs, has avoided electrical grid outages thus far because of this approach. Miners will decrease activity at the government's behest when there is an exceptionally high strain on the electrical grid at a given time (Chiu, 2022).

Lessons Learned

States have taken a wide variety of different policy approaches toward cryptocurrencies, and several tentative lessons can be learned from them. First, cryptocurrency activity cannot be prevented entirely, but more limited regulatory goals are feasible. Cryptocurrencies are designed to resist government control, and the tools needed to circumvent even the most advanced firewalls are far too accessible for an all-out ban to succeed. However, it is observed in the United States and elsewhere that more limited and targeted regulation of cryptocurrency use can meaningfully limit some of the adverse effects of cryptocurrency use (Nershi et al., 2022). Regulating cryptocurrency through exchanges, where most individuals will choose to interact with cryptocurrencies has already produced notable results and appears to be the most effective means of regulation (Nabilou, 2019). However, that cryptocurrency use in China has already surpassed the pre-ban levels suggests that complete cryptocurrency bans are not feasible. Second, decentralized cryptocurrencies are not presently an adequate substitute for national currencies. Their value is too volatile, and their use is not wide enough for them to replace fiat currencies (Alvarez et al., 2022). Central bank currencies may be a more effective solution, but this is yet to be seen.

While there is much to learn from the apparent successes and failures of the different policy approaches states have adopted for cryptocurrencies, it is also important to evaluate them carefully. While thus far, more limited regulatory efforts targeted at or working with cryptocurrency exchanges have been the most successful in achieving their goals, this may change. Cryptocurrencies are a new and constantly evolving technology, as are efforts to track cryptocurrency movements and deanonymize the blockchain. Changes in either could greatly affect the feasibility of any policy approach, making it exceedingly important for policymakers to stay informed on any technological developments related to cryptocurrencies to craft the most effective policies.

CONCLUSION

Do cryptocurrencies matter to political outcomes of interest, and if so, how? Cryptocurrencies may be the next evolution of currency, and while they bring many benefits, there is a litany of serious issues policymakers and scholars must understand to address.

First, cryptocurrency use is increasingly popular among hostile and illicit actors at all levels. Bitcoin is the default currency on the dark web and can be used to purchase a variety of illegal goods such as firearms, narcotics, child pornography, and bodily organs. However, the use of cryptocurrency for illicit means is not confined to purely criminal organizations. Terrorist organizations and rebel groups such as Al-Qaeda, ISIS, Boko-Haram, Russian separatist groups in Ukraine, and Hamas all use cryptocurrencies. Faced with increased scrutiny from state counter-terrorism financing efforts, these groups are turning to cryptocurrencies to solicit donations and fund violent attacks abroad. While cryptocurrencies have not become the default currency for terrorist groups, if they continue to show merit in helping them achieve their goals, the use of cryptocurrency by violent actors will only continue to grow.

Second, the illicit use of cryptocurrencies is also spreading among hostile states in the international community. States facing sanctions from the international community are implementing several cryptocurrency-based strategies to circumvent these sanctions or mitigate their effect. For example, in partnership with their BRICs allies, Russia is developing a shared cryptocurrency to serve as an alternative to the SWIFT global payment service. Similarly, Russia and Iran are developing cryptocurrency-based versions of their national currencies largely because they are perceived to be sanction-proof. Besides limiting the effectiveness of sanctions, Iran and Russia use their frozen oil assets to mine cryptocurrencies, effectively translating otherwise frozen assets into currency. Even North Korea has been able to profit off of cryptocurrencies, stealing millions annually in cryptocurrency-based schemes to help fund their regime.

The use of cryptocurrencies by hostile actors on all levels is concerning, but the domestic legal use of cryptocurrencies alone presents a host of challenges to policymakers. For example, cryptocurrency mining consumes a tremendous amount of electricity, and in extreme cases, it can cause power grids to fail. Also of concern is that cryptocurrencies are extremely volatile assets, prone to large price swings and various scams. In the worst-case scenario, individuals can lose large percentages of their life savings to these swings or scams.

In response to the domestic challenges of cryptocurrencies, states have adopted various policy approaches. El Salvador has sought to incentive the use of Bitcoin, even embracing it as an officially recognized currency. In contrast, China has banned the use of any cryptocurrency besides their central bank coin, the digital yuan. Neither state has experienced any observable success in achieving these large-scale cryptocurrency goals. China remains one of the largest cryptocurrency markets, and the use of Bitcoin in El Salvador is still relatively rare. In contrast, the United States has successfully achieved its much more limited policy goals.

Cryptocurrencies matter to political outcomes of interest. This chapter has provided evidence that their use is related to a host of meaningful outcomes states seek to control or prevent. The notion that cryptocurrencies are either irrelevant or a passing trend is no longer valid. Future research should be aimed at how cryptocurrencies affect outcomes of interest, not simply if they matter at all. Questions should be explored, such as 1) Are States that mine cryptocurrencies more likely to take action that incurs sanctions? Or 2) Are civil conflicts where one or more rebel groups last longer? Preexisting relationships based on finances should also be reexamined, considering cryptocurrencies' increasing use. Scholars will encounter difficulty finding quantitative measures of cryptocurrency use, but some options exist. The Singapore-based cryptocurrency analysis TripleA has estimated the percentage of each country's population using cryptocurrencies in 2021. Chainalysis, another cryptocurrency company, publishes annual rankings for each country's cryptocurrency use measured by trading volume and activity. Still, quantitative measures are challenging to come by and will be one of the primary obstacles in the study of cryptocurrency use. However, as this chapter has made clear, doing so is a more than worthwhile endeavor.

REFERENCES

Alvarez, F. E., Argente, D., & van Patten, D. (2022). *Are cryptocurrencies currencies? Bitcoin as legal tender in El Salvador*. https://www.nber.org/papers/w29968

Amick, S. (2022, June 17). *Third Largest Russian Oil Company Gazpromneft Partners With BitRiver To Mine Bitcoin*. Nasdaq.

Arias-Oliva, M., Pelegrín-Borondo, J., & Matías-Clavero, G. (2019). Variables influencing cryptocurrency use: A technology acceptance model in Spain. *Frontiers in Psychology*, *10*, 475. Advance online publication. doi:10.3389/fpsyg.2019.00475 PMID:30949085

Baur, D. G., & Dimpfl, T. (2021). The volatility of Bitcoin and its role as a medium of exchange and a store of value. *Empirical Economics*, *61*(5), 2663–2683. doi:10.100700181-020-01990-5 PMID:33424101

Bitcoin mining – An (un)surprising resurgence? (2022, May 22). University of Cambridge Judge Business School.

Chainalysis Team. (2022, August 29). *2 Million and Counting: How Dozens of Pro-Russian Groups Are Using Cryptocurrency Donations to Fund the War in Ukraine*. Chainalysis.

ChangK.HobbsW.RobertsM.Steinert-ThrelkeldZ. (2021). *Crisis is a Gateway to Censored Information: The Case of Coronavirus in China*. https://ssrn.com/abstract=3756577

Chiu, A. (2022, June 15). Limiting crypto helped the Texas power grid weather a heat wave. *The Texas Tribune*.

Chudinovskikh, M., & Sevryugin, V. (2019). Cryptocurrency regulation in the brics countries and the Eurasian economic union. *BRICS Law Journal*, *6*(1), 63–81. doi:10.21684/2412-2343-2019-6-1-63-81

Clarke, C. P. (2016). Drugs & Thugs: Funding Terrorism through Narcotics Trafficking. *Journal of Strategic Security, Vol. 9, No. 3. Special Issue: Emerging Threats*, *5*, 1–15. doi:10.2307/26473335

Clautice, T. (2019). *Nation State Involvement in Cryptocurrency and the Impact to Nation State Involvement in Cryptocurrency and the Impact to Economic Sanctions Economic Sanctions Nation State Involvement in Cryptocurrency and the Impact to Economic Sanctions*. https://digitalcommons.lasalle.edu/ecf_capstones

Clarke, C. P. (2017, July 25). *ISIS Is So Desperate It's Turning to the Drug Trade.* RAND.

Couger Team. (2021, April 21). *What Countries and States That Have Trialled Blockchain Voting Learned.* Medium.

Cox, C. (2022, September 20). *Treasury warns against Russia's efforts to evade sanctions with cryptocurrencies.* CNBC.

Crosby. (2016). *BlockChain Technology: Beyond Bitcoin.* Academic Press.

Dagres, H., & Slavin, B. (n.d.). *How Iran Will Cope with U.S. Sanctions.* https://www.atlanticcouncil.org/blogs/

de Vries, A., Gallersdörfer, U., Klaaßen, L., & Christian, S. (2022). Revisiting Bitcoin's carbon footprint. *Joule, 6*(3), 498–502. doi:10.1016/j.joule.2022.02.005

Deane. (2018). *The New Game: Cryptocurrency Challenges U.S. Economic Sanctions.* Academic Press.

di Pierro, M. (2017). What Is the Blockchain? *Computing in Science & Engineering, 19*(5), 92–95. doi:10.1109/MCSE.2017.3421554

Dion-Schwarz, C., Manheim, D., Johnston, P. B., & Rand Corporation. (2019). *Terrorist use of cryptocurrencies : technical and organizational barriers and future threats.* RAND.

Dorokhova, E., Belykh, T., & Koren'kova, G. (2021). Economic and Legal Aspects of Cryptocurrency Usage in Russia. In *Strategic Latency Unleashed: The Role of Technology in a Revisionist Global Order and the Implications for Special Operations Forces* (pp. 327–332). 10.2991/aebmr.k.210710.055

Dudely, S. (2021). Cryptocurrency: Will the Digitization of Currency Allow Malign Actors to Achieve Strategic Effects? In Strategic Latency Unleashed (pp. 269–282). Academic Press.

Dupuis, D., & Gleason, K. (2021). Money laundering with cryptocurrency: Open doors and the regulatory dialectic. *Journal of Financial Crime, 28*(1), 60–74. doi:10.1108/JFC-06-2020-0113

FDIC. (2021, December 17). *How America Banks: Household Use of Banking and Financial Services.* Federal Deposit Insurance Corporation.

Finklea, K. (2017). *Dark Web.* www.crs.gov

Foley, S., Karlsen, J. R., Putniņš, T. J., Goldstein, I., Jiang, W., Karolyi, A., Weber, M., & Easley, D. (2019). *Sex, drugs, and bitcoin: How much illegal activity is financed through cryptocurrencies?* https://www.emcdda.europa.eu/attachements.cfm/att_194336_EN_TD3112366ENC.pdf

Frost, L. (2020, May 19). ISIS's $300 million war chest is hidden in Bitcoin, says think tank. *Decrypt.*

Gauer, Kueshner, & Updegrave. (2022). *The 2022 Crypto Crime Report.* Academic Press.

Gould-Davies, N. (2020). Russia, the West and Sanctions. *Survival, 62*(1), 7–28. doi:10.1080/00396338.2020.1715060

Hasbi, A. H., & Mahzam, R. (2018). *Cryptocurrencies: Potential For Terror Financing?* www.rsis.edu.sg

Huo, Y. (2022). *The Effect of Government Policies on Cryptocurrency Market*. Academic Press.

Iddon, P. (2022, November 15). *How Iran is cashing in on cryptocurrencies to evade U.S. sanctions*. Arab News.

Irwin, A. S. M., & Milad, G. (2016). The use of crypto-currencies in funding violent jihad. *Journal of Money Laundering Control, 19*(4), 407–425. doi:10.1108/JMLC-01-2016-0003

Johnson, B. (2021, April 20). *ISIS Cyber Group Warns of Tracking Through Bitcoin Use*. Homeland Security Today.

Katzman, K. (2022). *Iran Sanctions*. Academic Press.

Kaur, R., & Kaur, A. (2012). Digital signature. *Proceedings: Turing 100 - International Conference on Computing Sciences, ICCS 2012*, 295–301. 10.1109/ICCS.2012.25

Keatinge, T., & Danner, K. (2021). Assessing Innovation in Terrorist Financing. *Studies in Conflict and Terrorism, 44*(6), 455–472. doi:10.1080/1057610X.2018.1559516

Kethineni, S., & Cao, Y. (2020). The Rise in Popularity of Cryptocurrency and Associated Criminal Activity. *International Criminal Justice Review, 30*(3), 325–344. doi:10.1177/1057567719827051

Kim, S. J. (2021). *North Korea's Sanctions Evasion and Its Economic Implications*. Academic Press.

Library of Congress. (2021). *Regulation of Cryptocurrency Around the World: November 2021 Update*. http://www.law.gov

Lyngaas, S. (2022, July 10). *Here's how North Korean operatives are trying to infiltrate U.S. crypto firms*. CNN Politics.

Macfarlane, E. (2021). Strengthening Sanctions: Solutions to Curtail the Evasion of International Economic Sanctions Through the Use of Cryptocurrency. *Michigan Journal of International Law, 42*(1), 199. doi:10.36642/mjil.42.1.strengthening

MacKenzie Sigalos. (2022, March 22). *Exxon is mining bitcoin in North Dakota as part of its plan to slash emissions*. CNBC.

Maharrey, M. (2018, August 6). *SWIFT and the Weaponization of the U.S. Dollar*. Foundation for Economic Education.

Mahdavieh, R. (2019). *Governments' Adoption of Native Cryptocurrency: A Case Study of Governments' Adoption of Native Cryptocurrency: A Case Study of Iran, Russia, and Venezuela Iran, Russia, and Venezuela*. http://library.ucf.edu

Makhlouf, F., & Selmi, R. (2022). *Do sanctions work in a crypto world? The impact of the removal of Russian Banks from SWIFT on Remittances Do sanctions work in a crypto world?* https://hal.archives-ouvertes.fr/hal-03599089

Malik, N. (2018). *How terrorists use encryption, the darknet, and cryptocurrencies*. Academic Press.

Marian, O. (2013). Part of the Banking and Finance Law Commons, Criminal Law Commons, Internet Law Commons, and the Tax Law Commons Recommended Citation Recommended Citation Omri Marian, Are Cryptocurrencies Super Tax Havens? 112 MICH. *Michigan Law Review First Impressions, 112*. https://repository.law.umich.edu/mlr_fiAvailableat:https://repository.law.umich.edu/mlr_fi/vol112/iss1/2

Mehrotra & Kharif. (2021, November 3). Ransomware H.Q.: Moscow's Tallest Tower Is a Cybercriminal Cash Machine. *Bloomberg*.

Meshulam, D., Hausfeld, K., Kinney, J., & North, E. (2022, February 15). *U.S. Department of Justice, aided by cryptocurrency exchanges, seizes over US$3.6 billion in stolen Bitcoin*. DLA Piper.

Motamedi, M. (2018, December 6). *U.S. sanctions accelerate Iran's blockchain drive*. Al-Monitor.

Nabilou, H. (2019). *How to Regulate Bitcoin? Decentralized Regulation for a Decentralized Cryptocurrency*. Academic Press.

Nadeem, M. A., Liu, Z., Pitafi, A. H., Younis, A., & Xu, Y. (2021). Investigating the Adoption Factors of Cryptocurrencies—A Case of Bitcoin: Empirical Evidence From China. *SAGE Open, 11*(1). Advance online publication. doi:10.1177/2158244021998704

NershiK.CollinsM.ElzaynH. S.GrayJ.HeverlyR.HoffmanD.LeviM.LittrellC.MansfieldE. D.NielsonD. RobertsM.Rodine-HardyK.SharmanJ.SimmonsB.WerbachK. (2022). *How Strong Are International Standards in Practice? Evidence from Cryptocurrency Transactions*. https://ssrn.com/abstract=4186244 doi:10.2139/ssrn.4186244

Nofer, M., Gomber, P., Hinz, O., & Schiereck, D. (2017). Blockchain. *Business & Information Systems Engineering, 59*(3), 183–187. doi:10.100712599-017-0467-3

Norry, A. (2017, October 11). *The History of Silk Road: A Tale of Drugs, Extortion & Bitcoin*. Blockonomi.

O'Leary. (2021, March 31). The bitcoin terrorists of Idlib are learning new tricks. *Wired*.

Park, J. (2021). *The Lazarus group the cybercrime syndicate financing the North Korea state*. Academic Press.

Parker, E. (2018). Can China Contain Bitcoin? *MIT's Technology Review, 121*(1), 61–65.

Perrin, A. (2021, November 11). *16% of Americans say they have ever invested in, traded or used cryptocurrency*. Pew Research Center.

Riley, J. (2021). The current status of cryptocurrency regulation in China and its effect around the world. *China and WTO Review, 7*(1), 135–152. doi:10.14330/cwr.2021.7.1.06

Robinson, T. (2021, May 21). *How Iran Uses Bitcoin Mining to Evade Sanctions and "Export" Millions of Barrels of Oil*. Elliptic.

Staff, B. B. C. (2022, June 27). *What are the sanctions on Russia and are they hurting its economy?* BBC News.

Statista Report: Cryptocurrencies. (2022). Statista.

Tassev, L. (2022a, January 22). *Iran to Pilot 'National Cryptocurrency,' Considers Blockchain Tech for Stock Market*. Bitcoin.Com.

Tassev, L. (2022b, June 26). *Iran Aims to Roll Out Pilot Version of Crypto Rial Within 2 Months*. Bitcoin.Com.

Teichmann, F. M. J. (2018). Financing terrorism through cryptocurrencies – a danger for Europe? *Journal of Money Laundering Control*, *21*(4), 513–519. doi:10.1108/JMLC-06-2017-0024

Thiemann, A. (2021). *Cryptocurrencies: An empirical view from a tax perspective JRC Working Papers on Taxation and Structural Reforms No 12/2021*. https://ec.europa.eu/jrc

Triple, A. (2022). *Cryptocurrency across the world*. TripleA.

Turak, N. (2021, May 26). *Iran bans bitcoin mining as its cities suffer blackouts and power shortages*. CNBC.

Turner, A., & Irwin, A. S. M. (2018). Bitcoin Transactions: A digital discovery of illicit activity on the blockchain. *Journal of Financial Crime*, *25*(1), 109–130. doi:10.1108/JFC-12-2016-0078

United States Department of Justice. (2020). Global Disruption of Three Terror Finance Cyber-Enabled Campaigns.

United States Postal Service. (2020). *Secure Voting System* (Patent No. 2020/0258338 A1). United States.

U.S. Department of Justice. (2020, August 13). *Global Disruption of Three Terror Finance Cyber-Enabled Campaigns Largest Ever Seizure of Terrorist Organizations' Cryptocurrency Accounts*. U.S. Office of Public Affairs.

Weimann, G. (2016). Terrorist Migration to the Dark Web. *Source: Perspectives on Terrorism, 10*(3). https://www.jstor.org/stable/26297596

Wertz, D. (2020). *Understanding U.S. and International Sanctions on North Korea*. www.ncnk.org

Whyte, C. (2018). Cryptoterrorism: Assessing the Utility of Blockchain Technologies for Terrorist Enterprise. *Studies in Conflict and Terrorism*. Advance online publication. doi:10.1080/1057610X.2018.1531565

Wronka, C. (2021). Digital currencies and economic sanctions: The increasing risk of sanction evasion. *Journal of Financial Crime*. Advance online publication. doi:10.1108/JFC-07-2021-0158

Xi, C. (2022). *The End of the War or the Commencement of Battle?* The Chinese University of Hong Kong Faculty of Law Research Paper No. 2022-15.

Yin, H. H. S., Langenheldt, K., Harlev, M., Mukkamala, R. R., & Vatrapu, R. (2019). Regulating Cryptocurrencies: A Supervised Machine Learning Approach to De-Anonymizing the Bitcoin Blockchain. *Journal of Management Information Systems*, *36*(1), 37–73. doi:10.1080/07421222.2018.1550550

Zahirah, R., & Ridwan, M. (2018). Kelompok Studi Mahasiswa Pengkaji Masalah Internasional The Utilization of Cryptocurrencies by the Terrorist Group as an Alternative Way of Hawala for Illicit Purposes Mahasiswa Ilmu Hubungan Internasional, Fakultas Ilmu Sosial dan Ilmu Politik. *Jurnal Sentris KSMPMI, 2*. https://dictionary.cambridge.org/dictionary/english/cryptocurrency

ENDNOTES

1 Several different hashing algorithms may be used, but the most popular is SHAH-256, which is best known as the hashing function for bitcoin. SHAH-256 is an evolved form of the SHAH-1 algorithm, developed by the United States National Security Agency (NSA) in the early 1990s to enable secure data transmission across the Internet. The primary difference between the two algorithms is the length of the encryption key. SHAH-256 has a 256-bit encryption key, while the key for SHAH-1 is only 160 bits. Hash functions are also used to maintain the public and private key systems for a given cryptocurrency; this will be discussed more in the next section.

2 For cryptocurrencies the paper changes the terms from buyer to initiator and from seller to receiver, as cryptocurrencies are often used for non-retail purchases, such as donations or remittances.

Chapter 10
A Proposal for Enhancing the Security and Privacy of Digital Currencies

Kazuo Takaragi
National Institute of Advanced Industrial Science and Technology, Japan

Takashi Kubota
https://orcid.org/0000-0001-6240-2994
Waseda University, Japan

ABSTRACT

Regarding digital currencies, the authors focus on 1) the security issues with the advent of quantum computers and 2) the privacy issues in relation with CBDC's wider collection of personal data. Many legal issues can be solved using technology. Thus, they propose central banks to adopt their privacy-enhancing technologies in cryptography, called "SK4SC," which performs KYC (know your customer) while keeping PKI (public key infrastructure)-registered trading partners anonymous. Security and privacy in digital currency can be trade-offs. For example, the AML/CFT effort requires the collection of personal information about innocent people, which can be detrimental to their privacy. On the other hand, excessive restrictions on the collection of personal information to protect privacy may weaken the AML/CFT. However, adopting SK4SC can resolve the trade-offs by anonymizing KYC, which is essential for the AML/CFT.

INTRODUCTION

While various private and national digital currencies, including cryptocurrencies such as bitcoins and CBDCs (Central Bank Digital Currencies), have rapidly developed, security and privacy concerns still exist. Usually speaking, personal freedom of being free from potential hazards is defined as security, while privacy is defined as being free from unwanted attention. In other words: a) Rapid advances in quantum computers will sooner or later break existing cryptography. At that time, the authentication

DOI: 10.4018/978-1-6684-6429-8.ch010

itself necessary for online payment is broken, so security is significantly reduced. Therefore, we need to switch from existing cryptography to quantum-safe cryptography. b) In AML/CFT (Anti-Money Laundering and Combating the Financing of Terrorism), privacy concerns remain as before, even if security is enhanced with quantum-safe cryptography. Therefore, it is necessary to strengthen both security and privacy by developing a new cryptographic application system that incorporates cryptography into an ingenious management method.

Regarding the privacy concern, let us consider a CBDC as an example. A CBDC that collects information on consumers has serious privacy implications. In the past, when transactions were digitized, there were many incidents in which a large amount of personal information was stolen instantaneously by so-called man-in-the-middle attacks by hackers. CBDC poses a similar threat from hackers. Moreover, according to Sheluchin (2020), establishing a CBDC would allow the central bank to collect more information on its consumers than ever before. These personal details and data could then be shared with law enforcement without needing a warrant, which currently require a warrant to investigate the spending habits of an individual. Now that the authorities have not decided how to obtain a warrant in the case of CBDC, there are concerns about the unintended exploitation of users' personal information when it comes into effect. The introduction of any centralized cryptocurrency through the central bank should be subject to scrutiny since it would likely lead to an unprecedented level of state surveillance.

As for the security concern, let us remind the issue of quantum-resistant cryptography by referencing the U.S. NIST (National Institute of Standard and Technology) homepage (NIST (2022)), which explains as follows:

In recent years, there has been a substantial amount of research on quantum computers – machines that exploit quantum mechanical phenomena to solve mathematical problems that are difficult or intractable for conventional computers. If large-scale quantum computers are ever built, they will be able to break many of the public-key cryptosystems currently in use. This would seriously compromise the confidentiality and integrity of digital communications on the Internet and elsewhere. The goal of post-quantum cryptography (also called quantum-resistant cryptography) is to develop cryptographic systems that are secure against both quantum and classical computers, and can interoperate with existing communications protocols and networks. The question of when a large-scale quantum computer will be built is a complicated one. While in the past it was less clear that large quantum computers are a physical possibility, many scientists now believe it to be merely a significant engineering challenge. Some engineers even predict that within the next twenty or so years sufficiently large quantum computers will be built to break essentially all public key schemes currently in use. Historically, it has taken almost two decades to deploy our modern public key cryptography infrastructure. Therefore, regardless of whether we can estimate the exact time of the arrival of the quantum computing era, we must begin now to prepare our information security systems to be able to resist quantum computing.

Though there are many other issues regarding digital currencies, we focus on two issues: security and privacy. First, with the advent of quantum computers, existing digital currency cryptography will be broken. Therefore, we overview the current situation, and consider a security-enhancing strategy to develop new quantum-resistant cryptography, which can serve as a reference for other countries. Second, we propose that digital currency providers, such as central banks issuing CBDCs adopt privacy-enhancing technologies in cryptography. While there are various technical mechanisms to ensure security and privacy, this paper attempts to make a more concrete proposal by taking "SK4SC," which we were involved in developing, as a specific example. SK4SC performs KYC (Know Your Customer) while keeping PKI (Public Key Infrastructure)-registered trading partners anonymous. Security and privacy in digital currency can be trade-offs. For example, the AML/CFT effort requires the collection of personal information about people, which can be detrimental to their privacy. On the other hand, excessive restrictions on the collection of personal information to protect privacy may weaken the AML/CFT. However, adopting our SK4SC can resolve the trade-offs by anonymizing KYC, which is essential for the AML/CFT.

Section I overviews security and privacy from a legal perspective and stresses that the legal approach has limits and technical solutions are needed. It is not the purpose of this paper to provide a legal analysis of these issues, but the legal issues surrounding security and privacy are many and varied. Therefore, as an introduction to the technical explanations that follow, we present examples of the current status and challenges of legal solutions.

Section 2 reviews research into the theory and application of high-performance cryptography, such as the one used in the cryptocurrency Zcash, which offers breakthroughs in reconciling security and privacy.

Section 3 illustrates the current situation of developing quantum-resistant cryptography from a technical perspective.

Section 4 describes how our SK4SC features can help enhance privacy in PKI.

Section 5 describes the functions provided by SK4SC and its performance when implemented in PC software.

Our original paper (Takaragi et al., 2023) included detailed descriptions of the technical specifications and mathematical proofs. However, considering the reader's needs, the main part of this article omitted these and explained the functionality of SK4SC in easy-to-understand narrative terms. The appendix presents the mathematical formulas for readers specifically interested in cryptography theory but can be ignored by the general reader.

Conclusion shows a concrete proposal. In this proposal, we show how SK4SC should be introduced and improved step by step, with an eye on the development of quantum computers.

Overview of Legal Aspects of Security and Privacy

As an introduction to the technical explanations that follow, we present examples of the current status and challenges of legal solutions. It is not the purpose of this section to provide a legal analysis of these issues, but we want to show that the legal approach has limits and a technical solution is needed.

Along with IT development and globalization, security enhancement and privacy protection in the digital space have become urgent issues in the context of digital money. For example, North Korean cybercriminals are reported to have attacked foreign cryptocurrency platforms. They extracted nearly $400 million worth of digital assets in 2021 (Chainalysis (2022). As legal countermeasures to such illegal hacks supported by law-abiding nations such as North Korea is limited, technical measures against hacking are more effective. Various sanctions are possible legal countermeasures against North Korean hacking. However, the hacking itself cannot be stopped, and various challenges remain in preparing evidence of the hacking and taking legal action to make the other party comply with it. In this regard, if there is a technological solution to prevent hacking, many legal challenges would not arise because the problem would not occur in the first place.

On the other hand, there have been some privacy concerns about the collection of personal information of consumers by their CBDCs as shown above in Introduction. In addition, some governments have also enacted laws, such as the U.S. Clarifying Lawful Overseas Use of Data (CLOUD) Act and the

Chinese new laws as shown below in subsection ii, enabling them to collect certain foreign data. The law is providing some solutions to these problems, albeit inadequate. However, the solution to the problem will be easier if technological solutions are used in conjunction with legal ones. The privacy issue arises because central banks in their own and foreign countries can collect large amounts of personal information when consumers use CBDC. Once again introducing a technical solution would prevent a legal problem from occurring.

Security Enhancement

Quantum Computers

With the advent of quantum computers soon, existing cryptography used in digital currencies can be broken according to the NIST explanation shown above in Introduction.

The rise of quantum computers will greatly affect various fields of our life and financial system. If quantum computers can be used to easily decipher the codes used for digital currency, then the digital currency can be easily stolen. If this happens, the legal system (e.g., ordinary transaction laws, financial regulations, taxation, AML/CFT, etc.) built on the premise of the current financial order, which has been securely supported by cryptography, will collapse. Therefore, developing new cryptographic technologies compatible with quantum computers is essential.

AML/CFT

Let us look at the Anti Money Laundering/ Countering the Financing of Terrorism (AML/CFT). According to the FACTI (2021) report, 2.7% of the world's annual GDP is already laundered by criminals, and quantum computers can greatly facilitate money laundering by breaking the current cryptographies. So far, each country's AML/CFT measures have been developed globally based mainly on the recommendations by the international body called the FATF (Financial Action Task Force). The FATF Recommendation does not directly address quantum computer measures, but even without considering such measures, many countries currently have many unresolved issues pointed out by the mutual reviews by the FATF member countries. For example, like other major countries including the U.S., China, Singapore, and Switzerland, Japan was also recognized as an "enhanced follow-up country" in the 4th FATF Mutual Review adopted in June 2021. As a result of various requests for improvement, public and private sector efforts, including legislation and system development, are proceeding at a rapid pace (Fukushima, 2022).

Still, the security countermeasures for the advent of quantum computers have been under research and development, as shown below in Section III. There is little room for legal solutions in this area except for promoting financial support, etc. Though preparing the perfect countermeasure is difficult as quantum-resistant cryptography is still under development, we show our second-best technological approach in the post-quantum computing environment in Conclusion.

Policy Coordination

Legal roles are limited if a technological revolution occurs, such as the emergence of quantum computers. However, in the area of CBDC policy coordination, it may be possible to facilitate it among the group of countries to some extent.

In October 2021, the G7 Finance Ministers and Central Bank Governors' Meeting released the "Public Policy Principles on Retail Central Bank Digital Currencies (CBDCs)," reportedly as a check on China's CBDC called e-CNY. The principles drafted by the G7 Finance Ministers emphasize that CBDC providers should have three important characteristics: 1) strict adherence to the rule of law (*Principle 2: G7 values for the International Monetary and Financial System should guide the design and operation of any CBDC, namely observance of the rule of law, sound economic governance and appropriate transparency.*); 2) strict compliance with FATF standards in AML/CFT (*Principle 6: Any CBDC needs to carefully integrate the need for faster, more accessible, safer and cheaper payments with a commitment to mitigate their use in facilitating crime.*); and 3) strong data privacy protections (*Principle 3: Rigorous standards of privacy, accountability for the protection of users' data, and transparency on how information will be secured and used is essential for any CBDC to command trust and confidence. The rule of law in each jurisdiction establishes and underpins such considerations.*).

These principles have the potential to facilitate policy coordination among G7 and some friendly countries. Yet, as explained in subsection ii, they also have political implications in restraining the Chinese Government Access to the Chinese CBDC known as e-CNY. Therefore, countries belonging to politically opposite groups to the G7 do not readily comply.

Privacy Protection

The legal protection of privacy has been strengthened in recent years. The traditional right to privacy was "the right to be left alone" by the mass media, but with the advent of computers, it has been transformed into "the right to control one's own information" concerning where personal information is collected (Okamura, 2021). Again, once privacy or personal data protection is ensured through technological solutions, legal solutions for failure to protect privacy become less necessary in the first place.

In the following, we will look at the legal situation in Japan, China, Europe, and the United States regarding personal data protection.

Japanese Law Reform

Japan has drastically revised its Act on the Protection of Personal Information (APPI) in 2020 and 2021 (see Asai, 2022b) for the English translation of the articles). The use of personal information is currently allowed to the minimum necessary except for legitimate reasons such as AML/CFT. The main points of the 2020 amendment, which is considered a drastic reform, are: (a) to make the responsibility of business operators to handle personal information heavier and raise the maximum amount of fines for corporations by referencing foreign moves such as the E.U. GDPR (General Data Protection Regulation); (b) to restrict the use of cookies; and (c) to establish a framework for data utilization called "pseudonymous processed information" to make it easier for companies to utilize data.

This pseudonymous processed information refers to information about an individual that has been processed so that a specific individual cannot be identified unless it is compared with other information, and for example, it corresponds to information that has been deleted from customer information that is personal (member ID, name, age, gender, service used). In Section IV, we will see a kind of pseudonymous information processing by our SK4SC model.

International Law Development

On the other hand, international rules on data flow are still developing due to the different policies on personal data in major countries. Let us see some examples.

E.U.-U.S. Privacy Shield

First, there has been a conflict between the E.U. and the U.S. over the cross-border movement of personal data. The E.U.-U.S. Privacy Shield was a legal framework for regulating transatlantic exchanges of personal data for commercial purposes between the E.U. and the U.S. However, on July 16, 2020, the European Court of Justice (ECJ) invalidated the E.U. Commission's decision on the adequacy of the data protection provided by the E.U.-U.S. Privacy Shield in its Case C-311/18, ECLI:EU:C:2020:559 ("Data Protection Commissioner v Facebook Ireland and Maximillian Schrems ("Schrems II"). Still, on March 25, 2022, the U.S. and the E.U. announced that a new data transfer framework called the Trans-Atlantic Data Privacy Framework had been agreed to, replacing the Privacy Shield (White House, 2022).

The above are legal issues regarding whether to recognize each other's legal rules in the E.U. and the U.S. If data privacy were sufficiently protected technically, it would not be an issue in the first place.

Data Localization

Second, some countries, such as China and Indonesia, seek to enclose important information within their borders (data localization). In contrast, other countries such as the E.U. and Japan aim for free circulation in order not to disrupt digital trade. Though the effectiveness of existing WTO (World Trade Organization) rules to discipline data localization measures remains inconclusive, Hodson (2018) appreciates the possible role of FTA (Free Trade Agreement) to strike an appropriate balance between facilitating businesses' growing need to transfer and store data across national borders while preserving governments' right to regulate in the public interest.

The above would still not be a legal problem in the first place if there were a technical mechanism that allowed only certain data to be data localized.

Government Access to Data

Third, there are increasing foreign concerns about recent domestic legislation in China, such as the National Security Law (NSL), the Cybersecurity Law (CSL), the Data Security Law (DSL), and the Personal Information Protection Law (PIPL) (see Asai (2022a) for the English translation of the articles). Such laws have significantly allowed state agencies to access information held by private companies (Article 53 NSL, Article 28 CSL, Article 35 DSL, Article 33, 34 PIPL, etc. For the details of broad Chinese Government Access to Data, see Asai (2022a), pp. 280-283). Extraterritorial application makes the above Government Access to Data more complicated. Foreign and Japanese companies outside of China are also subject to the PIPL for providing services to China (Article 3 (2) PIPL, *"The Law also applies to the proceeding of personal information of individuals in China that is conducted outside China in any of the following cases: (a) [The purpose of the processing is] to provide goods or services to individuals in China; (b) [The purpose of the processing is] to analyze and evaluate the activities of individuals in China; (c) Other cases as provided by laws or administrative regulations."* see Asai (2022a) p.136).

As a result, there are concerns that the personal information of foreigners may be unfairly collected by the Chinese government. So far, it seems that the Chinese government has not collected such foreign personal information. Once again, these issues would not be a problem if a technical mechanism allowed the government to access only certain types of data.

Domestic Compliance

Fourth, there are other concerns about domestic compliance with the above government access. For example, there is the possibility of conflicts with Japan's Article 24 of the APPI (*"When providing personal data to a third party in a foreign country, a business operator handling personal information must obtain the consent of the person himself / herself to approve the provision to a third party in a foreign country in advance, except in the cases listed in each item of paragraph 1 of the preceding Article."*), which restricts the provision of information to third parties located abroad. This can make providing service to China, where Government access is broad, difficult for Japan, where privacy from the Government is respected. A similar problem arises in the E.U. by the Article 46 (*"Transfers subject to appropriate safeguards: (1)...a controller or processor may transfer personal data to a third country or an international organization only if the controller or processor has provided appropriate safeguards, and on condition that enforceable data subject rights and effective legal remedies for data subjects are available; (2) the appropriate safeguards referred to in paragraph 1 may be provided for, without requiring any specific authorization from a supervisory authority, by: (a) a legally binding and enforceable instrument between public authorities or bodies...."*) of the GDPR regarding the sufficiency of cross-border transfers of personal information to third countries.

To address these legal issues, it is useful to incorporate technical mechanisms that automatically conform to laws and regulations. For example, in AML/CFT and financial regulation, Regtech, a technological mechanism that manages the automatic compliance of large volumes of transactions with detailed regulations, is actively employed.

Need for Technical Solutions

As we have seen above, a variety of legal problems can arise. Still, it is difficult to discipline them by law alone, especially in cross-border fields, and technical solutions greatly decrease such problems. Protecting the security and privacy of digital currencies, where transactions are instantaneous, should be done instantaneously and with high enforcement through technological solutions; legal solutions that take time to resolve and have limited international enforcement are insufficient. If we have technology that can ensure security and privacy to a certain level, we can automatically fulfill the following policy goals: a) to counter cyber-terrorism measures; b) to counter hacking with the advent of quantum computers; c) to comply with legal regulations in the development of CBDCs; d) to utilize privacy-protected data; and e) to curb foreign government access to personal information. As a result, the law would only need to function in areas that could not be handled technically and would make policy more effective than it is today.

In other words, many legal issues can be solved using technology. Therefore, in the following, we would like to seek technical solutions by explaining quantum-resistant cryptography and a privacy protection technology called SK4SC.

APPROACH BY HIGH-PERFORMANCE CRYPTOGRAPH

High-performance cryptography, such as zero-knowledge proofs applied to the cryptocurrency Zcash, has already been implemented in practice and is a predictable scientific advance in many fields. KYC (Know Your Customer) with AML/CFT (Anti-Money Laundering and Combating Financial Terrorism) in the financial system, including CBDC (Central Bank Digital Currencies), seems to be one of the predictable areas of its application. An example of a new function is the ability to redact with a range proof of personal information, such as age when submitting documentary evidence for KYC in customer eligibility screening.

In KYC, personal information is processed and disclosed to the other trader for them to prove the necessary attributes for the transaction. Personal information includes, for example, the contents of national ID cards, certificates issued by companies such as bank account balances, and cryptographic keys. Here, a trader ID is a collection of various trader attributes collected to ensure trustworthiness in transactions. Attribute proof is usually accomplished in practice by the issuer's organizational hierarchy. Hierarchization makes it possible to specify duties and data storage locations, which is useful for streamlining operations and protecting and managing personal information and other information that must be strictly adhered to within an organization. However, exposing the full chain of issuers for certificate validation raises privacy concerns, as it can expose sensitive information about the issuer's organizational structure and certificate owners. As an application of zero-knowledge proof, delegatable anonymous credentials (DAC) developed by IBM (Camenisch et al., 2017) solve this problem, hiding the full delegation chain and providing privacy in both delegation and presentation of certificates. This system is useful in preventing the spread of personal information within the organization since it is possible to conceal from other departments what information has been handled by the departments within the organization.

DAC belongs to the field of Anonymous Credentials (AC). AC is a signing method that allows group members to sign messages anonymously on behalf of the group. Among ACs, group signing (GS) is a method suitable for a simple structure that does not involve an intermediate administrator (Local CA) between a group representative (Root CA) and a group member (End signer). GS also provides a way to anonymously revoke a signer's public key (Bresson et al., 2001). Hierarchical GS (Fuchsbauer et al., 2009) is an extension of GS that allows a hierarchical tree of group managers. However, a party can act as either a manager or a user, but not at the same time. Hierarchical GS differs from DAC in that the group manager can de-anonymize the signature.

Another related concept is attribute-based signatures (ABS) (Maji et al., 2001). The concept of ABS also provides the ability to indicate that a user owns an attribute anonymously, but the purposes of ABS and DAC differ in several ways. ABS aims at different goals, as explained in (Maji et al., 2001). It is shown that ABS can be used to create an authentication system in which there is no central authority and no global coordination is required, except for cryptographic parameter settings (Okamoto et al., 2022). On the other hand, DAC requires a centralized trust anchor, but in return provides strong anonymity at the registration stage. ABS is well suited to describe complex logical relationships between attributes used for authentication but does not require strict anonymity at the registration stage.

Our work is most directly related to DAC, and in particular, Camenisch et al. (2017) stands out among DAC in terms of feasibility and security proof. Separately, (Bünz et al. (2018) developed a zero-knowledge range proof scheme that proves an attribute is within a certain numerical range rather than simply hiding the attribute. In this research, we use a scheme in which multiple organizations using

DAC issue different partial IDs to one transactor. That transactor presents the aggregate ID to the other transacting party to perform attribute proof.

As far as the authors know, there is no example of applying the technology that combines DAC and ZKRP to KYC, other than the original paper by the authors (Takaragi et al., 2022).

DEVELOPING QUANTUM-RESISTANT CRYPTOGRAPHY

Quantum computers are not only projected to be used in commercial applications but are also considered to be strategic technology for ensuring national security. The competition for their development is intensifying worldwide. In the U.S., the federal government has increased the QIS (Quantum Information Science) R&D budget from $793 million in FY2021 to $877 million in FY2022 (US Government, 2021). Quantum Information Science (QIS) is the synthesis of quantum mechanics and information theory. QIS research includes revolutionary new types of computers, sensors and networks that improve national prosperity and security. Additionally, the military (with an undisclosed budget) and corporations are investing heavily in this technology. Japan, Europe, and China are also investing heavily in the field. A powerful quantum computer will likely be realized relatively soon, with breakthrough phenomena often observed in intensely competitive development fields. In that case, the Rivest–Shamir–Adleman (RSA) algorithm and elliptic curve cryptography (ECC), which have been in use since the 1990s, will be broken eventually.

The U.S. NIST (National Institute of Standards and Technology) has indicated that the development of powerful quantum computers is more of an advanced engineering issue than the advanced theoretical basic research previously envisioned, and its realization time is imminent. Based on such prediction, NIST is planning to establish a draft of U.S. standards for cryptography that cannot be broken even by a quantum computer (referred to as quantum-resistant cryptography, or post-quantum cryptography) by 2024. The impact of the U.S. standard cryptography will be significant because it will become the de facto global standard.

With the announcement of NIST on July 5, 2022, standardization candidates were narrowed down to three digital signatures and one encryption scheme (see Table 1). All schemes have achieved practical performance in PCs, IoT devices, and other applications.

For example, CRYSTALS-Dilithium, which is considered to be the first candidate for signature, takes about 0.2 seconds for signature creation and about 0.04 seconds for signature verification on a commercially available personal computer (using Core-i7 6600U). CRYSTALS-Dilithium is about 20 to 100 times slower than conventional ECC.

Table 1. Candidates for the standard (2022)

Field	Algorithm	Type	Country of the submitting institution
Signature	CRYSTALS-DILITHIUM	Lattice	NL, DE, US, CH, FR
	FALCON	Lattice	US, FR
	SPHINCS+	Hash	NL, US, DE, TW, AT, CH, DK
Encryption	CRYSTALS-KYBER	Lattice	NL, DE, US, CH, FR

These candidates will be further reviewed and amended as necessary by 2024. After that, it will be published as a standard. However, the algorithm standardized this time cannot eliminate the risk of being unexpectedly broken after 2024. As a backup in case of such a situation, NIST plans to separately define a quantum-resistant signature method and encryption method that are not lattice type (NIST, 2022).

It should be noted that the mathematical formula for quantum-resistant cryptography is the culmination of accumulated cryptography research over many years, and contributions made by Japanese researchers are also significant (Table 2).

Table 2. Quantum-resistant cryptography, contributions made by Japan

Algorithm	A part of the academic contribution by Japan
Cryptographic methods in general	Secure (IND-CCA) cryptographic construction by Eiichiro Fujisaki and Tatsuaki Okamoto
Lattice	Safety evaluation through analysis by Tadanori Teruya, Kenji Kashihara, Yoshinori Aono, Eiichiro Fujisaki, and Ken Naganuma (Technical University of Darmstadt "Lattice Challenge")
Hash*	Improvement of the Merkle signature by Katsuyuki Okeya and Tsuyoshi Takagi

RSA and ECC, which have been in practical use since the 1990s, are still embedded in communication and information processing hardware and software worldwide at the time of this writing (July 2022). Although it is difficult to replace all these with quantum-resistant cryptography immediately, it is important to prepare for a change in cryptography without waiting for the announcement of the 2024 U.S. Standardization of quantum cryptography. In order to replace ciphers, first, various preparatory procedures such as changes in public and private rules occur. Furthermore, changing equipment and software also requires considerable time for testing, trial confirmation, etc., in addition to cost burden. The time is near for cryptography to be compromised, and these efforts should be undertaken as soon as possible in time. In fact, Google in the US has announced plans to realize a quantum computer with one million qubits by 2029. If this is achieved, it will bring us one step closer to breaking conventional encryption.

The U.S. standard for quantum-resistant cryptography to be established by NIST only covers the basic functions of digital signatures and encryption. However, the academic community is working to make cryptography even more sophisticated. Considering the past, in parallel with the practical use of basic functions such as RSA and ECC in the 1990s, more sophisticated cryptographic schemes have been developed. As an example of sophisticated cryptography, zero-knowledge proofs have already been put into practical use in the cryptocurrency Zcash, and are regarded as predictable scientific advances for social applications (National Academy of Sciences, 2022). Similarly, in the field of cryptocurrency Monero, zero-knowledge range proofs called Bulletproofs are applied to hide the connection between the sender and receiver. The academic community has proposed improving Bulletproofs to become quantum-resistant (Bünz et al., 2018). Furthermore, research is also underway to improve bilinear cryptography, known as pairing, which is useful in ID-based cryptography, delegatable anonymous credential systems, and others, to make it quantum resistant as well (Boneh et al., 2018). As a result, current zero-knowledge proofs and other high-performance cryptography will be improved to quantum-safe types, eliminating the risk of being broken by quantum computers.

So far, we have discussed the situation where public key cryptography should be compromised by the development of quantum computers. Unlike public key cryptography, the common key cryptography AES (FIPS 197), standardized in 2001, is difficult to break using quantum computers. This does not mean that the common key cryptosystem AES is secure enough. Public key cryptography has been established over a long period of time based on mathematical assumptions based on classical calculations (writing or calculators). Therefore, it was strong in classical calculation. However, there was no guarantee that it would also be strong in quantum computing. On the other hand, many of the currently popular symmetric-key ciphers, such as AES, did not take advantage of well-established mathematical problems over a long history based on classical computations. Rather, it exploited the difficulty of solving a mathematical puzzle that someone recently came up with. In other words, the mathematicians' verification time is short, so the mathematical puzzle could be solved at any time in the future. As a result, it may one day suddenly be broken by an attack method based on classical computation.

MEASURES TO ENHANCE PRIVACY PROTECTION

This section presents the challenges and improvements of the PKI. PKI is a technology for assuring the correspondence between a public key and its owner using a third-party organization called a Certification Authority (CA).

The mechanism of PKI is similar to that of the Japanese seal registration (*inkan toroku*), a registration system publicly endorsed since the Meiji Period (1868–1912). The seal registration certificate is a document on which the stamped image of the user's seal is drawn, and the relevant government agency seal (*inkan*) (as seen in the second row of Figure 1) is further stamped. In PKI, this is achieved through the public key cryptography. This means that the seal is a secret key, and the seal registration certificate is the public key with an endorsement (digital signature). The endorsement is generated using the seal of a higher authority (the higher-level secret key, the second row in Figure 1). The reliability of the top-level seal (*inkan*) (first row in Figure 1) is by definition not endorsed by the top-level authority but is indicated by methods other than digital certificates, such as a strict audit by an external organization, disclosure of the rules for operating the certification service, operational results, and name recognition.

Conventional PKI has used basic cryptography, such as RSA and ECC, whose usage began in the 1990s. However, these ciphers raise privacy concerns because the sender's personal information is disclosed to the receiver, as shown on the right side of Figure 1.

Figure 1. Privacy-enhanced PKI

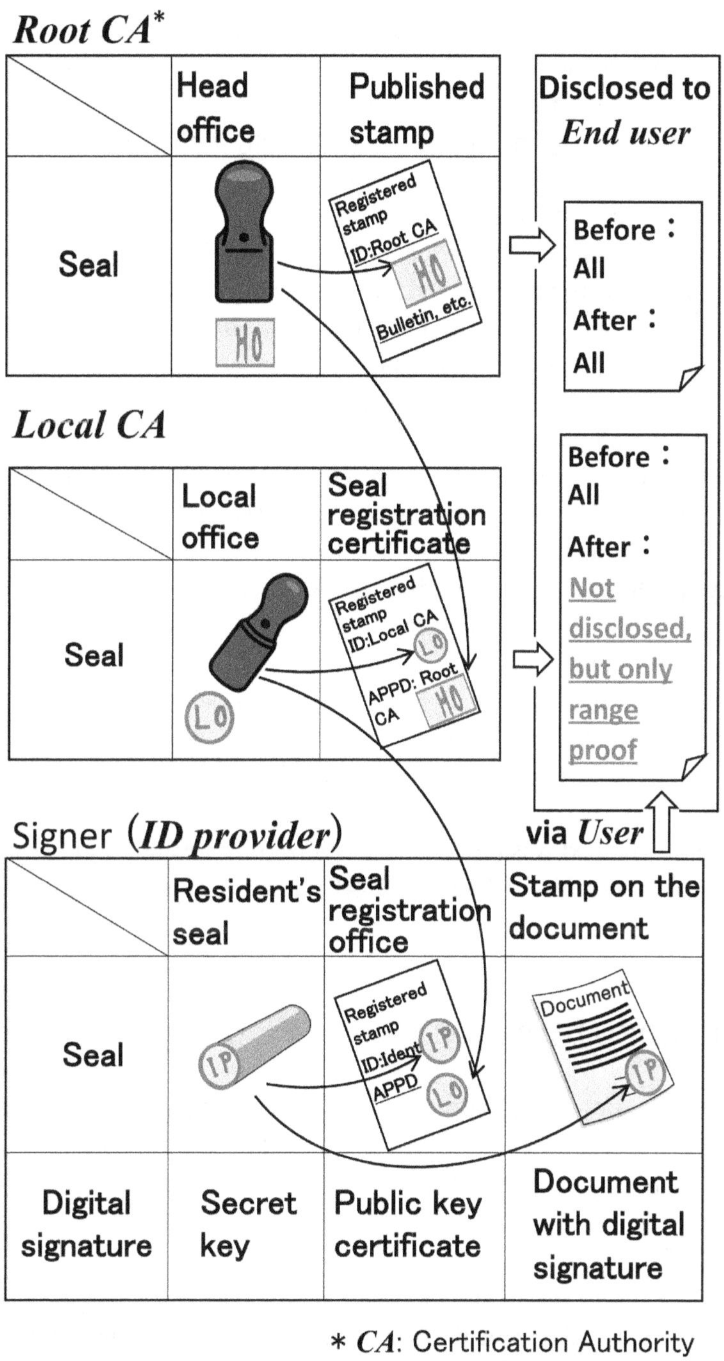

Suppose there is an identity provider that issues a third-party certificate for KYC to customers. The identity provider has one head office and multiple branches. Each branch uses its secret key to generate a digital signature on the document. At that time, the name and public key of the head office that delegated the signature to the branch will be known.

Identity proof is usually accomplished by a hierarchy of issuers in practice. However, exposing the full chain of issuers for certificate validation raises privacy concerns because it can expose sensitive information about the issuer's organizational structure and certificate owners.

The customer wants to hide the name of the issuing branches, i.e., *Local CA*s and *ID provider*, of the signature due to privacy concerns.

To address this issue, this study incorporates cryptography with new features such as Delegatable Anonymous Credentials (DAC) and Zero-Knowledge Range Proofs (ZKRP) instead of RSA or ECC. Then, the customer can prove to the business partner that any part of the document sent by the identity provider is within the range indicated by the inequality, without disclosing accurate data.

Aiming to put this mechanism into practical use in social systems, the authors, in cooperation with AIST, Waseda University, Shonan Institute of Technology, Hitachi, and SECOM, have developed a mechanism referring SK4SC (Secure Kernel for Supply Chains), which utilizes privacy-enhanced PKI to realize AAA based access control, i.e., authentication, authorization, and accounting, on the blockchain with a high level of security and privacy protection.

Access control with AAA has traditionally been achieved using authentication information centrally managed by a central computer or server operated by an ISP (Internet Service Provider) or some other entity. With SK4SC, users obtain the necessary KYC documents from multiple decentralized ID providers to prove that they are eligible users for trading. Those documents are in the form of redacted documents. Using a function called zero-knowledge proof, users can arbitrarily select symbols and numbers necessary for transactions from redacted documents and prove to the counterparty that they are within the scope of the inequality. Furthermore, if multiple redacted documents contain the same keyword, such as the master ID number *MA*,1, the user can prove that they are equal without revealing the exact value *MA*,1 to the counterparty (see bottom row in Figure 2). SK4SC provides the evidence needed while preserving privacy in such a way.

Using SK4SC, a basic method was developed to realize self-sovereign identity (SSI) management, which is required in KYC protocols for CBDCs. SK4SC exhibited good performance in many aspects, including the processing speed (see (Bogatov et al., 2020), (Bünz et al., 2018) for details).

FUNCTIONS PROVIDED BY SK4SC

This section introduces what functions SK4SC provides. A detailed explanation of the cryptographic formulas is outside the scope of this article, so we leave the detailed formulas to another document (Takaragi et al., 2022) and only give a brief explanation here.

Overview of SK4SC

Consider an environment in which transaction data is transmitted across multiple PKIs. Figure 2 shows a scheme for protecting privacy in KYC.

In the following description, when referring to the entity shown in Figure 2, such as Root CA, the entity name is italicized, such as Root CA.

Figure 2. System with national ID card

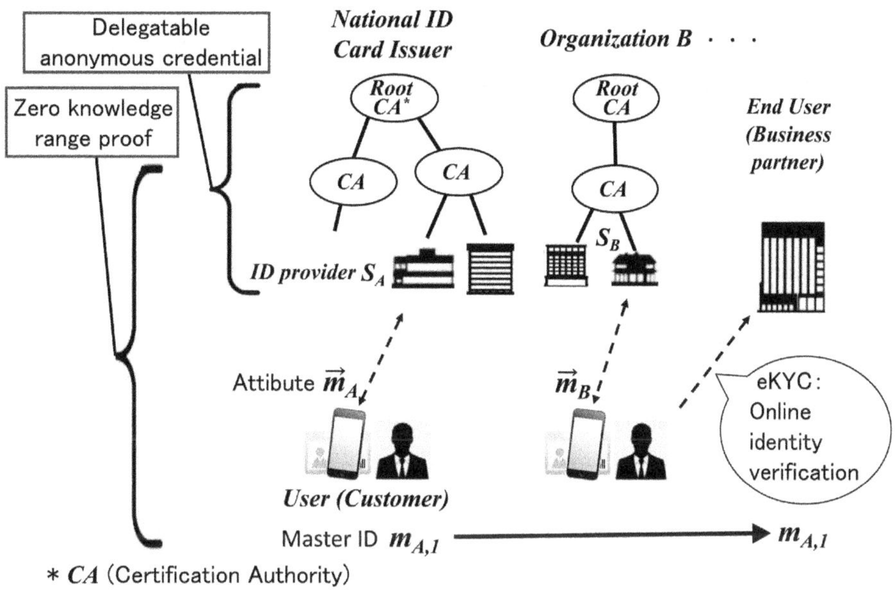

T*he uppe*r part of Figure 2 represents a hierarchical PKI to which DAC is applied. At the bottom, *Customer* at the end of the hierarchy uses zero-knowledge proof to create redacted documents with range proofs and send them to Business partner. Here zero-knowledge proof is a way for one person (prover) to tell another person (verifier) that a (usually mathematical) statement they have is true. It is a way of communicating that can be proved without conveying any knowledge other than that it is true.

The public key of *Root CA* (certificate authority) and the general formula for signature authentication are publicly available. However, the names of specific constituent departments within the organization, including unique identifiers for signatures and authorization, are not disclosed. The signature authentication formula is a narration of who makes what signature for what message and how it is authenticated in what environment. The narration can be described in natural language, and at the same time, the natural language description is converted into a kind of formal language description using a mathematical equation such as

$\mathscr{B}$ =NIZK{solution:equation}

Here, NIZK (Non-Interactive Zero-Knowledge proof) proves that the prover knows the solution x of the equation denoted after the colon, e.g. y=f(x), e.g., the national identification number on the document. The procedure outputs a value $\mathscr{B}$ that is not equal to the solution and proves that the prover knows the solution. In the process, even if the verifier is given the value $\mathscr{B}$, the concrete numerical value of the solution cannot be calculated back. As a result, the verifier establishes the fact that the prover knows the solution without being informed of the solution value itself. In other words, the verifier confirms that the prover (*ID provider*, *Customer*, etc.) knows the hidden numbers that satisfy the narration used for the authentication. This leads to *Business partner* executing the necessary KYC while hiding part of *ID provider*'s trade secret or *Customer*'s personal information.

Processing Flow

For simplicity of explanation, assume that there is only one *Local CA* and each *Local CA* and *ID provider* has only one attribute, which is its public key value.
The flow of the steps below is roughly as follows:

ID provider, which is the contact point of the organization for *Customer*, receives a delegation from the top of the organization (Root CA) and signs documents that prove the identity and attributes of *Customer*.

First, the procedure goes through steps 1 through 3. That is, the delegation of authority from the Root CA to the Local CA (Step 1, issuing certification 1). Further delegation of that authority from the Local CA to the *ID provider* (Step 2, issuing certification 2). The *ID provider* signs a document proving *Customer*'s identity and attributes (Step 3, Issuing proof of identity and attributes). DAC and ZKRP are used to sign those certifications and documents. Then, in those certifications and documents other than the Root CA's public key, the parts related to trade secrets or personal information will be handed over to *Customer* in a redacted form.

Next, in Steps 4 and 5, *Customer* can arbitrarily choose the redacted portion and prove to *Business partner* that the original content is within the range indicated by the inequality. Alternatively, instead of the inequality, *Customer* can disclose the exact content to *Business partner*. Moreover, *Business partner* establishes the fact through the proof that:

a. The document was signed by some *ID provider* subordinate to *Root CA* while keeping the attributes of *Local CA* and the *ID provider* undisclosed to the verifier.
b. Some parts of the document have been redacted.
c. Some of the redacted parts are shown to be within the numerical range indicated by the inequality.
d. The original content, even if it remains redacted, was signed with some acknowledgment by the *ID provider*.

Up to this point, processing by one ID Provider belonging to one organization has been described. In an actual PKI, one customer obtains attribute certificates from multiple ID providers belonging to multiple organizations. If the first organization is the National ID Card Issuer, then *the process* for obtaining other attribute certificates from other organizations is as follows:

a. Assume *Customer*'s master ID, i.e., *MA,1 is create*d by *ID provider* in *Organization A*.
b. Let $mA,1{=}gRA,1hMA,1$, $\vec{m}$ A= $(mA,1, ..., mA,nA)$
c. *ID provider SA* belonging to *Organization A* creates
$\sigma A = SIGN2(skA;\ \vec{m}\ A)$.
d. If the *ID provider SA* is a *National ID card issuer*, MA,1 of $mA,1{=}gRA,1hMA,1$ will be the personal identification number issued by the country for everyone. Also, RA,1 is used as a master key, and only *Customer* keeps it secret. And MA,1 is called the master ID. The *National ID card issuer* confirms the value of MA,1 by commitment and then creates (pkA, $\vec{m}\ A$, σA). It is put into a national ID card of *Customer*.
e. Similarly, the *ID providers* of *Organization B*, *Organization C*, ..., also follow the Step 3 above to create the signatures for *Customer* attributes (pkB, $\vec{m}\ B$, σB) (pkC, $\vec{m}\ C$, σC), ..., required for KYC.

f. When proving attributes to *Business partner* in Step 4, *Customer* first certifies that $\vec{m}$ *A*, $\vec{m}$ *B*, ... can be authenticated with the root public key of each *Organization*. Then, for each element of $\vec{m}$ *A*, $\vec{m}$ *B*, ..., proof of the attribute is performed.

Performance Analysis

The processing time of SK4SC is the total time of DAC and ZKRP. In previous DAC and ZKRP implementations, the scheme was implemented by using old/fragile cryptographic primitives. In other words, it exploits the difficulty of the elliptic curve discrete logarithm problem but is not quantum-resistant. The following processing performance is obtained:

a. An experimental implementation shows the result of running DAC using 128-bit secure elliptic curve (BN128) on PC (Intel Cascade Lake 3.1 GHz CPU) (Bogatov et al., 2020). As an example of the processing time required for one layer and one attribute, the processing time of the prover is 41 ms and the processing time of the verifier is 89 ms. According to this, their processing time increases in proportion to the number of layers, and each additional attribute increases the processing time by about 30%.
b. Another experimental implementation shows the result of running Bulletproofs using 128-bit secure elliptic curve (BN128) on PC (Intel i5-6300U, 2.40GHz CPU) (Bünz et al., 2018). As an example of the processing time required for one range proof, the processing time of the prover is 22.38 ms and the processing time of the verifier is 3.27 ms. According to (Bünz et al., 2018), it is expected to take k times the processing time to perform ZKRP for *mi* (*i* = 1, ..., *k*).
c. Due to the rapid development of quantum computers, high-performance cryptography using old/fragile cryptographic primitives, e.g., ECC, in the past may become obsolete in the next decade or so. A rough outlook is given on the processing performance when replacing the old/fragile cryptographic primitives with new/robust quantum-resistant ones. In order to make SK4SC quantum-resistant, research is progressing to replace primitives from old/fragile to new/robust (Boneh $_e$t. al., 2020) (Kim, 2021) (Cabarcas et. al., 2015). It will take more time for research and development to be completed. A rough estimate for making SK4SC quantum-safe is tens to hundreds of times more processing time than the numbers above for DAC and ZKPR, respectively.

CONCLUSION

As mentioned earlier, legal issues can be solved to some extent using new technology. Therefore, we overview the current situation of developing quantum-resistant cryptography and show a new privacy-enhanced system called SK4SC from a technical perspective. Our final goal is to show the necessary steps to implement SK4SC, referencing the advent of quantum computers.

Google plans to develop a quantum computer with one million qubits by 2029. If this is achieved, we will be close to realizing a quantum computer in the hundreds of millions of qubits that can break RSA and ECC. However, the new features of our SK4SC, Delegatable Anonymous Credentials and Zero-Knowledge Range Proofs, currently use Groth signatures and Bulletproofs cryptography. These ciphers are not resistant to strong quantum computers. Research on improving Groth signatures and

Bulletproofs to be quantum-resistant and their practical applications are in progress (Bünz et al., 2018; Boneh et al., 2018). The competition for the development of this technology is also highly active, and we will encourage the related parties to promote the development of an improved version that can be put to practical use within a few years at the earliest.

In the post-quantum computing environment, we propose the following approach:

a. The general framework of SK4SC, which realizes privacy-enhanced PKI on the blockchain using Delegatable Anonymous Credentials and Zero-Knowledge Range Proofs, should be presented in such forms as a whitepaper and released globally.
b. For the time being, Groth signature and Bulletproofs' cryptographic libraries (e.g., open source) are to be used to demonstrate POC (proof of concept).
c. Until the realization of a powerful quantum computer is in sight (currently, it is unknown whether it will be around 2030 or after 2040), trial applications and evaluations of SK4SC should be carried out.
d. The cipher should be updated when a working quantum-resistant version of Groth signatures and Bulletproofs is completed.

ACKNOWLEDGMENT

This research was supported by a grant from the SECOM Science and Technology Foundation. We would like to express our deepest gratitude to all those who have engaged in and contributed to this valuable discussion.

REFERENCES

Asai, T. (2022a), China Data and Personal Information Laws. Amazon.

Asai, T. (2022b). Japan Data Protection Law (2nd ed.). Amazon.

Bogatov, D., De Caro, A., Elkhiyaoui, K., & Tackmann, B. (2021). Anonymous transactions with revocation and auditing in hyperledger fabric. Cryptology and Network Security, 13099, 435–459. doi:10.1007/978-3-030-92548-2_23

Boneh, D., Glass, D., Krashen, D., Lauter, K., Sharif, S., Silverberg, A., Tibouchi, M., & Zhandry, M. (2020). Multiparty non-interactive *key exchange and more from isogenies on el*liptic curves. Journal of *Mathematical Cryptology,* 14(1), 5–14. doi:10.1515/jmc-2015-0047

Bresson, E., & Stern, J. (2001). Efficient Revocation in Group Signatures. In K. Kim (Ed.), Lecture Notes in Computer Science: Vol. 1992. PublicKey Cryptography. PKC2001. Springer. doi:10.1007/3-540-44586-2_15

Bünz, B., Bootle, J., Boneh, D., Poelstra, A., Wuille, P., & Maxwell, G. (2018). Bulletproofs: Short proofs for confidential transactions and more. 2018 IEEE Symposium on Security an*d Privacy (SP), 315-334. 10.1109/S*P.2018.00020

Cabarcas, D., Demirel, D., Göpfert, F., Lancrenon, J., & Wunderer, T. (2015). An Unconditionally Hiding and Long-Term Binding Post-Quantum Commitment Scheme. IACR *Cryptol. ePrint Arch., 628.*

Camenisch, J., Drijvers, M., & Dubovitskaya, M. (2017). Practical UC-secure delegatable credentials with attributes and their application to blockchain. CCS '17: Proceedings of the 2017 A*CM SIGSAC Conference on Computer and Communication*s Security, 683–699. 10.1145/3133956.3134025

Chainalysis Team. (2022, May 20). North Korean hackers have prolific year as their unlaundered cryptocurrency holdings reach all-time high. Chainalysis. Retrieved October 18, 2022, from https://blog.chainalysis.com/reports/north-korean-hackers-have-prolific-year-as-their-total-unlaundered-cryptocurrency-holdings-reach-all-ti*me-high/*

ECLI:EU:C:2020:559, Case C-311/18, Data Protection Commissioner v Facebook Ireland and Maximillian Schrems ("Schrems II"), 2020

FACTI. (2021). Financ*ial Integrity for Sustainable Development. Report of the High Level Panel on International Financial Acc*ountability, Transparency and Integrity for Achieving the 2030 Agenda, 25 February 2021. Available from https://www.factipanel.org/docpdfs/FACTI_Panel_Report.pdf

Fuchsbauer, G., & Pointcheval, D. (2009). Anonymous Consecutive Delegation of Signing Rights: Unifying Group and Proxy Signatures. In V. Cortier, C. Kirchner, M. Okada, & H. Sak*urada (Eds.), Lecture Notes in Computer Science*: Vol. 5458. Formal to Practical Security. Springer. doi:10.1007/978-3-642-02002-5_6

Fukushima S. (2022). FATF対日相互審査の教訓 [Lessons Learned from the FATF Mutual Review of Japan]. Journal of the Japanese Institute of International Business Law, 50(2).

Hodson, S. (2018). Applying WTO and FTA Disciplines to Data Localization Measures. World Trade Review, 18(4), 579–607. doi:10.1017/S1474745618000277

Kiayias, A. (Ed.). (2011). Topics in Cryptology - CT-RSA *2011. In The Cryptographers*' Track at the RSA Conference 2011, San Francisco, CA, USA, February 14-18, 2011. Proceedings. Springer Berlin Heidelberg. 10.1007/978-*3-642-19074-2*

Kim, S. (2021). Complete Analysis of Implementing Isogeny-Based Cryptography Using Huff Form of Elliptic Curves. IEEE Access. doi:10.1109/*ACCESS.2021.3128515*

Maji, H. K., Prabhakaran, M., & Rosulek, M. (2011). Attribute-Based Signatures. In A. Kiayias (Ed.), Lecture *Notes in Computer Science: Vol. 6558. Topics in Cryptology – CT-RSA 2011. CT-RSA 2011. Springer. doi:10.1007/97*8-3-642-19074-2_24

National Academy of Sciences. (2022), Cryptography and the Intelligence Community The Future of Encryption. The National Academies Press.

NIST. (2022). Call for Additional Digital Signature Schemes for the Post-Quantum Cryptography Standardization Process. Available from the NIST homepage at https://csrc.nist.gov/csrc/media/Projects/*pqc-dig-sig/documents/call-for-proposals-dig-si*g-sept-2022.pdf

Okamoto, T., & Takashima, K. (2020). Decentralized Attribute-Based Encryption and Signatures. IEICE Trans. Fundamentals, E103–A(1). doi:10.1587/transfun.2019CIP0008

Okamura, H. *(2021).* 個人情報保護法の知識 *[Overviewing the Personal Information Protection Act] (5th ed.). Nikkei Bunko Publi*sher.

Sheluchin, A. (2022, June 16). A national digital currency has serious privacy implications. The Conversation. Retrieved November 20, 2022, from https://theconversation.com/a-national-digital-currency-has-serious-privacy-implications-130520

Takaragi, K., Kubota, T., Wohlgemuth, S., Umezawa, K., & Koyanagi, H. (2023). Secure revocation features in EKYC - Privacy protection in Central Bank Digital Currency. IEICE Transactions on Fundamentals of *Electronics, Communications and Computer Sciences, E106.A(3)*, 325-332. Available from https://www.jstage.jst.go.jp/article/transfun/E106.A/3/E106.A_2022CIP0008/_pdf/-char/ja

The United States Government. (2022, March 25). United States and European Commission joint statement on Trans-Atlantic Data Privacy Framework. The White House. Retrieved November 20, 2022, available *from https://www.whitehouse.gov/briefing-room/statements-releases/2022/03/25/united-states-and-*european-commission-joint-statement-on-trans-atlantic-data-privacy-framework/

Treasury, H. M. (2021, October 14). G7 Public policy principles for retail central *bank digital currencies and G7 Finance Ministers and Central Bank governors' statement on cent*ral bank digital currencies and digital payments. Retrieved October 18, 2022, from https://www.gov.uk/government/publications/g7-public-policy-principles-for-retail-central-bank-digital-currencies-and-g7-finance-ministers-and-central-bank-governors-statement-on-cent*ral-bank*

*US Government. (2021). National quantum initiative supplement to the president's FY 2022 budget. Report by the subcommittee on quantum information science committee on science o*f the national science & technology council national quantum initiative supplement to the president's FY 2022 budget, United States Government, Dec. 2021. Available from https://www.quantum.gov/release-of-the-national-quantum-initiative-supplement-to-t*he-presidents-fy-2022-budget/*

APPENDIX I

Delegatable Anonymous Credential (DAC)

The formula for delegatable anonymous credential used in this research is outlined. See [Camenisch et al., 2017] for detail.

In a hierarchical PKI, suppose certificate authorities from layer 1 to layer L-1 issue a chain of public key certificates. At layer L, the end signer (*ID provider*) signs its message.

Let L be a number greater than or equal to 3.

The chain of public key certificates and the signature of the end signer's message is expressed as follows.

$$(\text{pk}0,\ \vec{a}\,1,\ \sigma 1) \rightarrow (\text{pk}1,\ \vec{a}\,2,\ \sigma 2) \rightarrow \ldots \rightarrow (\text{pkL-1},\ \vec{a}\,\text{L},\ \sigma\text{L})$$

For $i = 1, \ldots, L$-1, let the first element of $\vec{a}_i$ be pk_i. That is $a_{i,1} = \text{pk}_i$ and $(\text{pk}_{i-1}, \vec{a}_i, \sigma_i)$ is the public key certificate of i-th level signer.

For $i = L$, let $\vec{a}_i$ be End signer's message $\vec{m}_i$ such that $\vec{a}_i = \vec{m}_i$. The signature σ_L is authenticated with public key pk_{L-1}.

i. Let e be a bilinear map over groups $\mathbb{G}_1$ and $\mathbb{G}_2$:
   ① Each of $\mathbb{G}_1, \mathbb{G}_2, \mathbb{G}_T$ is a group of order p.
   ② e: $\mathbb{G}_1 \times \mathbb{G}_2 \rightarrow \mathbb{G}_T$ is a bilinear map.
   ③ $g1$ is the generator of $\mathbb{G}_1$, g_2 is the generator of $\mathbb{G}_2$, $e(g_1, g_2)$ is the generator of $\mathbb{G}_T$.
ii. Let n_i be the number of elements of vector $\vec{a}_i$ for $i = 1, \ldots, L$.
iii. When i is even,
   $pk_{i-1} \in \mathbb{G}_1$, $\vec{a}_i \in \mathbb{G}_2^{n_i}$, $\sigma_i = SIGN_2(sk_{i-1;}\ \vec{a}_i)$.
   Here $SIGN_2(sk_{i-1;}\ \vec{a}_i)$ is given as:
   ① Let $y_{2,1,\ldots,}\ y_{2,ni}$ be system parameters.
   ② Enter sk_{i-1} that satisfies $pk_{i-1} = g_1^{sk_{i-1}}$
   ③ $\rho \leftarrow \$Z_q^* \quad r_i = g_1^{\rho} \quad s_i = (y_{2,1} \cdot g_2^{sk_{i-1}})^{\frac{1}{\rho}} \quad t_{i,j} = (y_{2,j}^{sk_{i-1}} \cdot a_{i,j})^{\frac{1}{\rho}}$
   ④ $SIGN_2\,(sk_{i-1};\ \vec{a}_i) = (r_i, s_i, t_i 1, \ldots, t_i n_i)$
iv. When i is an odd number, switch $\mathbb{G}$ 1 and $\mathbb{G}$ 2.
   That is, $pk_{i-1} \in \mathbb{G}_2$, $\vec{a}_i \in \mathbb{G}_1^{n_i}$ and $\sigma_i = SIGN_1\,(sk_{i-1};\ a_i)$.
   Here $SIGN_1\,(sk_{i-1};\ \vec{a}_i)$ is given as:
   ① Let $y_{1,1}, \ldots, y_1 n_i$ be system parameters.
   ② Enter sk_{i-1} that satisfies $pk_{i-1} = g_2^{sk_{i-1}}$.
   ③ $\rho \leftarrow \$Z_q^* \quad r_i = g_2^{\rho} \quad s_i = (y_{1,1} \cdot g_1^{sk_{i-1}})^{\frac{1}{\rho}} \quad t_{i,j} = (y_{1,j}^{sk_{i-1}} \cdot a_{i,j})^{\frac{1}{\rho}}$
   ④ $SIGN_1(sk_{i-1};\ \vec{a}_i) = (r_i, s_i, t_i 1, \ldots, t_i n_i)$
   Authentication of signature σ_i using public key pk_{i-1} is given as:

v. When i is an even number, the equation for authentication is

$Verify_2\,(pk_{i-1};\,\sigma_i;\,\vec{a}_i) \rightarrow \{1, 0\}$: on input of pk_{i-1}, σ_i, $\vec{a}_i$, output the result of

$$e(pk_{i-1}, g_2) \cdot e(g_1, y_{2,1}) = e(r_i, s_i) \wedge_{j=1}^{n_i} e(pk_{i-1}, y_{2,j}) = e(r_i, t_{i,j}) \cdot e(g_1, a_{i,j})^{-1}$$

Here, underlined parts are variables other than system parameters.

vi. When i is an odd number, The equation for authentication is

$Verify_1\,(pk_{i-1};\,\sigma_i;\,\vec{a}_i) \rightarrow \{1, 0\}$: on input of pk_{i-1}, σ_i, $\vec{a}_i$, output the result of

$$e(g_1, pk_{i-1}) \cdot e(y_{1,1}, g_2) = e(s_i, r_i) \wedge_{j=1}^{n_i} e(y_{1,j}, p_{ki-1}) = e(t_{i,j}, r_i) \cdot e(a_{i,j}, g_2)^{-1}$$

vii. In traditional signatures, the verifier confirms that the authentication equations hold after all underlined values are also disclosed to the verifier. However, in this research, zero-knowledge proof is applied to prove that the authentication equations hold without disclosing some of the variables to the verifier. This is possible in the form of pairing product equations:

$$\prod_i \prod_j e\left(\alpha_i, \beta_j\right)^{z_{ij}} = 1_{\mathbb{G}_T}$$

where $z_i\, j \in Z_q$ are constants, $\alpha i \in \mathbb{G}_1$ and $\beta_j \in \mathbb{G}_2$ are group elements appearing as variables.

viii. The above $SIGN_1$ and $SIGN_2$ are called Sibling Signatures and have been shown to have the following unforgeability.

Definition (Unforgeability): No adversary with oracle access to $SIGN_1$ and $SIGN_2$ can create a signature that correctly verifies with $Verify_b$, if no $SIGN_b$ query was made for the same message. For every such $b \in 1, 2$ we call it unforgeability-b.

ix. Lemma: Recipient who obtains the above signature can change it to a degenerate signature $SIGN_2$ $(sk_{i-1};\,\vec{a}_i')$ and its proof $Verify_2\,(pk_{i-1};\,\sigma_i'\,;\,\vec{a}_i')$ for $\vec{a}_i' = (a_{i,1,\ldots,}\, a_{i,n'i})$ for any integer n'_i where $1 \leqq n'_i \leqq n_i$. Same for $SIGN_1$ and $Verify_1$.

APPENDIX 2: MATHEMATICAL EXPRESSION OF SK4SC PROCESSING FLOW

Step 1: Root CA to Local CA

a. Let sk_0 and pk_0 be the private key and public key of *Root CA*. pk_0 is known to all participants. This may be established using the same method as publicizing the highest level public key in a conventional PKI. Alternatively, the certificate of pk_0 may be issued as one node of a larger hierarchical PKI, or it may be due to cross certification between different security domains.

b. Let sk_1, pk_1 be the private key and public key of *Local CA*. *Root CA* creates the delegatable anonymous credential $cred_1$ using $SIGN_2$ as the signature creation function as follows:

   $cred_1 = (pk_0, pk_1, \sigma_1)$
   where $\sigma_1 = SIGN_2\,(sk_0; pk_1)$.

   The authentication of σ_1 is given by:

   If $Verify_2\,(pk_0; \sigma_1; pk_1) = 1$, then accept. Otherwise, reject. $Verify_2$ is an equation for digital signature authentication.

Note that the $Verify_2$ equation does not explicitly include the private key sk_0.
See APPENDIX 1 for the formulas for $SIGN_2$ and $Verify_2$.
Root CA sends $cred_1$ to *Local CA* through a confidential communication path.

Step 2: Local CA to ID Provider

a. Let sk_2, pk_2 be the private key and public key of *ID provider*. *Local CA* creates the delegatable anonymous credential $cred_2$ using $SIGN_1$:
   $cred_2 = (pk_1, pk_2, \sigma_2)$
   where $\sigma_2 = SIGN_1\,(sk_1; pk_2)$.
   The authentication of σ_2 is given by:
   If $Verify_1\,(pk_1; \sigma_2; pk_2) = 1$, then accept. Otherwise, reject.
   See APPENDIX 1 for the formulas for $SIGN_1$ and $Verify_1$.

b. *Local CA* sends $cred_1$ and $cred_2$ to *ID provider* through a confidential communication path.

Step 3: ID Provider to Customer

a. *ID Provider* creates a document that lists various attributes collected by *ID Provider* about *Customer*.
From that list, attributes that are potentially relevant to privacy are chosen. Then, they are sorted into those that are converted to Pedersen commitments (see there are k') and those that are not:
$\vec{M} = (M_1, ..., M_{k'}, M_{k'+1}, ..., M_k)$
The former attribute is translated into Pedersen's commitment, yielding a redaction. The attribute M_i of the latter is mapped to point *mi* on the elliptic curve in whatever way is appropriate for the application:
$\vec{m} = \left(g^{R_1} h^{M_1}, ..., g^{R_{k'}} h^{M_{k'}}, m_{k'+1}, ..., m_k\right)$
where *R*1, ..., *Rk*' are random numbers generated by either Identity Provider or *Customer*.
   i. There is a case where *ID provider* knows the value of M_i but should not know the value of R_i, such as the master key described later. In that case, *Customer* generates random number R_i and calculates $m_i = g^{R_i} h^{M_i}$ and passes m_i and g^{R_i} to *ID provider*. The *ID Provider* confirms that M_i is m_i's commitment by recalculation and incorporates it into $\vec{m}$.
   ii. There is a case, such as the Time Stamp Server described later, where the values of both M_i and R_i should not be disclosed to the Time Stamp Server. In that case, *Customer* passes m_i to the Time Stamp Server hiding the values of M_i and R_i. The Time Stamp Server incorporates m_i into $\vec{m}$.

b. The signed redaction *sred* on $\vec{m}$ is created as:
$sred = (pk_2, \vec{m}, \sigma_3)$
where $\sigma_3 = SIGN_2(sk_2; \vec{m})$. The authentication of σ_3 is given by:
If $Verify_2(pk_2; \sigma_3; \vec{m}) = 1$, then accept. Otherwise, reject.

c. *ID provider* sends $cred_1$, $cred_2$, *sred*, $M_1, ..., M_k$ and random numbers generated by *ID provider* among $R_1, ..., R_k$ to *Customer* through confidential communication path.

Step 4: Customer to Business Partner

a. *Customer* arbitrarily classifies M_i $(i=1, ..., k)$ into three.
① M_i $(i \in D)$ to be disclosed to *Business partner* where *mi* is also disclosed.
② M_i $(i \in ZKRP)$ which is on the right side of the Pedersen commitment $m_i = g^{R_i} h^{M_i}$, where M_i and R_i are not disclosed and *mi* is disclosed
③ M_i $(i \in Hidden)$ where the value of M_i is not disclosed to the verifier, nor is the value of m_i disclosed to the verifier.
- In case ②, let V_i and W_i be the lower and upper bounds such that $V_i \leqq M_i \leqq W_i$. Furthermore, in ②, when ZKRP for m_x is to be proved again and enhanced unlinkability is desired, one-time show approach can be taken:
   i. *Customer* requests to *ID provider* that random numbers are changed α times in advance for the same M_x to create multiple Pedersen commitments and add those in signature σ_3.

ii. *Customer* degenerates m_i ($i = 1, ..., k+\alpha$) to *mi* ($i = 1, ... k$) (see *Lemma* in APPENDIX 1). The discarded elements m_i ($i = k+1, ..., k+\alpha$) are deposited for the second and subsequent proofs of m_x.

Here achieving unlinkability, which is one of the requirements for anonymity, is one of the purposes of ZKRP. Anonymity is defined as the state in which users can use resources and services without disclosing their attributes. Unlinkability means that it is impossible to determine whether a plurality of sessions is by the same person. Linkability is the opposite of this.

- In case ③, if M_y in $m_y = g^{R_y} h^{M_y}$ is the master ID and classified as $y \in Hidden$, *Business partner* (verifier) verifies that *Customer* (prover) is the rightful holder of the National ID card without knowing the National ID number *My* itself Moreover, since the zero-knowledge proof value $\mathcal{B}$ changes randomly each time as shown below, unlinkability can be maintained no matter how many times this process is repeated. In other words, if the master ID is used in condition ③, multi-show can be performed while maintaining unlinkability.

b. *Customer* creates *NIZK* (Non-Interactive Zero-Knowledge proof) and sends it to *Business partner*.

$\mathcal{B} \leftarrow\$ NIZK\{pk_1, pk_2, \sigma_1, \sigma_2, \sigma_3,$
$\langle R_i, M_i\rangle_{i\in ZKRP}, \langle m_i, M_i\rangle_{i\in \text{Hidden}}:$
$Verify_2(pk_0; \sigma_1; pk_1) = 1,$
$Verify_1(pk_1; \sigma_2; pk_2) = 1,$
$Verify_2(pk_2; \sigma_3; \vec{m}) = 1,$
$\langle mi = g^{R_i} h^{M_i}, V_i \leqq M_i \leqq W_i\rangle i \in ZKRP\}$

Here, \$ tangent to $\leftarrow$ indicates that the output is obtained via a randomization process.

That means the values of $pk_1, pk_2, \sigma_1, \sigma_2, \sigma_3, \langle R_i, M_i\rangle i \in ZKRP$ and $\langle m_i, M_i\rangle i \in Hidden$ are hidden to the verifier while other values are disclosed to the verifier. And the verifier knows in advance the formula to the right of the colon.

The value of $\mathcal{B}$ is sent to the verifier (*Business partner*). First, the verifier is convinced that the three verification equations for $Verify_2$ and $Verify_1$ hold. That is, all variables appearing in the equations are arguments of the pairing product equations, making this proof plausible (see APPENDIX 1).

Second, see that the *elements* of $\vec{m}$, i.e. m_i, $i \in ZKRP$, are disclosed to the verifier. Bulletproofs [7] based on the Pedersen commitments $m_i = g^{R_i} h^{M_i}$, $i \in ZKRP$, is applied to prove $V_i \leqq M_i \leqq W_i$ without disclosing the values of R_i and M_i ($i \in ZKRP$) to the verifier.

Step 5: Business Partner

If the verification result of $\mathcal{B}$ is positive, *Business partner* convinces the following:

a. $\vec{m}$ was signed by some *ID provider* subordinate to *Root CA* while keeping the attributes of *Local CA* and the *ID provider* undisclosed to the verifier.
b. The plaintext M_i and R_i of m_i are disclosed for $i \in D$.
c. The range of plaintext M_i of m_i is $V_i \leqq M_i \leqq W_i$ for $i \in ZKRP$.
d. The plaintext M_i and m_i are hidden for $i \in Hidden$.

The above outlines the mathematical expression of SK4SC.

APPENDIX 3: PROOF OF NON-REVOCATION

The public key pk_A can be revoked at any time. The information registered in the national ID card expires due to various reasons such as loss of the card, change of address, family name and so on.

Here, if the registration information to be authenticated by the public key pk_A is to be revoked, the registration information is revoked by revoking the public key pk_A itself.

Let T_A be the time when the time stamp was stamped on the signature of the registration information. Let T_N be the time when the non-revocation of the public key pk_A is authorized by the *Non-revocation certification authority* S_N. In the case of $T_A \leqq T_N$, the public key pk_A and the registration information are determined to be non-revoked (see Figure 3).

Figure 3. Example non-revocation

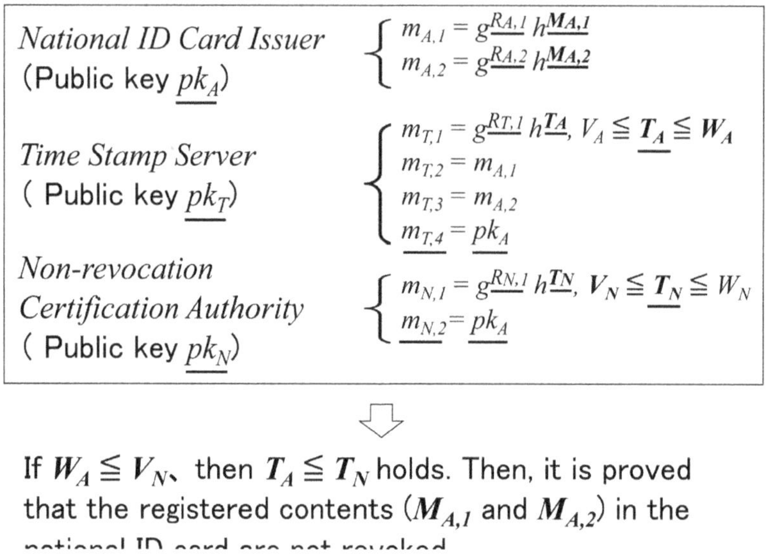

In order to realize this mechanism, *Organization T* that provide the time stamp service is placed in the horizontal direction of Figure 1. In addition, *Organization N* that issues a non-revocation certificate of a public key is set up.

The procedure is as follows:

a. Let (sk_T, pk_T) be the set of private key and public key of the *Time stamp server* S_T.
   Let (sk_N, pk_N) be the pair of private key and public key of the *Non-revocation certification authority* S_N.
b. When the *Time stamp server* S_T receives a time stamp request for $\sigma_A = SIGN_2(sk_A; \vec{m}_A)$, it authenticates the signature and creates the following time stamp.
   $\sigma_T \leftarrow\$ SIGN_2(sk_T; \vec{m}_A, g^R h^{T_A}, pk_A)$
   Here, R is a newly generated random number, and TA is the time when the time stamp is stamped.

Here, the signature layer i is an even number. Equation $Verify_2$ for authentication of σ_A is obtained according to the formula in APPENDIX 1:

$e(pk_A, g_{A,2}) \cdot e(g_{A,1}, y_{A,2,1}) = e(r_{A,1}, s_{A,1}) \wedge_{j=1}^{n_{A,i}} e(p_{kA}, y_{A,2,j}) = e(r_{A,i}, t_{A,i,j}) \cdot e(g_{A,1}, a_{A,i,j})^{-1}$

where $a_{A,i,j} = m_j$ for $j = 1, ..., n_{A,i}$. The first subscript of each symbol indicates the *Organization* name to which the signer belongs.

The equation $Verify_2$ for authentication of σ_T is obtained as

$e(pk_T, g_{T,2}) \cdot e(g_{T,1}, y_{T,2,1}) = e(r_{T,1}, s_{T,1}) \wedge_{j=1}^{n_{T,i}} e(p_{kT}, y_{T,2,j}) = e(r_{T,i}, t_{T,i,j}) \cdot e(g_{T,1}, a_{T,i,j})^{-1}$

where $a_{T,i,j} = a_{A,i,j}$ for $j = 1, ..., n_{A,i}$, $a_{T,i,j} = g^R h^{T_A}$ for $j = n_{A,i+1}$, $a_{T,i,j} = pk_A$ for $j = n_{A,i}+2$ and $n_{T,i} = n_{A,i}+2$

c. The *Non-revocation certification authority SN* is requested to issue a non-revocation certificate. If the non-revocation of the public key pk_A is confirmed, the non-revocation proof of the public key is created as

$\sigma_N = SIGN_2\,(sk_N\text{: } g^{R'}h^{T_N}\ ,\ pk_A)$

Here T_N is the time when the non-revocation proof is created.

The equation $Verify_2$ for authentication of σ_N is obtained as

$e(pk_N, g_{N,2}) \cdot e(g_{N,1}, y_{N,2,1}) = (r_{N,1}, s_{N,1}) \wedge_{j=1}^{n_{N,i}} e(pk_N, y_{N,2,j}) = e(r_{N,i}, t_{N,i,j}) \cdot e(g_{N,1}, a_{N,i,j})^{-1}$

where $a_N,i,1 = g^{R'}h^{T_N}$, $a_N,i,2 = pk_A$ and $n_N,i = 2$

d. Combine the above three $Verify_2$ equations into pairing product equations as shown in APPENDIX 1:

$$\prod_i \prod_j e(\alpha_i, \beta_i)^{z_{ij}} = 1_{\mathbb{G}_T}$$

Then by applying zero-knowledge proof, it is proved to the verifier that the three $Verify_2$ equations hold, i.e., the above three signatures are authenticated.

e. Then ZKR_P is applied.

Let n be the number of all Pedersen commitments thus obtained (see Figure 4).

Figure 4. Pedersen commitment with range proof

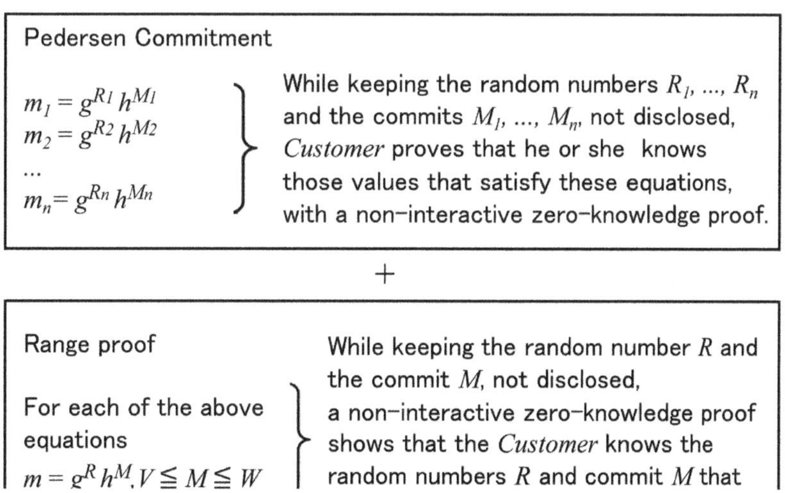

Each Pedersen commitment is associated with the evidence that one of the *ID providers* belonging to each *Organization* whose name is known to the verifier signed it. *Customer* then creates a zero-knowledge proof that he or she knows all the variables that satisfy these formulas. In addition, range proofs are given.

f. Note especially a_T, i, $j = g^R h^{T_A}$ for $j = n_{A,i}+1$ is signed by *Time stamp server* S_T and $a_{T,i,1} = g^{R'} h^{T_N}$ is signed by *Non-revocation certification authority* S_N. If $T_A \leqq T_N$ is proved, then it is proved that the signature has not been revoked. *Customer* wants to prove it. Then *Customer* selects the values V_A, W_A, V_N and W_N to satisfy $V_A \leqq T_A \leqq W_A \leqq V_N \leqq T_N \leqq W_N$ to give zero-knowledge proof $\mathscr{B}$ according to the *NIZK* formula in Step 4-b above.

g. So far, a method of giving a non-revocation proof was described. When giving a proof of revocation, a separate proof of revocation authority S_R and its private key / public key pair (sk_R, pk_R) are provided. The time T_R when the *ID provider*'s public key pk_A is revoked is signed by the *Revocation certification authority* S_R. If $T_R < T_A$ is verified by *NIZK*, then the signature (pk_A; σ_A; $\vec{m}_A$) is proved to be revoked.

APPENDIX 4: SECURITY ANALYSIS

The DAC used in this research is built from signature schemes and zero-knowledge proofs and is proven to be secure in the Universal Composability (UC) framework (Camenisch et al., 2017). In particular, the signatures have been proven to have unforgeability as defined in APPENDIX 1.

Another feature of this research, Bulletproofs, is a short non-interactive ZKRP that does not require a trustworthy setup. Bulletproofs is used to convince the verifier that the encrypted plaintext is well formed. For example, it proves that an encrypted number is within a certain range but reveals nothing else about that number. It has knowledge soundness and zero-knowledgeness (Bünz et al., 2018). In practice, it is most often implemented in non-interactive versions obtained using the Fiat-Shamir transform despite the lack of formal proof of security for this setting.

Achieving unlinkability, which is one of the requirements for anonymity, is one of the purposes of ZKRP. Anonymity is defined as the state in which users can use resources and services without disclosing their attributes. Unlinkability means that it is impossible to determine whether a plurality of sessions is by the same person. Linkability is the opposite of this.

In this research, DAC and ZKRP are combined to realize the function of non-revocation proof. This combination raises two issues from a security and privacy standpoint.

First, the time when ID provider created the signature is not disclosed, but the time range (V_A, W_A) stamped by the timestamp server is disclosed. Therefore, the fact that the signature was created prior to W_A is known to Business partner.

Second, the value of mi for Pedersen Commitment is known to the business partner, so it's reuse may break unlinkability. Regarding the former, in most uses of eKYC, when issuing a non-revocation certificate, it is customary to include the date of issue. So, for many uses of KYC, this is not a problem. As for the latter, unlinkability is enhanced as in Case 2 below by using the proof of mi only one-time and having ID provider issue a new mi' each time.

As security proof, assume that an attacker is looking at data sent from Customer to Business partner.

The attacker can get the values of ① M_i ($i \in D$), ② m_i ($i \in ZKRP$), $\langle V_i, W_i \rangle_{i \in ZKRP}$, pk0 and $\mathscr{B}$ as described in Step 4 of APPENDIX 2 at the first use of signature from ID provider. At this time, ① M_i $i \in D$), ② $\langle V_i, W_i \rangle i \in$ ZKRP and pk0 are obtained as statistically meaningful data that leads to linkability by the attacker. ② m_i ($i \in ZKRP$) and $\mathscr{B}$ are meaningless random number data for the attacker.

If the same signature from the ID provider is reused again, the attacker will know that the same m_i ($i \in ZKRP$) has been used again at a different time or place. This may be added as a significant statistical data for linkability.

Case 1: If they accept this, reuse the signature.

Case 2: If they desire enhanced unlinkability as described in Step 3 ②, change m_i ($i \in ZKRP$) to m_i' ($i \in ZKRP$) by *another ra*ndom number and then reuse it. As a result, statistical data significant for linkability does not increase except for ① M_i ($i \in D$), ② $\langle V_i, W_i \rangle i \in$ ZKRP and pk_0. eKYC users should check the possibility of such linkability and select the acceptable method (Case 1 or 2).

Chapter 11
Disruptive Technology of the 21st Century:
How Blockchain Can Improve Electoral Voting

Keisha Williams
Eastern Kentucky University, USA

ABSTRACT

Blockchain is arguably the most disruptive technology of the 21st century. Although most known for its role in cryptocurrency, blockchain is utilized in several innovative ways. For example, an artist's copyright, royalties, and licensing information can be stored using smart contracts, a blockchain-backed technology. This chapter will address the role of blockchain-backed technology throughout the voting process and how state government plans of action may differ as they attempt to regulate and adopt this innovative yet disruptive technology. Additionally, it will compare different e-voting platforms and their underlying frameworks, languages, and consensus protocols. Ultimately, this chapter reveals how electoral practitioners and policymakers have responded to regulating blockchain-backed technologies and how cryptographic framework and technological advancements directly influence the e-voting ecosystem.

INTRODUCTION

Cryptocurrency and e-Voting differ in utility. E-Voting systems need cryptocurrency to perform as a secure remote application. In the voting context, blockchain's immutability indicates that falsified voting would make it increasingly more challenging. For instance, each citizen would be given one cryptographic coin, like Bitcoin, assigned a unique hash number. Each validated voter would then be given a special wallet address and send their coin to the address of whichever political candidate they wish (Hayes, 2022). Perhaps blockchain's transparent and traceable nature would eliminate the need for human vote counting and the possibility of fraudulent activities.

DOI: 10.4018/978-1-6684-6429-8.ch011

So, what exactly is blockchain? In layperson's terms, blockchain is an immutable digital public ledger that utilizes cryptographic algorithms as security mechanisms. Although most known for its role in cryptocurrency, blockchain is used in several innovative ways. For example, a music artist's copyright, royalties, and licensing information can be stored using smart contracts, a blockchain-backed technology. Artists can utilize blockchain-backed technology to communicate directly with their fan base and have complete control over pricing and distribution, thus eliminating the intermediary.

Smart contracts, for example, are blockchain-backed computerized transaction protocols built into standardized contracts and traded with low transaction costs due to computerized analysis (Frankenfield, 2022). They can also act as an unconditionally trusted third party in conjunction with the immutability provided by a blockchain data structure and distributed consensus (Jafar et al., 2021). Consequently, an advantage of blockchain's decentralized capabilities is that all participants do not need an intermediary, as it acts as a self-executing line of code. Lastly, blockchain-backed technologies are increasingly becoming recognized as a secure voting alternative, providing traceable, transparent, and irreversible transactions. For example, in Iowa, policymakers recently approved a bill, SF541, to acknowledge blockchain smart contracts legally. Ultimately, the bill allows legal recognition of transfers and registrations developed using blockchain smart contracts. The bill, SF541, states that "a contract shall not be denied legal effect or enforceability solely because the contract is a smart contract or contains a smart contract provision." (Thomson, 2021).

In West Virginia, troops serving overseas can vote securely via a blockchain-backed voting application starting in 2019 (Reiff, 2020). Voatz, an e-Voting platform, is a smartphone-based voting system on blockchain to vote remotely and anonymously. Voters confirm their ideal candidates and themselves on the application. They give proof by an image and their identification, including biometric confirmation using a distinctive signature such as fingerprints or retinal scans (Reiff, 2020). Voatz addresses the timeliness and voter anonymity concerns associated with absentee voting. Depending on the success of the initial application launch, that state will plan to expand voter access to the voting applications to residents in all fifty-five counties. West Virginia's e-Voting effectiveness is discussed further in the case studies section of this chapter.

Unsurprisingly, with change comes pushback. The new concept of e-Voting can be viewed as unstable, as the most common association only extends to the market volatility cryptocurrencies face as a commodity. However, e-Voting platforms only incorporate cryptography as a security measure and don't face market volatility like cryptocurrencies. The crucial point to remember in contrasting cryptocurrency with e-Voting utility is that they both only use cryptography to record a ledger of transactions transparently (Hayes, 2022). So, cryptocurrencies like Bitcoin, or e-Voting platforms like Voatz, couldn't exist without an underlying cryptographic framework. Since the two are directly related, the technical development of cryptographic frameworks that support cryptocurrency naturally prepares for a more improved e-Voting ecosystem. Therefore, electoral practitioners and policymakers should guide how to regulate frameworks legally to gain support for improving e-Voting effectiveness. However, e-Voting improvement as a single priority may stand second to the legal framework development of cryptocurrencies.

FTX Scandal: The Next Enron?

Cryptocurrency exchange FTX founder, Sam Bankman-Fried, regarded as the Democratic party's second biggest donor, is facing what Bloomberg called the "biggest one-day collapse" it had ever seen among billionaires (Mollman, 2022). The total $39.8 million in campaign donations are causing some Democrats to weigh in on the issues around FTX. Senator Elizabeth Warren (D) tweeted on 11/17/2022, "The collapse of one of the largest crypto platforms shows how much of the industry appears to be smoke and mirrors. We need more aggressive enforcement, and I will keep pushing @SECGov to enforce the law to protect consumers and financial stability (Mollman, 2022). Sens. Elizabeth Warren (D) requests Sam Bankman-Fried to release the plethora of documents for a "complete and transparent accounting" of business practices and financial activities, which she hopes will reveal the blockchain-backed exchange discrepancies (Egan, 2022).

After a long day of testimony, as of December 14, 2022, Sam Bankman-Fried has been arrested. In response, Senator Elizabeth Warren (D) proposed a bipartisan bill to implement mandates on crypto money laundering. The Digital Asset Anti-Money Laundering Act addresses money laundering by joining the digital asset ecosystem into compliance with the current anti-money laundering system in the international financial system. Unambiguously, the Act would guide FinCEN in finalizing and implementing a rule proposed in 2020. The regulation requires banks and money services to keep records, file reports, and verify customer and counterparty identities linked to non-hosted wallets or ones in jurisdictions non-compliant with the Bank Secrecy Act (Egan, 2022).

It should be noted that the Bahaimin FTX exchange was not regulated by the SEC. According to Brad Garlinghouse, Ripple CEO, the problem is that the SEC failed to create regulatory transparency in the US. Therefore, many American investors, and 95% of trading activity, went offshore (PYMNTS, 2022). Ripple CEO Garlinghouse continued by saying companies lack guidance on functioning because firms in the US require a system that adopts trustworthiness and transparency. Using Singapore's regulatory framework as an example. "Singapore can regulate their crypto ecosystem because they "have the work experience and know all cryptocurrencies aren't securities." Co-founder Jeremy Allaire also said, "The lack of a clear and sound regulatory framework for US crypto markets has left people exposed to the supervisory powers of the Bahamas and who knows where for others" (PYMNTS, 2022).

Garrison, a Reuters columnist, alleges that while FTX ensnared US investors in its attractive yield-bearing accounts, it was a "Ponzi scheme" that took investor resources and shuffled them to related entities as a facade of liquidity. Allegedly, Bankman-Fried and FTX promoters intentionally conspired to mislead investors, violated Florida state laws requiring securities to be registered, and prohibited discriminatory business practices (Person & Godoy, 2022).

What have US regulators done to address such illegal financial turmoil? Regulators in the US responded to this historic "biggest one-day collapse" primarily through fraud detection, enforcement, anti-money laundering, and general criminal deterrence. Congress is mindful of the risks of cryptocurrency. Bitcoin is a recognized illegal harbor and facilitates nefarious activity like "donating to Wikileaks."

Recently, the Commodity Futures Trading Commission ("CFTC") made an example of an alleged Ponzi scheme composed through Gelfman Blueprint, Inc. and requested information on leverage and margin requirements from Coinbase concerning a June 2017 "flash crash" on its GDAX platform (Clements, 2018). In response, the CFTC has approved registration orders to LedgerX as derivatives clearing organization and exchange execution facility and will allow the self-certification of Bitcoin derivative products by the Chicago Mercantile Exchange and the CBOE Futures Exchange as well as Bitcoin binary options on the Cantor Exchange. Moreover, the CFTC pursued administration actions against BFXNA Inc., Coinflip, Inc. (Derivabit), and Francisco Riordan for functioning an unauthorized online facility connecting Bitcoin options counterparties (Clements, 2018).

Moreover, the Department of Justice pursued criminal charges in money laundering using cryptocurrency as a threat to drug enforcement measures. The Financial Crimes Enforcement Network ("FinCEN"), an extension of the US Treasury Department, issued supervision as applications are integrated into the Bank Secrecy Act with other anti-money-laundering laws for cryptocurrencies (Clements, 2018). This Act necessitates "administrators" or "exchangers" of virtual currency to be subject to "money service business" ("MSB") regulations and licensing (Clements, 2018).

Unfortunately, the collapse of FTX isn't the first of its kind. Although the circumstance isn't ideal for investors and can be anxiety-inducing, the crypto industry shows hope in the long run. Enron disrupted our lives momentarily but had no impact on the energy markets. Blockchain technology will undoubtedly make the American economy more competitive and productive if properly regulated and implemented (Clements, 2018). One thing is for sure before any real accountability can be issued, the US will have to reach a bipartisan agreement and perhaps a more stable framework. Congress should begin by passing bipartisan legislation that creates a sensible regulatory framework for digital stablecoins backed by the US dollar. Regulation is important because a well-regulated stablecoin backed by the US dollar and other high-quality, liquid assets could become the global default payment system. Considering the US already governs pricing in other commodities, why not cryptocurrency? FTXs' collapse can't cause us to abandon blockchain-backed technologies' great promise and potential.

Policymakers Ignite International Blockchain Regulation

The downfall of FTX reminds us of the dire need to regulate blockchain-backed technologies. Blockchain's strongest association is cryptocurrency. Policymakers can provide a comprehensive international framework to help unblur cryptocurrency and blockchain utility differences. Even the US Treasury is beginning to acknowledge the necessity of regulation on blockchain-backed technologies, now more significantly in the wake of the FTX scandal. Nellie Liang, the US Treasury Department's undersecretary for domestic finance, believes digital assets have the potential to reform payments fundamentally when implemented strategically. The regulation of stablecoins can catalyze a comprehensive digital asset framework (Hamilton, 2022). Stablecoins are cryptocurrencies combined with other currencies, commodities, or financial instruments (Hayes, 2022). Essentially, they provide a comparable substitute to high-volatility cryptocurrencies, like Ethereum and Bitcoin, which are known to fluctuate as a medium of exchange (Hayes, 2022). Nellie Liang speaks to its potential to affect the broader financial system. The US Treasury representative reiterates the organization's intentions to integrate it into the banking system without enduring the regulation of depository insurance.

Organizations that adopt stablecoin into their financial infrastructure will need to be regulated as a bank and require a new regulatory structure from Congress. Even so, traditional banking agencies that once recommended that issuers be regulated as banks now suggest fintech, blockchain-backed technologies to solve the depository insurance inquiry. "These are more than money market funds," Liang said of stablecoin issuers, and so they need novel regulations that deal with them as payments companies, too" (Hamilton, 2022). Although there are no specific details, a potential House Financial Services Committee bill has leaned toward setting up rules for both banks and nonbanks." These are more than money market funds," Liang said of stablecoin issuers. "So they need novel regulations that deal with them as payments companies, too" (Hamilton, 2022). Ultimately, dialogs are still embryonic and will require bipartisan policies. Could stablecoins used in a wholesale capacity solve cross-border payment issues? According to Michelle Neal, Federal Reserve Markets Group Head, one of the most challenging obstacles to cross-border payment processing involves time. Research efforts conducted by her company see promise in using a central bank digital dollar to speed up settlement time in foreign exchange markets. The Federal

Reserve is reportedly developing a stablecoin; however, like most blockchain-backed regulations, lack of support from elected officials often results in paralysis (Person & Derby, 2022).

Recent macroeconomic studies have proven regulatory ambiguity to hinder policymakers and lobbyists from achieving bipartisanship support. Therefore, it can be viewed as passive for continued postponement of regulation because it will allow systems to remain vulnerable to obsoletion and unable to detect illegal cryptocurrency activity (WEF, 2022). Matthew Blake, Head of Shaping the Future of Financial and Monetary Systems World Economic Forum, says, “Effective regulations are needed to help mitigate the risks associated with digital currencies while realizing the benefits” (WEF, 2022).

How can cryptocurrency and stable coin’s established regulatory structure influence e-Voting? The World Economic Forum has proposed recommendations for regulatory best practices for cryptocurrency and stablecoins, revealing improved e-Voting implications. The macroeconomic ecosystem views blockchain-backed cryptocurrency not as a replacement but as complementary services that can only improve existing systems. In this context, the WEF is proposing the following:

- An international classification framework. Regarding e-Voting, this can help voters avoid technical barriers such as VPNs and proxies, further adding to the e-Voting ecosystem.
- Coordinate with other governments to mitigate regulatory arbitrage and reduce the negative economic impacts of choices made by developed economies on emerging economies. Although e-Voing removes the need for paper ballots, human skill-related responsibilities are still required to run an effective digitized public service.
- Include crypto and stablecoins in monetary financial statistics. Transparent reporting will help reveal which politicians are accepting cryptocurrency and can help voters determine their ideal candidate.
- Consider these when designing regulations as economic projections become possible and qualitative assessments are completed. Policymakers should also account for the differences between types of cryptocurrency and stablecoins when developing rules.

Thus, WEF would agree that cryptocurrencies and stablecoins should be regulated in economies (WEF, 2022). Nevertheless, it should be noted that different regulatory approaches may have variable impacts in each country or jurisdiction. Regulatory frameworks should be created to back accountable innovation while responsibly handling risks, particularly risks that can disrupt the economic and financial systems as we know them. Perhaps this is the answer to gaining widespread digital asset framework adoption.

On March 9, 2022, President Biden signed an Executive Order outlining potential risks and ways to capitalize on the potential benefits of digital assets and their underlying technology. The Ensuring Responsible Development of Digital Assets EO addresses the development of frameworks and policy recommendations that advance the six key priorities identified in the EO: consumer and investor protection; promoting financial stability; countering illicit finance; US leadership in the global financial system and economic competitiveness; financial inclusion; and responsible innovation. (WhiteHouse. Gov). Ultimately, effective regulations are needed to help assuage the risks linked with digital assets while simultaneously realizing their benefits. In response to the Executive Order, the current digital asset framework is beginning to attract interest from various supporters across the political spectrum. Once the legal and political constraints have been established, the focus can shift toward discovering the technical limitations an e-Voting platform must respect.

Technical limitations may be found in the digital divide to access. If e-Voitng occurs, it is necessary to establish proper support for citizens without a computer or smartphone so they can still participate in the electoral voting process. Many elderly, homeless, disabled, and low-income citizens are known to be excluded when public services are digitized. Without effective digital technology democratization, institutions may face discriminatory allegations as many citizens would not have access to vote. Perhaps a step to accomplishing ease of access could begin with addressing the inequalities in internet access while incorporating a universal digital literacy learning initiative. All of which eliminate avoidable barriers to voting.

Policymakers would be well advised to make a bipartisan agreement on how they wish to regulate digital assets promptly. Technological advancements evolve at an accelerated rate, whereas policy requires processes that can be counterproductive to the current ecosystem's technical relevance. Suppose politicians from across the political spectrum work together promptly to establish effective regulation regarding digital assets. In that case, their collective expertise and creative ideas can help develop the frameworks for secure e-Voting systems.

Ultimately, FTX's downfall shouldn't frighten e-Voting advocates. As an unexpected byproduct, malicious crypto actors are forcing policymakers to galvanize and achieve bipartisan agreement on digital asset regulation; as a result, key characteristics of architecture and protections for blockchain-backed e-Voting platforms are being created.

Cryptography: Characteristics of Architecture and Protections

The concept of blockchain technology was introduced via Cryptography, the mathematical and computational practice of encoding and decoding data. (Seth, 2022). Following the Latin root rules, "crypto" means secret, and the Greek "graph" represents "to write," joined, creating the meaning "secret-writing." The blockchain-backed software supports cryptocurrency, provides transparency of the transactions and its participants, eliminates intermediaries, and enables protection from double-spending (Seth, 2022). Cryptography is malleable and has several business applications. Still, it remains consistent in the following offerings: providing network security through an immutable verification system, transparency as a public digital ledger, and decentralized control of generating new units. Decentralization supports the transfer of ownership between its users, eliminating the need for an intermediary.

Developers haven't always seen the value in cryptography. However, in 2008, Satoshi Nakama introduced the concept of "double spending" (Hayes, 2022), thus creating the blockchain system of verification. No one knows if Satoshi is a single person or a secret organization, as the name is considered a pseudonym. However, we can all agree that the disruptive technology developed by this person or group's ideas has remodeled many industries and will likely be around for the long haul (Clements, 2018). Several cryptographic characteristics are applied to e-Voting platforms, and most contribute to the security necessary for secure e-Voting.

Outlining the blockchain sequential steps and security mechanisms may better explain double-spending in cryptocurrency terms (Frankenfield, 2022). Firstly, a transaction's movement is distinguished by a hash, or encrypted number, which includes a timestamp, information from the previous block, and transaction data. Secondly, the block closes after being unanimously verified by the network via PoW. A new block containing the previous hash and a new timestamp is created in response. Lastly, a Bitcoin is awarded to the participant that verified the hash (Frankenfield, 2022).

In a hypothetical world where the concept of fully adopted blockchain-backed voting software exists simultaneously in a world of increasing vulnerability to cyber hacks, widespread voter manipulation would be a foreign idea. Thanks to pseudonym developer(s) Satoshi Nakama, double-spending attempts to mitigate these concerns (Frankenfield, 2022). Double-spending happens when someone alters a blockchain network and inserts a special one to reacquire a cryptocurrency. Double-spending is helpful for e-Voting frameworks since each coin is assigned a unique hash. Electoral practitioners can verify ballot count accuracy by confirming the absence of identical hash sequences.

Proof of Work is a blockchain consensus mechanism using computing power to verify cryptocurrency transactions and add them to the blockchain. Proof of work was initially created as a proposed solution to the growing problem of spam email. (Frankenfield, 2022). Hal Finney eventually adapted and developed the concept in 2004 using the SHA-256 hashing algorithm to secure digital money. Ultimately, Proof-of-Work contributes significantly to e-Voting security protocols.

Proof-of-stake is a cryptocurrency consensus mechanism for processing transactions and creating new blocks in a blockchain (Frankenfield, 2022).). Proof-of-stake adds another level of security to e-Voting. It changes the way blocks are verified using the machines of coin owners, so there doesn't need to be as much computational work done. The owners offer their coins as collateral—staking—for the chance to validate blocks and then become validators (Frankenfield, 2022). Validators are selected randomly to confirm transactions and validate block information. This system randomizes who gets to collect fees rather than using a competitive rewards-based mechanism like proof-of-work (Frankenfield, 2022). Ultimately, it comes down to PoW and PoS having similar attributes but two very different algorithmic approaches. (Frankenfield, 2022).

Practical Byzantine Fault Tolerance (PBFT) is a consensus protocol that allows a distributed system to reach a consensus even when a few nodes demonstrate malicious intent (Feng et al., 2018). For example, if a node shows an invalid block or transaction, every other node will recognize and reject it, blocking invalid transactions from participating on the blockchain. In the end, PBFT adds another level of security to its e-Voting platform.

SHA-256 stands for Secure Hash Algorithm 256-bit, and it's used for cryptographic security. Cryptographic hash algorithms produce irreversible and unique hashes (CryptoWallet, 2022). The larger the number of possible hashes, the smaller the chance that two values will create the same hash. Thus, SHA-256 is one of the most secure hashing functions on the market. The system is so safe that the US government requires its agencies to protect certain sensitive information using SHA-256s. Ultimately, SHA-256 only benefits its e-Voting platform by strengthening its security protocols.

It should be noted that SHA-256 is not an encryption method; rather, SHA-256 is a mathematical function. As mentioned earlier, for double-spending to occur, a secret block must be mined to outdo the original blockchain's creation. It would require supercomputing power because blockchain and consensus mechanisms move so quickly that the modified block would be outdated before it was accepted (CryptoWallet, 2022). Even if it was received, the network would still have passed up the information in the block and would reject it (Frankenfield, 2022). Although the double-spending risk has not been eradicated, blockchain-backed applications can help mitigate such concerns.

Bitcoin vs. Ethereum: Currency and E-Voting Strength

Bitcoin's chief utility is to be a commodity (Hooson, 2022). Taproot, introduced in 2022, is an example of Bitcoin's utility upgrade that enables smart contracts (Sharma, 2022). Taproots smart contracts' purpose is to streamline the processing of transactions and make them more efficient in speed and cost(Sharma, 2022). Multiple signatures can be batched together and validated (Sharma, 2022). Enabling signature aggregation incorporates Schnorr signatures, which are more secure and simple to implement. (Sharma, 2022) .

As the scope of Bitcoin continues to expand, an additional utility extension named the Bitcoin Lightning Network is being developed. The Bitcoin Lightning Network is a second-layer protocol that aims to take transactions off-chain to increase the network speed (Hooson, 2022).

Ethereum extends its utility scope to several business applications, including but not limited to cryptocurrency, DeFi apps, smart contracts, gaming, technology, and NFTs. Consequently, this feature broadens the opportunity of Ethereum's capabilities. Ethereum's programmable capabilities have positioned itself as a marketplace for financial services, games, and apps (Jeffries, 2022).

So, how can Blockchain-backed technology help prevent voter fraud, and how effective is this disruptive tool? Some would argue that this disruptive technology's value lies in its ability to upgrade archaic practices. In recent years, financial institutions have transformed their business application to ensure they can accept cryptocurrency, like bitcoin, because it maintains a more secure and transparent payment processing and banking solution than traditional banking systems. Even so, blockchain technology is the underlying concept that holds the potential to solve issues from banking to voting (Liebkind, 2022).

The power of blockchain lies within its nodes, which verify, process, and record all transactions throughout the chain (Liebkind, 2022). Therefore, blockchain isn't limited to financial transactions. Almost any type of data can be stored in blockchain. Capabilities such as these can be helpful to the county board of elections by providing a means to store voter demographic information securely. Additionally, it can solve the issues associated with absentee ballot voting as well as allow our military troops to participate in the voting process at a much greater rate (Reiff, 2020). Further, since a blockchain node cannot be duplicated, even a potential hacker with access to a private internet terminal cannot affect the chain of blocks fast enough to keep up with the speed of PoW or PoS.

The boom in blockchain-backed cryptocurrency development has made a global network of uniquely identified communities. Since its implementation in 2008, Bitcoin has developed several competitors. Ethereum, established in 2015, has become runner-up in the race for the highest value in cryptocurrency. Although they share similar cryptocurrency features, Ethereum currently takes the lead by facilitating and monetizing the utility of its network from smart contracts to decentralized applications (dApps).

Similarly, Bitcoin and Ethereum are powered by the principle of distributed ledgers and cryptography. Both are traded on online trade exchanges, stored in crypto-wallets, use blockchain-backed technology, and are decentralized (Reiff, 2022).

Examples of E-Voting Platforms

Powered by Bitcoin, and using a PoW census mechanism, Follow My Vote is a company with an online voting platform performed on the blockchain with a polling box audit skill to see democratic development in real-time (Jafar et al., 2021). They enable voters to cast their votes remotely and safely and vote for their political candidate. Lastly, they can use their identification to open the ballot box, locate their ballot, and check that it is correct and that the election results are accurate (Jafar et al., 2021).

Powered by Ethereum and using a PoS census mechanism, Polys is a blockchain-backed online voting platform backed by public crypto algorithms (Jafar et al., 2021). Kaspersky Lab powered, Polys supports the organization of polls by student councils, unions, and associations and helps them make students aware of electoral information. E-Voting with Polys leads to community productivity, improves contact with group leaders, and attracts new supporters. In the end, Polys aims to reduce time and money for local authorities, state governments, and other organizations by aiding them to concentrate on gathering and developing proposals (Jafar et al., 2021).

Powered by a Hyperledger Fabric framework, Voatz uses Python or Javascript as its coding language (Jafar et al., 2021). Unlike its predecessors, Bitcoin and Ethereum, Hyperledger Fabric is a private and permissioned network. In Ethereum, anybody can participate in the network at any time. But Hyperledger has a predefined community of participants, and access to the network is restricted only to them. Essentially, Hyperledger is not an organization, a cryptocurrency network, or a blockchain system. It does not support a cryptocurrency like bitcoin, but it works by providing the necessary infrastructure and standards for developing various blockchain-based systems and applications for industrial use (Voatz, 2023). This company launched a smartphone-based voting system via blockchain to vote remotely and anonymously and validate voting count accuracy.

According to the Follow My Vote experts, Bitcoins' framework may prove a viable option, except it wasn't built for third-party integrations (Follow My Vote, 2021). Consequently, developing applications on Bitcoin's Blockchain requires extensive effort to co-opt numerous structures of Bitcoin into purposes for which they weren't planned, making it challenging for an initial launch. Additionally, Follow My Vote considers the maintainability of Bitcoin in the future as an e-Voting platform and is reviewing a more unpopular functionality like Return By Free to upgrade their chains which allow for larger blocks (Follow My Vote, 2021).

As an e-Voting application, Ethereum tends to centralize validated efforts. Proof of Work incentivizes centralization (Follow My Vote, 2021). Centralized efforts can be seen as problematic as it tends to perpetuate a cycle of sensitive censorship outlining the vulnerabilities of the framework. Ethereum allows arbitrary code execution as part of core blockchain validation, and there are legitimate questions about the security, performance, and economic implications. Ethereum's proclivity near centralization is easier to recognize in more mature Bitcoin, where ten individuals control 95% of the hashing power, for example. As mentioned earlier, Proof of Work provides more security to its e-Voting platform. Ethereum is attempting many new ideas related to application development, but Follow My Vote experts know the challenges of designing a new type of blockchain. Moreover, Ethereum is, as of this writing, still a Proof of Work network (Follow My Vote, 2021).

As of mid-September 2022, the blockchain-backed tech solution company upgraded its consensus security mechanism from PoW to PoS. By upgrading, the company has "eliminated the need for energy-intensive mining and instead secures the network using staked ETH," the Ethereum Foundation said (Bizouati-Kennedy, 2022).

Voatz uses the Practical Byzantine Fault Tolerance (PBFT). The Practical Byzantine Fault Tolerance allows a distributed system to reach a consensus even when a few nodes demonstrate malicious behavior (Feng et al., 2018). Ultimately, this blockchain-backed security mechanism adds significantly to protection in e-Voting security.

CASE STUDIES

Estonia's I-Voting

Separating from the Soviet Union in 1996, the Estonian government was known to be an e-Government trailblazer (Authur, 2014). Ultimately, the newly formed government had limited access to citizen data and saw no other viable source but to fully adopt an e-government framework built utilizing blockchain-backed technologies. Thanks to Estonia's blockchain infrastructure and educational outreach efforts, by 2005, the country had integrated blockchain into its electoral voting process. Security experts quickly pushed back, saying that i-Voting shouldn't be used in the European elections. According to independent researchers, "the e-Voting application had security vulnerabilities that could lead to faked votes or totals" (Authur, 2014).

How is Estonia's i-Voting application structured? Like Follow My Vote and Polys, i-Voting is an e-Voting platform that utilizes elements of cryptography as a structured framework. The Estonian blockchain-backed technology uses tallying to safeguard the voting count, similar to the double postal envelope scheme used in postal voting (Pihlak, 2019). Using an e-ID, voters can digitally sign the digital outer envelope, and before the vote is counted, the system anonymizes the ballot by removing the digital outer layer. From there, votes are calculated on a computer without internal storage, internet, or network connection. Thus, the computer only contains a DVD drive and a smart card reader. The RAM disk is used as a processor and is turned off at the end of the vote-counting process. Once the count is complete, all information vanishes from the computer (Pihlak, 2019).

Estonia shares similar concerns, just as the US faces cyber security concerns due to the 2016 Russian hackers. More than anything else, strengthening the ideology of public trust should be the focal point if governments want to adopt a comprehensive e-Voting platform effectively. Fundamentally, it comes down to data governance and policy," Kaevats says, "currently the US needs something similar to GDPR [General Data Protection Regulation] in Europe. The US needs to up their game; otherwise, the potential likelihood of building trust in a society is not good" (Fillion, 2020). When launching anything disruptive in society, the level of success is primarily determined by the faith citizens have in their electoral voting process (Jafar et al., 2021).

Additionally, absentee ballot as a voting measure has been found to delay vote counting, exacerbating inefficient, traditional voting systems. With cyber security fears about the US elections being manipulated, several engineers recommend a "fiat Accompli" framework approach (Fillion, 2020). Adapting to cultural norms is especially important because government officials should be mindful of cultural nuances and state and local level jurisdiction to integrate effectively (Fillion, 2020).

During the 2019 European Parliament elections, nearly 44% of its citizens voted online (Pihlak, 2019). Overall, Estonia's e-Government was instrumental in helping the country fight COVID-19, giving citizens more to do online and familiarizing them with digital services. As an unintended byproduct, Estonians could evade the several problems that came with COVID-19 (Arthur, 2014). Since the coronavirus outbreak, the Estonian system has been convenient regarding time and public health: it allows people to exercise their democratic rights without risking their health. Although Americans are on the fence about online voting, the Estonian i-Voting platform gives hope, especially when electoral integrity in the US is becoming a pressing issue. However, experts say the US will need to build a stronger foundation of public trust before its citizens can securely cast their ballots from the tips of their fingers. Ultimately, the effectiveness of implementation is determined mainly by the people's faith in the election process (Jafar et al., 2021).

Perhaps the US can begin launching smaller campaigns to help better manage mistakes, or possibly the answer is becoming more transparent about the security mechanisms embedded within their digital asset frameworks. With the proper security mechanisms, like the SHA-256 algorithms, blockchain-backed technologies can solve several threats that reveal themselves in an e-Voting system and many other potential cyber risks. Moreover, blockchain-backed technology eliminates the concern of internet manipulation because although a hacker may have access to an internet terminal, it doesn't have access to the hash nodes. US voters can vote confidently, knowing their data will not be revealed, and officials can have an almost instantaneous identity verification thanks to the SHA -256 hashing algorithm (CryptoWallet, 2022). In the end, there isn't enough quantitative, if any, evidence to prove blockchain-backed technologies are ineffective or vulnerable to cyber threats.

West Virginia's E-Voting: Voatz

In 2018, the nation's first mobile voting pilot for Uniformed Service & Overseas (UOCAVA) voters was launched in West Virginia (Moore & Sawhney, 2019). Using Voatz, an e-Voting platform, authenticated registered voters could receive, mark and submit a secret ballot of the correct style from their hands using smartphone devices. Essentially, each submitted vote was encoded and stored on a geographically distributed and redundant network of blockchain servers managed by the two largest cloud infrastructure providers (Moore & Sawhney, 2019). During the polls closing time, each vote was printed at the county and tabulated on federally certified tabulation equipment. Ultimately, post-election audits were performed on each vote submitted via smartphones.

West Virginia, Secretary of State Mac Warner's goals were to ensure UOCAVA voters can use their smartphones to improve voting convenience and security and lower the burden on county clerks. While Warner's goals were largely met, the first pilot identified some gaps that the Secretary's office and the clerks asked to be implemented in time for the second pilot (Moore & Sawhney, 2019). These included:

Independent Security Evaluation and Post-Election Audits: Four independent security auditors were retained to conduct penetration testing and review the iOS and Android source code, a blockchain infrastructure, and the vendor's corporate procedures. With a voter-centric model in mind, the voters' confidence in the privacy and security of their ballots and the auditability of the process and results were key focal points. Accordingly, the vendor was asked to propose a method of performing post-election audits.

Scalable Election Definition: The ballot styles were programmed manually in the Primary Election. Going into the General election with an uncertain number of participating counties, creating the ballot styles had to be done programmatically. In time for the General Election pilot, ballot definition files of the 999 ballot styles from the primary voting system were produced programmatically. Blockchain infrastructure – to increase performance and security, the number of blockchain servers doubled from 16 to 32, evenly split over multiple geographical locations across the US between the two largest cloud infrastructure providers.

Ability to Spoil a Ballot: While the voter could verify their ballot in the first pilot, the ability to alter a ballot was not available. A process was devised for the second pilot whereby a second ballot could be issued to voters, and only their last ballot would be counted.

Automatic Preparation of Tabulatable Ballots: In the first pilot, ballots had to be manually transcribed, which worried the clerks. An additional capability was developed to print ballots, which automatically could be inserted directly into ballot tabulating machines.

After collecting the results, it was discovered that of the 24 participating counties, 183 voters submitted FCPA forms requesting by "email or online," and 160 (87%) completed the application download (Moore & Sawhney, 2019). Research also revealed that while no data was collected on the 23 (13%) voters who requested but did not download, the likely reasons include: did not have an eligible smartphone, choosing email instead, being in a country that was not enabled in the Apple or Google Play stores. However, of the 160 voters who finished the download, 147 (92%) completed the authentication process (Moore & Sawhney, 2019). In part, the research concludes that institutions should seek to improve overall voter engagement. For example, have the app automatically issue a reminder if the voter hasn't voted by Election Day (Moore & Sawhney, 2019).

Political Climate in LA and New Jersey

Recently, policymakers and politicians have pushed for regulating blockchain-backed digital assets, and several other key stakeholders are becoming increasingly aware. According to a report conducted by Harris Poll, nearly 37% of US voters are considering their representative political stance on cryptocurrency issues, and 81% agree there should be more precise regulation of the industry (Martin, 2022). Since key stakeholders slowly acknowledge the value of blockchain-backed technology, just how holistic is its presence in this decade? The US 2022 midterm election had several blockchain-backed implications. Blockchain was involved as a voting function and used as a funding source to support political allies via cryptocurrency.

Earlier this year in LA, The Fair Political Practices Commission voted to approve new rules allowing donations of digital currencies. Political candidates can accept donations via cryptocurrency only if they convert it into US dollars immediately. They must use a registered cryptocurrency processor to handle the digital financial transaction, which will collect each benefactor's name, address, occupation, and employer (AP, 2022). In other states like New Jersey, political candidates are prohibited from receiving cryptocurrency as a gift. The General Assembly approved a package that provides the structure and regulation needed for digital assets, cryptocurrencies, and blockchain-backed technologies. Additionally, the package prohibits elected officials from accepting non-fungible cryptocurrency s as gifts and demands the state create a blockchain integration program (Donyéa, 2022). New Jersey's "Digital Assets and Blockchain Regulations" Act seeks to improve tourism as lawmakers strategically position themselves in the cryptocurrency industry.

With the US 2022 midterm elections behind us, several voting analysts expect 2022 will be the year policymakers formulate a plan to regulate the digital asset class. Suppose the House and Senate can come to a bipartisan agreement. In that case, we can expect a crypto bill to be tacked onto a government funding package or other necessary legislation (Versprille, 2022). Policymakers are set to establish a framework for stablecoins and one more seeking to resolve jurisdictional issues (Versprille, 2022). The debate may go well into 2023, which can be alarming since blockchain-backed technology in the US currently lacks a comprehensive regulatory framework. Consumers aren't protected from cyber "Ponzi Schemes without a well-defined regulatory system" (Versprille, 2022).

Once policymakers can see the full impact of blockchain-backed technologies in electoral voting, they will understand how the overall development of the cryptographic ecosystem can only add to the e-Voting ecosystem. Consequently, combining effective regulation with a universal acceptance of blockchain's secure capabilities galvanizes competing institutions to push for improved regulation as key societal stakeholders. How can society prepare for increased dependence on digitized public service sectors? US electoral practitioners can prepare voting participants by promoting a universal blockchain learning initiative to address the current inequities of our societal digital divide. By doing so, electoral practitioners can expect e-Voting engagement levels to increase due to having gained sufficient knowledge of blockchain-backed applications thanks to the universal blockchain learning initiative. Fortunately, blockchain is evolving toward 3.0 as technology improves and decentralized applications are developed. Electoral practitioners should prepare a web 3.0 learning guide to maintain public trust, as blockchain-backed technologies provide the security protocols necessary for e-Voting success.

Will Blockchain Be Disruptive Like ARPANET?

Before blockchain is fully adopted into US society, the gradual progress in improving existing systems may take decades, according to HBR (Harvard, 2022).

Web3, or web 3.0, is being touted as the future of the internet. The vision for this new, blockchain-backed web includes cryptocurrencies, NFTs, DAOs, decentralized finance, and more (Stackpole, 2022). It offers a read/write/own version of the web, in which users have a financial stake and more control over the web communities to where they belong (Stackpole, 2022). Web3 promises to transform the online experience as dramatically as PCs and smartphones did. It is not, however, without risk. Some companies have entered the space only to face a backlash over the environmental impact and financial speculation (and potential for fraud) that comes with Web3 projects. And while blockchain is offered as a solution to privacy, centralization, and economic exclusion concerns, it has created new versions of many of these problems (Stackpole, 2022). Companies must consider the risks and the benefits before fully adopting any framework.

The gradual swelling wave of enthusiasts characterizes the transition from Web 2.0 to Web3 by internet services and mobile apps rebuilt on decentralized blockchain technology. Web3 descriptions regularly include an extensive spectrum of innovative technology like Cryptocurrency, DAOs, and digital assets like NFTs or non-fungible cryptocurrency (Stackpole, 2022). Social communities like gaming, the Metaverse, and augmented and virtual reality are associated with Web3 because some virtual worlds rely on blockchain-backed digital assets. "Web3 is repackaging some particular technologies," McCullough told CBS News. "Blockchain went down a tech culture cul-de-sac, and consumers got tired of the hype. Traditional crypto never became currency; NFTs became this cultish thing, and VR has been 'the next big thing for decades. Web3 is a brand that knits all these ideas into a plausible whole" (Patterson, 2022).Web3 isn't just some trending word. Web3 uses blockchains, cryptocurrencies, and NFTs to give authority back to the users in the form of ownership (Ethereum, 2022.)

While still embryonic, Web3 comprises a few core principles: decentralization, permissionless, native payments, and trustless. Decentralization refers to ownership that gets distributed amongst its builders and users—equally, e-Voting's verification by participation protocol. The criterion, often called desirability, makes it possible to assess whether or not a single voter engaged in the election. Permissionless refers to equal access to Web3, and everyone is included. Comparably, e-Voting can eradicate traditional barriers to voting, like internet access inequities and absentee voting discrepancies, further improving electoral voting. Native payments allow Web3 to utilize cryptocurrency for digital transactions instead of relying solely on archaic banking practices. Naturally, the development of cryptocurrency directly impacts the e-Voting ecosystem since each e-Voting platform requires cryptocurrency to function securely.

Lastly, trustless means operating economical and incentive-based mechanisms removing intermediaries. Similarly, e-Voters can avoid intermediaries since they can vote remotely without waiting in line.

Providing ownership over digital assets, Web3 uniquely improves internet-based utility (Ethereum, 2022). With blockchain verification mechanisms, gamers can maintain their digital assets. Thus, preventing gamers from losing their in-game purchase items or position if the game gets deleted (Ethereum, 2022). Additionally, since gamers own it, they're legally allowed to sell it, and the market for that is enormous.

Censorship resistance seeks a proper balance between platforms and helps keep your data on a blockchain. For example, if you leave a social media platform, you get to take your reputation with you. In e-Voting, you get to vote for your ideal candidate in secrecy. In Web 2.0, content creators had to put their trust in platforms not to switch up the rules; however, censorship resistance is native to the Web3 platform.

DAOs provide cryptocurrencies that perform like shares in a company. Ultimately, DAOs allow cryptocurrency holders to use smart contracts as they coordinate the company's decentralized platform as owners and make decisions about its future. Cryptocurrency users vote on how resources get spent, resulting in the code performing the voting conclusion.

Signing in on multiple accounts will soon be a nuance of the past. Web3 lets you control your digital identity with an Ethereum address and ENS profile. Consequently, the Ethereum address provides sign-up ease requiring only one login across secure, censorship-resistant, and anonymous platforms. Traditional payment systems have been known to exclude people lacking banking access. Like DAOs, Web3 utilizes Ethereum cryptocurrency to enable participants to exchange payments securely and doesn't require an intermediary.

Accordingly, educational initiatives must be implemented since Web3's disruptive presence is a fundamentally different and new paradigm. Web 1.0 had a similar education drive in the late 1990s as the world needed to prepare for the contemporary exponents of the world wide web (Ethereum, 2022). Transitioning from web 2.0 to web 3.0 will not be easy, but having educational initiatives to inform Web2 users is essential for its success. Ethereum.org caters to Web3 instruction through its Translation Program, which translates Ethereum into as many coding languages as possible (Ethereum, 2022). Counterarguments concerning specific attributes of Web3, like centralization pushback, are beside the point. It's equivalent to saying the internet is terrible because OnlyFans or Twitter exists. Web3 isn't tricky, but it is different and requires much examination. Indeed, technical advances in cryptography extend to the e-Voting ecosystem since web 3.0 and e-Voting platforms both use blockchain-backed technologies. Perhaps innovative applications such as web 3.0 will accelerate the development of blockchain-backed technologies critical for the e-Voting ecosystem.

Estonia has shown the world how to prepare society for digitized public services. For a sustainable e-Voting platform, the safety of remote participation must be feasible, and society must be adequately prepared as electoral voting becomes digitized.

E-Voting Conclusion

Technology changes over time, and so too, does its framework. The principles that hold the essence of democratic integrity can be maintained while adding an extra layer of security with blockchain-backed technologies. Citizens of a democratic society participate in the voting process because it is viewed as their right to give a voice to their values. E-Voting can remedy where traditional violations of individual rights and freedoms are known to exist. The decentralized architecture can be used to run and support a voting system that is open, fair, and independently verifiable.

Fortunately, e-Voting platforms are gaining a position in the minds of American citizens. E-Voting replicates traditional electronic voting solutions, providing digitally distributed, non-repudiation, and security protection characteristics (Jafar et al., 2021). Not only does blockchain-backed e-Voting have the potential to cut printing and paper costs by nearly 50%, but it also decreases labor costs associated with managing polling stations (Pihlak, 2019). E-Voting would also help reduce public health and environmental concerns during COVID-19 while increasing the accuracy of democratic decisions.

Traditionally, paper-based voting systems were utilized to help citizens build faith and trust in their electoral voting process (Jafar et al., 2021). Policymakers must ensure a universal blockchain learning initiative is integrated into their launching strategy if they expect their citizens to accept their ideas. With transparent security mechanisms, a solid IT infrastructure, and a comprehensive regulatory framework, US citizens may entertain learning more about blockchain's fluidity (Jafar et al., 2021).

It is recommended that more exploration of the scalability of blockchain-backed e-Voting systems is executed. In general, blockchain performs well when there are few users. However, when the network is overcrowded, e-Voting platforms run the risk of spiking usage and time consumption. Voting institutions may want to discuss this during their initial consultation with their e-Voting platform representative to determine the best pricing models based on their organization's service region scope. Since e-Voting is fairly new, this could be the reason the research field is inadequate. Voting institutions are well advised to start with a smaller-scale pilot launch to better manage any errors that may arise. By doing so, voting institutions can plan to modify any discrepancies and plan to relaunch on a larger-scale launch.

Further, electoral practitioners should ensure all citizens have access to e-Voting platforms. Practitioners can start by creating pathways to equal internet access. Politically speaking, bipartisan agreement should remain solution-oriented if policymakers wish to regulate cryptographic frameworks swiftly. Technological advancements evolve at an accelerated rate, making the threat of cyber threats riskier, too. It can be argued that the FTX scandal blurs the lines of cryptocurrency and blockchain utility. However, blockchain advocates can passionately reiterate how cryptocurrency and e-Voting platforms are only alike in using cryptography to record a ledger of transactions transparently. Unlike cryptocurrencies, e-Voting platforms don't face market volatility since they are not commodities.

On an individual level, in ten years, blockchain will likely be dinner table politics where families seemingly grapple with how they've been able to adapt to everyday life with blockchain (McCormick, 2022). Electoral practitioners should ensure citizens understand how blockchain seeks to improve existing systems, like democratic voting systems, not replace them.

REFERENCES

Associated Press. (2022). Californians can soon use cryptocurrencies to donate to state and local candidates. *Los Angeles Times*. Retrieved December 10, 2022, from https://www.latimes.com/california/story/2022-07-21/california-to-allow-cryptocurrency-campaign-donations

Authur, C. (2014). Estonian e-voting shouldn't be used in European elections, say security experts. *The Guardian.* Retrieved November 29, 2022, from https://www.theguardian.com/technology/2014/may/12/estonian-e-voting-security-warning-european-elections-research

Bizouati-Kennedy, Y. (2022). Ethereum Merge: 3 winners and 2 losers. *GOBankingRates.* Retrieved November 18, 2022, from https://www.gobankingrates.com/investing/crypto/ethereum-merge-3-winners-and-2-losers/?utm_campaign=1185176&utm_source=nasdaq.com&utm_content=20&utm_medium=rss

Clements, R. (2018). Assessing the Evolution of Cryptocurrency: Demand Factors, Latent Value, and Regulatory Developments. *Michigan Business & Entrepreneurial Law Review.* Retrieved December 2, 2022, from https://repository.law.umich.edu/cgi/viewcontent.cgi?article=1079&context=mbelr

CryptoWallet. (2021). What is SHA-256? *CryptoWallet.* Retrieved December 6, 2022, from https://cryptowallet.com/glossary/sha-256/

Donyéa, T. (2022). New Jersey legislature moves to make state's cryptocurrency rules 'industry-friendly'. *WHYY.* Retrieved November 14, 2022, from https://whyy.org/articles/new-jersey-legislature-industry-friendly-cryptocurrency-rules/

Egan, M. (2022). Elizabeth Warren unveils bipartisan bill to crack down on crypto money laundering. *CNN.* Retrieved December 14, 2022, from https://www.cnn.com/2022/12/14/business/elizabeth-warren-bipartisan-crypto-crackdown/index.html

Egan, M. (2022). First on CNN: 'greed and deception.' Elizabeth Warren demands Sam Bankman-fried and FTX turn over a trove of Records. *CNN.* Retrieved December 3, 2022, from https://www.cnn.com/2022/11/17/investing/ftx-sbf-elizabeth-warren-crypto/index.html

Ethereum.org. (n.d.) What is web3 and why is it important? *ethereum.org.* Retrieved December 6, 2022, from https://ethereum.org/en/web3/

Feng, L., Zhang, H., Chen, Y., & Lou, L. (2018). Scalable dynamic multi-agent practical byzantine fault-tolerant consensus in permissioned blockchain. *MDPI.* Retrieved December 20, 2022, from https://www.mdpi.com/2076-3417/8/10/1919#:~:text=Although%20the%20PBFT%20algorithm%20is,low%20consensus%20speed%20and%20scalability

Fillion, S. (2020, August 25). If Estonia figured out online voting 15 years ago, why can't the US? *Forbes.* Retrieved December 9, 2022, from https://www.forbes.com/sites/stephaniefillion/2020/08/24/if-estonia-figured-out-online-voting-15-years-ago-why-cant-the-us/?sh=6301f0f015ab

Follow My Vote. (2021). Blockchain comparison, a closer look at Bitcoin, BitShares, and Ethereum. *Follow My Vote.* Retrieved December 10, 2022, from https://followmyvote.com/blockchain-comparison-a-closer-look-at-bitcoin-bitshares-and-ethereum/

Frankenfield, J. (2022a, October 13). What is proof of work (POW) in blockchain? *Investopedia*. Retrieved December 9, 2022, from https://www.investopedia.com/terms/p/proof-work.asp

Frankenfield, J. (2022b, October 15). Understanding double-spending and how to prevent attacks. *Investopedia*. Retrieved November 17, 2022, from https://www.investopedia.com/terms/d/doublespending.asp

Frankenfield, J. (2022c, October 20). What does proof-of-stake (pos) mean in crypto? *Investopedia*. Retrieved December 11, 2022, from https://www.investopedia.com/terms/p/proof-stake-pos.asp

Frankenfield, J. (2022d, September 16). What are smart contracts on the blockchain, and how do they work? *Investopedia*. Retrieved November 15, 2022, from https://www.investopedia.com/terms/s/smart-contracts.asp

Grace. (2022, July 22). Top Metaverse use cases & applications. *LimeChain*. Retrieved November 18, 2022, from https://limechain.tech/blog/top-metaverse-use-cases

Hamilton, J. (2022, July 18). US Treasury open to nonbanks issuing Stablecoins, official says. *CoinDesk Latest Headlines RSS*. Retrieved November 19, 2022, from https://www.coindesk.com/policy/2022/07/18/us-treasury-open-to-nonbanks-issuing-stablecoins-official-says

Hayes, A. (2022, October 4). Stablecoins: Definition, how they work, and types. *Investopedia*. Retrieved December 4, 2022, from https://www.investopedia.com/terms/s/stablecoin.asp#:~:text=Stablecoins%20are%20cryptocurrencies%20that%20attempt,a%20commodity%20such%20as%20gold

Hayes, A. (2022a, September 28). Who is Satoshi Nakamoto? *Investopedia*. Retrieved December 5, 2022, from https://www.investopedia.com/terms/s/satoshi-nakamoto.asp

Hayes, A. (2022b, December 19). Blockchain facts: What is it, how it works, and how it can be used. *Investopedia*. Retrieved February 15, 2023, from https://www.investopedia.com/terms/b/blockchain.asp

Hooson, M. (2022, September 28). Bitcoin vs. Ethereum. *Forbes*. Retrieved December 5, 2022, from https://www.forbes.com/advisor/investing/cryptocurrency/bitcoin-vs-ethereum/

Iansiti, M., & Lakhani, K. R. (2017). The truth about blockchain. *Harvard Business Review*. Retrieved November 15, 2022, from https://hbr.org/2017/01/the-truth-about-blockchain)

Jafar, U., Aziz, M. J. A., & Shukur, Z. (2021, August 31). Blockchain for Electronic Voting System-review and open research challenges. *Sensors*. Retrieved November 11, 2022, from https://www.ncbi.nlm.nih.gov/pmc/articles/PMC8434614/

Jeffries, S. (n.d.). *Ethereum vs. bitcoin: Which crypto is better?* Nasdaq. Retrieved December 8, 2022, from https://www.nasdaq.com/articles/ethereum-vs.-bitcoin%3A-which-crypto-is-better

Liebkind, J. (2022, July 18). How blockchain technology can prevent voter fraud. *Investopedia*. Retrieved December 8, 2022, from https://www.investopedia.com/news/how-blockchain-technology-can-prevent-voter-fraud/

Marr, B. (2022, October 12). Web3 and the future of Music. *Forbes*. Retrieved January 13, 2023, from https://www.forbes.com/sites/bernardmarr/2022/07/13/web3-and-the-future-of-music/?sh=6e0d96402a29

Martin, M. (2022, November 8). Crypto issues are a factor for 37% of US voters, study finds. *Forbes*. Retrieved November 15, 2022, from https://www.forbes.com/sites/mitchellmartin/2022/11/01/crypto-issues-a-factor-for-37-of-us-voters-study-finds/?sh=630c18fd341d

McCormick, P. (2022). The Web3 debate. *The Web3 Debate*. Retrieved December 2, 2022, from https://www.notboring.co/p/the-web3-debate?s=r

Mollman, S. (2022). Oops. Sam Bankman-Fried's implosion took down Democrats' second-biggest donor with it as the party gears up to regulate crypto. *Fortune*. Retrieved December 11, 2022, from https://fortune.com/2022/11/10/sam-bankman-fried-ftx-joe-biden-democratic-party-second-biggest-donor/

Moore, L., & Sawhney, N. (2019). *The West Virginia Mobile Voting Pilot - Under the Hood*. Retrieved December 20, 2022, fromhttps://sos.wv.gov/FormSearch/Elections/Informational/West-Virginia-Mobile-Voting-White-Paper-NASS-Submission.pdf

Patterson, D. (2022). Explaining web3: From the Blockchain and crypto to nfts and the Metaverse. *CBS News*. Retrieved December 4, 2022, from https://www.cbsnews.com/news/web3-blockchain-crypto-nft-metaverse-explainer/

Person, & Derby, M. S. (2022). Ny Fed official: Digital Dollar Could Speed Foreign Exchange Settlement. Reuters. Retrieved December 7, 2022, from https://www.reuters.com/markets/currencies/ny-fed-official-digital-dollar-could-speed-foreign-exchange-settlement-2022-11-04/

Person, & Godoy, J. (2022). FTX's Bankman-Fried, Tom Brady and other celebrity promoters sued by crypto investors. *Reuters*. Retrieved December 6, 2022, from https://www.reuters.com/legal/ftx-founder-bankman-fried-sued-us-court-over-yield-bearing-crypto-accounts-2022-11-16/

Pihlak, H. (2019). *I-voting – the future of elections?* e-Estonia. Retrieved November 19, 2022, from https://e-estonia.com/i-voting-the-future-of-elections/

PYMNTS.com. (2022). *Crypto execs call for clearer regulations in wake of FTX collapse*. Retrieved December 15, 2022, from https://www.pymnts.com/cryptocurrency/2022/crypto-execs-call-for-clearer-regulations-in-wake-of-ftx-collapse/

Reiff, N. (2020). Some US military overseas will vote via blockchain. *Investopedia*. Retrieved November 9, 2022, from https://www.investopedia.com/news/some-us-military-overseas-will-vote-blockchain/

Reiff, N. (2022a). Which industries will blockchain technology disrupt next? *Investopedia*. Retrieved November 1, 2022, from https://www.investopedia.com/tech/which-industries-will-blockchain-disrupt-next/

Reiff, N. (2022b). Bitcoin vs. Ethereum: What's the difference? *Investopedia*. Retrieved November 26, 2022, from https://www.investopedia.com/articles/investing/031416/bitcoin-vs-ethereum-driven-different-purposes.asp

Seth, S. (2022). Explaining the crypto in cryptocurrency. *Investopedia*. Retrieved November 27, 2022, from https://www.investopedia.com/tech/explaining-crypto-cryptocurrency/

Sharma, R. (2022a). Non-fungible cryptocurrency (NFT): What it means and how it works. *Investopedia*. Retrieved November 24, 2022, from https://www.investopedia.com/non-fungible-cryptocurrencys-nft-5115211

Sharma, R. (2022b). Bitcoin's taproot upgrade: What you should know. *Investopedia*. Retrieved November 4, 2022, from https://www.investopedia.com/bitcoin-taproot-upgrade-5210039

Stackpole, T. (2022). What is web3? *Harvard Business Review*. Retrieved November 14, 2022, from https://hbr.org/2022/05/what-is-web3

Thomson, G. (2021). Iowa House approves bill to legally recognize blockchain smart contracts. *Cointelegraph*. Retrieved December 15, 2022, from https://cointelegraph.com/news/iowa-house-approves-bill-to-legally-recognize-blockchain-smart-contracts

Versprille, A. (2022). US midterm elections may have big crypto implications. *Bloomberg.com*. Retrieved December 1, 2022, from https://www.bloomberg.com/news/newsletters/2022-11-03/us-midterm-elections-may-have-big-crypto-implications

World Economic Forum. (2022a). Slow moving regulatory decision making for cryptocurrency not economically favourable, New World Economic Forum Study finds. *World Economic Forum*. Retrieved December 9, 2022, from https://www.weforum.org/press/2022/07/slow-moving-regulatory-decision-making-for-cryptocurrency-not-economically-favourable-new-world-economic-forum-study-finds

World Economic Forum. (2022b). The macroeconomic impact of cryptocurrency and stablecoins. *World Economic Forum*. Retrieved November 8, 2022, from https://www.weforum.org/whitepapers/the-macroeconomic-impact-of-cryptocurrency-and-stablecoins

Chapter 12
The Vision of Freedom and the Potential of Blockchain

Kazushi Yamamoto
Tashkent State University of Law, Japan

ABSTRACT

This chapter examines freedom and blockchain. Hegel stated that world history is the progression of freedom consciousness. While modern law and the state have facilitated the realization of freedom, blockchain technology has facilitated global, autonomous decentralization. This chapter defines freedom and explores how blockchain aids in its promotion and examines why central governments of developed countries opposed Libra, a global currency concept based on blockchain. While the central government intervened to crush Libra, this chapter shows that blockchain has vast possibilities for evolution of freedom. From the perspective of the absolute, blockchain and modern law and the state, while in conflict, will promote human freedom dialectically. This chapter uses the concepts such as publicness, autonomy, setback, external third party, facts and norms, primary rules, law and facts, family, bodily extension, constitutionalism, objective reality, absolute spirit, and so on. Also, the theories of Hayek, Habermas, H.L.A. Hart, Wittgenstein, Hegel, and so on will be utilized in this chapter.

INTRODUCTION

Will Blockchain Promote Human Freedom?

In this book chapter, Hegel's teachings on world history as the progress of the consciousness of freedom are discussed. However, what does "freedom" truly mean? In this chapter, freedom is defined as the realization of perfection, the alignment of ideas with reality, and the fulfillment of one's ideal life as a human being with an individual body. This chapter explores how modern law and the state serve as the stage for this notion of "freedom" and introduces the concept of "publicness" as a way to achieve it. The chapter then examines how technology, specifically Blockchain, can either conflict with or support the realization of "freedom" as a public platform. Through conceptual reconstruction and philosophical examinations, the research explores the implications of Blockchain and questions the need for an

DOI: 10.4018/978-1-6684-6429-8.ch012

external third party in facilitating joint activities. Then, the chapter discusses the concept of power and its implications and ultimately suggests a vision of "freedom" from the perspective of our human condition as individual beings existing in space-time with our bodies. This chapter aims to shed light on the complex relationship between Hegelian philosophy, modern institutions, technology, and human autonomy in the pursuit of "freedom."

This book chapter argues that utilizing an autonomous, self-regulated, decentralized, and non-centralized system through Blockchain can potentially facilitate the "freedom" of humanity in the world. While Hegel's teachings limited the medium for realizing "freedom" to the state, our exploration reveals the possibility of achieving "freedom" without relying solely on the state. From the perspective of absoluteness, we will examine how modern law, the state, and Blockchain interact and respond to each other, developing the concept of "freedom" dialectically. By shedding light on this dynamic relationship, this chapter aims to provide insights into the potential of Blockchain as a transformative force in pursuing human "freedom" beyond traditional institutional frameworks.

THE PROGRESS OF THE CONSCIOUSNESS OF FREEDOM IN WORLD HISTORY

Hegel once said, "World history is the progress of the consciousness of freedom." He found in the nation the instrument by which this freedom is phenomenally realized. The world spirit can be manifested as a national (ethnic) spirit. However, the state exists only in empirical reality. Hegel calls for the state as a procedure for phenomenally manifesting the impulse of "freedom," which is the true nature of the world's spirit. In this way, Hegel's concept of state is a tool for manifesting the impulse of freedom.

Hegel found in the state the tool for the development of world history. The point in question can be explored through the technological tool of the Internet. Therefore, a discussion of this point will be provided first.

What is "Freedom"?

Some may ask for a definition of the concept of "freedom." If "freedom" were to be discussed on an academic level, it would be impossible to do so in an essay as short as this one. However, since this chapter discusses "freedom," it would benefit the reader to give a specific definition. Therefore, this chapter briefly describes the envisioned concept of "freedom" below.

First, human beings can conceive of "infinity" and "perfection" conceptually but not literally. It appears in human beings as the awareness that "something exists objectively." For example, let us assume the concept of "circle" (the awareness that "something exists objectively" is not limited to it). "Circle" cannot be realized. Because line segments and curves always bring about a situation in which "points" and "dots" are separated without overlapping in phenomena. Therefore, a conceptual "circle" cannot be realized in a phenomenon. However, is the concept of "circle" useless? No, it is not. Because it is possible to conceive of a conceptually perfect "circle," the concept of "circle" can be made into a phenomenon (although it cannot be realized perfectly). For example, "justice" cannot be perfectly realized, but we can "assume" that such a concept is real. It is analogous to the relationship between the freedom to realize individual "rights" and the existence of "justice." The conceptually perfect "circle" does not correspond to the "line" that represents the phenomenalized "circle" of reality.

This chapter argues that the perfect "reality" of the "circle" is objectively real, and it assumes that human beings have the innate ability to recognize these perfect concepts. Furthermore, it holds that perfections such as "circle" or "justice" exist. This note assumes that humans can recognize these perfect concepts or value ideas and believe them honestly. Therefore, the idea of the "world" or "universe" is also assumed to be real and can be recognized by man. When compared to perfect concepts like a circle, the degree to which perfection constitutes the concept of the world is weaker, as the concept of the world consists of diverse empirical elements in any case. In this case, man's empiricist or inductive perception of the world based on his daily five senses is different from his deductive perception of the concept.

Second, men try instinctively to realize a conceptually perceivable perfect world, such as "the world" or "the circle," in the phenomenal empiricist world. Therefore, since man can recognize "the world," if his sphere of action is empirically limited to, for example, a nation, a city, or a village, he will try to realize his recognition of "the world" as his sphere of self-potentiality in reality. This is the source of the desire for order formation based on global autonomous will. By this, "freedom" is presented here as the ability and power to realize a perfect concept (e.g., "circle") or ideological value (e.g., "justice"), which I assume exits a priori, in the phenomenal world, or the connecting power to complete the procedure between them.

The following is supplementary information. It is interesting to note whether the concept of perfection or the ideological value presented above, which is assumed to be deductively real, is related to the idea of the reality of the human "soul," for example. If they are related, the discussion on the relation between the idea of "Idea" and the "soul" will be more powerful, as Plato says.

It would also open the discussion of the union of the "soul" and the "universe." This analysis interprets that the "promotion of the consciousness of freedom in world history" in Hegel's conception of the "Absolute Spirit" may also result from the union of the objectively real "Absolute Spirit" and man's inner spiritual capacity.

In conclusion, this chapter does not assert that establishing world order can be accomplished by either man's explicit cognitive reasoning or by his autonomous will. One can comprehend the philosophical standpoint advanced here by examining the entirety of the chapter rather than the brief description provided above. It is worth noting that Plato may have already presented a similar argument as stated above, that the conceptually perfect "circle" does not correspond to the "line" that represents the phenomenalized "circle" of reality. Another related idea can be found in Morigiwa's work (Williams, 1985/2020, pp. 467-468).

FREEDOM BASED ON MODERN LAW AND THE STATE, AND FREEDOM WITHOUT MODERN LAW AND THE STATE

It is well known that the Internet revolution has progressed since the 1990s. The information revolution has advanced in recent years through platforms such as IoT and AI or autonomous computing mechanisms. Such technologies are not public goods provided by the state but private goods provided by private companies as exclusive and competitive corporate services. Citizens can enjoy the convenience of information services by paying for them.

Two notable features can describe the concept of public goods, non-excludable and non-rivalrous. The purpose and principle of Big Tech are to pursue the private interest on a system of private property; it is difficult to determine that what they are providing as a service is public goods.

However, the following points are interesting to consider. Big Tech has created and is contributing to a public social space, a socially common knowledge platform, global social communication platforms, and a global system of payments and settlements. In this sense, we can see Big Tech: GAFAM from the point of view of "Publicness" or "Public sphere" (Sasaki & Kim, 2002, p. 6).

By reframing the analysis of GAFAM (Google, Apple, Facebook, Amazon, Microsoft) through the lenses of "Publicness" and the "Public sphere," a third perspective emerges that goes beyond the conventional dichotomy of "public" (i.e., government) and "private." This novel perspective recognizes that governments and private companies like GAFAM contribute to creating "Publicness" or the "Public Sphere." Drawing on Habermas' discourse theory, this chapter will delve into the intricacies of this concept, shedding light on the nuanced ways in which GAFAM's value in the public world can be understood beyond traditional categorizations of public and private. In other words, the state does not provide the means or platform to promote "freedom" or realize "freedom" in information exchange, which information technology facilitates, especially in communication.

When the media realizes "freedom," the platforms are created by private corporations that provide private properties to each country's citizens globally, and each country's citizens realize integrated communication through information services. Citizens are provided means by private corporations in the US in the case of GAFAM, so a situation has formed where it cannot be said that the modern law and states are only the agents that realize "freedom."

This discussion that follows presents two arguments. One is that "freedom" is successfully realized within the framework of law and the state in the modern era. The other is envisioning the realization of "freedom," not necessarily premised on the framework of modern law and the state as a requisite for "freedom."

REACTIONS OF THE GOVERNMENT AUTHORITIES OF THE MODERN LAW AND STATE IN THE LIBRA CASE

The creation of Libra invoked immediate responses from different governmental bodies and institutions to this blockchain-focused business partnership. Libra was a project aimed at building a globally autonomous circulating digital currency by private organizations. In response, central governments in advanced countries, mainly France and Germany, reacted. These governments' responses to the technology can be taken as movements in the former direction to realize "freedom." What perspective should be used to interpret the differences in methods between these two directions?

The following is a detailed examination of this phenomenon. One specific issue is the right of a nation to issue its currency. Friedrich Hayek, who condemned controlled or government-planned economies out of respect for freedom, also emphasized competition in the market and criticized the state's monopoly on the right to issue currency.

For example, he considered Keynesian demand-creation policies during recession times problematic because they would lead to inflation eventually and, thus, to a decline in the purchasing power of the people or an uneven distribution of wealth (Hayek, 1976/2020). Hayek was critical of the centralized government's planned control of the economy from the perspective of freedom. Because he trusted the self-generating order, he allowed the central bank to issue currency. However, he also allowed the private sector to issue its currency to promote free competition in the currency itself. This is thought to have been planned to promote free competition in the currency itself (Ueda, 2022, p. 225).

It is important to note that each country's government and central banks intervened in a thoroughly confrontational manner in response to this coalition concept. First, let us summarize the characteristics of Libra. In June 2019, the stablecoin "Libra" attracted worldwide attention when the plan was announced, with Facebook Inc. (now Meta Inc.) taking the lead.

The central government's response to the right to issue currency in modern states is particularly important. Libra has provoked strong reactions in many countries worldwide, such as fierce opposition in France and Germany, where it is typically seen as a threat to the monetary sovereignty of the central government. See below for a discussion of the French and German ministers' condemnation of the Libra from the perspective of monetary sovereignty.

It is essential to think about the relationship between the global currency trail of Libra and central governments. The French government response, by Singh (Singh, 2019, September 13), was "French Finance Minister Bruno Le Maire on Thursday said the country cannot authorize the development of Facebook's proposed Libra cryptocurrency on European soil as it threatens the "monetary sovereignty" of governments."

In other words, Libra would interfere with the currency issuance power within the sovereignty of modern nation-states.

"The Minister of the Economy firmly demands that Facebook does not exceed its prerogatives as a private company. The ability to issue securities, to build up a reserve, and to be a lender of last resort: none of this is conceivable," said Bruno Le Maire. According to him, money, "an attribute of the sovereignty of States," must "remain in the hands of States and not of private companies that respond to private interests." (Lesaffre, 2019, June 18).

Moreover, we can see the response of German Financial Minister Olaf Scholz (Reuters, 2020, December 8):

In a statement after a video conference of G7 finance ministers discussing cryptocurrencies, Scholz said that relaunching Libra under its new name Diem was only a cosmetic change. "A wolf in sheep's clothing is still a wolf," he said. "It is clear to me that Germany and Europe cannot and will not accept its entry into the market while the regulatory risks are not adequately addressed." He added, "We must do everything possible to ensure the currency monopoly remains in the hands of states."

For a less direct concern about the Libra from a public goods perspective by the Governor of the Bank of Japan, although less direct than the German and French ministers mentioned to the right, see the following. Kuroda Haruhiko, the Governor of the Bank of Japan (Kuroda, 2019, December 4):

"The Global Stablecoin aims to stabilize by tying its value to the value of the legal tender of each major country, and is an international public good - a basket of public goods - of each country." From the perspective of the monetary sovereignty of governments based on modern law and the state, strong opposition was made in developed countries, explicitly at the level of finance ministers internationally. In other words, Libra's global currency concept, established by a group of so-called private companies, threatened national monetary sovereignty and was not tolerated.

Libra's philosophy was that global currencies and financial infrastructures should be designed and governed as public goods. It should be designed as a global public good, which challenged the international monetary system based on the right of each country to issue its currency, causing intense alarm among national financial authorities.

The concept of a global order system has been introduced previously, as it has already been proposed in world justice (Inoue, 2012). On the other hand, the fact that it is backed by the legal tender system in each country suggests that it could not resist the intervention of financial regulators in the US, just as the military intelligence system backed GAFAM. Conversely, both GAFAM, at least Meta and Libra, are based on the premise of a system of freedom and justice based on modern law and the state, and the world as a federation of nations based on confidence in that premise. As such, Nakanishi suggests (2019, July 17) that they are only practicing the realization of freedom within the base of modern law and the state.

THE IDEA OF FREEDOM WITHOUT OR IN OPPOSITION TO MODERN LAW AND THE STATE

It is possible to posit that "freedom" can be manifested by not necessarily assuming the modern law and state or in opposition to them.

For example, in "The Structural Transformation of Publicness" by Jürgen Habermas, he describes forming a civic public sphere through communication in the so-called private sphere, such as cafes, newspapers, salons, dinner parties, etc. The so-called private sphere thus established is called the "public sphere." It can be said that Habermas sees the possibility of the development of "freedom" in the "publicness" of the so-called private, public sphere that is established in this way. As Habermas himself was aware, the concept of the "public sphere" is often used in the development of "freedom" in the global space through the Internet, especially in the creation of free speech space (Habermas, 1962/1973).

In the realm of facts, global communication through the Internet space, global transfer of information and goods, and global order as a settlement system through the formation of credit have already been established beyond the realm of the modern state or across the jurisprudence of the state.

Nevertheless, as mentioned earlier, whether it is GAFAM or a credit card company that enables credit creation and payment communication for international payments, it is a joint-stock company based on private ownership, and the condition for its existence is the modern state based on constitutionalism, which is premised on private ownership.

In contrast, creating credit through blockchain technology and forming a public sphere through communication technology based on blockchain technology is different (Nakamoto, 2009). From the perspective of a mechanism to realize "freedom," a network using blockchain technology is an autonomous public system that does not require a manager or, from the perspective of a national organization, a centralized third-party organization. The principle fundamentally differs from the conventional approach of forming a global "public sphere" with the Internet as the mediating term. To summarize Yukio Noguchi's and others' arguments (Noguchi, 2017, pp. 21-67):

Credit creation does not require a government or a third party.

It eliminates the need for credit backed by the centrally governed authority of law and the state in the modern era:

The technical characteristics of Blockchain may be summarized in the following three points. In other words, it is a composite technology created by combining three elemental technologies: "P2P (Peer to Peer) type network technology," "consensus building mechanism" (Noguchi, 2017, pp, 42-53), and "cryptography" (hash function, public key cryptography, and digital signature) (Ito, 2018, March). What is revolutionary about the system characterized by these elements is that it eliminates the need for a centralized governance structure, thereby enabling decentralized autonomous organizations, and also that it is tamper-proof and does not require the creation of a trust, which is maintained by a mechanism that requires a manager, such as a country or a company. It also allows people to conduct various transactions directly with the world. It ensures the truthfulness of factual proof without separating facts and norms (Noguchi, 2017, pp. 195-215, 235-292).

This chapter emphasizes the following points regarding these technical features and their phenomenal and practical benefits. The first is the nature of autonomous organizational operations, or DAO, or "decentralized autonomous organization," which operates in the Blockchain. Unlike GAFAM, a traditional centralized organization, the autonomous decentralized network established by Blockchain does not have a centralized information gathering system by a centralized management system. Therefore, it cannot be centrally controlled by a management system based on the assumption of a modern state and its laws and state system.

DEFINITION, FUNCTIONS, AND IMPLICATIONS OF THE BLOCKCHAIN

The fundamental differences between internet service providers such as GAFAM and Blockchain and the nature of Blockchain from the point of view of DAO were pointed out. Section four of GAFAM explores how these characteristics are made possible by the specific technicality of Blockchain. Given that the characteristics in this section also enable autonomy, which is one of the essential elements of freedom, let us take a closer look.

Definition

According to the Japan Blockchain Association (2016, October 3), the definition of "blockchain" is "a protocol, or implementation of a protocol, used by an unspecified number of nodes containing Byzantine faults and converges the probability of consensus reversion with time to zero." In a broader sense, blockchain is a technology with a data structure that can quickly detect manipulation using digital signatures and hash pointers and where the data has high availability and integrity due to distribution across multiple nodes on a network.

Furthermore, according to the Harvard Business Review, "block" is (Iansiti & Lakhani, 2017, January–February)the technology at the heart of Bitcoin and other virtual currencies. Blockchain is an open, distributed ledger that can record transactions between two parties efficiently and in a verifiable and permanent way.

Blockchain is a system that records transactions in a public ledger. Compared to conventional information systems, Blockchain is characterized by the facts that (1) records are open to the public, (2) it is operated in a decentralized manner, and there is no administrator, (3) operation costs are low, and the system does not go down, (4) it is not necessary to trust the organization that is the business entity (Noguchi, 2017, p.40).

In Blockchain, instead of such a controlling entity, a network of computers, called P2P, controls the correctness of transactions. A blockchain is not "out there," but you, the person running the code, are part of the Blockchain (Yamazaki, 2016, p. 243). In essence, it is a chain of data structures connected like a chain, generating units of data called "blocks" (Yamazaki 2016, "Document 4-3-6 Blockchain," p. 244).

Another defining characteristic of blockchain is proof of Work, PoW. PoW is the value incorporated within each block. That is, as the name implies, "proof of a great deal of calculation." Each block also contains a "hash value" calculated from the previous block by a specific calculation. This hash value means that if even one bit of the data on which the calculation is based is changed, the calculation result will be a completely different value (Yamazaki, 2016, p. 244). Therefore, to tamper with a block added to the chain once, all subsequent blocks must be discarded, and all PoW calculations for each block performed over time must be redone. It is said that falsification is not possible in reality. Saito states the reason as follows(Saito 2014, May 16, pp. 3-4): When other miners confirm the correctness of the content of a mined block, the block in question becomes the new end of the Blockchain. In other words, the block becomes the new end of the Blockchain, with successive blocks following it. Since the contents of a block depend on the hash value obtained from the contents of the previous block, mining must be redone after the block in question to tamper with past transactions. The difficulty of mining is tuned so that, on average, a miner succeeds once every 10 minutes, relative to the computing power of all miners, making it extremely difficult for some malicious miner to overtake the pace of legitimate miners and justify his or her tampering.

Satoshi Nakamoto describes blockchain's cryptocurrency feature as a purely peer-to-peer version of electronic cash that would allow online payments to be sent directly from one party to another without going through a financial institution. Digital signatures provide part of the solution, but the main benefits are lost if a trusted third party is still required to prevent double-spending. The solution proposed to counter the double-spending problem is using a peer-to-peer network. The network timestamps transactions by hashing them into an ongoing chain of hash-based proof-of-work, forming a record that cannot be changed without redoing the proof-of-work (Nakamoto, 2009, "Abstract," p.1).

Implications

This section will examine the various aspects of blockchain implementation in different domains, including society, economy, state, and law.

An example of a social implementation case is the "Proposal for the Social Implementation of Blockchain - Following the Revision of the New Law on Cryptographic Assets" by the Japan Association of New Economy (2019, July 30, p. 8). The potential applications of Blockchain technology are diverse and include areas such as small payments for copyright and usage fees collection in the private sector, management of local currencies and corporate rewards points by both government and private sector, proof of ownership for land rights by the government, data linkage between government agencies, supply chain management for recording the distribution process from raw materials to retailing in the private sector, for use in the sharing economy for recording transaction status of idle assets and sharing evaluations/reviews in the private sector, automatic execution of transactions between M2M machines for automatic transactions between AI/agents, and instantaneous revenue sharing in the private sector. The market size of Blockchain in Japan is estimated to reach 67 trillion yen and globally 300 trillion yen by 2030 (Japan Association of New Economy, 2019, July 30, p. 9).

NOTABLE FEATURES OF BLOCKCHAIN

This section discusses notable characteristics of blockchain technology.

One notable characteristic is the issue of the authenticity of the information. Even in the case of nations and giant corporations, information falsification, including fraud, is often made public, and Blockchain solves this problem of information authenticity. Internet technology cannot solve this problem, and this blockchain technology will drastically improve the quality of trust formation in the global space.

Second, there are cases in which Blockchain is replacing the functions of the state itself, including the management of official documents. This is typically seen in Estonia, known as "e-Estonia." Here Antonio De La Iglesia says, "The case of Estonia explores several blockchain-based socio-technical objects to identify the narratives that have emerged in Estonia, a perfect example to research this topic as the alleged first national infrastructure integrating blockchain." At the same time, as Iglesia points out, it is worth remembering that blockchain systems used at the governmental level in Estonia are based on private blockchains and reflect a centralized approach to distributed recording technology (Iglesia, 2022, April 18). Moreover, the case of the e-Governance Academy (eGA) serves as an ideal solution for promoting a sustainable society, e-Government, e-Democracy, cyber security, and related areas. It is an institution "Founded in 2002, the e-Governance Academy (eGA) is a center of excellence: a joint initiative of the Government of Estonia, Open Society Institute (OSI) and the United Nations Development Programme" (The e-Governance Academy (eGA), 2018). The establishment of e-Government in Estonia has contributed significantly to the development of e-services. Uuno Vallner, who played a significant role in the invention and creation of various e-services, immensely impacted the widespread usage of such services in Estonia today. "For example, the world's first public sector website was completed under his leadership: the official website of the Estonian Government, back in 1993." (The e-Governance Academy (eGA), 2018, February 21). Further, there is a consultant group helping clients to do business registrations in Estonia. This includes Start a Cryptocurrency Business, Estonia e-Residency application support, and so on (Healy Consultants Group, 2023).

Another case is in the United Kingdom, where blockchain technology was introduced in the National Archives (Noguchi, 2018, pp. 253-260). Matters that do not require interpretation, such as those under the jurisdiction of the judicial clerk and registration, are modern legal areas that would be perfectly fine to be converted to Blockchain. See the introduction of Noguchi's for a similar purpose, but regarding notaries. The economic rationality from the user's perspective regarding fee income, etc., is also discussed here. (Noguchi, 2018, pp. 261-266).

The point is that facts and norms must match. This is the topic discussed below.

Third, regarding the provision of public goods related to information and credit, the good public provision provided by the state system, which is based on modern law, would be substituted and better. We have already mentioned that it is possible to replace the right to issue money from the state. It is possible because of this functionality that regulators and financial authorities in many countries have decided to impose sanctions on Libra.

THE LEGAL AND POLITICAL-PHILOSOPHICAL DIMENSIONS OF MATTERS BROUGHT ABOUT IN THE BLOCKCHAIN

The conceptual framework of "facts to be recognized" does not exist in blockchain technology. In other words, facts cannot be falsified in blockchain technology, and it cannot be assumed that facts will deviate from authenticity or that norms and facts will be separated.

Given this critical property in Blockchain, the impact of blockchain technology on the concept of "freedom" in contemporary society will be examined from a political-philosophical perspective. In today's developed countries, many safety nets are provided to citizens. For example, consumer protection laws regulate consumers who purchase defective products online and provide fraud remedies to protect private property. States have more or fewer mechanisms for redistributing resources through state intervention in private property. In this sense, the modern state must exist to substantiate justice in society law can be seen as a tool for substantiation or implication of this idea of justice in society. In blockchain technology, however, it is impossible to falsify information, which means there is no such thing as "fraudulent information" in the first place. The absence of a manager also means that the state, as a mandatory executor of the law, or state officials, who are responsible for the state, does not exist. In addition, the absence of management also implies that there is no state as an enforcing body of law or state officials as personnel in charge of the state. Thus, in the example mentioned above, it is possible that corporate information and fraud against unscrupulous consumers cannot be assumed conceptually, at least in the public system itself. It is reasonable to assume that an "autonomous" global society will be constituted. It is not envisioned here that a fair contract can be secured because of the third-party or ex post facto sanction of the state.

The person's credibility is secured by the information in the system, and any civil action cannot be "set back" by assuming a norm apart from the fact. Thus, ideas like the following need not be assumed: Securing the authenticity of the fact by some third-party position or mechanism requires external proof. Alternatively, to grant credit to the consumer by a third-party credit granting company to grant credit to the debtor in the case of a credit card transaction.

In simpler terms, in a world where individuals are accustomed to individuality from birth, blockchain technology can create a global mechanism where information credibility is ensured autonomously by the system and its users. This allows people to live as individuals without relying on third-party organizations like the state or centralized entities. This world would guarantee "freedom" without external intervention, respect for private rights, and alignment with the normative concept of libertarianism. For an example of libertarianism, see Nozick, 1992. The concept of "open software" referred to in this chapter does not necessarily assume the existence of private rights on the Internet or Blockchain. It is said that blockchain technology itself is derived from libertarian political thought (Ennis, 2021, September 14), and it is possible to point out the ideological homogeneity between the two.

REBUTTALS AND DOUBTS

Why is it necessary to establish "rights" if a new public sphere is to be formed?

During the rise of the Internet, there were discussions on software development with two distinct approaches. One approach was to offer software products as commodities under private ownership, such as Microsoft's model. The other approach was to structure software as open software, allowing contributions from anyone for its evolutionary development. In the latter case, it is essential to note that contributions to the software are not considered private rights in principle, as seen in the collaborative development of free and open-source software like Linux. This approach is based on the idea that improvements should be made to public goods through voluntary contributions. These two approaches represent different philosophical schools of thought with distinct perspectives on the merits and demerits of "private rights."

The former perspective is rooted in the philosophical affirmation of private property rights, often supported by theories like John Locke's labor ownership theory (1690/2010, Section 27). On the other hand, the latter perspective emphasizes the public goods view of open resources, as seen in concepts like "commons-based peer production (CBPP)" presented by Yochai Benkler, a professor at Harvard Law School. This perspective does not exclude private property at a principled level, but it does not assume that most participants prioritize private interests based on private rights. Instead, private rights are typically associated with the family, while public arguments are made within the domain of the state without bringing in the private nature of the family. The modern world can be seen as a tension between the family, private rights within the family (as discussed by Engels in 1884/1965), and the state as a modern manifestation of the public world in ancient Greece, the Polis.

On the Possibility of a World Where Facts and Norms do not Detach

On the possibility of a world where facts and norms do not detach, some say that the concept of "right" was established by people who detached themselves from "possession," such as the aristocrats, in the history of the emergence of the concept (Koba, 2017).

In essence, the concept of "rights" cannot be separated from "possession" regarding blockchain technology. The facts recorded on the blockchain are guaranteed authentic and possess normative value. This realization aligns with the theory of the normativity of facts, where facts and norms are perfectly matched. Blockchain technology can eliminate the need for confirmations or verifications of general acts, such as information in contracts, thereby creating a world where facts and norms are integrated and accepted.

However, it is essential to note that while the existence of an afterlife or soul cannot be ruled out, our existence in the universe is typically associated with having a physical body.

Suppose the existence in the universe with a body is a human condition. It raises questions about whether the constitutional requirement of constitutionalism, as a secondary rule, is still necessary in the context of blockchain technology. This is because, even in the ethical dimension of life, we do not consider the birth of life itself as something governed by the market. Instead, we assume that our lives will be reproduced and sustained within a private order based on the family system, where the concept of family and having a body are inherent. This forms the foundational structure of constitutionalism.

Therefore, when considering the conception of a shared network in open resources, it can be argued that private rights may continue to be viewed as a global communication tool based on the principles of private ownership unless the core concept of a shared network in open resources can effectively exclude the element of second-order rule, which is a form of an amendment to constitutional principles within the realm of constitutionalism.

Alternatively, there could be an order that does not require the fundamental principle of constitutionalism, the guarantee of private rights under the Constitution, such as communism, if the ideology in open source is thoroughly implemented (Engels, 1884/1965).

On the Possibility of Securing "Trust" as a Self-Generating Order

This is related to 8-2 described above, but the fact that "set-backs" are not required to form "trust" has significant social implications. As mentioned earlier in the definition of "set-back," whether money or credit creation (e.g., credit creation through the provision of credit information by a third party about consumers in credit card transactions), facts and norms are not detached. All information is stored in the Blockchain, which is used as the basis for creating "credit." This is also groundbreaking because it is an entirely different way of creating "credit" from how it is secured and maintained in modern society. This point relates to what has been mentioned above: everything is completed through direct actions that do not require "setbacks" to secondary rules. It can be said that the Blockchain is a mechanism to establish this "belief" only by Intrinsic to the primary act.

Humanity From the Viewpoint of "Freedom" of Autonomy

This section will examine the concept of freedom through the lens of autonomy. The structure of society with autonomy will also be discussed, particularly from the perspective of libertarianism, as well as the potential for a global social order facilitated by the Blockchain. Additionally, the possibility of this construction will be explored concerning the political philosophy that prioritizes free will or "private autonomy" in legal contexts (Morimura, 2020, pp. 231-254, with a specific focus on pp. 253-254) and whether it is overly simplistic and naïve in its perception of human nature.

Hart states that "the original form of the power conferring rule," such as contracts and wills, "is also the basis of the moral system of promises" (1961/1976, p. 106). This order explains the establishment of the institutional guarantee of the promise of the recognition of private autonomy (Morimura, 2020, p.238). In other words, Morimura explains that the moral sense that a promise should be kept is common, and the fact that such a sense is shared in the act itself rather than in the contract itself or the natural state is why contracts are usually kept (2020, p. 239).

Let us examine whether society can be constructed with only primary rules. For example, the redistribution of resources in liberalism, through the second principle, through state power to those born with disabilities, is also a social implication of the redistribution of resources using power or justice. The essence of the idea would be appropriate, and one can see the ideas of conceiving a second principle of justice in this direction in the works of Rawls (Rawls, 1971/2010).

Therefore, it might be necessary to make some reservations about the idea that primary rules or private autonomy alone can be applied to all the principles of social composition. Alternatively, it is necessary to consider in principle what kind of social assistance is possible for those who are born with disabilities, especially those who are born with disabilities without their own autonomous will.

In Wittgenstein's view of language, which is the background of Hart's legal theory, it can be understood that in the latter period, he abandoned the mapping-theoretic method of ascertaining "certainty" utilizing a matching relation between facts and the world in the earlier period and carried out his philosophy inherent in the communication itself. In this case, the difference between the two is whether he tries to prove "certainty" externally (early philosophy) or whether he "jumps in" in communication without a such external proof (in the case of hesitation or life, he is forced to "jump in" as it can only be said). This assumes that there can be situations where there is no other "extrinsic material" intrinsic to the situation to guide one's judgment. It is a question of one's attitude toward life: there is no other way to live (late philosophy).

However, even if a system is created where facts and norms align flawlessly on the Blockchain, not all facts existing in the Blockchain contain values. In addition, in communication, "meaning is conveyed," some naturalistic "entity" is assumed (and possibly related to the consideration of cosmological dimensions). In particular, the fact that this dimension, which may not be wholly resolved between the two, can be assumed to constitute one of the dimensions of the existence of law and the state, and, to refer back to the beginning of this chapter, is a possible realm of metaphysical existence, such as Hegel's "Absolute Spirit."

THE MAIN ELEMENTS ARE THE REMAINING SUBJECTS OF CONSIDERATION BUT WILL BE LEFT FOR FURTHER STUDY

Power

Throughout this chapter, our focus has been on exploring the potential of blockchain technology in realizing "freedom" on a global scale. In doing so, we need to consider how power is understood, which is also a topic discussed in libertarianism. When discussing the possibilities of a global "public sphere" or the potential of a decentralized autonomous organization (DAO), we must pay attention to specific points. Specifically, we need to critically examine whether the concept of "power," whether centralized or otherwise, is detrimental to the idea of a self-governing civic order. The implicit assumption of seeking civic publicity through blockchain technology suggests that centralized power, such as the constitutional power established by law and the state in the modern era, may not be ideal. For instance, in the case of global communication giants like GAFAM, the assumption is that the current state-based power is not conducive to building a better world. Even projects like Libra, which attempted to explore new possibilities through blockchain, positioned themselves as opposing the power of law and the state. The assumption is that a world without centralized managers or executives, enabled by blockchain, may be more favorable for creating a better world for people. However, we must critically examine whether this assumption is valid, as it provides essential insights into how power is positioned to realize and manifest the value of "freedom."

As indicated at the beginning of this chapter, the proposition advanced is that the various social implications of Blockchain will facilitate the realization of human "freedom" in the history of the world. However, this does not necessarily mean that the existence of "power" is evil. In the dimension of social concepts in Blockchain or the Internet, various themes, such as anarchism and anarcho-capitalism in the context of political philosophies, are presented, and sometimes they argue that the modern state established by law and the exercise of its power can be seen as "evil." Is this so? Focusing on the dimensions of our bodies can effectively examine this question, which is examined in the next section.

In the pursuit of life, people exercise their bodies to realize their values. In this sense, the body functions as a "medium" for realizing the value each person can achieve and manifest in this world. Similarly, modern law and the state also function as public "media" that establish individuals as equal and free subjects to realize the value of each individual. How freedom is realized in Blockchain may differ, but the spirit of seeking "freedom" among people remains the same. In other words, the approach may differ, but the underlying desire for "freedom" is consistent.

What "freedom" is being promoted?

At the start of this chapter, the discussion focused on "freedom." Therefore, the question arises about what type of "freedom" is expanded using blockchain technology.

As this chapter aims to contribute to the world's evolution in a meaningful way, it is suggested that promoting a sense of freedom through the concept of "physical extensibility" is one approach through which blockchain technology can help achieve this goal. This is because individuals exist in the physical world with a physical body, which anchors them to the present moment and a specific location in space and time. As a result, individuals are compelled to navigate online spaces, including those facilitated by blockchain technology such as the Internet or DAOs, within the confines of their physical existence.

Humans come to live in this phenomenal universe with a physical body. On the other hand, people can connect to the dimension of absolute infinity from their ability to conceptually deduce or to assume the infinite. Nevertheless, humans can capture the infinite on the conceptual level. In this phenomenal world, they live with a physical body; in a sense, they are forced to be relative in space-time. It is not appropriate to describe this as being forced. However, in any case, this forces us to feel a relative inconvenience in that the absolute realm of deductive conceptuality is not always fully phenomenalized, and this inconvenience. This sense of nonfreedom is something that humans would like to achieve a breakthrough in some way.

The concept of "freedom" in the context of a global, autonomous, and decentralized order brought about by blockchain technology is evident. If a physical body did not initially bind individuals and could move freely through time and space solely with their consciousness, then "freedom" may not hold the same significance. Paradoxically, it can be argued that the condition of our existence in this world, where we unavoidably experience the relative constraints of having a physical body, gives rise to the intrinsic potential of the "freedom" proposed in this chapter. While it may be challenging to establish a definitive concept at a deductive level, we can transcend the "inconvenience" of the relativity of space-time and experience the essence of "free" activity in a global space.

The fundamental condition of our existence in the universe is that we are physical beings. This corporeal dimension is why the social space facilitated by global blockchain technology is so alluring to us. However, it also highlights the inherent limitations of the social vision that can be realized through blockchain technology. Private goods cannot adequately provide various dimensions of public goods. For instance, aspects such as birth into the world, which is a public good, are facilitated by societal norms and regulations. While private companies can employ blockchain technology to provide physical goods and services such as sewage management, it cannot create the entire dimension that constitutes the conditions for human existence. For example, issues related to life and death, and the ethical considerations around topics like test-tube babies, are beyond the scope of blockchain technology alone. Therefore, it is crucial to consider how individuals can ensure the assurance of their "life" and "death" beyond the limitations of blockchain technology.

Therefore, considering the physical dimension, it appears possible to clarify the dimension in which the need for public goods provided by the modern state becomes apparent. However, the condition in our universe of having a body may cause a feeling of detachment between ideas and phenomena.

CONCLUSION

The introduction of this chapter posed the question: "Will Blockchain promote human freedom? "The answer is a resounding yes. The reasons supporting this answer will be presented later in the chapter. Before delving into those reasons, a review of what has been explored thus far.

The first section confirmed that Hegel stated that world history is the progress of the consciousness of freedom. This chapter responded to the request for a definition of freedom. Despite organizing the concept into several elements, freedom can be condensed into the expression of the manifestation in the phenomenal world of a universally real perfection or ideal reality. It is also to autonomously bridge the above-stated gap between reality and phenomenalization in the life of the "this I" nobody else, to autonomously realize bridging between the Ideal and the reality through my mind and body.

The second section discussed the concept of publicness was introduced. From the perspective of publicness, the so-called private enterprise, as a platform for realizing freedom, can be treated like modern law and the state. Two methods of considering the achievement of freedom were presented: one based on the existence of modern law and the state and another that does not necessarily depend on them.

The third section explored the topic of free competition in money, drawing upon Hayek's critique of centralized control within the modern state. Libra tried to build a global economic platform and was condemned, especially by the central governments of the modern states of Germany and France.

In the fourth section, the argument presented was based on Habermas' idea of achieving freedom without relying on modern law and the state. Once again, the concept of publicness was reconsidered concerning freedom. With its DAO concept, the Blockchain is unique because it does not require a centralized third party.

The fifth section defined Blockchain and its functions.

The sixth section pointed out the most notable features of the Blockchain as the authenticity of the information, the fact that it replaces state functions such as the case in Estonia, the conformity in information management that does not require a consistent interpretation of facts and norms, and the provision of public goods related to information and trust.

In the seventh section, the characteristics of the Blockchain were discussed in terms of its ability to facilitate autonomous transactions between two parties without requiring the involvement of an external third party. This quality enabled transactions without "set-backs" because the Blockchain does not rely on external factors. Furthermore, a political and philosophical connection was drawn between libertarians and the Blockchain.

The eighth section explored the evolution of software from the Internet to the Blockchain and was examined from two perspectives: private ownership and open software. John Locke's theory was discussed concerning the former, while Yochai Benkler's views were presented. Engels' theory was introduced, highlighting the role of the family in establishing private property. The section also examined the "facts and norms matching" in the Blockchain from the perspective of the law theory. It is theoretically correct. However, that law has the potential of a cognitive dimension detached from facts. Therefore, if a principle in which facts and norms are in complete agreement is realized, then one of the conditions of communism may be complete. Since there is no need for "set-back" to "external third parties" on the Blockchain, there is no need to return to secondary rules in the creation of "trust" or "credit." It can be said that it is a mechanism that creates a "belief" inherent only in primary actions.

The perspective of autonomy was examined at the end of this section to structure society, and its relationship to the concept of freedom was explored. The potential for the Blockchain to create a global social order was discussed, drawing upon the views of libertarians. The concept of "private autonomy" in jurisprudence was analyzed, along with the moral institution of promise described by H.L.A. Hart. Morimura's view that keeping promises is essential for contractual compliance was also considered. However, doubts were raised about whether such actions alone could distribute goods to those with natural disabilities.

Consequently, Rawls' theory of justice in liberalism was briefly discussed. Additionally, Wittgenstein's view of language was introduced, highlighting the importance of the context in which actions occur: "jumping into the situation." Nevertheless, there may be a dimension of reality beyond what can be captured in descriptive language, which is connected to the existence of law and the state and could align with Hegel's concept of the "absolute spirit."

The following issues were examined in section nine. Since Blockchain emphasizes the characteristics of a Decentralized Autonomous Organization (DAO), questions arise about whether power is evil. This proposition was doubted.

The examination in the ninth section focuses on the type of "freedom" Blockchain promotes. It was concluded that Blockchain promotes "physical extensibility" as a form of freedom. We exist in this universe with our bodies, and that may be the cause of our wants and the basis of our precious lives, autonomies, and rights. However, our physical existence also prevents us from fully grasping the idea or reality itself. Paradoxically, the condition of our existence in this world, where having a body is a relative inconvenience, creates the potential for the "freedom" proposed in this chapter.

The ninth section further elucidates how the specific dimension of corporeality is what makes the social space facilitated by global blockchain technology appealing to us. However, it also underscores, to some extent, the inherent limitations of the social vision that can be actualized through blockchain technology. The fact that we possess physical bodies in the universe can create a sense of detachment between abstract ideas and tangible phenomena. This work clarifies that Blockchain can create an autonomous, self-regulated, decentralized, non-centralized system. Moreover, Blockchain can facilitate the "freedom" of humanity in the world.

The evolution of "freedom" can be seen from the point of view of our condition as human beings existing in this universe, and it is our bodies. Paradox is that it is the reason why we take it as truth that Blockchain is offering us the sense of the expanding of our "freedom" from the point of view of "physical extensivity," however, it is at the same time, the cause that we cannot exercise our "freedom" perfectly to conduct our autonomous power to realize the unification between the existing perfect idea and the realization of the idea into this world physically.

The Blockchain offers a system that allows us to create things through a DAO, eliminating the need for external centralized governance. This is why the central governments of France, Germany, and the United States of America tried to interfere with the activities of Libra. Libra tried to create an autonomous private company association that gathered the global financial system, including the currency system. Although Libra withdrew, many attempts have emerged to apply the philosophy of Blockchain and manifest blockchain technology into this physical world.

One may ponder how to make sense of the diverse phenomena. From the standpoint of the concept of absolutes, both blockchain technology and the modern legal system and state can be seen as proponents of freedom. While blockchain emphasizes decentralization and autonomy, the modern state acts as a centralized entity for governance. The modern state has attempted to resist the emergence of blockchain-based global currency initiatives, as it challenges the traditional third-party trust and credit creation systems. Nevertheless, individuals strive for increased freedom through global autonomy rather than relying solely on centralized entities.

Therefore, returning to the last question in the introduction:

From the perspective of absoluteness, do not modern law and the state, and the Blockchain, interact and respond to each other, evolving "freedom" dialectically? While Blockchain, modern law, and the state dialectically contradict each other, will they not contribute to the evolution of the "freedom" of humankind of the world?

Due to limitations in the length of this chapter, only a small portion of the issues have been thoroughly examined. However, numerous aspects still need to be explored in-depth and could be the subject of future discussion. For example, the active political participation aspect of freedom, namely the active public opinion formation by individuals in the public sphere and the active participation in exercising power through decisions made in public opinion, could only be verified a little. This operational aspect of freedom in politics is essential in analyzing the concept of freedom.

Moreover, even with the addition of this political activity, which is also a condition of human existence, more than the analysis of freedom is needed. The analysis of the concept of freedom in this chapter is only a proposal from one perspective of the concept of freedom in terms of autonomy for this chapter. Therefore, although this chapter deals with freedom, it is difficult to claim that sufficient verification has been achieved.

Postscript

It is possible that Blockchain, like the Internet, may contribute to the evolution of culture and civilization in human history, particularly in terms of the way "freedom" is realized, despite encountering numerous challenges. Although it may be seen as a normative statement, the author draws this conclusion based on specific events that occurred during the writing of this chapter. Specifically, before the author's commencement of writing, FTX filed for Chapter 11 of the Federal Bankruptcy Code on November 11, 2022, and went bankrupt. While this chapter may have reached its conclusion due to numerous significant constraints, many aspects still require further verification and discussion.

REFERENCES

Engels F. (1965). *Kazoku・Shiyuzaisan・Kokka no kigen: Lewis H. Morgan no kenkyu ni kanren shite* [The origin of the Family, Private Property, and the State: About the study of Lewis H. Morgan] (S. Tobara, Trans.). Iwanami Shoten. (Original work published 1884)

Ennis, P. J. (2021, Sep 14). Ethereum's Political Philosophy Explained. *CoinDesk*. https://www.coindesk.com/markets/2021/07/09/ethereums-political-philosophy-explained/

Habermas, J. (1973). *Koukyosei no kozo tankan* [The Structural Transformation of the Public Sphere: An Inquiry into a Category of Bourgeois Society] (S. Hosoya & M. Yamada, Trans.). Miraisha. (Original work published 1962)

Hart, H. L. A. (1976). *Hou no gainen* [The concept of law] (M. Yazaki, Trans.). Misuzu-shobo. (Original work published 1961)

Hayek, F. (2020). *Kahei hakkou jiyuuka ron kaiteiban – Kyousou tsuuka no riron to jikkou ni kansuru bunseki* [Denationalisation of Money: The Argument Refined] (A. Murai, Trans.). Nikkei BP Classics. (Original work published 1976)

Healy Consultants Group. (2023). *Estonia cryptocurrency/bitcoin business set up in 2023*. https://www.healyconsultants.com/estonia-company-registration/cryptocurrency-solution/

Iansiti, M., & Lakhani, K. R. (2017, January–February). The Truth About Blockchain. *Harvard Business Review*. https://hbr.org/2017/01/the-truth-about-blockchain

Iglesia, A. D. L. (2022, April 18). Decentralizing what? When blockchain's perspectives clash. *The P2P Model*. https://p2pmodels.eu/decentralizing-what-when-blockchains-perspectives-clash/

Inoue T. (2012). *Sekai Seigi Ron* [World Justice Theory]. Chikuma--shobo.

Ito, M. (2018, March). *Kumikomi blockchain no shogeki (Jou)* [The impact of embedded blockchain (Volume one of the two)]. Takaoka Toko Co., Ltd. https://uquest.tktk.co.jp/embedded/outline/outline10.html

Japan Association of New Economy. (2019, July 30). Blockchain no shakai jissou ni muketa teigen ~Ango shisan no shinpo kaisei wo ukete [Proposals for the social implementation of Blockchain - Following the revision of the new law on cryptographic assets]. *Naikakufu of Japan*. https://www8.cao.go.jp/kisei-kaikaku/kisei/meeting/wg/toushi/20191220/191220toushi03.pdf

Japan Blockchain Association. (2016, October 3). "Blockchain no teigi" wo koukai shimashita ["Definition of Blockchain" is now available] *JBA Nihon blockchain kyokai*. https://jba-web.jp/news/642

Koba, A. (2017). *Roma hou annai – Gendai no houritsuka no tameni* [Introduction of Roman law - for contemporary lawyers]. Keiso Shobo.

Kuroda, H. (2019, December 4). *Kessai no innovation to chou ginko no yakuwari -Stable coin ga nagekaketa mondai* [Payments Innovation and the Role of Central Banks- the problems posed by stable coins]. Bank of Japan. https://www.boj.or.jp/about/press/koen_2019/data/ko191204a1.pdf

Lesaffre, C. (2019, June 18). Facebook va créer sa monnaie: "Nous allons demander des garanties", prévient Bruno Le Maire [Facebook will create its own currency: "We will ask for guarantees", warns Bruno Le Maire]. *Europe, 1*. https://www.europe1.fr/economie/facebook-va-creer-sa-monnaie-nous-allons-demander-des-garanties-previent-bruno-le-maire-3905215

Locke J. (2010). *Touchi niron kanyaku* [A complete translation of the Second Treatise of government] (Kato T. Trans.). Iwanami Shoten. (Original work published 1689)

Morimura, S. (2020). *Houtetsugaku ha konnani omoshiroi* [Philosophy of law is so interesting]. Shinzan-sha.

Nakamoto, S. (2009). Bitcoin: A Peer-to-Peer Electronic Cash System. *Satoshi Nakamoto Institute*. https://bitcoin.org/bitcoin.pdf

Nakanishi, T. (2019, July 17). Libra 'Kakkoku kisei ni shitagau' Bei kouchou kai de hihan ya kenen [Libra 'subject to national regulations,' criticisms and concerns at U.S. hearing]. *Nihon Keizai Shinbun Sha*. https://www.nikkei.com/article/DGXMZO47418380X10C19A7000000 /

Noguchi, Y. (2017). *Blockchain Kakumei – Bunsan jiritsu gata shaki no shutugen* [Blockchain Revolution - Emergence of a decentralized and autonomous society]. Nihon Keizai Shuppan-sha.

Noguchi, Y. (2018). Kasou tsuuka ha dounaruka - Bubble ga owari, atarashii shinka ga hajimaru [What will happen to virtual currency? Bubble ends, new evolution begins]. *Diamond*.

Nozick, R. (1992). *Anarchy・Kokka・Utopia* [Anarchy, State, and Utopia] (I. Shimazu, Trans.). Bokutaku-sha. (Original work published 1974)

Rawls, J. (2010). *Seigiron* [A Theory of Justice] (T. Kawamoto, Trans.). Kinokuniya Shoten. (Original work published 1971)

Reuters. (2020, December 8). Facebook's renamed cryptocurrency is still 'wolf in sheep's clothing': German Finance Minister. *Reuters*. https://www.reuters.com/article/uk-g7-digitial-facebook/facebooks-renamed-cryptocurrency-is-still-wolf-in-sheeps-clothing-german-finance-minister-idUKKBN28H202

Saito, K. (2014, May 16). *Bitcoin ni okeru transaction, sono tensei to eikyo* [Transactions in Bitcoin, their ability to exhibit and impact]. WIDE Technical-Report in 2014. http://member.wide.ad.jp/tr/wide-tr-ideon-bitcoin-transaction2014-00.pdf

Sasaki, T., & Kim, T. C. (Eds.). (2002). *Koukyo Tetsugaku <6> Keizai kara mita koushi mondai* [Public philosophy <6> Public and private issues from the point of view of the economy]. Tokyo Daigaku Shuppankai.

Singh, C. (Ed.). (2019, September 13), France Finance Minister Calls Facebook Libra a Threat to 'Monetary Sovereignty'. *News18*. https://www.news18.com/news/tech/france-finance-minister-calls-facebook-libra-a-threat-to-monetary-sovereignty-2307541.html

The e-Governance Academy (eGA). (2018). *eGA*. https://ega.ee/about-us/

The e-Governance Academy (eGA). (2018, February 21). Let's get acquainted: Uuno Vallner, the father of e-Estonia. *eGA*. https://ega.ee/news/uuno-vallner-father-estonias-estate/

Ueda, K. (2022). *Kinyu System no Keizaigaku* [Economics of Financial Systems]. Nihon Hyoronsha.

Williams, B. (2020). *Ikikata ni tsuite tetsugaku ha nani ga ieruka* [Ethics and the limit of philosophy] (Y. Morigiwa & K. Shimokawa, Trans.). Chikuma-shobo. (Original work published 1985)

Yamazaki, S. (2016). *Blockchain gijutu no shikumi to kanosei* [How blockchain technology works and its potential]. Internet hakusho 2016 20 nen tokubetsu kinen gou [Internet White Paper 2016 20 Year Special Anniversary Issue]. Impress A&D. https://iwparchives.jp/files/pdf/iwp2016/iwp2016-ch04-03-p243.pdf

KEY TERMS AND DEFINITIONS

Autonomy: Refers to the ability or characteristic of self-determination and self-governance, where individuals or organizations can make decisions and act based on their own purposes and values. Those with autonomy can make decisions and take actions without relying on external constraints or directives, taking responsibility for themselves, thinking for themselves, and achieving self-improvement, growth, and evolution. It is a crucial element for the concept of freedom and an important concept that can also answer the question of why freedom is important.

Big Tech: The term used to refer to a group of huge IT (Information Technology) companies with dominant global influence. These companies are primarily based in the United States and provide popular internet services and products worldwide. The representative Big Tech companies include GAFAM: Alphabet (Google), Amazon, Apple, Meta (Facebook), and Microsoft.

Blockchain: Blockchain is a decentralized digital ledger technology that connects transaction records into a chain of blocks, preventing tampering or fraud. Its immutability makes it highly reliable and does not require a centralized system. While it is well known for its use in cryptocurrency transactions, it is expected to be applied in various fields. Please refer to section 5 of this chapter for details.

External Third Party: A term coined by the author in this chapter. The concept is considered essential in traditional public systems such as modern law and the state and credit payment systems such as banks and credit cards. In blockchain, conceptually, this is not necessary. This difference highlights a key distinction between blockchain and these traditional systems regarding credit and trust.

Freedom: In this chapter, this means autonomously manifesting an objectively existing, perfect concept in the phenomenal world through one's own body or the public system, according to the individual's values and beliefs. Please refer to Section 1 for a more detailed definition by the author.

Libra: A global payment system proposed by Facebook (now Meta) and others. The issuing organization included about 30 companies and organizations such as major US payment companies Visa, Mastercard, and PayPal. The system is based on blockchain technology. Unlike other cryptocurrencies, such as Bitcoin, Libra is a type of stablecoin that can be exchanged for a specific ratio of dollars, euros, and other currencies, ensuring a stable value.

Publicness: In this chapter, from the point of view of freedom, publicness is a fundamental concept for re-examining the position of modern law and the state, as well as that of so-called private enterprises, based on the nature of the goods they provide.

Setback: A term coined by the author in this chapter. An attitude or act of deciding whether to act in some action, such as a contract, by checking it against a reference or norm from an external, third-party position considered "certain." In this chapter, this concept is treated critically in terms of both the truth that, theoretically, "certainty" cannot be ensured by referring to something external in a particular situation and the reality in life that one has no choice but to "jump in" according to what one believes in the situation in question.

Section 5

Innovations in Social Media and Data Technologies

Chapter 13
The Disruptive Nature of Short Video-Based Social Media Platforms in China:
The Case of TikTok

Jingwen Wu
University of Kentucky, USA

ABSTRACT

Can short video-based social media platforms disrupt economic markets and upend government censorship? Short videos have become the vanguard of social media. This chapter examines how this modern information communication technology (ICT) disrupts government regulation and the market as a whole in China, using TikTok as an example. Specifically, it argues that a video social media platform with unlimited content and low usage thresholds, combined with citizen's lack of legal awareness, makes it easy to expose privacy, encounter fraud, and infringe upon intellectual property. These factors are exasperated by TikTok's video format since video-based content is more difficult to censor in a timely manner. Consequently, TikTok poses challenges for government regulation and disrupts government censorship. The chapter concludes by discussing policy implications. Specifically, it proposes the implementation of modernized laws and regulations to tackle the distinct challenges and risks presented by cutting-edge technologies.

INTRODUCTION

What is the potential disruptive nature of short video based-social media platforms? Information Communication Technology (ICT) disrupts both government regulation and the capital markets in China by challenging the existing norms and creating a new dynamic that requires both regulatory adaptation and market innovation. The rise of short video-based social media, such as TikTok, has led to significant changes in the market and government policies in China. This chapter looks at these changes from two perspectives: the disruption to the government's regulatory function and the disruption of the traditional

DOI: 10.4018/978-1-6684-6429-8.ch013

market. To understand the disruptive nature of short video-based social media, TikTok is used as a case study, as it is the most widely used short video-based social media platform. "Short, Small and Refined" content allows people to obtain large amounts of information in the shortest amount of time (Bondy et al., 2022; Lee, 2021; Shutsko, 2020; Wang et al., 2019). This format enables people with low literacy levels to understand what is being said by watching videos (Shutsko, 2020). In recent years, there has been a marked increase in the number of individuals engaging with short videos in China, due to the low educational threshold required for both production and consumption (Bondy et al., 2022; Shutsko, 2020).

Short video-based social media platforms offer unique engagement opportunities compared to traditional social media platforms like Facebook or Twitter. These platforms emphasize user-to-user interaction and cross-platform sharing, leading to higher levels of user engagement and greater potential for social impact (Campbell et al., 2014; Figenschou & Fredheim, 2020; Wiley et al., 2022). The increased engagement facilitated by short videos can potentially influence government regulations by amplifying the voices of marginalized communities and providing a platform for social change. The unique features of micro-vlogging, such as the emphasis on user engagement and cross-platform sharing, make short video-based social media content more influential than other forms of social media content. This is demonstrated through factors such as emotional appeal, increased potential for virality, higher levels of user engagement, accessibility for a wider range of users, and increased opportunities for creative and innovative content creation. These features increase level of engagement—by exposing users to more diverse perspectives and engaging members in useful dialogue—which can lead to greater social impact (Wiley et al., 2022). In one example, nonprofits more effectively engage with stakeholders when micro-vlogging (Wiley et al. 2022).

However, the advancement of ICT presents complex challenges for governments regarding both the cost of employing repressive measures and the cost of regulation. ICT's ability to enhance the transparency of information can lead to increased public discontent and heighten the probability of protests (Little, 2016). Effective communication technology empowers the general public to organize and act collectively, which poses a potential threat to authoritarian regimes by issuing popular demands for a change of course to a particular policy action taken by the government (Masterson, 2015). Little (2016) shows that effective communication technology increases the likelihood of protests, given that it increases dissatisfaction with the regime by making information transparent, and enables people to know when, where, and how potential protests will be held.

For instance, the Chinese government policies have been altered to reflect citizens' preferences as a result of social media attention (Masterson, 2015). Given that ICT promotes political participation and people prefer more democratic policies (Lupu, 2014), it is likely that the development of ICT will result in an increase in democratic levels (Martin, 2013, 2014, 2015; Rojas and Puig-i-Abril, 2009; Yamamoto et al., 2018). However, the development of ICT also affords authoritarian governments the means to employ digital surveillance to detect and suppress dissent, thereby reinforcing their grip on power (Gohdes, 2014; Xv, 2021). Since authoritarian regimes lack voting channels, Xv (2021) argues that digital surveillance provides the regime with information about citizen preferences, which it uses to repress citizens. Specifically, dictators, unaware of the true anti-regime sentiment within the population, use digital surveillance to identify radical opponents and prevent coordinated uprisings. ICT, therefore, creates two possible outcomes: problems for governments in revealing their wrongdoings in an immediate manner (Mergel, 2012; Palmer and Perkins, 2012; Warschauer 2001; Poreh 2012); and empowering governments to more rapidly and precisely oppress citizens.

Furthermore, the costs of regulation and monitoring compliance with laws and regulations have increased, due to the complexity of technology and the need to keep up with industry changes. What do we expect with the developments of the short-video platform in terms of those costs, specifically TikTok? The short-video industry has undergone a decade-long development and has had a profound impact on the Chinese market, characterized by the rise of various products. Several short video platforms have launched and failed over the last decade, with TikTok becoming the largest and most popular, especially in the Chinese market (Serrano et al., 2021). Data from the *Pew Research Center* show that China has the largest number of smartphone owners, followed by India, and the United States (See Figure 1). The popularity of smartphones combined with applications for video production, editing, and sharing attracted a user base of more than one billion Internet users in China as of December 2021, with 934 million of them using short video platforms (CNNIC). The growth in users on short video platforms is accelerating, with platforms seeing more than a 600 percent increase in 2019 (iiMedia). Capitalization of the short video market reached 140.83 billion yuan in 2020 (20.42 billion USD), representing a 70 percent increase over the previous year.

This chapter argues that short video social media platforms (e.g., TikTok) disrupt the market in a variety of ways. First, it raises the issue of legal regulations in copyright and fraud protection. The two-fold question on copyright concerns original protections afforded to short videos and whether editing existing videos infringes upon the original works. Second, short video social media platforms created a live streaming business model that involved fraud, which ultimately resulted in millions of problems with online shopping and market competitiveness.

Existing literature examines how traditional social media such as Twitter influence policymaking by studying the United States (Mergel, 2012; Schwoerer, K., 2019). Despite the extensive literature on how traditional social media influence government policy, there is little discussion on how authoritarian governments deal with short video type ICTs. This chapter aims to fill this gap in the literature by exploring the impact of short video-based social media on government regulation and the market in China.

This chapter explores the impact of short video-based social media videos on government regulations, with a focus on China. It starts by providing an overview of Chinese cyber regulations and the complexities of the Chinese market. Next, the chapter discusses the impact of mobile social media on Chinese society and the challenges faced by governments in regulating video content. The rise of short video social platforms, including TikTok, and their impact on the market is also discussed. Finally, the research explores the difficulties of regulating short video platforms, the relationship between ICT and government accountability in authoritarian regimes and offers some suggestions for consideration.

Figure 1. Cross-sectional smartphone ownership population (2020)
Notes: *Data collected from Pew Research Center. The unit of the population that appears on the graphs as millions.*

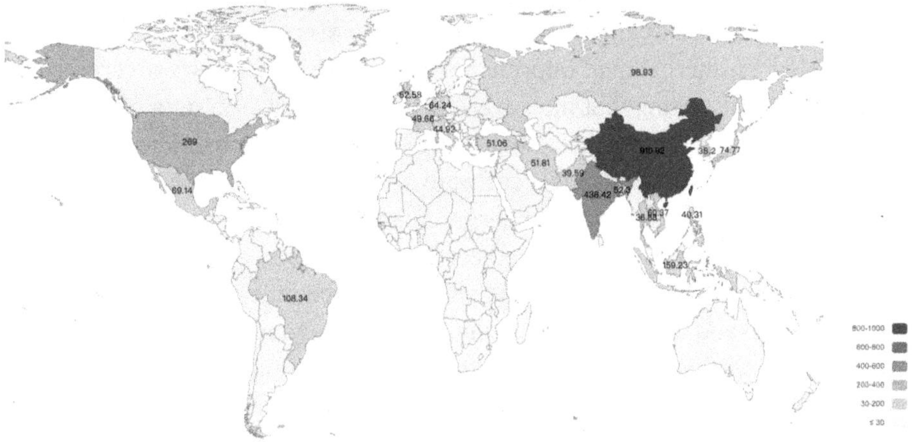

Understanding Chinese Cyber Regulation

Chinese cyber regulation refers to the legal and policy framework that has been established by the Chinese government to govern and regulate the Internet and other forms of online communication within China (Liang and Lu, 2010; Hong and Goodnight, 2020). The development and evolution of the regulatory framework in China has been shaped by the country's unique political, cultural, and economic context (Creemers, 2017; Hong and Goodnight, 2020l Lindsay, 2015; Wu 2007; Masterson, 2020). The first significant development in Chinese cyber regulation occurred in the late 1990s, when the Chinese government began taking steps to control and regulate the Internet as it became more widely available in the country (Cheung and Zhao, 2013; Endeshaw 2004; Rusakova et al., 2021; Tsui, 2003). Among the measures taken were the establishment of the China Internet Network Information Center, which coordinated Internet-related activities and issued licenses to Internet service providers (ISPs). As part of the government's efforts to regulate online activity, several laws and regulations have been introduced, including the Cyber Security Law, the Network Security Law, and the Cyber Crime Prevention Law (Endeshaw, 2004). To oversee and enforce cyber regulations, the Chinese government has established a few regulatory agencies and bodies in addition to these legislative and technical measures. The Cyberspace Administration of China is responsible for formulating Internet policy, while the Ministry of Public Security investigates and prosecutes cybercrimes (Cai, 2018; Johnson, 2018; Li, 2016).

Chinese cyber regulation has continued to evolve and expand in response to a variety of factors, including the growth of the Internet, the increasing importance of the Internet in the economy and society, and the emergence of new technologies and challenges (Clayton et al., 2006; Winter and Lindskog, 2022). In recent years, the Chinese government has taken several steps to further tighten control over the Internet and other forms of online communication, including the implementation of the Great Firewall of China, a system of technical measures that is designed to block or restrict access to certain foreign websites and platforms (Robert, 2020; Winter and Lindskog, 2022). The Chinese government has a long history of regulating and controlling the Internet within its borders. The government in China initially censored and blocked certain websites and content during the early days of the Internet. However, the government has recognized the need for comprehensive cyber regulations as the Internet has become increasingly important for economic and social development in China. Specifically, social media companies must create two separate servers, one for domestic Chinese citizens and one for overseas Chinese citizens, in compliance with their countries' regulations. As a result, the overseas version of TikTok has published a report on the problem of international information security risks. Chinese government officials have raised public concerns by asserting that the government will access user data on these platforms from foreign countries, since the data of Chinese social media companies is regulated by the government, at a time when several of China's social media software companies are on the rise internationally. International companies are also at risk from Chinese regulations on social media and the Internet, which creates potential disruption to international markets (Suciu, 2022).

As China embraces the immense commercial potential of the Internet, it is also enforcing state control over the Chinese web and its usage (Hachigian, 2001). Allowing a wide-ranging platform helps to protect the commercial goldmine from becoming politically hazardous, recognizing that an unregulated network would give power to the citizens (Hachigian, 2001). However, a win in the cyber world cannot be definite, as the Internet is unable to offer its complete commercial advantage when controlled by government agencies (Hachigian, 2001). Consequently, China has been investing a considerable amount of money to guarantee that their interests are defended by strategic regulation over the cyber environment (Interesse, 2022). The Chinese government's approach to cyber regulation has been dominated by censorship and control, along with the promotion of domestic Internet companies and the protection of national security. There have been several criticisms and pushback from both domestic and international actors, however the Chinese government has demonstrated little interest in adjusting and evolving its approach to cyber regulation in response to the increasing importance of the Internet to economic and social development in the country. Rather, those complex situations encourage the government to become more strategic (Noesselt, 2014).

Unraveling the Complexity of the Chinese Market

The Chinese market has a long and complex history, dating back to ancient times when the country was ruled by various dynasties. In recent history, China underwent significant economic and political changes, including the establishment of the People's Republic of China in 1949 and the implementation of economic reforms in the late 1970s by Deng Xiaoping (Chevalier and Lu, 2009; Qian, 2017; Weber, 2021).

During the 1980s and 1990s, China experienced rapid economic growth and became a major player in the global economy. This growth was largely driven by the country's manufacturing sector, which benefited from cheap labor and export-oriented policies (Bell et al., 1993; Knight and Song, 2005; Tung and Worm, 2001). In recent years, China has undergone significant economic and social changes, including a shift towards a more consumption-driven economy, increased urbanization, and the rise of a middle class (Sicular et al., 2022; Yang et al., 2011). Today, China is the world's largest market for a variety of products and services, including automobiles, electronics, and luxury items.

The Chinese private sector has experienced significant growth in recent years, becoming increasingly competitive in the global marketplace as a result. Government policies, technological advances, and market access have enabled these companies to achieve success. Several government policies have been especially beneficial, providing access to credit and other forms of financing, as well as tax incentives for private companies. Technology has played a major role in improving efficiency and modernizing the production processes of Chinese companies. Furthermore, Chinese companies have been able to tap into expanding global markets, which has further led to their expansion and growth. As a result of these developments, Chinese private companies have become increasingly competitive, and many are now playing a major role in the international economy (Garnaut et al., 2012; Gregory and Tenev, 2001; Lardy, 2014).

Despite these changes, the Chinese market remains characterized by significant government intervention and state-owned enterprises, along with a complex regulatory environment (Lin et al., 2020; Wang et al., 2004). State-owned enterprises (SOEs) are enterprises owned and operated by the Chinese government. In addition to providing a variety of goods and services, they are primarily engaged in the energy, finance and infrastructure sectors. SOEs have played an integral role in China's economic growth, facilitating the industrialization and modernization of the country. They also have a significant presence on international markets, making them a major player in international trade. Government regulations affect SOEs on a day-to-day basis, and the state has a significant impact on their activities. It has been criticized that SOEs are overstaffed and inefficient in the use of resources despite their importance (Lin et al., 2020). Despite these challenges, SOEs remain a major part of China's economy.

The Impact of Mobile Social Media on Chinese Society: A New Era of Communication

The proliferation of technology, digitalization, and network connectivity has led to a significant transformation in the modalities of communication on social media platforms within the Chinese context. In the 90s, phone calls replaced pen pals, and by the turn of the century, writing blogs became a popular trend. Social applications platforms like Tianya, Sina blogs, and RenRen were widely used (Masterson, 2020). Furthermore, the emergence of posting platforms like Tie Ba, which allow people to post comments online, and forums have made communication much simpler than ever before. New forms of socialization emerged, such as sharing with those you do not know. Then smartphones took over computer-based connections, leading to a growth in mobile apps and microblogs (Sina Weibo), QQ, and WeChat flourished (Masterson, 2020). China's social media network is undergoing a diverse range of developments, and people are increasingly relying on it to access information. The sight of *low-headed people* who are constantly looking down at their phones has become a common occurrence as more and more people turn to social media. Innovations in technology have led to more effective and affordable mobile devices and applications. Social functions have become the standard of various contemporary Internet services and applications. The most intuitive embodiment of this development is the growth of mobile social media users. At the end of 2010, the total number of mobile Internet users in China was 288 million, of which 930 million users use various instant messaging services including social media through mobile phones in 2020, accounting for 99.8 percent of mobile Internet users.[1]

Due to the ever-increasing reach of mobile social media, the fundamental dynamics of social life are being reinterpreted on a large scale. This has resulted in significant alterations to Chinese society, ranging from the way in which people interact to how they entertain, work, trade and provide services. Mobile social media redefines interpersonal communication and reconstructs contemporary Chinese community relations, work style, family pattern in the form of friends' networks and group chat. Changes in government are also made to adapt to innovations. It is well known that the Chinese government censors outside and inside information, allowing only information that is in the interest of the regime to maintain stability within the country. Those disruptive innovations have profoundly changed the way governments respond to their citizens. Digital technology has enabled governments to collect information from citizens and to govern more effectively. For instance, in the past ten years, it has become a norm for government agencies at all levels to open official accounts on mainstream social networking platforms such as Weibo and WeChat to promote their propaganda and create channels for citizens to express their concerns at all levels of governments (Li et al., 2022). As a result, this became an effective way for the Chinese government to govern and a means for gauging the general public's preferences as an authoritarian regime where there was no way of determining the "median voter preference." Governments find it more difficult to censor information due to the divergence of information. Specifically, in place of conventional methods such as blogging or wording posts, video features on social media (so called short video) have become extremely difficult for them to censor (Xiang, 2019).

We live in a media-driven society today. As compared to traditional media, social media has the advantage of timely updates. Users can receive the information released by the outside world at the first instance, make comments in real time according to the information content, and gain access to historical information more conveniently and faster (Appel et al., 2020; Kaplan and Haenlein, 2010; Miller et al., 2016). Aside from counting their popularity, social media can also change media forms in real time in accordance with users' preferences, such as from pictures and words to videos and audio. Social media is designed to share and allow users to express themselves more freely while reducing browsing costs. Furthermore, because of the characteristics of Internet connectivity, sharing, commenting, and browsing can all be tracked and accounted for.

Tencent WeChat and Sina Weibo became the most popular social media sites in China after 2009. Tencent launched WeChat on January 21st, 2011, to provide instant messaging for intelligent terminals. WeChat enabled users to send free voice messages, videos, pictures, and text messages across multiple networks (Figure 2 the first column). It also allowed users to share streaming media content and location-based social plug-ins using services such as Shake, Friends Circle, Public Platform, and Voice Notepad. Users were able to add friends and follow the platform by shaking, searching for numbers, existing phone numbers, scanning QR codes, and following people nearby. WeChat had the feature of communicating only with friends' networks (comparable apps such as What's App), while also also allowing users to share content with friends only. Despite the features of WeChat, the disruptive potential for the social media platform is low. This is because the level of information dissemination and accessibility is limited due to how people primarily use WeChat, which is to send text messages. People normally use WeChat to send text messages, which is a convenient and popular way of communication. It is easy for the Chinese government to censor text messages sent by WeChat, which implies that it is easy for the government to regulate the content that is being spread. However, because WeChat is only used within a relatively small circle of people, there is less of a risk involved in causing a delay in the government's censorship and regulation. Therefore, it is generally assumed that WeChat is a secure platform for communication, as the government's ability to monitor it is limited.

Figure 2. Comparison of Chinese social media: Features and disruptive levels
Notes: *The above logos are software icons, representing WeChat, Weibo, and TikTok from left to right.*

	WeChat	Weibo	TikTok
Features	• Communication is limited to friend's network. • Sharing content can only be seen by friend's network.	• Weibo adopts a socialization model based on opinion leaders. • Allow followers engage with influencers by commenting on their content.	• The threshold for joining is low (literacy is not required). • Allow audience parallel interaction. • Communication through video content creation, and circle does not limit to friends.
Disruptive Level	• **Low.** • Finite network. • Text communication. • Easy to censor and control by government.	• **Medium.** • Spread widely by opinion leaders. • Celebrity effect requires quick censorship.	• **High.** • Infinite network. • Spread widely among all groups. • Censorship includes accurately reviewing images, videos, and messages.

The Chinese social network Weibo is comparable to Twitter in terms of its functionality (Figure 2 the column 2). The Chinese term Weibo is the abbreviation for micro-blogging, which is a broadcast social network platform used to share short and real-time information. Weibo is a sharing and communication platform. It pays attention to timeliness and randomness. While micro-blogs can express thoughts and trends at any time, blogs focus more on sorting out what you have seen, heard, and felt over time. Weibo differs from traditional blogs because it only displays 150 Chinese words rather than a full article like a blog. This, in turn, increases social engagement by reducing the cost of information accessibility. Furthermore, Weibo adopts a socialization model based on opinion leaders, allowing all followers to engage with influencers by commenting on their content. Thus, the celebrity effect is very common in Weibo. Celebrity effect refers to "the ability of famous people to influence others" (Stevenson, 2010). Weibo allows users to follow celebrities or influential people that they are interested in, and they are more likely to be influenced by these individuals. It is likely that the government's untimely management of influencers' content will result in troubles. For instance, most of the influencers in China have millions, or even billions, of followers. The influence of influential people can create a great deal of trouble for the government in some cases (Masterson, 2020). The inappropriate remarks of a celebrity may have a significant impact on society. An example of the phenomenon known as "Weibo celebrity effect" can be seen in the case of Li Jiaqi, also known as China's Lipstick King. Li has gained a tremendous amount of influence on Weibo, with over 35 million followers. Every time Li mentions a product on his page, it usually results in an upsurge in sales and brand awareness; one brand even reported a 3,600 percent rise in sales after Li publicized their product (Li, 2016). Weibo's comments feature allows users to interact with followers who follow the same person, so the network is not limited to friends. Based on its capacity to extensively propagate information, Weibo is considered to possess a medium level of disruptive potential.

TikTok is a social media app that allows users to create and share short videos set to music. It is different from other social media platforms in a few key ways: First, it is video-oriented. TikTok is primarily focused on video content, whereas other social media platforms may have both text and photos. Additionally, it is oriented towards music. Videos on TikTok are usually set to music, and users may create videos by selecting from a variety of popular songs or by creating their own audio. Having such a strong emphasis on music sets TikTok apart from other social media platforms. The biggest difference is that short video based platform such as TikTok is algorithm-driven. TikTok uses algorithms to recommend content based on the interests and viewing history of its users. This makes it a unique platform compared to other platforms that require users to follow specific accounts or hashtags to discover content. In China, at the end of 2016, social media appeared in the form of short video clips, mainly on the platform of TikTok. In 2021 it became one the most popular app in China. It is the fastest growing social software for Chinese users (QuestMobile, 2021).[2] Figure 3 shows the comparison trend of the daily active users of different apps over the period 2017-2021. As can be seen, TikTok has experienced the greatest increase since 2018. It is famous for its maximum video duration of fifteen seconds. The TikTok platform allows users to upload videos in real time and add special effects and filters as needed. Combined with cheerful rhythm music, it enhances the visual impact for the audience. TikTok is currently one of China's most popular short video platforms with an average daily activity of 150 million and an average monthly activity of over 300 million.

TikTok is considered a high-level disruptive technology because it has revolutionized the way people share and create content (Jia and Liang, 2021). The app's user-friendly interface and creative tools have enabled an unprecedented level of user engagement, allowing users to quickly create and share short-form video content. This has allowed businesses to reach new audiences with their products and services in an innovative way, while also allowing influencers to reach larger audiences and drive sales. Additionally, TikTok has enabled a whole new type of content to be created and shared, from lip syncing and dancing videos to comedic skits and educational content. This has changed the way people consume and interact with digital content and has allowed new types of content to gain popularity.

Furthermore, Video based content is much harder for governments to censor than text-based content like WeChat, QQ due to the nature of the content itself. Video content is much more complex, and censorship is typically easier when dealing with text. For instance, TikTok's disruptive nature creates many drawbacks as a short video publishing platform with almost no threshold. For example, in 2018, a video released by a young girl accidentally captured her mother bathing behind her back, and such a video is visible in the public's field of view through review. The video was released at 22:06 on June 19th, and it was deleted one hour and 40 minutes later when a user reported it. Although the number of broadcasts was only 216 times, this incident also exposed the drawbacks of TikTok in the supervision mechanism. Excessive pursuit of the immediate release of information leads to the omission of users' published content. In such an era of rapid information dissemination, once some bad/private content appears in the public media, it is likely to lead to situations that are out of the creator's control. However, due to the low entry threshold of TikTok, different users will frequently publish short new video content, which brings great challenges to auditors. Personal information, address information, and banking information can also be disseminated in the same manner.

Video content is made up of both audio and visual components, and censoring it requires a much deeper understanding of both the audio and visual elements. Governments must analyze the visuals to determine if they violate any laws or regulations, as well as the audio to determine if any offensive language is being used. This is a much more complex task than censoring text-based content, which can simply be scanned for words or phrases that are deemed offensive or illegal. Another problem with censoring video-based content is the sheer amount of data that is contained in it. With text, the number of words or phrases that must be read is relatively small, and as a result of binary digitization, these texts can be scanned nearly instantly upon uploading.

The magnitude of data present in video-based content is substantially greater than that in text-based content, as it encompasses both auditory and visual components. This necessitates the allocation of significant resources for the analysis of each frame of the video to determine any potential violations of laws or regulations, thereby rendering censorship of video-based content more challenging. Furthermore, the swift and extensive sharing of video-based content exacerbates the difficulty of censorship. Governments may be able to effectively censor text-based content owing to its relatively limited scale, but the vastness of data in video-based content renders censorship more arduous. As such, governments may face challenges in censoring video-based content, which may result in its reach to a broader audience than intended.

A prominent example of this occured during COVID-19 pandemic Shanghai lock down. Shanghai is China's economic center and one of the most developed cities in China. Shanghai's municipal government has often been regarded as among the most efficient and the "people's livelihood-oriented" government. Shanghai is home to the most educated and wealthy people in China, as well as many foreign companies with the highest level of foreign employment. The people in the region tend to obtain more information, and they possess a relatively thorough understanding of the democratic system. Political oppression appears to be resilient. Accordingly, Shanghai's pandemic lockdown was exposed to the world in a way it has never been exposed before. It is rare for Chinese social media users to criticize the government in large numbers. However, during the Shanghai lockdown, many Shanghai residents took to social media to voice their concerns about their poor living conditions. Residents complained that they cannot find food or must wait a long time for vegetables, meat and eggs provided by the government. The government also introduced measures to prevent virus carriers from leaving their homes, such as installing electronic alarms on residents' doors and forcibly evacuating residents to other locations so they can be disinfected. The government typically responded to these complaints by refuting rumors or censors. Thus, a large majority of residents recorded and post videos on the Internet using short video platforms showing government officials kicking in their doors, ruining their finances, and swearing at them. This caused heated criticism by citizens of the government. Video showing these unreasonable practices was widely circulated on the Internet. An anonymous video entitled "The Voice of April" provides an account of the lockdown in Shanghai. A black-and-white picture of Shanghai's skyline is accompanied by sounds recorded over the past month: a resident pleaded with the government for materials; a son pleaded for the hospitalization of his sick father; a grassroots cadre explained that she also felt tired and could not help the frustrated caller.

An anonymous user's video that was posted on a short video social media site garnered a significant amount of attention before it was removed. To circumvent censorship, citizens turned the video upside down, embedded it in other pictures, and used the dubbing to send the video to unrelated videos. For instance, they have created a video in which SpongeBob can be seen sitting in his room watching television that displays the video. Short videos are used by citizens to circumvent censorship and attract greater attention. Even though the government has increased the intensity of video censorship, available technology has not been able to keep pace with creative ways users circumvent the censors. Consequently, the short video based social media platform has created a challenge for government regulation regarding censorship (Yuan and Duo, 2018).

In addition to government censorship, it also causes disruption in laws and regulations resulting in challenges for the government. Short videos are capable of meeting fragmented time utilization and social needs of the public, as they possess the following characteristics: short time, fast transmission, short production cycle, a low threshold, and strong public participation. Despite the significant social value it creates, copyright protection issues, such as more complicated infringement methods, inadequate legal protections, and relatively backward industry supervision, result in an increased incidence of network infringement and piracy. With the huge number of short videos available, finding effective ways to use copyright law to protect the original content and resolving conflicts between original and second creations is inevitable. Short videos are subject to two types of copyright issues. The first question is whether the short videos themselves are protected by copyright; the second is whether the short videos produced and edited by existing videos infringe on the original work. Some short video producers have arbitrarily re-edited the original videos of other people into short videos and uploaded them to the platform to increase their traffic and advertising profits in a short amount of time and at a low cost. For instance, the Intermediate People's Court in Xi'an, China ruled in favor of Tencent in a copyright infringement case against TikTok on October 26th, 2022. The court ordered TikTok to remove the infringing videos and compensate Tencent with 32 million yuan (approximately 4.97 million USD) and 426,900 yuan (approximately 65,935 USD) in litigation fees.[3]

Besides influencing Chinese government regulation, it has also disrupted foreign government regulation. TikTok has faced controversy and scrutiny from governments around the world due to concerns about data privacy and national security. In the United States, the Trump administration threatened to ban the app in 2020, citing national security concerns related to the app's Chinese ownership and the possibility that user data could be accessed by the Chinese government. One way in which TikTok has disrupted government regulation is by challenging traditional approaches to data privacy and national security. The app's global reach and the complex nature of its ownership and data storage have made it difficult for governments to regulate and monitor its operations.

TikTok has disrupted government regulation by raising new and complex issues related to data privacy and national security, and by forcing governments to adapt to the rapidly evolving social media and technology landscape.

Figure 3. Chinese social media daily active users (2017-2021)
Notes: *Data collected from Statista. Unit in millions.*

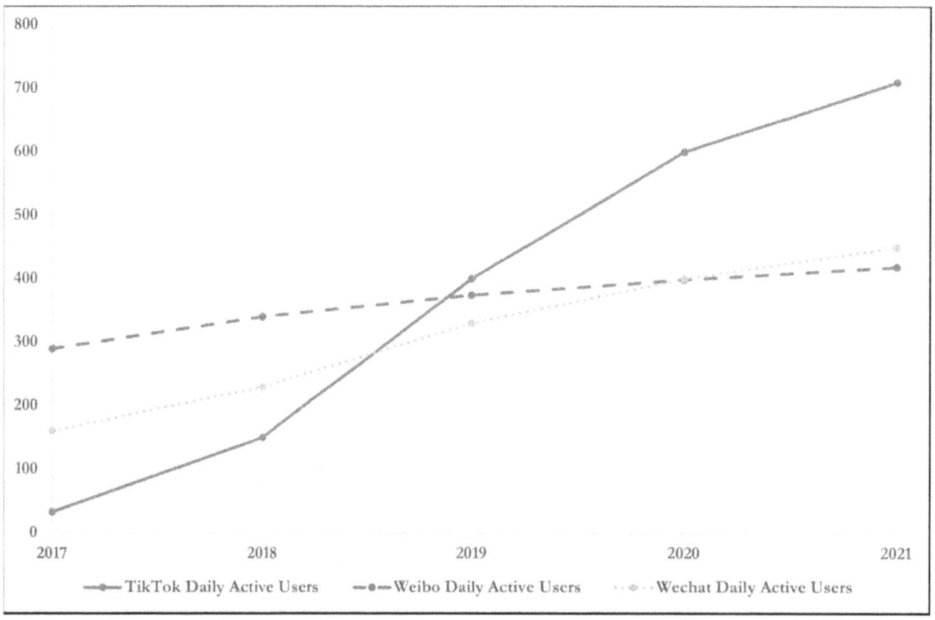

The Disruption of Short Video Social Platforms: How TikTok is Changing the Market

One way in which TikTok has disrupted the market is by providing a new platform for content creation and distribution. Many users have gained significant followings on the app and have even been able to turn their TikTok fame into successful careers, such as by promoting products or services or by creating and selling their own merchandise. These short video platforms offer more than uploading a video of duration from 15 seconds to 15 minutes; they also offer live streaming (Webcasting; Zhibo; 直播). Webcasting refers to the entertainment practice of broadcasting instant images on the Internet because of the growth of online audio-visual platforms. The biggest difference between live broadcast and uploading video is that the audience can interact with the anchor in real time through messages or bullet comments (Dan mu; 弹幕). This is because the anchor can adjust the program content or please the audience in time according to the audience's response. Live broadcasting as an industry grew in a straight line during the period of the epidemic. As a result of China's lock-down policy, people had social needs to interact directly with people. This prompted the live broadcast industry to develop rapidly to meet those needs. In 2019, the sales of live delivery goods in China reached nearly $61 billion, and this figure doubled in 2020.[4] According to the BBC report, Viya, a popular Chinese live broadcaster, announced to sell a commercial rocket launching service on Alibaba Taobao for a price of 40 million yuan (about 5.64 million US dollars) and 800 people paid an online deposit of 500,000 yuan (about 72,885 US dollars).[5]

A webcasting room is characterized by social interaction between netizens and each other, as well as between netizens and anchors. This interaction creates a sense of adrenaline that may cause some consumers to engage in impulse purchases. Additionally, the sudden outbreak of the epidemic has served as a boost and catalyst for the development of live selling goods. Likewise, the range of commodities is expanding, ranging from traditional clothing, cosmetics, to local products, houses, etc. They can pay directly on these live broadcast platforms. Furthermore, posting comments also allows one to ask for service in addition to purchasing products. For instance, one may request that the broadcaster perform a show for a fee. These characteristics have brought great influence on the economy.

Webcasting is also frequently subjected to false propaganda. Considering the low barriers to entry, most anchors lack relevant legal knowledge, and the timeliness of live webcasts forces anchors to describe products on the spot, which can lead to false information. This is likely to disrupt the fairness of the market, especially for competitors. Moreover, it is difficult for people to know the actual state of goods by watching videos on the Internet in the live broadcast room. This feature is frequently used by criminals to sell counterfeit goods. Since the methods for seeking rights and interests through live broadcast in online shopping have yet to be developed, many individuals hesitate to assert their rights and interests. Furthermore, live broadcast rooms usually place millions of orders for goods in a short time, which makes it difficult for the government to regulate such orders. Regardless of whether these rights are protected, the government is necessary to assist this enormous group in defending their rights. The situation is also problematic.

Data fraud is also a concern. As live delivery has gained popularity, data such as viewership volume, fans, comments, and sales have become key elements of industry competition. There are some platforms and businesses that try to increase their sales volume through online celebrity effect, so "faking bills," "buying fans" and "faking comments" are the unspoken rules. On certain e-commerce platforms, the sale of fraudulent services is still prevalent and priced differently. These services are created to mimic actual data streams and evade detection by the platform's monitoring system. Data fraud not only deceives businesses, misleads consumers, but can also destroy the ecology of the entire live broadcast industry. To this extent, the government's ability to supervise is limited.

Another way in which TikTok has disrupted the market is by changing the way that users consume content. The app's algorithm curates a personalized feed for each user based on their interests and the content they engage with, which has made it easier for users to discover new and relevant content. This has also made it easier for content creators to reach new audiences and gain visibility.

In addition to its impact on content creation and distribution, short video based social platforms have disrupted the advertising industry by offering a new platform for businesses to reach and engage with consumers. The app's targeting capabilities and the ability to create sponsored content have made it an attractive option for brands looking to reach a younger and more digitally savvy audience.

Short video social platforms have disrupted the market by offering a unique and innovative platform for content creation, distribution, and advertising, as well as by changing the way users interact with social media content.

Censoring Social Media: Navigating Government Regulations and Control on Short Video Platforms

Chinese government quickly reached a deal with TikTok to censor videos that contradict its views. In 2019, researchers tried to upload the historical video that shows young people standing up to block a line of PLA tanks during the Tiananmen crackdown, June 4th, 1989. One of the videos was uploaded with an account registered in Australia, but only the account holder can see the content. In fact, the integration of private enterprises, including social media companies, into government and party offices has been mutually beneficial. The government has implemented policies that support the growth of these enterprises, such as providing financial assistance to small and medium-sized enterprises. The government has implemented a negative inducement strategy for larger businesses, where non-compliance with regulations could result in loss of their current privileged status with government policy support. For instance, Alibaba faced sanctions from the government for engaging in monopolistic practices. Hence, the government has exercised greater regulatory oversight over these enterprises, driving social media companies to develop censorship technology to meet government requirements. In 2019, a clearer regulation on short video management was introduced. On Jan 9th, the China Network Audiovisual Program Service Association issued the "Network Short Video Platform Management Specification" (hereinafter referred to as "Specification"). As a result of these regulations, the government was now in a position to better control public opinion in the future.

The Specification includes six points:

1. The network platform for short video service shall possess the relevant qualifications stipulated by laws and regulations such as the License for Information Network Communication of Audiovisual Programs (AVSP) and conduct business strictly within the scope specified in the license.
2. The network short video platform should actively introduce mainstream news media, party, government and military organizations and other institutions to open accounts and improve the supply of positive and high-quality short video content.
3. Network short video platforms should establish a system for editor-in-chief content management.

4. Short video platforms on the network implement content review by the platform before posting anything for public viewing. Content review is required for all short videos broadcast on the platform, including the title, introduction, barrage, comments, and other program contents.
5. Platforms developing short video services should assemble a team of auditors with strong professional abilities and high political quality. Radio and television administrative departments should train auditors at or above the provincial level, and the number of auditors should match the number of short videos uploaded and broadcast. Ideally, there should be more than 1/1000 auditors for every short new video broadcast on this platform.
6. Failure to comply with this specification shall result in legal consequences.

ICT and Government Accountability: A Paradoxical Relationship in Authoritarian Regime

Given its disruptive nature, can ICT improve citizen's ability to hold governments accountable? The use of communication technology is being seen as an effective tool for holding governments accountable. First, it lowers the cost of information, enabling citizens to access more political information. In particular, the development of communication technologies has enabled one to learn more frequently about government wrongdoing (Gil de Zúñiga and Chen, 2019). Second, the public now has a wider range of means to report government wrongdoing (Gil de Zúñiga and Chen, 2019). As an example, the government creates official accounts and mini programs on social media platforms for citizens to post complaints and suggestions. By using this approach, the government can gain a better understanding of citizens' preferences. For instance, all levels of the Chinese government have established network channels and opened government apps for receiving complaints. It is of particular importance in authoritarian regimes, where there are no elections through which the government can obtain information about the preferences of the people regarding policy. Third, communication technology enables citizens to overcome collective action problems, thereby reducing the cost of revolution and posing a threat of not reacting to citizens' concerns. Consequently, the regime would be more likely to respond to their citizens as communication technology develops. Combined, this implies that, communication technology has the potential to increase government accountability as it develops.

That said, governments are strategic actors. Authoritarian regimes care about regime legitimacy. Different countries consider legitimacy differently and reinforce legitimacy strategically. In general, legitimacy comes in two forms. In democratic societies, citizens consider government legitimacy as having enough representatives, holding fair elections, separating power, or meeting people's policy preferences etc. In other countries, government legitimacy is measured by the government's ability to provide sufficient security (Mittiga, 2021). Mittiga (2021) considers the former types of legitimacy that were stated above as contingent legitimacy and the latter as fundamental legitimacy. Governments use both types of legitimacy to maintain political stability. I expect that when one salient issue undermines government legitimacy, the government will be more responsible for addressing other issues (contingent legitimacy) to restore fundamental legitimacy. COVID-19 is an example of a salient issue. The fundamental legitimacy of COVID-19 was undermined by restricting people's free movement. Owing to fears of inadequate hospital capacity and other potential risks, the government remains steadfast in its pursuit of a zero COVID policy, keeping the lockdown measures in place until December 2022.

With limits on free movement undermining the first types of legitimacy, will the Chinese government be more likely to use other methods (i.e., increase government response to other societal concerns) to enhance legitimacy? Currently, China is undergoing a period of economic development and social transformation, resulting in a wide range of social problems, social security emergencies, and social welfare crises. Internet popularity has led to an ever-growing number of Internet users, and the Internet has become a platform for people to share their views and opinions. Recent years have seen the development of a variety of social media, slowing down the pace of government management. As a result, managing the network environment presents many challenges to the government. Is the government more responsive to other concerns during COVID-19 regulations? The answer to this question seems to be no. Historically, China has adopted a strategy of selective responses, which implies that they only address concerns that maximize their rewards. For instance, the literature suggests that Chinese respondents are more likely to answer questions on their platforms on single tasks and issues that can be easily handled (Su and Meng 2016); they are more likely to censor online comments related to collective action problems (King, Pan and Robert 2017; Chen Pan and Xv, 2016); and the Chinese government is increasingly responding to climate change related issues (Mittiga, 2021). In addition, more strategies can be used to divert people's attention from COVID policy. They typically do this by expanding their propaganda online and reporting the worse off conditions internationally. Thus, instead of undermining legitimacy, people will become more tolerant of arbitrary national policies as their nationalism strengthens. During China's COVID lockdown, social media reported that COVID cases were on the rise, particularly in the United States. In a survey conducted during COVID, 1,000 people were interviewed; they found that the Chinese were more critical of the United States and more trusting of the Chinese government. Another example is a large government-related scandal that took place recently in Henan during COVID. Lockdown policy brought many economic issues. Banks refuse to give citizens money they have saved in their accounts. As a result, large numbers of people demonstrate on the street, which becomes the largest protest under high political pressure. The story was covered on a variety of media platforms where blooming news promotes Nancy Pelosi's visit to Taiwan and condemns the US for failing to respect China's sovereignty. This reinforces nationalism and divergent public attention for addressing diminished legitimacy. Similarly, a democratic transition was expected in Middle East countries following the Arab Spring. It is in such instances that social media becomes a powerful tool for citizens to organize protests and overcome the collective action challenge. Despite this, a lesson that can be drawn from the Arab Spring is that authoritarian governments are soon to learn how to use technology to their advantage.

Technology shutdowns and restrictions are not limited to China. Russia, for instance, Internet regulator Roskomnadzor announced on March 4, 2022, that Twitter and Facebook would be blocked along with TikTok upload would be prohibited. On the 14th, Instagram was added to the banned list. In addition to restricting social media, Russian authorities have also constrained several new websites, such as BBC, Voice of America, Radio Free Europe. In 2021, Cuba shut down Internet access for 176 hours to restrict mass mobilization in response to the government's poor handling of the COVID-19 pandemic.[6] Myanmar, after military power takes over, they blocked the Internet off and on, along with banning certain social media platforms during certain times and to maintain monopoly on information. Since governments need time to adapt to the development of science and technology, they generally have strategic means to control domestic information flows in a way that favors them, especially in authoritarian governments where the people hold the government less accountable. There may not be a more accountable government as communication technology develops. Figure 4 shows the China Freedom Index, which shows that all indicators, including civil liberties, freedoms of expression, and political rights, have decreased significantly. After 2019, there has been a significant decrease.

To conclude, the development of new technologies in China is not likely to result in increased government accountability. This is because Chinese government has a well-established track record of limiting access to information that they consider sensitive or contrary to their interests. Additionally, the widespread use of technology in China has provided the government with even more tools and resources to monitor and control the flow of information. Instead, these technologies will likely be used to further strengthen the government's censorship apparatus and enable them to gather more information about citizens, suppress dissenting voices, and maintain control over the flow of information, making it more difficult for citizens to hold the government accountable.

Figure 4. China freedom index (2013-2022)
Notes: *Freedom house data. The graph shows China's freedom trends in civil liberty rights, freedom of expression rights, political rights, and the aggregate index of all these factors (note significant decrease after 2019).*

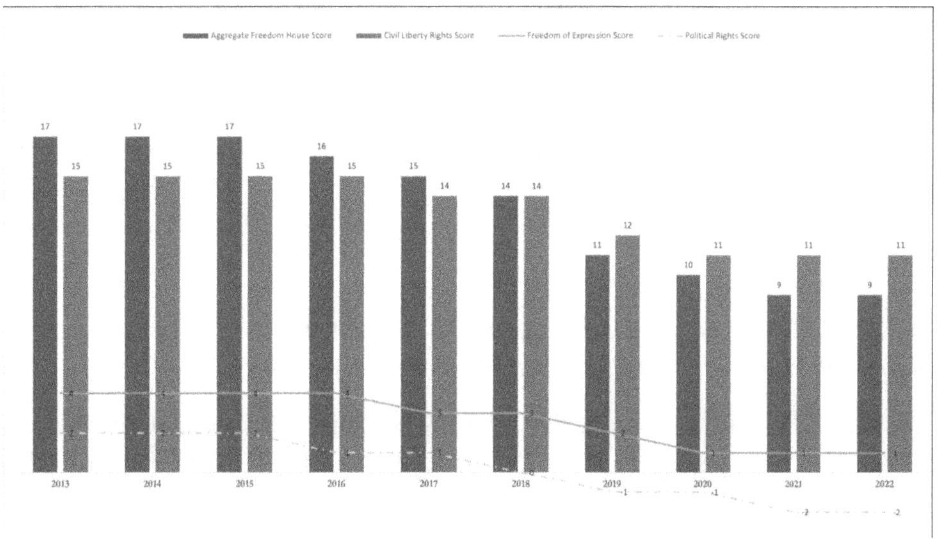

CONCLUSION

This chapter examines the disruptive nature of ICTs with a special focus on "short video-based" social media platforms. The complexity of science and technology increases the difficulty of government management. It is also important that laws and regulations remain current. The development of short video platforms is not limited to China. Now, many platforms have launched short video applications one after another. For example, Facebook now has a short video function.

Unlike other social media platforms, TikTok provides unlimited and unrepeated content. The content will be continuously updated as you slide down. In lieu of manual information retrieval and filtering, the algorithm is recommended as a method of program delivery for platforms such as TikTok that support short video clips. This application analyzes the types of videos that users might be interested in, and then recommends those videos to the user interface as they "slide down" the viewing screen. A platform with unlimited content and low usage thresholds, as well as people's lack of legal awareness, make it easy to expose privacy, encounter fraud, and infringe upon intellectual property.

The implications for policymakers are complex and varied. The disruptive impact of technology requires governments across global to adapt in multiple ways. New technologies present unique challenges and risks that must be addressed by regularly updating laws and regulations. To protect individuals' personal information, governments may enact laws regarding data privacy or cyber security (Akrivopoulou, 2013; Howard, 2010). Governments may also establish regulatory agencies or task forces to enforce technology-related laws and regulations, while providing legal training to regulatory agencies. Furthermore, they may engage in international cooperation and negotiations related to technology issues, such as data privacy and cyber security.

Governments should also invest in research and development to adapt to disruptive technologies in order to remain competitive with technological advancements. Additionally, this may require funding for research and development, partnering with private companies to develop cutting-edge technologies, or supporting technology startups and incubators. As part of their responsibilities, governments may also seek to educate and inform the public about new technological risks and advantages. Further, they may provide resources and support to help individuals and businesses adapt to technological developments. Overall, the constant evolution of technology and social media requires both citizens and governments to adapt. Leaving these advancements unaddressed is not a viable option and any responses must be carefully assessed in terms of the impact on democratic principles and the market economy.

REFERENCES

Akrivopoulou, C. M. (2013). *Digital Democracy and the Impact of Technology on Governance and Politics: New Globalized Practices: New Globalized Practices*. IGI Global. doi:10.4018/978-1-4666-3637-8

Appel, G., Grewal, L., Hadi, R., & Stephen, A. T. (2020). The Future of Social Media in Marketing. *Journal of the Academy of Marketing Science*, *48*(1), 79–95. doi:10.100711747-019-00695-1 PMID:32431463

Bell, Kochhar, & Khor. (1993). *China at the Threshold of a Market Economy.* International Monetary Fund.

Bondy Valdovinos, Zeng, & Wikstrom. (2022). *TikTok: Creativity and Culture in Short Video*. John Wiley & Sons.

Bower, J. L., & Christensen, C. M. (n.d.). *Disruptive Technologies: Catching the Wave*. https://www.academia.edu/download/59516552/disruptive_technologies20190604-89196-pd4gyy.pdf

Cai, C. (2018). Global Cyber Governance: China's Contribution and Approach. *China Quarterly of International Strategic Studies*, *4*(1), 55–76. doi:10.1142/S2377740018500069

Campbell, D. A., & Lambright, K. T. (2020). Terms of Engagement: Facebook and Twitter Use among Nonprofit Human Service Organizations. *Nonprofit Management & Leadership*, *30*(4), 545–568. doi:10.1002/nml.21403

Campbell, D. A., Lambright, K. T., & Wells, C. J. (2014). Looking for Friends, Fans, and Followers? Social Media Use in Public and Nonprofit Human Services. *Public Administration Review*, *74*(5), 655–663. doi:10.1111/puar.12261

Chen, J., Pan, J., & Xu, Y. (2016). Sources of Authoritarian Responsiveness: A Field Experiment in China. *American Journal of Political Science*, *60*(2), 383–400. doi:10.1111/ajps.12207

CheungA. S. Y.ZhaoY. (2013). *An Overview of Internet Regulation in China*. doi:10.2139/ssrn.2358247

Chevalier, M., & Lu, P. X. (2009). *Luxury China: Market Opportunities and Potential*. John Wiley & Sons.

Clayton, R., Murdoch, S. J., & Watson, R. N. M. (2006). Ignoring the Great Firewall of China. In Privacy Enhancing Technologies (pp. 20–35). Springer Berlin Heidelberg. doi:10.1007/11957454_2

Creemers, R. (2017). Cyber China: Upgrading Propaganda, Public Opinion Work and Social Management for the Twenty-First Century. *Journal of Contemporary China*, *26*(103), 85–100. doi:10.1080/10670564.2016.1206281

D'Arienzo, M. C., Boursier, V., & Griffiths, M. D. (2019). Addiction to Social Media and Attachment Styles: A Systematic Literature Review. *International Journal of Mental Health and Addiction*, *17*(4), 1094–1118. doi:10.100711469-019-00082-5

Dragu, T., & Lupu, Y. (2021). Digital Authoritarianism and the Future of Human Rights. *International Organization*, *75*(4), 991–1017. doi:10.1017/S0020818320000624

Dunaway, J., & Lawrence, R. G. (2015). What Predicts the Game Frame? Media Ownership, Electoral Context, and Campaign News. *Political Communication*, *32*(1), 43–60. doi:10.1080/10584609.2014.880975

Endeshaw, A. (2004). Internet Regulation in China: The Never-ending Cat and Mouse Game. *Information & Communications Technology Law*, *13*(1), 41–57. doi:10.1080/1360083042000190634

Figenschou, T. U., & Fredheim, N. A. (2020). Interest Groups on Social Media: Four Forms of Networked Advocacy. *Journal of Public Affairs*, *20*(2). Advance online publication. doi:10.1002/pa.2012

Garnaut, L. S., Yao, Y., & Wang, X. (2012). *Private Enterprise in China*. ANU Press.

Geyser, W. (2021). *What is TikTok? The fastest growing social media app uncovered*. Influencer Marketing Hub. Retrieved from https://influencermarketinghub.com/what-is-tiktok/

Gil de Zúñiga, H., & Chen, H.-T. (2019). Digital Media and Politics: Effects of the Great Information and Communication Divides. *Journal of Broadcasting & Electronic Media*, *63*(3), 365–373. doi:10.1080/08838151.2019.1662019

Gohdes, A. R. (2020). Repression Technology: Internet Accessibility and State Violence. *American Journal of Political Science*, *64*(3), 488–503. doi:10.1111/ajps.12509

Gregory, N., & Tenev, S. (2001). *The Financing of Private Enterprise in China*. Finance & Development. https://www.elibrary.imf.org/view/journals/022/0038/001/article-A005-en.xml

Griffiths, J. (2019). *The Great Firewall of China: How to Build and Control an Alternative Version of the Internet*. Zed Books Ltd. doi:10.5040/9781350225497

Guidry Jeanine, P. D., Waters Richard, D., & Saxton Gregory, D. (2014). Moving Social Marketing beyond Personal Change to Social Change: Strategically Using Twitter to Mobilize Supporters into Vocal Advocates. *Journal of Social Marketing*, *4*(3), 240–260. doi:10.1108/JSOCM-02-2014-0014

Hachigian, N. (2001). China's Cyber-Strategy. *Foreign Affairs*, *80*(2), 118–133. doi:10.2307/20050069

Hong, Y., & Thomas Goodnight, G. (2020). How to Think about Cyber Sovereignty: The Case of China. *Chinese Journal of Communication*, *13*(1), 8–26. doi:10.1080/17544750.2019.1687536

Howard, P. N. (2010). *The Digital Origins of Dictatorship and Democracy: Information Technology and Political Islam*. Oxford University Press. doi:10.1093/acprof:oso/9780199736416.001.0001

Interesse, G. (2022). China's Cybersecurity Industry: A Market Analysis. *China Briefing News*. October 26, 2022. https://www.china-briefing.com/news/chinas-cybersecurity-industry-a-market-analysis/

Jia, L., & Liang, F. (2021). The Globalization of TikTok: Strategies, Governance and Geopolitics. *Journal of Digital Media & Policy*, *12*(2), 273–292. doi:10.1386/jdmp_00062_1

Johnson, J. S. (2018). China's Vision of the Future Network-Centric Battlefield: Cyber, Space and Electromagnetic Asymmetric Challenges to the United States. *Comparative Strategy*, *37*(5), 373–390. doi:10.1080/01495933.2018.1526563

Kaplan, A. M., & Haenlein, M. (2010). Users of the World, Unite! The Challenges and Opportunities of Social Media. *Business Horizons*, *53*(1), 59–68. doi:10.1016/j.bushor.2009.09.003

King, G., Pan, J., & Roberts, M. E. (2017). How the Chinese Government Fabricates Social Media Posts for Strategic Distraction, Not Engaged Argument. *The American Political Science Review*, *111*(3), 484–501. doi:10.1017/S0003055417000144

Knight, J., & Song, L. (2005). *Towards a Labour Market in China*. OUP Oxford. doi:10.1093/0199245274.001.0001

Lardy, N. R. (2014). *Markets Over Mao: The Rise of Private Business in China*. Columbia University Press.

Lee, G. (2021). *Chinese Short Clip Video Case Study: Focusing on TikTok*. Korea Digital Publishing Society., doi:10.30580/kdips.2021.18.2.24

Li, H. (2016). The Weibo Celebrity Effect on Luxury Brands Marketing in China: An Analysis of Burberry's Weibo Marketing Strategy. *International Journal of Business and Commerce*, *5*(1), 1–11.

Li, X. (2016). *Regulation of Cyber Space: An Analysis of Chinese Law on Cyber Crime*. International Journal of Criminal Justice Sciences., doi:10.5281/zenodo.56225

Li, Y., Chandra, Y., & Fan, Y. (2022). Unpacking Government Social Media Messaging Strategies during the COVID-19 Pandemic in China. *Policy and Internet*, *14*(3), 651–672. doi:10.1002/poi3.282

Li, Y., Guan, M., Hammond, P., & Berrey, L. E. (2021). Communicating COVID-19 Information on TikTok: A Content Analysis of TikTok Videos from Official Accounts Featured in the COVID-19 Information Hub. *Health Education Research*, *36*(3), 261–271. doi:10.1093/her/cyab010 PMID:33667311

Liang, B., & Lu, H. (2010). Internet Development, Censorship, and Cyber Crimes in China. *Journal of Contemporary Criminal Justice*, *26*(1), 103–120. doi:10.1177/1043986209350437

Lin, K. J., Lu, X., Zhang, J., & Zheng, Y. (2020). State-Owned Enterprises in China: A Review of 40 Years of Research and Practice. *China Journal of Accounting Research*, *13*(1), 31–55. doi:10.1016/j.cjar.2019.12.001

Lindsay, J. R. (2015). The Impact of China on Cybersecurity: Fiction and Friction. *International Security*, *39*(3), 7–47. doi:10.1162/ISEC_a_00189

Little, A. T. (2016). Communication Technology and Protest. *The Journal of Politics*, *78*(1), 152–166. doi:10.1086/683187

Lorentzen, P. (2014). China's Strategic Censorship. *American Journal of Political Science*, *58*(2), 402–414. doi:10.1111/ajps.12065

Lupu, Y. (2015). Legislative Veto Players and the Effects of International Human Rights Agreements. *American Journal of Political Science*, *59*(3), 578–594. doi:10.1111/ajps.12178

Martin, J. A. (2013). Closing gaps in international knowledge and participation: News attention, online expression, and the 2010 Haiti earthquake. *Mass Communication & Society*, *16*(3), 417–440. doi:10.1080/15205436.2012.713149

Martin, J. A. (2014). Mobile media and political participation: Defining and developing an emerging field. *Mobile Media & Communication*, *2*(2), 173–195. doi:10.1177/2050157914520847

Martin, J. A. (2015). Mobile News Use and Participation in Elections: A Bridge for the Democratic Divide? *Mobile Media & Communication*, *3*(2), 230–249. doi:10.1177/2050157914550664

Masterson, J. R. (2015). Social Media in China: The Double-Edged Sword. In *Revolutionizing the Interaction between State and Citizens through Digital Communications* (pp. 238–263). IGI Global. doi:10.4018/978-1-4666-6292-6.ch013

Masterson, J. R. (2020). Chinese Citizenry Social Media Pressures and Public Official Responses: The Double-Edged Sword of Social Media in China. In *Digital Transformation and Its Role in Progressing the Relationship Between States and Their Citizens* (pp. 139–181). IGI Global. doi:10.4018/978-1-7998-3152-5.ch007

Mergel, I. (2012). The Social Media Innovation Challenge in the Public Sector. *Information Polity*, *17*(3,4), 281–292. doi:10.3233/IP-2012-000281

Miller, D., Sinanan, J., Wang, X., McDonald, T., Haynes, N., Costa, E., Spyer, J., Venkatraman, S., & Nicolescu, R. (2016). *How the World Changed Social Media*. UCL Press. doi:10.2307/j.ctt1g69z35

Mittiga, R. (2022). Political Legitimacy, Authoritarianism, and Climate Change. *The American Political Science Review*, *116*(3), 998–1011. doi:10.1017/S0003055421001301

Noesselt, N. (2014). Microblogs and the Adaptation of the Chinese Party-State's Governance Strategy. *Governance: An International Journal of Policy, Administration and Institutions*, *27*(3), 449–468. Advance online publication. doi:10.1111/gove.12045

Palmer, N. A., & Perkins, D. D. (2012). Technological Democratization: The Potential Role of ICT in Social and Political Transformation in China and Beyond. *Perspectives on Global Development and Technology*, *11*(4), 456–479. doi:10.1163/15691497-12341236

Poreh, K. (2012). A Revolution in Communications: Democratization and the Growth of Cyberspace. *SAIS Review (Paul H. Nitze School of Advanced International Studies)*, *32*(2), 217–220. doi:10.1353ais.2012.0019

Qian, Y. (2017). *How Reform Worked in China: The Transition from Plan to Market*. MIT Press. doi:10.7551/mitpress/8098.001.0001

Roberts, M. E. (2020). *Censored: Distraction and Diversion Inside China's Great Firewall*. Princeton University Press.

Rojas, H., & Puig-i-Abril, E. (2009). Mobilizers Mobilized: Information, Expression, Mobilization and Participation in the Digital Age. *Journal of Computer-Mediated Communication*, *14*(4), 902–927. doi:10.1111/j.1083-6101.2009.01475.x

Rsakova, E. P., Inshakova, A. O., & Frolova, E. E. (2021). Legal Regulation of Internet Courts in China. In *Modern Global Economic System: Evolutional Development vs. Revolutionary Leap, 1515–21*. Springer International Publishing. doi:10.1007/978-3-030-69415-9_167

Schwoerer, K. (2019). Individuals' Use of Twitter to Discuss Freedom of Information in the United States: A Social Network Analysis. *The Journal of Civic Information*, *1*(2), 45–59. doi:10.32473/joci.v1i2.119010

Serrano, M., Carlos, J., Papakyriakopoulos, O., & Hegelich, S. (2020). Dancing to the Partisan Beat: A First Analysis of Political Communication on TikTok. In *12th ACM Conference on Web Science* (pp. 257–66). Association for Computing Machinery.

Shutsko, A. (2020). User-Generated Short Video Content in Social Media. A Case Study of TikTok. In *Social Computing and Social Media. Participation, User Experience, Consumer Experience, and Applications of Social Computing* (pp. 108–125). Springer International Publishing. doi:10.1007/978-3-030-49576-3_8

Sicular, T., Yang, X., & Gustafsson, B. (2022). The Rise of China's Global Middle Class in an International Context. *China & World Economy*, *30*(1), 5–27. doi:10.1111/cwe.12400

Stevenson, A. (2010). *Oxford Dictionary of English*. OUP Oxford.

Su, Z., & Meng, T. (2016). Selective Responsiveness: Online Public Demands and Government Responsiveness in Authoritarian China. *Social Science Research*, *59*(September), 52–67. doi:10.1016/j.ssresearch.2016.04.017 PMID:27480371

Suciu, P. (2022). Is TikTok Really A National Security Threat? *Forbes Magazine*. https://www.forbes.com/sites/petersuciu/2022/11/18/is-tiktok-really-a-national-security-threat/

Tsui, L. (2003). The Panopticon as the Antithesis of a Space of Freedom. *China Information*, *17*(2), 65–82. doi:10.1177/0920203X0301700203

Tung, R. L., & Worm, V. (2001). Network Capitalism: The Role of Human Resources in Penetrating the China Market. *International Journal of Human Resource Management*, *12*(4), 517–534. doi:10.1080/713769653

Wang, X., Xu, L. C., & Zhu, T. (2004). State-Owned Enterprises Going Public The Case of China. *Economics of Transition*, *12*(3), 467–487. doi:10.1111/j.0967-0750.2004.00189.x

Wang, Y.-H., Gu, T.-J., & Wang, S.-Y. (2019). Causes and Characteristics of Short Video Platform Internet Community Taking the TikTok Short Video Application as an Example. *2019 IEEE International Conference on Consumer Electronics - Taiwan (ICCE-TW)*, 1–2. 10.1109/ICCE-TW46550.2019.8992021

Warschauer, M. (2001). Singapore's Dilemma: Control versus Autonomy in IT-Led Development. *The Information Society*, *17*(4), 305–311. doi:10.1080/019722401753330922

Weber, I. M. (2021). *How China Escaped Shock Therapy: The Market Reform Debate*. Routledge. doi:10.4324/9780429490125

Wiley, K., Schwoerer, K., Richardson, M., & Espinosa, M. B. (2022). Engaging Stakeholders on TikTok: A Multi-level Social Media Analysis of Nonprofit Microvlogging. *Public Administration*, (April), padm.12851. Advance online publication. doi:10.1111/padm.12851

Winter, P., & Lindskog, S. (n.d.). *How the Great Firewall of China Is Blocking Tor*. https://www.usenix.org/sites/default/files/conference/protected-files/winter_foci12_slides.pdf

Wu, X. (2007). *Chinese Cyber Nationalism: Evolution, Characteristics, and Implications*. Lexington Books.

Xiang, Yu. (2019). User-Generated News: Netizen Journalism in China in the Age of Short Video. *Global Media and China*, *4*(1), 52–71. doi:10.1177/2059436419836064

Xu, X. (2021). To Repress or to Co-opt? Authoritarian Control in the Age of Digital Surveillance. *American Journal of Political Science*, *65*(2), 309–325. doi:10.1111/ajps.12514

Yamamoto, M., Kushin, M. J., & Dalisay, F. (2018). How Informed Are Messaging App Users about Politics? A Linkage of Messaging App Use and Political Knowledge and Participation. *Telematics and Informatics*, *35*(8), 2376–2386. doi:10.1016/j.tele.2018.10.008

Yuan, D., & Guo, L. (2018). Video-based social media platforms and censorship circumvention: The case of China. *China Media Research*, *14*(3), 1–13.

Yuan, Z., Wang, G., & Khor, N. (2011). The Rise of the Middle Class in the People's Republic of China. SSRN *Electronic Journal*. https://studycrumb.com/alphabetizer doi:10.2139/ssrn.1806769

ENDNOTES

1 China Internet Report, 2021 by the China Internet Network Information Center (CNNIC).

2 "TikTok and the Rise of Short Video in China" report by QuestMobile, 2021.

3 Detail information can be found at: https://finance.sina.com.cn/stock/hkstock/ggscyd/2022-11-04/doc-imqqsmrp4868078.shtml

4 "Live Broadcasting in China: A Look at the COVID-19 Effect on Consumer Spending." We Are Social, 21 May 2020, wearesocial.com/blog/2020/05/live-broadcasting-in-china-covid-19.

5 It is primarily used to launch low orbit satellites and was developed by a subsidiary of China Aerospace Science and Industry Corporation.

6 Gillan, M. (2021, April 29). How Governments Around the World are Using Internet Shutdowns to Cope with the Coronavirus Crisis. The Diplomat. https://thediplomat.com/2021/04/how-governments-around-the-world-are-using-Internet-shutdowns-to-cope-with-the-coronavirus-crisis/

Chapter 14
The Social Credit System in China

Martin Woesler
Witten/Herdecke University, Germany

ABSTRACT

This chapter introduces the social credit system (SCS) as one of the Chinese government's disruptive-innovative answers to control its citizens and organizations to secure the retention of political power. Based on source texts and surveys, "external control" and "extrinsic motivation" are identified as core principles of SCS. The unique qualities of the SCS are described in the context of the existing surveillance and control systems in China. The example of dropouts is discussed. The impact of artificial intelligence on communication, the efficiency of the SCS, its effects on everyday life in China are described. The SCS violates the UNESCO agreement, of which China is a signatory. Also the European Union has passed an Artificial Intelligence Ethics Act. The chapter shows the limits of the system and that it is not transferable to liberal democratic societies. However, IT companies and interest groups in data capitalism have also developed small-scale credit systems for economic profit and political manipulation.

LITERATURE REVIEW, RESEARCH QUESTIONS, AND METHODOLOGY

This chapter describes, analyzes, classifies, and explains one of China's disruptive innovative responses to the new possibilities of digital surveillance and external control. Many of the existing papers on the Social Credit System take sides in favor of (mostly mainland Chinese contributions) or against the introduction of an SCS. However, they often consider the starting point of the SCS as 2014 (State Council 2014) and they do not account for the official documents as written in Chinese. The book chapter in hand does not take sides, but points out the sources of the SCS from the 1990s and describe the SCS, what its propagated and intended goals are, how it was and is implemented, and whether SCS has reached the propagated and real goals. For the analysis of the Social Credit System (SCS) in this book chapter, the relevant accessible texts of laws, regulations, and official announcements were translated from Chinese into English (Woesler and Warnke, 2023), along with numerous official Chinese and Western media reports. The author spent about a month in China each year since the introduction of the SCS in 2020,

DOI: 10.4018/978-1-6684-6429-8.ch014

and nearly the entire year in 2022, interviewing Chinese citizens and expats and conducting on-site surveys. As such, this study applies a mixed-methods approach. It differs from others research which does not discuss the SCS itself, but instead, discusses an image of the SCS conveyed in Western media.

An authoritarian regime cannot rely solely on repressive measures. Repression measures can be costly. They face a greater risk of undermining regime legitimacy, so the leading group in authoritarian regimes adopts them strategically. Repressive measures are only used when regime leadership considers the repressive payoff to be greater than alternative methods (Chen 2016, King 2013, King 2014, King 2017, Mittiga 2021). This book chapter argues that China has skipped the stage of control society as suggested by Foucault and directly entered a combination of discipline society and control society with the digitalized state capitalism. While Western control societies get along without punishment, China uses the parts of both the discipline societies and control societies. Therefore, repressive measures are used both, directly with individuals as well as in the form of public shaming as a deterrence for bad public actors. And the Social Credit System makes the repressive measures (with a tight control system claiming to be able to identify anybody in China within 1 second, claiming to know the personality of the users better than family members and encouraging denunciation) the new normal, challenging even the last resorts of free thinking, free choices and anonymity.

The social credit system is a comprehensive evaluation system of people living in China (Chinese and expatriates) and of Chinese living abroad. About 20% of the data is collected from people while organizations existing in China (companies, associations, government agencies, representative offices, etc.) account for about 80% of the data.

The most important goal of the SCS is to keep the ruling party in power by controlling the people. Propaganda claims that the most important goal is to increase trust in all areas of society. This goal is to be achieved by means of a comprehensive system of evaluation, reward and punishment based on a behaviorist and cybernetic worldview. Since its introduction in 2000, the Social Credit System does not seem to have had much influence on how far the population trusts its government. Since 2020 other topics like the measures of the government to fight the pandemic dominated the development of trust. However, the power of the government is stronger than in 2000, with many new restrictive laws like the new data security law of 2021, which declares that all data (including all data of citizens) ever touching a Chinese server belongs to the government (not the citizen) and the lifting of term limits for the president in October 2022.

The specified reason for the introduction from 2000 (documented in a speech by President Jiang Zemin in 2002, see Woesler et al. 2019) was a lack of trust in the economy claimed by Lin Junyue in 1999: Based on his experience with the rating systems in the U.S. financial sector, which he had gathered at U.S. universities, Lin Junyue proposed to transfer the American system of credit histories and credit scores, which created trust in the U.S., to China and extend it to all areas of life, i.e., to turn a financial instrument into an instrument for overall assessment. In this way, trust could be increased not only in the financial sector, but in society as a whole.

Lin also claimed that with the SCS less people would be imprisoned, however, this goal was not achieved. (Lin 2015, Louvet 2019) The relatively low number of 1.4 million prisoners in 2000 was not reached any more afterwards, with an average of 1.6 to 1.7 million since 2005 (https://www.prisonstudies.org/country/china).

How does the SCS work? It collects the following information from individuals and organizations: Payment behavior for loan installments, credit cards, incidental costs, court costs, income tax (Blomberg 2021) or tendering behavior, verdicts in labor court cases and corruption trials, traffic behavior, birth control, public transport payments, academic honesty, honorary positions, certificate of good conduct, parental reverence, online shopping behavior, and behavior in social media.

Figure 1. A systematic illustration by the Wall Street Journal
Source: McAllister (2019)

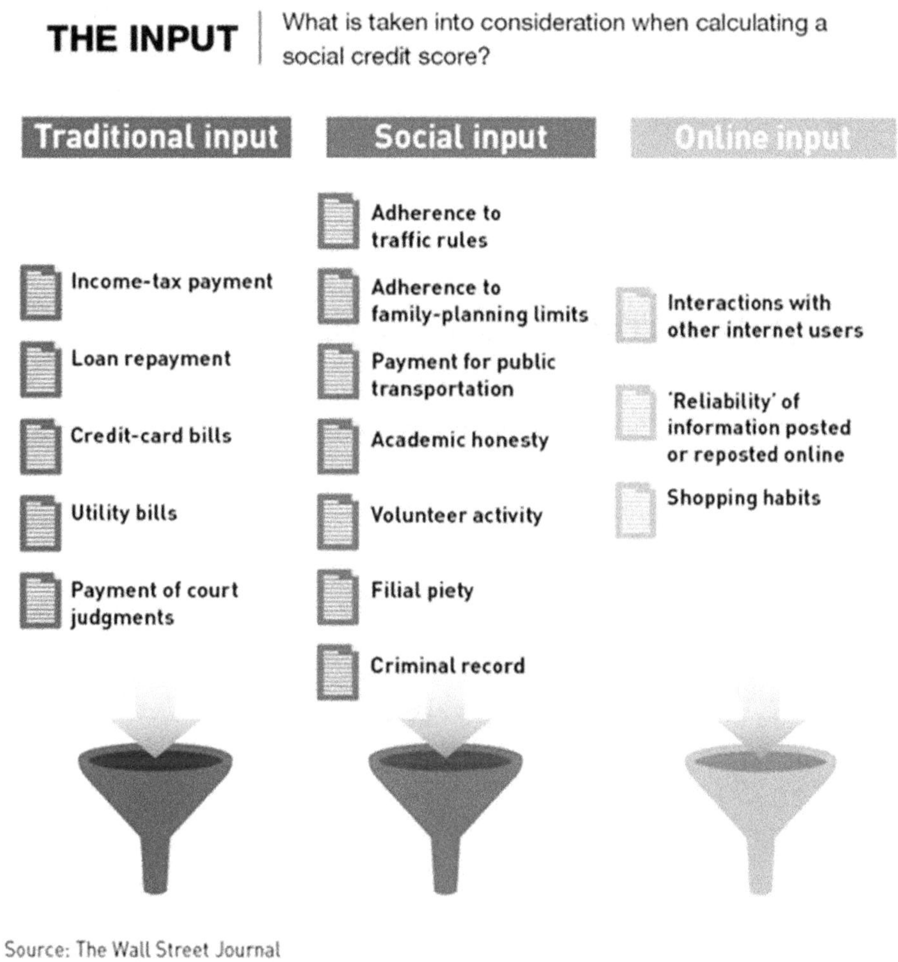

This data is collected centrally by the National Development and Reform Commission. The Commission oversees three agencies that operate the platforms on which the information is processed. First, the National Public Credit Information Center and the People's Bank operate the public portal of CreditChina (https://www.creditchina.gov.cn). Lists appear here with the full names of thousands of people who have been restricted from air and train travel, for example. Negative entries from companies, such as those that lost court cases, can also be seen. The Reform Commission continues to oversee the State Information Center. This maintains the non-public National Credit Information Sharing Platform. Finally, there is one data collection carried out independently of the Reform Commission, namely by the

State Administration for Market Regulation: it operates the public platform "National Enterprise Credit Disclosure System" (EUCCC 2019).

The information about companies can also be viewed unofficially with the "Sky's Eye" app.

The information is analyzed and evaluated; for example, with regard to tax behavior, one receives a rating in the categories A (very good) to D (poor), with a corresponding negative rating, e.g., because one had traffic violations, in the area of high-speed train and air travel, one is listed with full name on the above-mentioned portal and can then no longer purchase high-speed train and airline tickets. Other sanctions include refusal to issue visas to certain countries such as Singapore, etc. Advantages are obtained in the case of positive classification with regard to renting rental cars (no deposit may be required here), preferential treatment when dealing with authorities, etc.

Figure 2. Generating credit score with algorithm
Source: McAllister (2019)

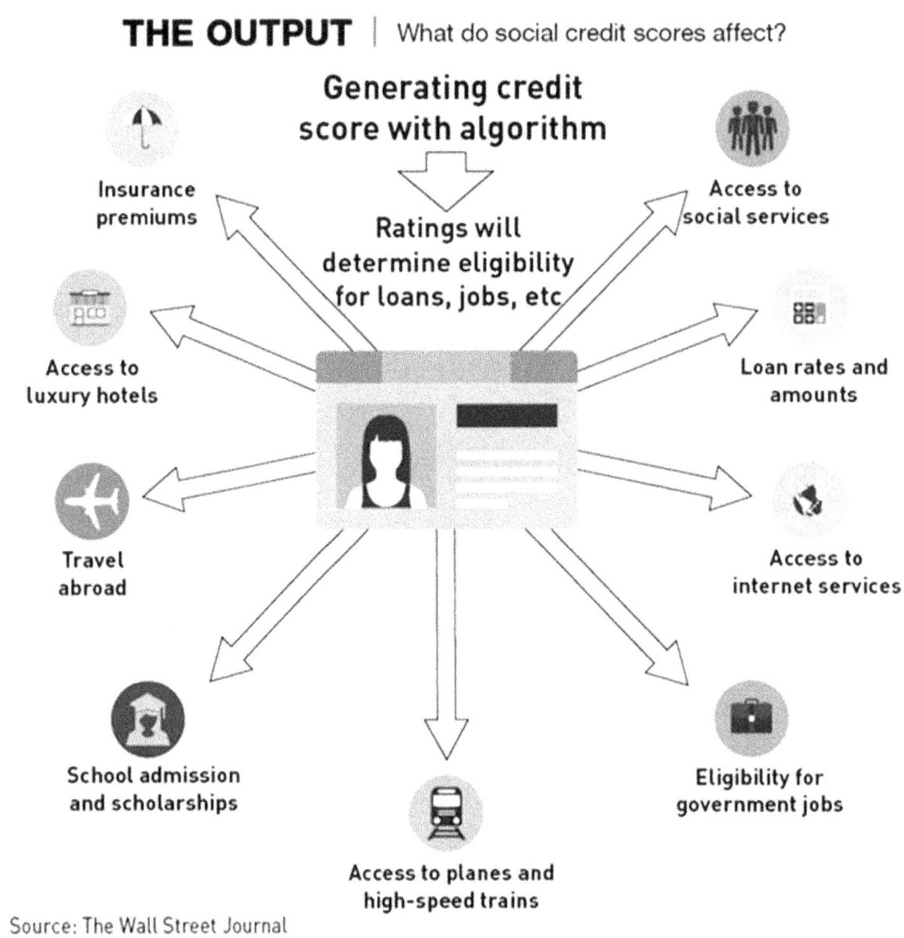

Companies receive blacklist entries for (labor) court cases brought against a company displayed on the above-mentioned website.

The research questions this chapter addresses are: Is this rating system an effective instrument of external control? Where does the system reach its limits? Is this system transferable to liberal democratic societies?

THE SYSTEM-STABILIZING USE OF INTERNET TECHNOLOGIES IN AUTHORITARIAN REGIMES

One disruptive development in communications technology is the possibility of increased monitoring of users and the evaluation of their data, as demonstrated by the personality profile of users on Facebook: Even just a few "likes" allow more precise statements to be made about a person than family members could do.

In liberal democracies, these evaluations are limited to protect personal rights, including privacy and personal data protection. In data capitalism, these limits are leveraged by large Internet corporations such as Google, Facebook, Amazon, Netflix, etc., and also by small app developers through general terms and conditions that users agree to without taking notice, because they either don't care to read the user terms or are unaware of the often far-reaching tracking performed by the app. In addition, there are companies like Cambridge Analytica (closed in 2018) that specialize in manipulating users based on their personality profile. This is achieved through priming, nudging, fake news, bots, filter bubbles, conspiracy theories, etc., i.e., influencing their voting decision for Trump or Brexit or their purchasing decision. There are interest groups behind this that need short-term electoral success or pursue economic interests. The characteristic of social media to spread fake news several times faster and more effectively than the truth plays into the hands of these attempts at manipulation, leading to a polarization of societies and the perception of dissenters as enemies.

When the Internet was introduced to Asia around the beginning of the 1990s, it appeared to be a threat to autocratic dictatorships such as China's, which build their power on a synchronization of the media with propaganda and lifelong indoctrination. Here alternative narratives to the filter bubble were offered to the user, who was apparently anonymous on the Internet. An attitude of "lip service" has developed among citizens of such regimes. Meanwhile, in the 2020s, autocratic dictatorships and authoritarian regimes use the Internet, connected technologies (such as user identification), and surveillance technologies to spy on users. The goal is to look behind the "lip service", behind the explicit communication and get to the indirect, implicit opinion, conviction, belief, values, thoughts of users.

Even brain scanners are used in China to check on the reliability of employees.

Figure 3. Brain scanners
Source: Chen (2018)

While they technically can only register if someone falls asleep or has strong emotions like being very sad or very happy, they actually achieve the targeted effect of compliance due to their deterring effect.

Knowing that the devices you are wearing have such functionality also has an effect on individual behavior and psycho-emotional disposition. Shanghai technology company Deayea has confirmed that train drivers on the Shanghai-Beijing line, for example, regularly wear these brain scanners. (Chen 2018). Future study on the effects of these brain scanners on the mental and physical health of these train drivers will illuminate further the tension between state control and autonomy of the individual.

Figure 4. Worker in a Shanghai factory with work helmets, in which brain scanners were implemented
Source: Chen (2018)

In China, about 6.5% of the budget is spent on domestic security. This money has been used to expand communications and surveillance technology over the past decades. This government funding of tech corporations has led to the global supremacy of Chinese communications companies. In addition, surveillance by cameras inside and outside buildings has been expanded. Facial recognition can also identify people wearing a medical face mask, leaving only their eye areas exposed; other technology recognizes people by their gait. Chinese police boast that they can determine the location of any person in China in less than a second. Chinese cell phones are so cheap that Africa is adopting this technology and using it for communication and to handle cashless payments, for example. This data passes through Chinese servers, so users of Chinese cell phones with Chinese SIM cards cannot access Google or Facebook, for example, even when the user is travelling in Europe.

According to Chinese data laws, all data that is on or passes through Chinese servers belongs to the Chinese state. The State uses it to document, evaluate, manipulate and control society. A typical example that has been uncovered was, that local officials manually switched the health code on cell phones of people from green to red, who were on their way to a demonstration. With a red code indicating they were probably infected with COVID, these want-to-be-demonstrators complied and were taken into isolated quarantine on suspicion of corona virus infection and then (although at that time a 1-week quarantine was prescribed) were deported back to their home region next day by bus. (Gan 2022)

In a general comparison to Europe and the US, a huge part of Chinese and African citizens lacks awareness, except in a few exceptional cases, that revealing their data, such as when paying for a product via QR code, could be detrimental to them. In fact, however, this makes it possible to create a movement profile, for example, even decades later.

People who openly oppose the regime and call it a dictatorship disappear before the eyes of social media cameras and their family members are taken into clan custody or put under house arrest. (Anon. 2018b). Threatening to harm family members is standard practice in authoritarian regimes. Algorithms monitor social media and delete posts and videos that show certain regime-critical traits - keywords and basically any kind of hype, as alternative words often develop to escape censorship.

THE DISRUPTIVE NEW QUALITY OF THE SOCIAL CREDIT SYSTEM

The centrally planned economy has lost competition to the liberal market economy. In China, another form of economy became more successful than these two systems, it was state capitalism. In this system, Manchester capitalist conditions prevail (exploitation, prohibition of unions, prohibition of strikes, etc.) and the state increases its control over companies the more successful they become. Careful attention is paid to ensure that the increased control does not affect the economic power of the company, but at the same time all the benefits of successful companies are siphoned off, such as user data in the case of communications or social media companies.

In China, the narrative is that the centrally planned economy was unsuccessful because of the human factor, being visible in such phenomena as reporting embellished numbers upward so that the wrong quantities were produced and the need for innovation or diversification was not recognized in time. Digitalization has boosted disruptively the means for central governments to surveil the needs and supplies of the market. With almost complete monitoring, the human factor now appears to be eliminated and market needs can now be analyzed by evaluating actual conditions. A whole reporting chain is abandoned and with it the human reporting errors. This is another variation of cybernetics, which had failed in Chile with CyberSin during 1970-1973 (Lavanderos 2022).

The new disruptive quality of the Social Credit System is that all available data, including data from the various authorities, from all citizens and from all organizations is centrally aggregated using unique identification numbers, and thus a person who has not filed his/her tax return on time and has been placed in category D for the 10% worst taxpayers becomes a low performer. The lowest performers lose their job, may be placed under house arrest, may have their face and personal data - such as full name, address, etc. - exposed to public shame on large LED boards in busy places, and the person's children may be excluded from certain schools. So, a new quality is the central aggregation of data. Another quality is double punishment for the same 'wrongdoing', since the person was already fined for the traffic or tax violation. Liu Hu, a Chongqing based investigative journalist who exposed a corrupt cadre on a social media channel was ordered by a court to pay a fine for this 'reputation damage', at the same time he was put under house arrest, and therefore could not perform his job as an investigative journalist since his accounts were deleted on various social media platforms. Furthermore he is no longer allowed to buy airline and high-speed train tickets (according to *People's Daily* (Anon. 2018a), the issuance of 10.888 million airline tickets and 4.164 million train tickets to low-performers has been denied). Other documented punishments are, that a warning tone sounds from the cell phone when the person approaches a public place or someone who calls the person receives a warning message. The third new quality is clan detention. This occurs when children are no longer allowed to go to their school because of their parent's 'misbehavior' and thereby suffer disadvantages. This can also occur when people are friends with an accused criminal and are lowered in their Social Credit System evaluation because of this friendship. So here, in addition to the legal system, a parallel punishment system is created. This second system runs under the slogan "Make every step more difficult for low-performers." Daily newspapers gleefully show cartoons in which these low-performers have their feet concreted in a bucket.

The Social Credit System had the propagated goal of increasing trust among the population, in the economy, state, party and government. (State Council 2014) In fact, it has led to more distrust by rewarding denunciation and known abuses of the system. The second level of meaning of the word "credit" in the Social Credit System, financial creditworthiness, has also not led to a reduction in loan defaults: In fact, an assessment that rewards, among other things, party loyalty has been found to lead to more loan defaults. This was at least the case in the model experiment in Rongcheng, where previously "unbanked" citizens were certified as creditworthy through loyalty and afforded a car loan, but had no experience with installment payments and therefore defaulted. That the approach of positively valuing party loyalty in a credit rating system has failed is indirect evidence that party loyalty itself is rather detrimental to creditworthiness. Indeed, China is one of the most corrupt and restrictive systems in the world (Source: Transparency International). Today, Rongcheng is not using punishments any more, but rewards. (Brussee 2022)

In assessing organizations, the Social Credit System also revealed that it is politically manipulated. In a field test, companies that are economically integer but have fallen out of favor with the political leadership and state-owned enterprises that have been rocked by corruption scandals were elicited from the Social Credit System website, with the result that corrupt state-owned enterprises like CCCC or PolyProperty had a clean slate there and the successful but disagreeable private enterprises like Alibaba or Wanda had numerous blacklisted entries.

Another stated goal of the father of the Social Credit System Lin Junyue was to relieve the prisons, since the low performers would already be led back to the right path by their neighbors and friends (Lin 2019). In fact, the prison population has not declined since the Social Credit System was introduced.

Criticism of the system is not permitted. At best, Chinese journals contain essays by lawyers who do not criticize the system per se, but rather discuss problems with its implementation and give advice on how the system can be made even more legally secure and effective. Western experts like Victor Brussee 2022 argue in their "all-clear" pleas that there was not one but many Social Credit Systems in China, that very different actors were involved, that there was a lack of technical implementation, that much was still done by hand, that only people who had been convicted by a court are sorted into the low-performer group, and that China was so large and had different regions. The criticism may have its justification, but the Social Credit System itself has been successfully implemented, almost every person has his or her ID card, social credit number, every organization has its social credit number, and even if the Social Credit System runs more in the background. It is effective, as evidenced by the denial of millions of tickets with the reason to have a low score or being blacklisted. It is not the only control system in China, of course, and how important it will become in concert with the other control systems remains to be seen. After all, in terms of its overall assessment approach, it is designed to play a leadership role within the control systems.

EFFECTS OF THE SCS FELT TODAY AND THE ROLE OF ARTIFICIAL INTELLIGENCE IN THE SCS AND IN COMMUNICATION IN GENERAL

Any form of assessment of a person or organization is scored. Under certain conditions, scoring is accepted, such as grading in school, etc.

What is special about Social Credit scoring is that individual grades are assigned, i.e., overall grades, which in turn can have an impact on all areas. So, what might it look like if all areas of daily life depended on a score? The Health Code offers a good insight into the effects of scoring in everyday life. The author of this article has himself experienced that the rules established (yellow health code, 7 days quarantine in a center in Canton, then green code for travel home, then yellow code for 3 days domestic quarantine, then green code for freedom of movement) were simply manually controlled and arbitrarily undermined by locally responsible authorities such as the street committee ("hygiene unit"). Thus, on the 10th day of quarantine, the original yellow code became red again and the author had to start the procedure again. When asked, the local authorities did not explain why they had manually switched the code to red, rather he was only given instructions on how he now had to do daily PCR tests for a week to then have a chance to be switched back to yellow and finally back to green. Without a green code, one could not use transportation, enter stores or buildings in China in November 2022.

That the collected data is first used for the purpose of power maintenance and not for the public, can be seen with China's handling of Covid data. While at the early stage, reporting was suppressed and later figures were adjusted, which is visible at sharp rises in the reported number of infections, when the Zero Covid measures were lifted overnight on December 7, 2022 (after mass tests almost no tests were available any more, and almost no medication), the Central Committee was informed about 249 mil. infected people (Ritchie 2022), while the reports to the World Health Organization were 4.128 infections, which is an underreporting of 60.320 times, and on January 15, 2023, the reports were totally abandoned. During the three years of the Zero Covid policy, not enough new hospitals were built, not enough emergency beds and ventilators were bought and no Western vaccine was approved for the Chinese market. Since everybody knew that the Zero Covid policy could not go on forever, because China is connected with the world, such preparations would have been necessary for the time after the policy. In December 2022 and January 2023, the often unvaccinated old and vulnerable people died in the hallways of the overcrowded hospitals and the death toll reached new heights. Because people were discouraged to document Covid as a cause of death, even the number ofC victims will remain unknown. Only the extra mortality rates, which experts estimated in December 2022 at up to one or two million for the years 2022-2023 reflect the tragedy (Steenhuysen 2022).

The core of the Social Credit System is the external control of citizens and organizations. By allowing the political leadership to set the criteria for evaluation itself, the complex reality is artificially reduced to a few aspects that are evaluated. So there is a risk that such evaluation will be unfair, that it will not take important factors into account, that errors may arise or that the system itself will be manipulated. In addition, disincentives arise; for example, in Rongcheng, it was offered that citizens could improve their score by donating blood or money to an institution designated by the party. Citizens might be forced to post a positive comment about government policies on social media when they want to leave their homes. The Social Credit System uses fear, nudging, social distancing, operant conditioning, etc. to force constant action against one's beliefs. This is psychologically stressful. The link between psychological distress and mortality has been scientifically proven (Reisinger 2015; Holt-Lunstad 2015).

According to Luhmann 1997 and Baecker 2007, man has passed through the communication epochs 1.0 (orality), 2.0 (writing), 3.0 (printing) and 4.0 (digitalization). In this process, Luhmann divided the exchange with the machine into surface (human-machine communication through interfaces such as keyboard, screen, etc.) and depth (the inner life of the machine, which is closed to the people). With the numerous smart surveillance devices in our environment (when teaching at Peking Normal University, the author was surrounded by three bullet cameras, the students by nine), we are in a communication epoch "Virtual Communication 5.0" in which we have already been swallowed up in the depth of the machine itself. While technology supported human communication up to epoch 4.0, actual communication in epoch 5.0 is no longer between humans, but between machines. Although communication between humans is often still the starting point for the actual communication, namely when a communication is recorded and evaluated by algorithms for keywords, classified by comparison with Big Data, and when it leads to decisions and actions on the basis of these correlations (not causalities). The author was told that the teacher, who had previously been sitting on the job, was dismissed due to politically unwelcomed remarks in class. The actual act of communication, namely the dismissal decision, is made by algorithms. Here, humans have the role of being passive research objects and recipients of decisions that even the programmers of the algorithms can no longer explain, and they also play the role of fine-tuners. For example, in China's SCS it was decided to assign the categories A (well-behaved taxpayer) and D (tax evader) not on the basis of absolute standards, but relatively, depending on the total. In this case, 10% was specified, so the best 10% were assigned category A, the worst 10% category D. In this way, the screw can be tightened even in the short term, because people who are in category D will try to avoid the mistakes, but still 10% will be assigned category D. At the beginning of the external control, people may still find the specifications annoying but bearable and comply, but precisely because they comply, the automatism raises the bar higher and now they have to make even more of an effort to avoid slipping into category D. This is a cold progression.

The tightly woven web of surveillance and the Social Credit System leaves no loopholes. When restrictions on access to businesses became too strict under the zero-covid policy, Chinese media reported that people who had gone underground for various reasons voluntarily turned themselves in to the authorities.

What happens to people who no longer want to pay lip service and constantly act against their own will? A young woman named Dong filmed herself spraying a poster of the state president with black ink. Shouting epithets directed at President Xi, she was giving heed to her anger at the dictatorship, calling him names forbidden in China. To make the security authorities' job easier, she also gave her exact location in the video. – She stated: "Xi Jinping, I am waiting here for you to catch me." At 3 p.m., she posted photos of uniformed men outside her door, after which she and her Twitter account disappeared from the scene (Anon, 2018b).

Even great notoriety does not protect against disappearance without a trace, as there are numerous cases of disgraced heads of corporate empires doing just that. Jack Ma (head of Alibaba) once made critical remarks about the "pawnshop mentality" of Chinese banks. This alone was enough to make him disappear. (Peach 2021) People who bring sexual assaults of high cadres to public attention, such as a Chinese tennis player Peng Shuai, have also been muzzled (Fan 2021).

It seems difficult to leave the system; emigration would be an option, but Chinese can only apply for a passport when they have concrete intentions to leave and are often denied the document. Should a critical person succeed in leaving the country, the family is often held back as collateral.

Meanwhile, UNESCO (UNESCO, 2021) and also the EU (EU, 2021) with its agreements and laws on Ethics of Artificial Intelligence (AI) have explicitly opposed the use of AI in Social Credit Systems as it poses a high risk. Even if China has agreed to the UNESCO agreement, it cannot be assumed - based on experience with similar agreements, such as human rights issues - that China will abolish its Social Credit System.

DISCUSSION: ADOPTION OF THE SOCIAL CREDIT SYSTEM IN LIBERAL DEMOCRACIES

As has been shown with the many examples and experiences with the SCS, it is a disruptive control system of citizens and organizations, therefore a tool to manage a country (see Hoffmann 2017). What are the implications of liberal countries accepting the use of SCS? This discussion leads us back to the topic of the disruptive nature of technology as a whole.

According to the inventor of the social credit system, Lin Junyue, the goal was also to export the system to France and other "capitalist countries": "then there would never have been the yellow vest movement in France." Lin Junyue claims that Cambodia, Sri Lanka, Poland, and Chile have already signaled interest (Lin 2019). However, the system has not yet been exported for societies as a whole, but in specific sectors. Canada uses an assessment from the social credit system as a component for selecting Chinese citizens for Canadian visas. In Europe, there are small model experiments such as a 'voluntary' bonus points system in Bologna starting in 2023.

The Foresight Value Study on the Future of Germany for the German Federal Ministry of Education and Research (BMBF) says of one of six possible scenarios, "In the 2030s, a digital, participatory negotiated points system will begin to provide incentives for behavioral change in Germany" (BMBF, 2020, p. 123).

What conditions would the SCS have to meet in order to be accepted in liberal democracies?

First of all, this system would have to be introduced on a voluntary basis, and that alone is where it fails, because the moment some volunteers agree to be evaluated and thereby gain advantages over others, these others are placed in a worse position. Incentives are created to participate, which in turn undermine the principle of voluntariness. Then the rating categories would have to be defined by the users themselves and it would have to be possible to change them. Finally, once a rating has been given, it would have to be possible to reverse it, for example if the person complains that the rating was incorrect. (In China, the complaint has to be directed not to the upper-level authorities, but to the authorities who were responsible for the score in the first place.) Then a person would have to have the right to know how the classification was decided upon. This poses a problem in that programmers of the algorithms that make the ratings cannot explain how the algorithms arrived at their decisions. Finally, a person would have to have the right to decide whether to inspect the result of the scoring, personally or if it would be made public. Furthermore, it would have to be regulated that punishments do not lead to double punishments (this can hardly be ruled out systemically; in fact, the system is designed to do so). As for bonuses for high scores, these should be individually variable, since the same bonus may be differently valuable to different people.

According to Foucault 1975, societies were initially disciplinary societies and developed into control societies. He used the metaphor of the "panopticon" prison tower, from which the guard could monitor only one prisoner at a time. On this basis, all prisoners complied. In China, there is a combination of both, and the new control society also uses direct disciplinary measures. Moreover, in China, the guard can no longer just monitor only one prisoner at a time, but with the help of technology, all prisoners are actually monitored 1:1 by algorithms again, as in the disciplinary society. Nonetheless, as in the control society, the Social Credit System of course also exercises indirect power in that prisoners now keep each other in check (comparison, denunciation, isolation of low performers). With the fear of denunciation, the Social Credit System undermines trust in social relations and has the opposite effect than it was originally intended for, namely to enhance trust in social relations. Describing the impact on the average citizen, Chen Shengzhang from Fulushan village, where social credit system test was launched, said, "Whatever we do, we think about our credit score" (Dorloff, 2017). In principle, breaking the law or disloyal behavior is also more difficult to carry out in the Social Credit System, since monitoring takes place everywhere and at all times, so to speak. The fact that this leads to psychological and possibly health impairment can be seen in China in the situations when monitoring becomes too tight.

Not under the Social Credit System, but with the use of similar surveillance technology and scores, in this case green and red codes on cell phones that are centrally recorded and managed (health codes under the Covid pandemic), the dynamic zero Covid policy in China in 2022 restricted the movement of people in their flats, in entire neighborhoods, sometimes for months (tourists were stranded in Xinjiang for half a year), so that people were not allowed to leave their homes, even if there was just one suspected case in one city district. During this time, incidents such as supply shortages, forcibly euthanized pets, separation of children and their parents, and deaths in house fires that could not be extinguished in time due to the roadblocks occurred. Numerous businesses were forced to close, supply chains were disrupted and the economy collapsed. The people's displeasure vented itself in demonstrations with white slips of paper held high (without text, conviction was less likely – everyone knew what was supposed to be written there anyway) and occasional outbreaks of violence. On December 8, 2022, the permanent QR code scans for location notification at every store, building and subway entrance was abolished with short notice. In other words, even if enormous pressure is put on citizens and they risk heavy fines if they protest, people can be externally controlled only to a certain extent. The regime's reaction of eliminating location scans without replacement was too radical from a medical point of view, as it allowed the pandemic that had previously been kept out of China to spread rapidly. True, there were even more far-reaching relaxations in Europe and America, but those countries also had effective vaccines and more resilient health systems. In China, instead, propagandistically glossed-over but ineffective dead vaccines and a low number of intensive care beds with ventilators are now facing a massive wave of infections. Therefore, China was hit hard in December 2022 with a high mortality rate and especially since there was also a risk of long-COVID with the more harmless but more contagious Omicron variant. However, since China discouraged to reveal the cause of death in the certificates and stopped reporting figures to the WHO, the exact dimension of this human tragedy remains a secret.

OTHER EXAMPLES AND SURVEYS RELEVANT TO ANSWERING THE RESEARCH QUESTIONS

The research questions are repeated here: "Is this rating system an effective tool for third-party control? Where does the system reach its limits? Is this system transferable to liberal democratic systems?" Based on our study, these questions can now be answered.

The evaluation system is an effective instrument of external control. However, it is only one of a complex system of instruments. More immediately, other instruments have the effect of scissors in the head: People wonder whether the person they are talking to is an informer, and whether the camera pointed at them is recording something that will later be used against them. The constant indoctrination and evaluation by mentors, party committees, etc. trains citizens in China to be at least adept at lip service. Even survey results from China should be taken with a grain of salt, as surveys are conducted in a non-free setting. As for inner convictions, at least the directed and channeled indignations toward ostensible insults, disadvantages or aspirations to supremacy by neighboring Asian countries or superpowers like the U.S. actually appear to correspond to inner convictions, a consequence of the homogenized media and filter bubbles. People in China have developed a capacity for suffering because of the frequently changing political direction, depending on which party faction is in power, and they put up with much more than people raised in individualistic societies, for example. The taboo reappraisal of political mistakes of the past is also part of this; if the violent suppression of the democracy movement in 1989 is defined and tabooed as a "failed counterrevolution," this narrative prevails in the next generation at the latest; in a survey of 90 master's students in November 2022, the majority of students could relate anything to this date. Model regions that have tried different versions of the Social Credit System showed that most citizens conform to the system, i.e., they willingly allow themselves to be externally controlled. Replacing credit scoring by Social Credit Scoring in Rongcheng has led to more credit defaults, and the punishments for low scores have been viewed negatively by the population, so the system there now only generates rewards, i.e., uses more carrots than sticks. However, the millions of train and airline tickets denied are being hailed in China's mouthpiece media as a success of the Social Credit System. Versions of the SCS deployed nationwide, such as the Sesame Credit Score by Alibaba, which won a bid to a private sector implementation of the SCS. It shows that cell phone-savvy youth and young adults are changing their behavior to get a better score. This was revealed in a November 2022 survey of 90 bachelor's and master's students conducted by the author. The SCS has also led to behavioral changes among companies. Entrepreneurs pay more attention to ensuring that their suppliers have a good score and that their own employees do not post anything that destabilizes the system. However, when the pressure becomes too great, for example when the civil liberties of too many people are restricted too much at the same time, fellow Chinese citizens also take to the streets. However, the demonstrations and outbreaks of violence at the Covid roadblocks were few in number. The protest did not go beyond holding up white sheets of paper, and was also immediately met with a wave of arrests, as Chinese police can recognize faces even with masks. As elaborated above, at least the protests have led to a short-term reaction from the government. The obligation to constantly scan "location QR codes" in combination with the "health code" was abolished on December 8 and almost all travel restrictions were lifted shortly after.

Where does the system hit its limits? First, it took until 2020 for all people to have ID card numbers, which are their social credit numbers, and for all organizations to have a unique social credit number. Then the data collection set-up is still ongoing, with much of the data initially being collected manually and only gradually being switched to automatic collecting. Since citizens are also not told their scores (only blacklist entries are published, and a concrete score of points is only visible in private offshoots of the system such as Sesame Credit), public awareness of the system is low. However, the train and plane blacklists are published with full names of people on http://creditchina.gov.cn. Also black or red-list entries of all organizations registered in China can be accessed on this platform. Citizens and organizational leaders seem to allow themselves to be externally controlled by the system, but not indefinitely, as the protests against Corona measures show.

Second, gamification like this presents the danger for loopholes and beneficiaries, individuals who specialize in exploiting the rules of the system for personal gain. In China, the Social Credit score for companies was manipulated to favor state-owned firms over private ones. In addition, journalists who exposed corruption among cadres, like Liu Hu from Chongqing, were put under house arrest and laid off by the Social Credit System (ABC, 2018, min. 10:30). Personal scores from other systems, such as the Corona Health Code, were manipulated by local governments to prevent individuals from demonstrating against expropriations in the provincial capital: They were given a short-term "red" score on the Corona app, were seemingly quarantined, but were actually deported to their home regions.

Is this system transferable to liberal democratic systems? In principle, systems of external control are also known in liberal democratic societies, such as priming and nudging, the popping up of individualized advertisements as well as the awarding of school grades, review results for the submission of scientific papers, a score for financial creditworthiness, prizes and awards for outstanding performance, etc. In addition, private consumption-oriented bonus systems are also common in liberal democratic societies, such as customer loyalty cards, discount promotions, promotions like to get a hot dog when you accept to get a vaccination, etc. Major differences to the Chinese model are usually, that they are voluntary in nature, that they focus on rewards instead of punishments, that they offer the possibility to have ratings explained, to challenge them and to have them deleted. Overall, the initiators for such external control systems in liberal democratic societies tend to be business enterprises as well as political interest groups that want to manipulate an election, for example. In China, the initiator is the government, which has a legitimacy deficiency due to the lack of genuine free elections and resorts to various systems, such as that of external control of citizens – for the goal of maintaining power. Thus, the system in China is more comprehensive. In principle, scoring always represents a reduction in the complexity of reality, and thus sets standards defined by an interest group that are intended to lead to a change in behavior and, systemically, thereby promote false incentives and undesirable developments. People who are forced to act against their own convictions over a longer period of time tend to become depressed and suffer health disadvantages leading to shorter life expectancy (cf. in work context: von der Oelsnitz 2014).

CONCLUSION

China is successfully using the Social Credit System to exert external control over its citizens and individuals. Within the Chinese government's numerous instruments of external control (surveillance, indoctrination, etc.), however, the SCS represents only an extremely small component. External control primarily serves to maintain the power of the ruling political class. The system is abused in China both to disadvantage private companies and to cover up corruption. Sufferers are citizens who, in part through no fault of their own, suffer loss of employment, house arrest, clan detention, double punishment for the same thing, and so on. Survey results showing high acceptance of the system among the Chinese population (Kostka, 2018) are not valid because of the unfree setting. Statements that classify the system as "ineffective" due to low levels of automation, regional differences, low awareness, and a lag in implementation are also not valid, since the system is already banning millions of people from access to high-speed trains and planes. In fact, disincentives and undesirable developments in the Social Credit System are systemic.

The system in this comprehensive form ("super scoring", Gapski, 2021) is not transferable to liberal democratic societies. Social Credit Systems based on algorithms, among other things, are ethically reprehensible. Nevertheless, even in liberal democratic societies, incentive and external control systems are already being used for smaller economic or political interest groups.

REFERENCES

ABC. (2018). Exposing China's Digital Dystopian Dictatorship. *Foreign Correspondent*. https://www.youtube.com/watch?v=eViswN602_k

Anon. (2018a). The social credit system under the microscope: The credit society is coming, are you ready? *People's Daily*. http://www.xinhuanet.com/2018-06/04/c_1122931164.htm

Anon. (2018b). Young woman goes missing after filming herself splashing ink on poster of Xi Jinping. *The Shanghaiist*. https://medium.com/@shanghaiist/young-woman-goes-missing-after-filming-herself-splashing-ink-on-poster-of-xi-jinping-669
4a646e98f

Baecker, D. (2007). *Studien zur nächsten Gesellschaft*. Suhrkamp.

Blomberg, Marianne v.: Regulation by Shaming - The Social Credit System as a pillory for environmental sinners, tax debtors, lockdown ignoramuses and other lawbreakers, unpublished paper at the conference "Social Credit System" Lüneburg 2021.

BMBF. (2020). *Future of People's Values in Our Country*. https://www.vorausschau.de/SharedDocs/Downloads/vorausschau/de/BMBF_Foresight_Wertestudie_Langfassung.pdf?__blob=publica
tionFile&v=1

Brussee, V. (2022, Feb 11). *China's social credit score – Untangling myth from reality*. https://merics.org/en/opinion/chinas-social-credit-score-untangling-myth-reality

Chen, J., Pan, J., & Xu, Y. (2016). Sources of Authoritarian Responsiveness: A Field Experiment in China. *American Journal of Political Science*, *60*(2), 383–400. doi:10.1111/ajps.12207

Chen, S. (2018). Forget the Facebook leak: China is mining data directly from workers brains. *South China Morning Post*. https://www.scmp.com/news/china/society/ article/2143899/forget-facebook-leak-china-mining-data-directly-workers-brains

Dorloff, A. (2017). Sozialkredit-System. China auf dem Weg in die IT-Diktatur. *Deutschlandfunk*. https://www.deutschlandfunk.de/sozialkredit-system-china-auf-dem-weg-in-die-it-diktatur-100.html

EU. (2021). *The AI Act*. https://artificialintelligenceact.eu/wp-content/uploads/2021/08/The-AI-Act.pdf

EUCCC. (2019). *European Union Chamber of Commerce in China and Sinolytics. The Digital Hand. How China's Corporate Social Credit System Conditions Market Actors*. https://www.sinolytics. de/wp-content/uploads/2019/08/Sinolytics_The-Digital-Hand-How-Chinas-Corporate-Social-Credit-System-Conditions-Market-Actors.pdf

Fan, W. (2021). Peng Shuai: Who is the Chinese Tennis Star, and Where is She? *Wall Street Journal*. www.wsj.com/articles/peng-shuai-what-to-know-about-the-missing-chinese-tennis-star-11637324061

Foucault, M. (1975). *Surveiller et punir*. Gallimard.

Gan, N. (2022). China's bank run victims planned to protest. Then their health codes turned red. *CNN*. https://edition.cnn.com/2022/06/15/china/china-zhengzhou-bank-fraud-health-code-protest-intl-hnk/index.html

Gapski, H., & Packard, S. (2021). *Super-Scoring? Datengetriebene Sozialtechnologien als neue Bildungsherausforderung*. Kopaed Verlag.

Heikkillä, M. (2021). *China backs UN pledge to ban (its own) social scoring. UNESCO member countries endorse plan for 'ethical AI,' calling for clear red lines*. https://www.politico.eu/article/china-artificial-intelligence-ai-ban-social-scoring-united-nations-unesco-ethical-ai/

Hoffman, S. (2017). Programming China. *Merics China Monitor*, *44*, 1–12.

Holt-Lunstad, J. (2015). Loneliness and Social Isolation as Risk Factors for Mortality: A Meta-Analytic Review. *SAGE Journals*. https://journals.sagepub.com/doi/10.1177/1745691614568352?url_ver=Z39.88-2003&rfr_id=ori:rid:crossref.org&rfr_dat=cr_pub
%20%200pubmed

Jiang, Z. (2002). *Comprehensively build a society of moderate prosperity, open a new chapter in the pursuit of socialism with Chinese characteristics* [Speech to the XVI Chinese Communist Party Congress]. https://www.fmprc.gov.cn/web/ziliao_674904/zyjh_674906/t10855.shtml

King, G., Pan, J., & Roberts, M. E. (2013). How Censorship in China Allows Government Criticism but Silences Collective Expression. *The American Political Science Review*, *107*(2), 326–343. doi:10.1017/S0003055413000014

King, G., Pan, J., & Roberts, M. E. (2014). Political Science. Reverse-Engineering Censorship in China: Randomized Experimentation and Participant Observation. *Science*, *345*(6199), 1251722. doi:10.1126cience.1251722 PMID:25146296

King, G., Pan, J., & Roberts, M. E. (2017). How the Chinese Government Fabricates Social Media Posts for Strategic Distraction, Not Engaged Argument. *The American Political Science Review*, *111*(3), 484–501. doi:10.1017/S0003055417000144

KostkaG. (2018). *China's Social Credit Systems and Public Opinion: Explaining High Levels of Approval*. https://ssrn.com/abstract=3215138 doi:10.2139/ssrn.3215138

Lavanderos, L. (2022). From cybersin to cybernet. Considerations for a cybernetics design thinking in the socialism of the XXI century. *AI & Society*, *37*(3), 1279–1292. doi:10.100700146-021-01354-2

Lin, J. (2015). *Lin Junyue - Office of the Central Committee of the Communist Party of China for Network Security and Informatization ONSI*. http://www.cac.gov.cn/2015-10/28/c_1116956896.htm. (Chinese)

Lin. (2019). *Interview. Louvet, Sylvain. Monitored: seven billion under surveillance*. Arte docu.

Louvet, S. (2019). *Überwacht: Sieben Milliarden im Visier. Arte Doku*. https://www.arte.tv/de/videos/083310-000-A/ueberwacht-sieben-milliarden-im-visier/

Luhmann, N. (1997). *Die Gesellschaft der Gesellschaft* (Vol. 1). Suhrkamp.

McAllister, K. (2019). To Raise Your Credit Score, Start by Understanding It. A beginner's guide to understandding how your credit score actually works. *Wall Street Journal*. https://www.wsj.com/articles/to-raise-your-credit-score-start-by-understanding-it-11548010800

Mittiga, R. (2021). Political Legitimacy, Authoritarianism, and Climate Change. *The American Political Science Review*, *116*(3), 998–1011. doi:10.1017/S0003055421001301

Peach, S. (2021). *Why did Alibaba's Jack Ma disappear for three months?* https://www.bbc.com/news/technology-56448688

Reisinger Walker, E. (2015). Mortality in Mental Disorders and Global Disease Burden Implications A Systematic Review and Meta-analysis. *JAMA Psychiatry*. https://jamanetwork.com/journals/jamapsychiatry/fullarticle/2110027

Ritchie, H. (2022). Leaked notes from Chinese health officials estimate 250 million Covid-19 infections in December: reports. *CNN*. https://edition.cnn.com/2022/12/23/china/china-covid-infections-250-million-intl-hnk/index.html

State Council. (2007). Stellungnahmen des Generalsekretariats des Staatsrates zum Aufbau des Sozialkredit-Systems. *Generalsekretariat des Staatsrates, 17*. http://www.gov.cn/zhengce/content/2008-03/28/content_1923.htm

State Council. (2014). *Bekanntmachung des Staatsrats: Grundriss des Aufbauplans für ein Sozialkredit-System (2014-2020)*. http://www.gov.cn/zhengce/content/2014-06/27/content_8913.htm

Steenhuysen, J. (2022). New COVID model predicts over 1 million deaths in China. *Reuters*. https://www.reuters.com/world/china/new-covid-model-predicts-over-1-mln-deaths-china-through-2023-2022-12-17/

UNESCO. (2021). *Report of the Social and Human Sciences Commission (SHS) 41 C/73 22 November*. https://unesdoc.unesco.org/ark:/48223/pf0000379920#page=14

von der Oelsnitz, D. (2014). Die frustrierende Organisation: Ungeschicktes Job Design und forcierte Entfremdung. In *Die auszehrende Organisation* (pp. 89–112). Springer Gabler. doi:10.1007/978-3-658-05307-9_5

Woesler, M. (2022). Digitalisierung und Kybernetik in China: Das Sozialkredit-System. In T. Klenk, F. Nullmeier, & G. Wewer (Eds.), *Handbuch Digitalisierung in Staat und Verwaltung*. Springer VS. doi:10.1007/978-3-658-23669-4_101-1

Woesler, M., & Warnke, M. (Eds.). (2023). *Sozialkybernetik in statu nascendi. Die Entstehungsgeschichte des chinesischen Sozialkreditsystems – Quellentexte zum chinesischen SCS*. Matthes & Seitz.

Woesler, M., Warnke, M., Kettner, M., & Lanfer, J. (2019). The Chinese Social Credit System: Origin, political design, exoskeletal morality and comparisons to Western systems. *European Journal of Chinese Studies*, *2*, 7–35.

Chapter 15

Data Protection Burdens in the United States and China:
A Focus on Privacy Data and Data Resources Security Governance Structure

Jin Kyung Lee
University of Cincinnati, USA

ABSTRACT

The expansion of personal information and data resources use is raising the need for many countries to respond. Moreover, the expansion of technologies and new technologies developed based on data resources is causing responses through legislation in each country. Through this process, the government and domestic actors share the burden of protecting data resources, which is implemented through the security governance structure. The United States and China are developing different security governance structures for personal data and data resources. In the United States, the burden of protecting personal data is shared primarily between the government and data processors. In China, the burden of protection of personal information and data resources is shared mainly between the government, data controllers, and individuals.

Personal information data generated in cyberspace is linked to economic benefits, and more directly linked to national security. Because of this, countries need to respond to the new technological environment to protect their citizens and national interests, incorporating the internet space and the personal information it generates.

In light of this, state-level intervention by the creation and management of personal information protection data is in accordance with companies that withhold personal information, then conduct production activities based on it. Therefore, each country's political and economic characteristics are reflected in factors for the personal information data management strategy of the government.

This chapter aimed to analyze the data resources formed by the US and China, focusing on how they manage personal information data, along with their structural characteristics. The analytical framework was security governance, and the primary analysis target was both countries' legislation and its influence

DOI: 10.4018/978-1-6684-6429-8.ch015

on personal information data resources. At the end of this chapter, the differing personal information data protection and security frameworks of the United States and China are evaluated. The final understanding of the analysis and potential analysis topic are presented in conclusion.

FRAMEWORKS AND SUBJECTS OF STUDY

The framework that is mainly going to be used in this study is security governance. Under security governance, this study followed government security around personal data in the United States and China, plus data resources in the larger category. The comparison is meaningful in showing that both the United States and China are responding to the development of data resources, including personal information data and the technological environment that utilizes these resources.

One might ask whether comparing the US and Chinese legislation is methodologically feasible. However, the comparison between the two is highly feasible because both countries have proposed a set of regulatory frameworks for personal information data. These regulations also have value in being a national asset.

To clarify, what this study intended to examine is how the governing authorities legislate statute and regulations on the structure of personal information data, and what influences domestic actors have. The subject of this study was limited to the current state of legislation protecting personal information in the United States and China. Specifically, in the case of the United States, the study followed federal legislative attempts for personal information data and data resources and state level legislation enactment. In the case of China, Personal Information Protection Law and Data Security Law were analyzed.

The primary purpose was to evaluate the structure of security governance as shown in the overarching legislation of the two nations, and the rights granted to domestic actors through their governing systems. The study focused mainly on 2020 and following years when privacy-related legislative activity in the United States and China became prominent.

It is understandable by examining the legalization of personal information data protection in the United States and China that both governments are trying to protect personal information data with the private sector.

LITERATURE REVIEW

Security Governance

Security governance branches into several concepts, which is the main framework of this study. The United Kingdom's National Cyber Security Center has clearly defined the phrase. According to this organization's definition, security governance is a decision-making coordination framework that operates to approach security (National Cyber Security Center, 2016). If this system is successful, security can be pursued effectively.

By a more academic approach, the concept of security governance has been generally developed and applied in international politics. Since the end of the Cold War, as the security environment evolved, the stakeholders involved in the pursuit of security have diversified. As a result, actors under security governance that are mainly state-centered have developed to include the state and non-state actors, along with public and private actors (Krahmann, 2003).

In the conceptual framework of security governance, security encompasses issues in traditional and non-traditional domains. This is where multilateralism and government logic are the basis of the formation of security governance. However, since the term is difficult to make concise, several conceptual and empirical studies are ongoing (Ceccorulli & Lucarelli, 2013).

Research on the importance of private actors in the security governance field has also been conducted. In particular, the role of private actors in traditional security fields, like the military, where national influence is relatively strong have been the focus of these studies. In addition, there is a notable attempt to evaluate the various impacts of national security interventions in the private sector under security governance (Westermeyer, 2013).

Some studies have also presented the expanding influence of the security governance structure in the private security sector. For example, private military security companies gradually expanded and took charge of security areas that had been controlled solely by the states. Through this process, private security companies had a significant influence on the national security governing structure (Bures & Carrapico, 2018).

The concept of security governance relating to data resources was carried out in this study, and more practical research is likely to be needed in the future. In light of this view, the conceptual ambiguity between data resources and security governance frameworks within the literatures in this section does not have good standing in the history of literature as it is fairly new in the cyberspace.

Nevertheless, some literature studies the relationship between open data and security governance. As open data is used more often, a wider variety of actors will be involved in security governance frameworks to participate in decision-making for data management. Because of this notion, various actors refer to stakeholders such as the government, civil society, and corporations (Kassen, 2022).

Based on these research trends, security governance for data resources in this study can be understood as the formation of a framework in which multiple actors are involved in decision-making on data resources, including personal information data. The key participants in decision-making regarding security governance include individuals, businesses, and governments.

The major participants in data resource security governance in the United States and China, which future researchers can consider, are these same individuals, companies, and governments. Moreover, two aspects of data resource utilization and protection will be examined for decision-making. To summarize, data resource security governance in this study refers to a structure that determines the scope of data resource protection-utilization centered on the individual, corporate, and government legislative measures on data resources.

Figure 1. Key actors of data resource security governance structures

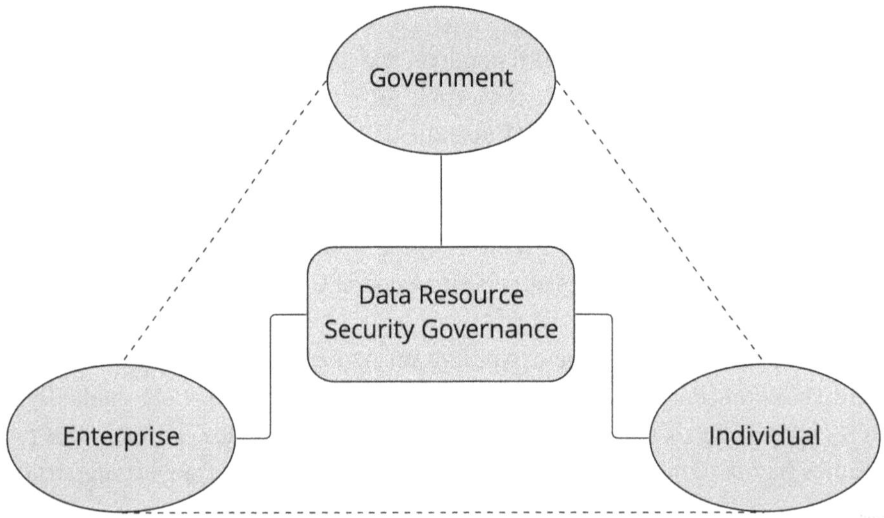

The above figures show essential participants in the security governance structure and their roles. As described above, countries, companies, and individuals are important participants in the security governance structure. The three actors make decisions about data resource management in this structure, while the state grants individuals and companies responsibilities and rights to data resources by legislation.

The framework was defined based on the theoretical basis of the existing security governance structure. It also serves as a theoretical framework that will identify the security governance structure of the United States and China to be simultaneously conducted. Through this theoretical framework, it is expected to understand how the US and China approach personal data management in different ways.

The European Union's General Data Protection Regulation (GDPR)

The European Union's GDPR is the world's most stringent privacy and security law. The legislation was enacted by the European Union, and it protects privacy obligations for those citizens under the European Union's discretion. Violating the law will result in sanctions in the form of substantial fines (*What is GDPR, the EU's new data protection law?*, 2018). A look at the case of the GDPR is meaningful because it sets out a framework for the imposition of comprehensive legal obligations on personal data and imposes strict responsibilities.

Also, since the enactment of GDPR, the United States and China followed in crafting legislation and enacting it to manage personal information data. It can be seen that this law has impacted the development of a general framework for personal information data protection and the development of security governance in several other countries.

The GDPR has a structure that guarantees the right of personal information to individuals while granting obligations to the personal information controller, which uses personal data. In particular, the principle of transparency that must be guaranteed in handling personal information, the minimum level of personal information processing, and the principle of accountability are specified, and similar contents are confirmed in the United States and China legislation. Accordingly, the application of GDPR to security governance dealing with personal data is significant and should be assessed. (McGavisk, n.d.).

Privacy Data Security Governance in the US: Focusing on Duty to Data Processors

It has been recognized that data privacy in the United States has become closely related to national security. This is based on concerns about the nature of new technologies, such as artificial intelligence, along with China's national technology policies. In particular, since data resources are essential for developing and utilizing new data-based technologies, privacy data is also recognized as an essential resource for the pursuit of national security (Williams, 2020). Therefore, several legislative attempts and enactments surrounding privacy data are a start to analyzing the US governing structures used in providing data privacy. This is provided at the federal level and the state level.

In the United States, there are no federal-level laws to protect privacy data. Instead, it is reasonable to view that the protection of privacy data in the United States has been accomplished by sectoral legislation (Klosowski, 2021). Representative laws include the Privacy Act, the Health Insurance Portability and Accountability Act, and the Child Online Privacy Act. These laws provide some protection of personal information in different areas of the cyberspace.

It was typical for each state in the United States to have a privacy law for each sector, like the nature of federal legislation. Still, after the European GDPR was enacted, several states introduced general privacy laws, starting with California. Many states are actively seeking to create comprehensive privacy laws to protect their own residents.

In addition to federal-level laws, the US federalist system separates the power of the national and state governments, but they have worked together on this issue. As a result, each state's legislation relating to protecting personal information online also tended to be made by the sector. However, in recent years, as the need for more cohesive management of personal information data has grown, several states in the United States are introducing privacy laws. For example, California, Virginia, and Colorado have enacted general privacy laws.

Of course, since federal and state-level laws play different roles, and each state has its unique legislative powers, it would be an overstatement to say that there is a single privacy-related legal system across the United States. However, federal and state laws in the United States are in proper balance, and the two objects form a cooperative relationship. Based on these relationships, the United States grew into a country with solid national power based on each state's diversity and innovation. Therefore, analyzing each state's laws dealing with data resources and privacy data is also reasonable compared to China's legislation.

In the case of China, which will be discussed later, the central government has the highest authority. It is implementing data protection legislation that is applicable to all of China, so it may not be appropriate to compare the US federal law with that of China in a strict comparative methodology. There needs to be more room to compare the two since they are so different when it comes to the separation of their government powers.

On the other hand, considering the independent legislative system of each state in the United States, the legislative activities related to privacy data can be understood in the context of the government authority in the United States that make comprehensive rules for the protection of personal information. This analysis that attempts to explore the efforts to protect personal information data in the United States by its legal activities, seems to be accompanied by a simultaneous examination of relevant federal and state laws in the United States.

Federal Level Legislation and Privacy Data Security Governance

Recently, as the opinion that federal law should protect personal information has became more popular, Congress making several moves to respond. This is based on the recognition that data resources are connected across various fields and that the development of the industry needs to protect and utilize personal information effectively (Allos, 2022). Attempts by the U.S. Congress to enact privacy-related federal legislation have primarily involved creating protective legislation and an agency responsible for overseeing personal information management.

S.2134: Data Protection Act (2021)

First of all, the Data Protection Act, introduced in 2021, advocates the establishment of an independent body to govern the management of specific high-risk data. It also regulates the collection, processing, and sharing of personal data. This institution has the authority to oversee and the responsibility of regulation privacy data by relevant legislation, while carrying out activities related to protecting personal information. There are high-risk data measures that include matter related to the large-scale collection of personal information and biometric data that can identify individuals.

The bill's initiative appears to primarily define "Big Tech" companies that handle personal data on a large scale as data controllers. If adopted, the bill would also provide stronger privacy protections for individuals in the United States from commercial entities that use their personal information.

Independent entities responsible for data processing may receive regular reports that particularly come from large commercial entities that collect and utilize large-scale data. These commercial entities can also be inspected regularly. In other words, the bill provides that the agency responsible for protecting personal information will be empowered to prescribe rules and to handle cases that violate the protection of personal information (Gillibrand, 2021).

The bill overall establishes an agency dedicated to protecting personal information. The bill's end goal of seeking centralized supervision from the federal government that is formed and upheld by said government. However, since some states have independently adopted general laws related to personal information, the state laws are expected to take precedence in this case.

No normative assignments related to the protection of national interests are found in this initiative, provided by the legislation related to the protection of data resources and privacy in China, which will be analyzed later. Even though the bill suggests that the agency responsible for processing personal information data can, if necessary, collaborate with national security-related organizations such as the National Security Agency, this appears to be an insignificant connection of national interest and security-privacy data resources than China's relevant laws.

S.1494: Consumer Data Privacy and Security Act of 2021

The bill, in seeking to protect consumers' privacy in the United States, aims to set clear federal-level standards for privacy. Currently, several states in the United States are enacting legislation to protect consumers at the state level. Therefore, different legal standards exist for each state, and the goal is to provide a uniform standard at the federal level to overcome the various limitations that come from this.

The legislation also mandates that companies that collect and process large volumes of personal data take steps to protect personal data in a more effective manner. It also requires companies to continuously develop and apply technologies related to personal information protection in their work.

S. 2499: SAFE DATA Act

The Setting an American Framework to Ensure Data Access, Transparency, and Accountability Act provides a framework that addresses several rights to the personal information of US consumers. The main content is to prohibit companies from processing or transferring consumer personal information without consent from those whose information has been collected. In addition, while collecting and processing personal information, companies must transparently disclose information to consumers and notify consumers when changes occur. The bill also requires that companies conduct a privacy impact assessment to understand how much damage they can cause to consumers.

The legislative attempts by Congress to protect personal information presented above have not been signed to become official law. However, it is noteworthy that the government imposes regulations on the collection and use of personal information on the personal information controller. The information controller are the companies, which the law is most directed towards. In other words, the federal government is sharing the burden of protecting personal information data with enterprises by imposing various regulations on companies regarding the responsibility of the higher authorities to protect the personal information and data of the people. Since the government's privacy-related legislative activity is expected to continue, it will be meaningful to note whether the US government can successfully produce and uphold data privacy laws for the people.

State-Level Legislation and Privacy Data Security Governance

In the past, each state government in the United States has not implemented comprehensive privacy-related legislation at the federal level. However, since the European Union's GDPR enactment, each state has begun to engage in comprehensive legislative activities dealing with personal information protection. California, Virginia, and Colorado have enacted privacy-related legislation and are implementing it.

California Consumer Privacy Act

California is the first state in the United States to enact and enforce comprehensive privacy laws. The first legislative attempt was made with the California Consumer Privacy Act of 2018. After several amendments and legislative overhauls, the California Consumer Privacy Act came into effect on July 1, 2020 (California Department of Justice, 2018).

California's consumer privacy laws are complex, reducing the burden on businesses that handle personal information while placing new burdens on oversight. First, the bill imposes substantial state-level restrictions on sensitive information in a more specific nature towards personal information. A new phrase, sensitive information, holds that there is information that can identify an individual, including medical information. This shall be protected by the bill.

According to this law, consumers can request correction of personal information, refuse to make decisions using personal information, and refuse to share or use sensitive information. In other words, it legally guarantees individual consumers the right to control their personal information.

In particular, this bill introduces and proposes the GDPR principles of the European Union and reflects minimizing the processing of personal information as big data, restrictions on the purpose of use of personal information, and the retention period of personal information in the draft.

Companies that utilize the collection and use personal information must create a contract specifying their obligations to protect personal information by this Act. Companies should actively protect personal information by appropriately implementing and developing security technologies and procedures essential for personal information protection. In other words, the state of California places a more significant burden on companies that use personal information to protect that information they withhold on their residents.

The most crucial point of this bill is creating a personal information supervision agency dedicated to collecting and using personal information. It is the first personal information supervisory authority established in the United States. The government agency is known as the California Privacy Protection Agency, will be able to determine the number of sanctions against data processors who violate California consumer privacy laws. In addition, following the rights guaranteed by state-level law, sanctions can now be imposed on companies processing personal information. In other words, in a situation contrary to or violating this law, personal information controllers may be subject to sanctions composed mainly of fines from the agency in charge.

However, it can also be seen in this Act that the number of companies affected by the Act has been reduced by applying the criteria for selecting personal information controllers subject to the Act in a lenient manner. Because of this, the bill has a multi-layering aspect, which reduces the number of companies subject to it while enforcing sanctions with substantial obligations and the legal basis on companies.

Virginia Consumer Data Protection Act

The Virginia Consumer Data Protection Act of 2021, passed by the Virginia State Legislature, is expected to take effect in 2023. This bill coincided with the start of the enforcement of the California Consumer Privacy Act in the case of Virginia, but it was the second in enactment after California. Therefore, in the case of Virginia, the content of the GDPR and California Consumer Privacy Act has been inherited in terms of similarity of legislation time (*Consumer Data Protection Act*, 2021).

The Virginia Consumer Data Protection Act applies primarily to those who manage and process personal information when doing business in Virginia or providing services to residents of Virginia. In this case, the application of the act excludes non-profits, universities, and government agencies.

The state of Virginia has a broader definition of personal information data than the state of California. It includes more information that relates to or can be linked to an individual. On the other hand, encrypted, non-personally identifiable, or publicly available information will not be included in the subject of personal information under the Act. However, race, ethnicity, religion, health information, and child information are classified as sensitive information and can only be used with the individual's consent.

The Virginia Consumer Data Protection Act classifies companies that make decisions about using personal information as controllers and companies that process personal information as processors. The bill specifies that the law applies to both of these actors. By this law, especially in the case of controllers, personal information processing must be decided in consideration of the purpose of using personal information. Personal information cannot be processed without the user's consent.

The Act also provides several rights that individuals who are sources of personal information have to protect their privacy. Typically, an individual can check if a company processes personal information, request a correction in case of inaccurate information, request deletion, and refuse to process personal information.

Virginia's bill does not discuss establishing a separate, centralized organization that takes complete control of personal information as in California. Thus, an independent body does not deal with law violations. This leaves the Virginia Attorney General with the enforcement powers. In this case, the primary means of sanctions is the imposition of fines. Virginia enacts legislation that obliges businesses to manage personal information more strictly. Moreover, the law gives individuals the right to be protected as the subject of personal information.

Personal information data is viewed as a subject that needs to be managed through comprehensive legislation. There is a recognized need for a comprehensive approach at the state level for the development of technology using personal information data and the various security changes that follow. In other words, the increase in the use of personal information and the development of technology have led to a response from the state government's legislation.

Colorado Privacy Act

Colorado is the third state in the United States to comprehensively enact legislation to protect personal information, after California and Virginia. The Colorado Privacy Act was also enacted in 2021 like California and Virginia. The three states made a comprehensive, state-level movement to protect personal information data simultaneously (Protect Personal Data Privacy, 2021).

Colorado's privacy protection law adopts the concept of a personal information controller and processor as in Virginia, so it can be seen that the concept related to personal information processor presented in the European Union's GDPR is an influence on this legislation.

Personal information data covered by this measure is data that can identify individuals. To Virginia, non-identifiable data and information already disclosed to the public is not apart of the personal information the measure intends to cover.

The concept of personal information data Information such as race, ethnicity, religion, health information, and biometric information corresponds to sensitive information that must be specially protected. Accordingly, through this Act, the subject of personal information has the legal right to receive protection as sensitive information.

The legislation gives consumers fundamental rights to personal information. First, individuals can have tangible access to verify that the controller processes personal information. Errors in personal information can be corrected, and personal information can be deleted or moved independently. Finally, if the personal information controller does not wish to use personal information, he/she also has the right to revoke consent to use personal information.

Aspects of the obligations imposed by the legislation are more focused on controllers and processors using personal data. Therefore, personal information users, i.e., controllers and processors, who mainly collect and use personal information in Virginia, must approach the privacy policy with transparency. This is a characteristic that is also seen in California and Virginia, and it has also been represented in Colorado.

Also, the personal information controller must specify for what purpose the personal information is collected and processed. Particularly, in collecting personal information, this can be done in a minimized range. The minimum level means the level that is reasonably necessary. In addition, the personal information controller cannot recycle personal information without consent. Finally, to protect personal data, controllers must continue to make their data protection efforts.

The Act also obliges personal data controllers to assess the impact of the use of personal data. It can be seen that the obligation to conduct an impact assessment in processing targeted advertisements or sensitive information using personal information is strictly imposed through this Act. Colorado could impose penalties in the form of fines on data controllers and processors for violating the law.

General Assessment of US State Privacy Laws in Security Governance Perspectives

Table 1. Key points of data resource security governance structures in California, Virginia, and Colorado

	California	Virginia	Colorado
Key actors in the privacy data resource security governance structure	State governments, businesses, and individuals	State governments, businesses, and individuals	State governments, businesses, and individuals
The rights of the information subject	Granting the right to know, access, delete and correct personal information data	Granting the right to know, access, delete and correct personal information data	Granting the right to know, access, delete and correct personal information data
Obligations of Personal Information Processing Enterprises	Obligation to collect and process data, minimize personal data processing, prohibit sensitive information processing, and data security measure	Obligation to collect and process data, minimize personal data processing, prohibit sensitive information processing, and data security measure	Obligation to collect and process data, minimize personal data processing, prohibit sensitive information processing, and data security measure
Use of terms for data processors	Business	Controller, Processor	Controller, Processor
The presence of a central agency	Introduced	Not introduced	Not introduced

The table above a key to understanding data resource protection legislation in California, Virginia, and Colorado. State governments, businesses, and individuals become important actors in the data resource security framework in all three states. The government guarantees several rights to the data of individuals who are the data subjects. The government as well imposes obligations on businesses that collect and process personal information data.

Apart from the above commonalities, California uses the term business instead of data collector and processor separately from Virginia and Colorado. California also has established a central agency that manages data resource utilization, unlike the other two states.

The link between the US state government's comprehensive privacy protection movement and the security governance framework is not a traditional discussion aspect, so a more detailed analysis is needed in the future. However, the environment in which the subject of personal information data becomes an essential means of creating productivity for companies and the development and use of the technology that supports it continues to be attempted. It seems that the need for comprehensive regulation in the US state governments has been brought up and considered to the point that legislation is in the works.

If the mechanisms of resource provision, resource handling, and strengthening of regulations formed because the subject of personal information, the persons concerned with personal information processing, and the government's ability to protect their data, it is expected to contribute to the formation of a stable industrial environment related to personal information in the respective states. This is shown by states' mandated technological efforts to protect their personal information on businesses and the setting of regulations. In other words, state governments share the burden of protecting personal information with companies by giving them the duty to protect personal information resources.

The privacy laws of California, Virginia, and Colorado do not directly address the pursuit of national or state-level security. However, this series of legislation suggests that the state legislation is responding to changes in the technological environment based on the use of personal information that will change in the future.

Also, each state's legislation does not provide the same regulatory framework. For example, in the case of California, the established centralized agency can be fully involved in collecting and using personal information. These separate agencies are not seen in Virginia and Colorado. The level at which personal information must be protected is not specified equally. As such, the rights and regulations granted by the state governments are not unified, but the direction toward stabilization of the personal information environment has been a similar notion by these states. Moreover, the stabilization of this environment leads to the provision of a stable environment in a broader sense. That is the pursuit of a stable security environment on a nationalized level.

Furthermore, the laws of these states do not take the approach that requires that personal information be protected on a normative level because personal information assets are directly related to national security for individuals and personal information controllers, as is the case with China's legislation, which will be analyzed later. Instead, it takes the form of protecting the rights of individuals as subjects of personal information and granting duties to companies to protect personal information.

In conclusion, several states in the United States are enacting general legislation for the integrated management of personal information. Through such legislation, they are responding to changes in personal information data and the technological environment withholding the data. The state governments are sharing the burden of personal information protection with companies, which are the personal information controllers at the state level.

Privacy Data Security Governance in China: Focusing on Duty to Data Processors and Personals

China is actively collecting and using various personal information data of its citizens. Therefore, accidents related to personal information leakage also occur frequently, and such an environment calls for the need to manage data resources comprehensively (Qin et al., 2022).

In the late 2000s, China began designing a path for social development based on informatization. As the scope of the usage of technologies such as cloud computing, the mobile Internet, and big data expands, network security and informatization have become increasingly essential values in modern Chinese society. In particular, China has strategically used informatization to improve domestic living standards and develop social systems, so protecting it has become an important duty (Du et al., 2018).

In the past, there was no uniform and personal information protection law in China. However, as it was continuously discussed that the protection of personal information is directly related to the national interest, the Personal Information Protection Act was enacted in 2021.

China has also enacted the Data Security Act for regulating data resources' management and supervision. The bill expresses that the Chinese government will actively intervene for reasons of protecting national security in data exchanges outside China's territory, and providing overall supervision of data resources generated within China.

China's Personal Information Protection Law General Information

According to Article 4 of China's Personal Information Protection Law of 2021, personal information refers to various types of information relating to an identifiable natural person. Also, personal information processing is an act that includes the collection, storage, use, processing, transmission, provision, disclosure, and deletion of personal information (Creemers & Webster, 2021).

Article 28 defines sensitive personal information as biological characteristics, religion, status, medical information, financial information, and personal information of minors under 14.

Principles of Personal Information Management in Accordance with China's Personal Information Protection Law

The bill contains principles that should be followed regarding handling personal information in China. For example, Articles 5, 6, and 7 explain that the processing of personal data must comply with the principle of transparency in a lawful and just manner.

Articles 13 and 17 stipulate cases where personal information can be processed, along with procedural obligations. For example, personal data may only be processed with the individual's consent, which is contractually necessary and fulfilling a legal obligation. Also, personal data can be used to supervise public interest or opinion. Also, for this purpose, the legal status or route of the person handling personal information and the procedure for the obligation to fulfill the notification to the individuals who'd data is being held are legally identified.

The legislation also states that individuals have a right to know about processing their personal data. The citizens also have the ability to make decisions, and object to other people's involvement in their personal information without the individual's consent.

The Role Burden of the Government and Domestic Actors Identified in China's Personal Information Protection Act

China's privacy data-related legislation gives the government independent authority over personal information data. This is confirmed in Articles 18, 26, and 35. In this provision, the Chinese government presents that state agencies may use personal information without notice or action to individuals in emergencies involving legal liability or national interests. In other words, through these provisions, the Chinese government provides individuals with transparency, relating to processing personal information. In contrast, the government takes the lead and provides exclusive access to personal information in cases related to national interests.

Where the government's monopoly authority over processing personal information is further specified is prominent in the law's principle of providing personal information abroad. The Chinese government strictly controls and manages the overseas transfer of personal information specified in Articles 38, 39, 40, and 42. First, when a personal information controller intends to provide personal information to Chinese citizens abroad, it must pass a safety evaluation and personal information protection certification from the relevant government department. In addition, the transfer and use of personal information abroad possible only when the Chinese law or the department in charge permits it.

Also presented in Articles 40 and 42, personal information data generated in China is domestically stored. When it is transferred abroad, it must undergo a government stability evaluation. Moreover, the government may prohibit using personal information directly related to China's national or public interest. Those who are prohibited from using personal information are managed separately, and the use of personal information is restricted by publicizing this information to the citizens of China.

These activities of the Chinese government are overseen by a department with uniform authority in protecting and managing personal information. The management department manages personal information protection while managing and coordinating various policy enforcement rules. Related activities include establishing standards and rules for the protection of personal information, establishing new rules in response to the increasing use of sensitive information and new technologies, such as facial recognition and artificial intelligence, and promoting network-based identification services for Chinese citizens.

An Overall Evaluation of China's Personal Information Protection Law From a Shared Perspective of Security Burden

China's personal information protection law is a plausible example of how new technology affects the government. The Chinese government is actively seeking standards for newly developed and applied technologies such as, again, artificial intelligence and facial recognition technology. China includes sensitive information, medical information, and financial information in the category of personal information data. Additionally, standards are given by law to individuals who are the source of personal information or users who want to use personal information.

In cases directly related to national interests or security, the government may be an exception to the procedural transparency or duty of disclosure guaranteed by this bill. This means that the state has the right to provide a personal information data management framework in the state-individual relationship represented by the government to the citizen. If the state prioritizes the rights of individuals, and the source of personal information, that would contend a completely different measure. However, in the case of China, the state can exercise powers beyond individual rights when it is necessary to use personal data for national interest and security stability.

In the case of data, controllers appear to be obliged by the Chinese government to actively comply with the government-provided personal data management framework and standards or rules. By obligating the protection of personal information, the government can take measures such as fines, confiscation of profits, suspension of business, and revocation of business licenses if the personal information controller fails to do so. If they fail to do so, they will be held legally responsible. In addition, if the personal data processing party is a corporation, the government may intervene in the audit. It may not allow the corporation to appoint a personal data protection officer.

Therefore, China's Personal Information Protection Act provides a legal basis for a transparent management procedure for personal information, while also imposing restrictions on personal information controllers and businesses using personal information like protection and management for individuals.

When personal information and national interest are significantly related, the state has a priority right to engage in personal information management beyond the procedural transparency presented above. Along with this, personal information managers are obliged to comply with the management standards set by the government, and failure to protect personal information may result in substantial sanctions from the state.

China's legislative body implemented the Data Security Act in September 2021. This can be seen as a legislative activity around the same time as countries worldwide are proceeding with various legislation to regulate and manage the transfer of data resources abroad.

Scope of Data According to China Data Security Act

Data under the Data Security Act is more comprehensive than the subject of management under the Personal Information Protection Act. Under this act, data means any record or resource created electronically or in other manners. The act of processing this data resource broadly includes the act of collecting, storing, using, processing, providing, and disclosing data (DigiChina, 2021).

Also, through Article 3, data security effectively protects data through several different measures, while enabling legitimate use of it. In Article 21 of the Data Security Act, the term "national core data" is used, which means that data related to national interests, security stability, and public interest are resources that the state should intensively manage and supervise. Therefore, national data is considered critical and is subject to special protection and management by the government.

General Principles of Data Sovereignty Confirmed in China Data Security Act

The data security law clarifies that any individual or group that utilizes data resources can be involved in China's safety and national interests. In response, Article 1 explains that rather than completely restricting the data, it is furthering national productivity by facilitating data openness and use.

In addition, Article 8 of the bill stipulates that activities using data in the bill must be directed toward complying with national laws and protecting the public interest a along with the ethical standards of professionals. It means that the use of data is in the context of the regulations that the state imposes on domestic actors. Therefore, as shown in Article 35 of the Act, when the national security agency collects data to protect national security, according to the law, it can be interpreted that the relevant organization or individual is obliged to cooperate.

The bill also states that the Chinese government will have supreme authority over data transfers abroad. According to Article 36, Chinese nationals or organizations cannot transfer data stored in China to foreign judicial-related organizations without the authorities' permission.

General Evaluation of China's Data Security Act in Terms of Security Governance

Articles 19 to 23 of this bill shows how the data safety management system will be implemented. First, to establish a safety management system according to the class and type of data, the state recognizes the importance of data for national development, classifies data according to the degree of decisiveness in national rights and interests, and actively protects data. Each local government also sets the target of critical data protection and proceeds with special protection. Second, the state actively manages data with a centralized, authoritative data safety and risk assessment, information sharing, monitoring, and warning system. Furthermore, a safety review system will be established to manage data safety and export, and government-led safety screening will be conducted. There will be a separate management system applied and managed in the case of data exported overseas.

The bill clarifies what obligations the Chinese government imposes on domestic actors on data resources. The state gives domestic actors the responsibility of managing data resources for the national interest and security stability. This management duty means to faithfully perform various procedures proposed by the state. For example, the classification of data resources according to their importance to national interests and examinations.

If those responsible do not fulfill their obligations related to monitoring and evaluation activities for data safety management, the state can impose strong sanctions. Penalties include the execution of fines, suspension of business, and cancellation of data transaction licenses.

China's data protection law can be seen as a form in which the state is deeply involved in using and managing data resources, giving domestic actors the duty to protect data resources. This also presents a confirming of the state's goal to select and manage essential data linked to national interests proactively. In other words, the state can exercise its authority to select data resources essential for the national interest and security environment.

Through this bill, the Chinese government proposes several institutional frameworks necessary to manage and supervise the use of data so that domestic participants can follow it. This can be confirmed by providing a government-led institutional management framework across all fields of data utilization certification, evaluation, and management.

Therefore, the Chinese government has acknowledged that data resources are tied to national interests, and incorporated legislation to provide a will to actively intervene to pursue a stable security environment and improve national productivity. It can be interpreted that the development of new technologies motivated the Chinese government to establish new rules and norms and give them to domestic actors. Through this, domestic actors involved in data resource processing adhere well to the state's norms; the state manages data resources in a centralized form.

Differences in the Structure of Privacy Data Security Governance in the United States and China

Figure 2. Two types of the U.S. privacy data security governance structure

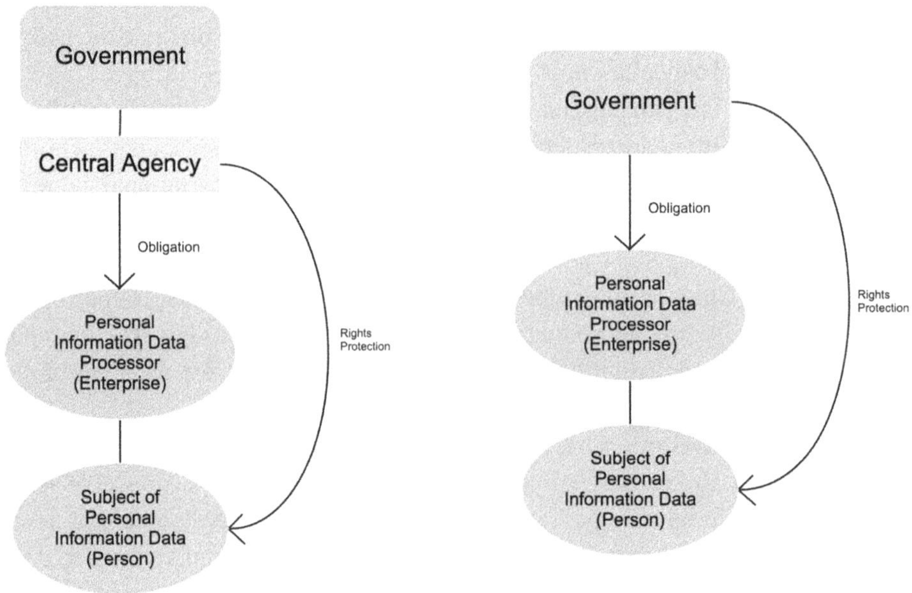

As shown in Figure 2, In the case of the United States, the state government imposes obligations on the data controller within the privacy data security governance structure. Individuals who are the subject of personal information are the priority party in receiving protection of their personal information. The existence or non-existence of a central agency that oversees the state's personal data involvement varies from state to state. For example, California has a central agency, but Virginia and Colorado do not have separate agencies for data protection. Accordingly, the United States shares the burden of privacy data protection between the data processor and the state government in this structure.

Figure 3. China's security governance structure for personal data and data resources

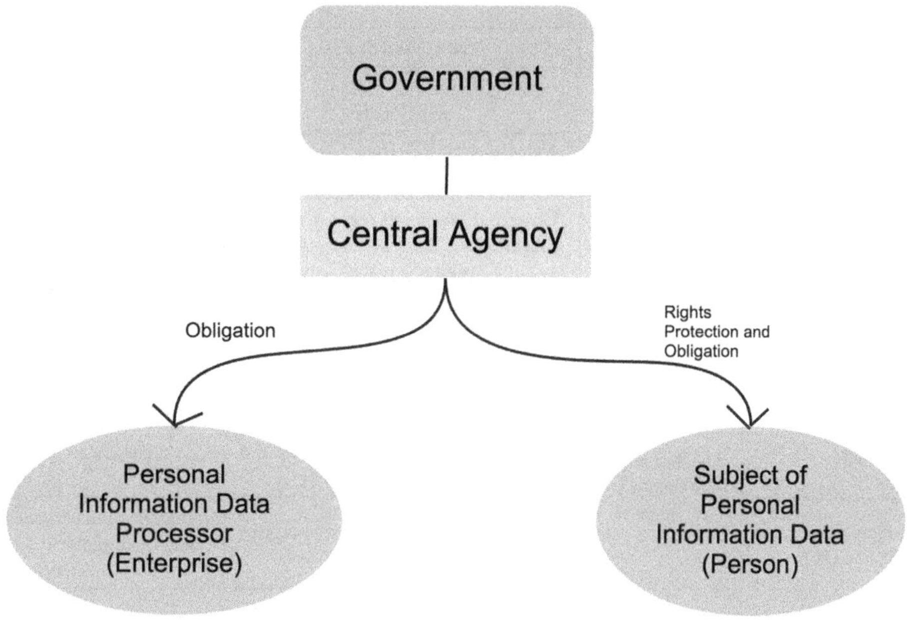

As shown in Figure 3, in China, the government imposes obligations on data controllers and individuals within the privacy data security governance structure. In this case, the normative purpose of national security and the pursuit of interests is reflected. At the same time, the right to protect personal information is granted to the individual. In addition, a central agency oversees the protection and management of personal information data from the government. Accordingly, in China, both the subject of personal information and the personal information controller bear the obligation to protect personal information data granted by the government. Also, in the case of China, the same structure works not only for personal information data but also for general data resources.

Table 2. Commonalities and differences in data security governance structure in the United States and China

		The United States	China
Main actors of the data security governance structure	The commonalities between the two states	State governments, businesses, and individuals	Governments, businesses, and individuals
Recognizing the need to manage data resources		Awareness of securing data resources exists.	Awareness of securing data resources exists.
Obligation to data collectors and processors (businesses)		Government imposes regulations and obligations on data resource collectors and processors.	Government imposes regulations and obligations on data resource collectors and processors.
Protection of the rights of individuals who are data subjects		The government grants rights to individuals subject to personal information data.	The government grants rights to individuals subject to personal information data.
The central agency for management	The differences between the two states	No central agency exists for privacy data resource management. (Exceptionally, California has a central agency.)	The central agency exists for privacy data resource management.
Obligation to individuals who are data subjects		Individuals are not obligated to protect data resources.	Individuals are obligated to protect data resources if national security protection is a priority.

The above table summarizes the commonalities and differences between the United States and China regarding the data resource security governance structure. First, the United States and China commonly include governments, businesses, and individuals as essential actors in the data resource security governance structure. Both countries recognize the need to systematically manage data resources and recognize that keeping data resources safe is essential for industry and national development. Third, both countries impose regulations and obligations on companies that collect and process data resources, mainly for data resource protection, management, and minimal use. Finally, both countries are pursuing the protection of rights to individuals who are privacy data subjects.

Meanwhile, the two countries show differences in the existence of central agencies and the obligation to individuals. First of all, in the case of China, there is a central agency for data resource management. However, no federal agency has been established in the United States. In the case of California, a central agency was unusually established, and further analysis is needed on this cause. China is taking a more specific approach to the situation where national security is prioritized, in which case individuals are obligated to cooperate with the government to provide personal information to the government. This regulation is not confirmed in the case of the United States.

The difference in the data resource security governance structure between the United States and China may also be comparable from a chronological perspective. In particular, even in modern China's political institution system, the Chinese Communist Party has an authoritarian representative character for state management and is with the national system (Duan, 2019). Therefore, it can be analyzed that the nature of right-wing representation has been applied even in China's data resource security governance structure. A new explanatory framework can become applicable if research on data resource security continues in the future.

CONCLUSION

The use of data resources is expanding. The development of new technologies based on data resources presents new tasks for the governments of superpowers. Specifically, data resources are becoming increasingly important for national security, economic benefits, and productivity enhancement. The government or legislative authorities are implementing various responses regarding personal information data among data resources. At this time, a structure for protecting personal information and data resources is presented as a security governance framework. Identification and analysis of these structures are expected to help devise a national-level strategy to utilize data resources, including personal information data, efficiently (Craddock, 2022). In addition, it is meaningful in that it is possible to understand how the response of the government and legislative authorities will develop in an environment where new technologies, especially data-based new technologies, are expanded and developed.

The United States and China together form a competitive landscape in the field of data security governance. The nations are creating different types of data security governance structures. In the United States, data protection obligations are primarily imposed on data controllers at the state level. Therefore, it shows the characteristics of a security governance structure in which the state government and the personal information processor jointly pursue the protection of data resources. The government imposes obligations on individuals and data processors in China through centralized agencies. This duty emphasizes the essence of national security and the pursuit of protecting said security.

The United States and China each have different personal information data security governance structures. Whether the United States can complete legislation on federal-level comprehensive privacy data and general data resources has not yet been presented. Along with this, more analysis is required to understand the relationship between security governance and data resources.

The results of this chapter provide a positive point for new but research-intensive issues such as data resources. Governments need new resources to manage them well for the benefit and stability of the country and thus form a unique governance framework for each country. The examples discussed represent the United States and China's development of data resource security governance frameworks, just in different forms.

In the case of the United States, each state has launched a bill to manage privacy data and is developing it uniquely. The case of California, which created a central management agency, needs to be studied more closely in the future. As in the case of China, if the government imposes an obligation on individuals to protect resources, it can be studied later whether this pattern is confirmed in other authoritarian countries.

REFERENCES

Allos, A. F. (2022, September 21). American Data Privacy and Protection Act Moves to The House. *The National Law Review*. https://www.natlawreview.com/article/american-data-privacy-and-protection-act-are-we-finally-getting-federal-data-privac
y

Bures, O., & Carrapico, H. (Eds.). (2018). Security privatization: How non-security-related private businesses shape security governance. Springer International Publishing. doi:10.1007/978-3-319-63010-6

California Department of Justice. (2018, October 15). *California Consumer Privacy Act (CCPA)*. State of California - Department of Justice - Office of the Attorney General. https://oag.ca.gov/privacy/ccpa

Ceccorulli, M., & Lucarelli, S. (2013). Conceptualizing multilateral security governance. In The EU and multilateral security governance (pp. 11–25). doi:10.4324/9780203095430

Consumer Data Protection Act. (2021, February 4). Virginia's Legislative Information System. https://lis.virginia.gov/cgi-bin/legp604.exe?211+ful+SB1392ES1

Craddock, P. (2022, September 19). EU Cyber Resilience Act: Cybersecurity Obligations for Connectable Hardware and Software Products Including IoT. *The National Law Review*. https://www.natlawreview.com/article/eu-cyber-resilience-act-cybersecurity-obligations-connectable-hardware-and-software

Creemers, R., & Webster, G. (2021, August 20). *Translation: Personal Information Protection Law of the People's Republic of China - Effective Nov. 1, 2021*. DigiChina. https://digichina.stanford.edu/work/translation-personal-information-protection-law-of-the-peoples-republic-of-china-effective-nov-1-2021/

DigiChina. (2021, June 29). *Translation: Data Security Law of the People's Republic of China (Effective Sept. 1, 2021)*. DigiChina. https://digichina.stanford.edu/work/translation-data-security-law-of-the-peoples-republic-of-china/

Du, P., Yang, D., & Yu, S. (2018). *The development of E-governance in China: Improving cybersecurity and promoting informatization as means for modernizing state governance*. Springer.

Duan, D. (2019). On authoritarian political representation in contemporary China. *Politics and Governance*, *7*(3), 199–207. doi:10.17645/pag.v7i3.2119

Gillibrand, K. E. (2021, June 17). *S.2134 - 117th Congress (2021-2022): Data Protection Act of 2021 (2021/2022)* [Legislation]. https://www.congress.gov/

Kassen, M. (2022). Open data governance as a theoretical concept: A stakeholder and institutional analysis. In M. Kassen (Ed.), *Open data governance and its actors: Theory and practice* (pp. 1–28). Springer International Publishing. doi:10.1007/978-3-030-92065-4_1

Klosowski, T. (2021, September 6). The state of consumer data privacy laws in the US (and why it matters). *Wirecutter: Reviews for the Real World*. https://www.nytimes.com/wirecutter/blog/state-of-privacy-laws-in-us/

Krahmann, E. (2003). Conceptualizing security governance. *Cooperation and Conflict*, *38*(1), 5–26. doi:10.1177/0010836703038001001

McGavisk, T. (n.d.). *The positive and negative implications of GDPR in the workplace*. TDS. Retrieved September 22, 2022, from https://www.timedatasecurity.com/blogs/the-positive-and-negative-implications-of-gdpr-in-the-workplace

Moran, J. (2021, April 29). *S.1494 - 117th Congress (2021-2022): Consumer Data Privacy and Security Act of 2021 (2021/2022)* [Legislation]. https://www.congress.gov/

National Cyber Security Centre. (2016, September 23). *Introduction to security governance*. https://www.ncsc.gov.uk/collection/risk-management-collection/governance-cyber-risk/security-governance-introduction

Protect Personal Data Privacy. SB21-190, Colorado General Assembly, 2021 Regular Session (2021). https://leg.colorado.gov/bills/sb21-190

Qin, A., Liu, J., & Chien, A. C. (2022). Police database in China was long left unsecured. *The New York Times*.

Westermeyer, G. (2013). *The impact of private actors on security governance: An analysis based on German ISR capabilities for ISAF*. Springer VS Wiesbaden., doi:10.1007/978-3-658-02230-3

Wicker, R. F. (2021, July 28). *S.2499 - 117th Congress (2021-2022): SAFE DATA Act (2021/2022)* [Legislation]. https://www.congress.gov/

Williams, R. D. (2020, December 1). To enhance data security, federal privacy legislation is just a start. *Brookings*. https://www.brookings.edu/techstream/to-enhance-data-security-federal-privacy-legislation-is-just-a-start/

Wolford, B. (2018, November 7). *What is GDPR, the EU's new data protection law?* https://gdpr.eu/what-is-gdpr/

KEY TERMS AND DEFINITIONS

Data Protection Burden: A phenomenon in which the burden incurred to protect data resources such as personal information is shared between the government and domestic actors (including individuals, companies, organizations, etc.).

Data Resource: It mainly refers to digital information assets, actors that provide data resources, and actors who utilize data resources.

Data Security Governance: Formation of a framework in which multiple actors are involved in processing decision-making on data resources, including personal information data.

Compilation of References

Abboud, G., & Lunelli, G. (2015, April). Ativismo judicial e instrumentalidade do processo: Diálogos entre discricionariedade e democracia. *Revista De Processo*, *242*, 21–47.

ABC. (2018). Exposing China's Digital Dystopian Dictatorship. *Foreign Correspondent*. https://www.youtube.com/watch?v=eViswN602_k

Abramowitz, A. (2012). *The Polarized Public? Why Our Government is So Dysfunctional*. Pearson Longman.

Acemoglu, D., Lelarge, C., & Restrepo, P. (2020, May). Competing with robots: Firm-level evidence from France. *AEA Papers and Proceedings. American Economic Association*, *110*, 383–388. doi:10.1257/pandp.20201003

Ahn, M. J., & Chen, Y. C. (2020, June). Artificial intelligence in government: potentials, challenges, and the future. In *The 21st Annual International Conference on Digital Government Research* (pp. 243-252). 10.1145/3396956.3398260

Ahuja, K., Chandra, V., Lord, V., & Peens, C. (2022, June 21). *Ordering in: The rapid evolution of Food Delivery*. Retrieved February 15, 2023, from https://www.mckinsey.com/industries/technology-media-and-telecommunications/our-insights/ordering-in-the-rapid-evolution-of-food-delivery

Akrivopoulou, C. M. (2013). *Digital Democracy and the Impact of Technology on Governance and Politics: New Globalized Practices: New Globalized Practices*. IGI Global. doi:10.4018/978-1-4666-3637-8

Alexopoulou, S., & Pavli, A. (2021). 'Beneath This Mask There is More Than Flesh, Beneath This Mask There is an Idea': Anonymous as the (Super) heroes of the Internet? *International Journal for the Semiotics of Law-Revue internationale de Sémiotique juridique, 34*(1), 237-264.

Allos, A. F. (2022, September 21). American Data Privacy and Protection Act Moves to The House. *The National Law Review*. https://www.natlawreview.com/article/american-data-privacy-and-protection-act-are-we-finally-getting-federal-data-privacy

Allyn, B. (2023, January 25). *A robot was scheduled to argue in court, then came the jail threats*. NPR. https://www.npr.org/2023/01/25/1151435033/a-robot-was-scheduled-to-argue-in-court-then-came-the-jail-threats

Alvarez, F. E., Argente, D., & van Patten, D. (2022). *Are cryptocurrencies currencies? Bitcoin as legal tender in El Salvador*. https://www.nber.org/papers/w29968

Amado, G. (2022, August 18). *Empresários bolsonaristas espalham fake news contra Dom e Bruno e atacam gays, jornalistas e TV Globo; leia zaps.* Metrópoles. Retrieved September 17, 2022, from https://www.metropoles.com/colunas/guilherme-amado/empresarios-bolsonaristas-espalham-fake-news-contra-dom-e-bruno-e-atacam-gays-jornalistas-e-tv-globo-leia-zaps

American Bar Association. (2020). https://www.americanbar.org/groups/professional_responsibility/publications/model_rules_of_professional_conduct/model_rules_of_professional_conduct_table_of_contents

American Bar Association. (2022). *Free Legal Answers.* https://abafreelegalanswers.org/

American Bar Association. (2022). *Profile of the Legal Profession 2022.* https://www.americanbar.org/content/dam/aba/administrative/news/2022/07/profile-report-2022.pdf

Amick, S. (2022, June 17). *Third Largest Russian Oil Company Gazpromneft Partners With BitRiver To Mine Bitcoin.* Nasdaq.

Amour, J., Parnham, R., & Sako, M. (2022). Article: Augmented Lawyering. *University of Illinois Law Review*, 71.

Anon. (2018a). The social credit system under the microscope: The credit society is coming, are you ready? *People's Daily.* http://www.xinhuanet.com/2018-06/04/c_1122931164.htm

Anon. (2018b). Young woman goes missing after filming herself splashing ink on poster of Xi Jinping. *The Shanghaiist.* https://medium.com/@shanghaiist/young-woman-goes-missing-after-filming-herself-splashing-ink-on-poster-of-xi-jinping-6694a646e98f

Ansolabehere, S., & Shaw, D. (2016). Assessing (and fixing?) Election Day lines: Evidence from a survey of local election officials. *Electoral Studies*, *41*, 1–11. doi:10.1016/j.electstud.2015.10.010

Anthopoulos, L., Reddick, C. G., Giannakidou, I., & Mavridis, N. (2016). Why E-Government Projects Fail? An Analysis of the Healthcare.gov Website. *Government Information Quarterly*, *33*(1), 161–173. doi:10.1016/j.giq.2015.07.003

Appel, G., Grewal, L., Hadi, R., & Stephen, A. T. (2020). The Future of Social Media in Marketing. *Journal of the Academy of Marketing Science*, *48*(1), 79–95. doi:10.100711747-019-00695-1 PMID:32431463

Apple. (n.d.). *Siri.* https://www.apple.com/siri/

Arias-Oliva, M., Pelegrín-Borondo, J., & Matías-Clavero, G. (2019). Variables influencing cryptocurrency use: A technology acceptance model in Spain. *Frontiers in Psychology*, *10*, 475. Advance online publication. doi:10.3389/fpsyg.2019.00475 PMID:30949085

Asai, T. (2022a), *China Data and Personal Information Laws.* Amazon.

Asai, T. (2022b). *Japan Data Protection Law* (2nd ed.). Amazon.

Associated Press. (2022). Californians can soon use cryptocurrencies to donate to state and local candidates. *Los Angeles Times.* Retrieved December 10, 2022, from https://www.latimes.com/california/story/2022-07-21/california-to-allow-cryptocurrency-campaign-donations

Aten, J. (2021). *The Cat Lawyer Zoom Fail Is Hilarious Mostly Because It Could Happen to Any of Us.* Inc.com.

Atkeson, L. R., Alvarez, R. M., & Hall, T. E. (2015). Trust in Election and Trust in Government: Why Voter Confidence Differs from Other Measures of System Support. *Election Law Journal*, *14*(3), 207–219. doi:10.1089/elj.2014.0293

Authur, C. (2014). Estonian e-voting shouldn't be used in European elections, say security experts. *The Guardian.* Retrieved November 29, 2022, from https://www.theguardian.com/technology/2014/may/12/estonian-e-voting-security-warning-european-elections-research

Autonomous Vehicles Confront a Political Viability Crisis. (n.d.). *Baron Public Affairs*. Retrieved July 13, 2022, from https://baronpa.com/library/autonomous-vehicles-confront-a-political-viability-crisis/

Ávila, F. D. (2022, March 4). *Repositório Institucional da Universidade Federal de Sergipe - RI/UFS: O estado de exceção no Supremo Tribunal Federal: análise da atuação do Tribunal como poder soberano nos posicionamentos judiciais e extrajudiciais da última década com enfoque na condução do Inquérito no 4781.* Retrieved September 5, 2022, from https://ri.ufs.br/handle/riufs/15101

Aziz, H., & Shah, N. (2021). Participatory budgeting: Models and approaches. In T. Rudas & G. Péli (Eds.), *Pathways Between Social Science and Computational Social Science* (pp. 215–236). Springer. doi:10.1007/978-3-030-54936-7_10

Baecker, D. (2007). *Studien zur nächsten Gesellschaft*. Suhrkamp.

Balkanweb.com. (2022a, August 10). *Zbardhet vendimi: Nga ky vit leksionet në Universitetet publike me detyrim. Udhëzimi mbi frekuentimin e mësimit në auditore.* Balkanweb.Com. https://www.balkanweb.com/zbardhet-vendimi-nga-ky-vit-leksionet-ne-universitetet-publike-me-detyrim-udhezimi-mbi-frekuen timin-e-mesimit-ne-auditore/

Balkanweb.com. (2022b, October 18). *Bojkoti: "Universiteti, të jetë aktor në treg", Rama i përgjigjet pedagogëve: Koha kur qeveria rrit pagat në mënyrë uniforme se kërkojnë Sindikatat, ka mbaruar.* balkanweb.com. https://www.balkanweb.com/bojkoti-universiteti-te-jete-aktor -ne-treg-rama-i-pergjigjet-pedagogeve-koha-kur-qeveria-rrit-pagat-ne-menyre-uniforme-se-kerkojne-sindikatat-ka-mbaruar-b ashke-me/

Baraniuk, C. (2013). The civic hackers reshaping your government. *New Scientist*, *218*(2923), 36–39. doi:10.1016/S0262-4079(13)61625-5

Barbosa, O. P. A., & Saracho, A. B. (2018). Estado democrático de direito – Superação do estado liberal e do estado social. *Revista Jurídica Luso-Brasileira*, *4*(5), 1305–1317.

Barrio, F. (2023). *El uso de inteligencia artificial en los procesos judiciales y el impacto en los Derechos Humanos*. La Ley, Thomson Reuters.

Barroso, L. R. (2009). Judicialização, ativismo judicial e legitimidade democrática. *[Syn]Thesis, 5*(1), 23–32.

BARROSO. (2011). *Luis Roberto. O Controle de Constitucionalidade no Direito Brasileiro*. Saraiva.

Baur, D. G., & Dimpfl, T. (2021). The volatility of Bitcoin and its role as a medium of exchange and a store of value. *Empirical Economics*, *61*(5), 2663–2683. doi:10.100700181-020-01990-5 PMID:33424101

BEA. (2021, September 30). *Wage and salary of U.S. employees, by industry 2020*. bea.gov. Retrieved October 7, 2022, from https://apps.bea.gov/iTable/iTable.cfm?reqid=19&step=2#reqid=19&step=2&isuri=1&1921=survey

Beard, C., & Beard, W. (1930). *The American Leviathan*. McMillian.

Bell, Kochhar, & Khor. (1993). *China at the Threshold of a Market Economy.* International Monetary Fund.

Bellan, R. (2022, April 26). *GM expects to spend $2B on Cruise in 2022*. Retrieved May 25, 2022, from https://techcrunch.com/2022/04/26/gm-expects-to-spend-2b-on-cruise-in-2022/

Bennis, W. G. (Ed.). (1970). *American bureaucracy*. Transaction Publishers.

Bergmann, H., Geissler, M., Hundt, C., & Grave, B. (2018). The climate for entrepreneurship at higher education institutions. *Research Policy*, *47*(4), 700–716. doi:10.1016/j.respol.2018.01.018

Berk, R., Heidari, H., Jabbari, S., & Scutari, M. (2017). Machine learning for parole decision making: A review of the literature. *AI & Society*, *32*(4), 591–610.

Bischoff, L. A. (2022, March 2). Ukraine conflict: Ohio politicos push Russian divestment, OSU cancels food delivery robots. *The Columbus Dispatch*. Retrieved August 9, 2022, from https://www.dispatch.com/story/news/2022/03/03/ohio-state-russia-ukraine-divest-pensions-robots-meal-delivery-yandex/9350250002/

Bishop, B. (2008). *The Big Sort: Why the Clustering of Like-Minded America is Tearing Us Apart*. Houghton Mifflin.

Bishop, C. M. (2006). *Pattern recognition and machine learning* (1st ed.). Springer.

Bitcoin mining – An (un)surprising resurgence ? (2022, May 22). University of Cambridge Judge Business School.

Bizouati-Kennedy, Y. (2022). Ethereum Merge: 3 winners and 2 losers. *GOBankingRates*. Retrieved November 18, 2022, from https://www.gobankingrates.com/investing/crypto/ethereum-merge-3-winners-and-2-losers/?utm_campaign=1185176&utm_source=nasdaq.com&utm_content=20&utm_medium=rss

Block, F. L., & Keller, M. R. (2015). *State of innovation: the US government's role in technology development*. Routledge. doi:10.4324/9781315631905

Blomberg, Marianne v.: Regulation by Shaming - The Social Credit System as a pillory for environmental sinners, tax debtors, lockdown ignoramuses and other lawbreakers, unpublished paper at the conference "Social Credit System" Lüneburg 2021.

Bluhm, M. (2022, November 5). *Biden's hugely consequential high-tech export ban on China, explained by an expert*. Retrieved December 17, 2022, from https://www.vox.com/world/2022/11/5/23440525/biden-administration-semiconductor-export-ban-china

BMBF. (2020). *Future of People's Values in Our Country*. https://www.vorausschau.de/SharedDocs/Downloads/vorausschau/de/BMBF_Foresight_Wertestudie_Langfassung.pdf?__blob=publicationFile&v=1

Bogatov, D., De Caro, A., Elkhiyaoui, K., & Tackmann, B. (2021). Anonymous transactions with revocation and auditing in hyperledger fabric. Cryptology and Network Security, 13099, 435–459. doi:10.1007/978-3-030-92548-2_23

Bogatyreva, K., Edelman, L. F., Manolova, T. S., Osiyevskyy, O., & Shirokova, G. (2019). When do entrepreneurial intentions lead to actions? The role of national culture. *Journal of Business Research*, *96*, 309–321. doi:10.1016/j.jbusres.2018.11.034

Bondy Valdovinos, Zeng, & Wikstrom. (2022). *TikTok: Creativity and Culture in Short Video*. John Wiley & Sons.

Boneh, D., Glass, D., Krashen, D., Lauter, K., Sharif, S., Silverberg, A., Tibouchi, M., & Zhandry, M. (2020). Multiparty non-interactive key exchange and more from isogenies on elliptic curves. *Journal of Mathematical Cryptology*, *14*(1), 5–14. doi:10.1515/jmc-2015-0047

Bonnefon, J.-F., Shariff, A., & Rahwan, I. (2016). The social dilemma of autonomous vehicles. *Science*, *352*(6293), 1573–1576. doi:10.1126cience.aaf2654 PMID:27339987

Bonney, R., Shirk, J. L., Phillips, T. B., Wiggins, A., Ballard, H. L., Miller-Rushing, A. J., & Parrish, J. K. (2014). Next steps for citizen science. *Science*, *343*(6178), 1436–1437. doi:10.1126cience.1251554 PMID:24675940

Boose, J. H., & Gaines, B. R. (1989). Knowledge acquisition for knowledge-based systems: Notes on the state-of-the-art. *Machine Learning*, *4*(3-4), 377–394. doi:10.1007/BF00130720

Bower & Christensen. (1995). *Disruptive technologies: Catching the wave*. Academic Press.

Bower, J. L., & Christensen, C. M. (n.d.). *Disruptive Technologies: Catching the Wave*. https://www.academia.edu/download/59516552/disruptive_technologies20190604-89196-pd4gyy.pdf

Boyadjieva, P., & Slantcheva, S. (2007). Public Perceptions of Private Universities and Colleges in Bulgaria. In S. Slantcheva & D. C. Levy (Eds.), *Private Higher Education in Post-Communist Europe: In Search of Legitimacy* (pp. 201–222). Palgrave Macmillan. doi:10.1057/9780230604391_11

Brainard, L., Boland, K., & McNutt, J. G. (2018). The Advent of Technology Enhanced Leaderless Transnational Social Movement Organizations: Implications for Transnational Advocacy. (185-). In J. G. McNutt (Ed.), *Technology, Activism and Social Justice in a Digital Age*. Oxford University Press.

Braue, D. (2022, July 14). Will schools ever learn cybersecurity? *Cybercrime Magazine*. Retrieved October 4, 2022, from https://cybersecurityventures.com/will-schools-ever-learn-cybersecurity/

Brazil. (1941). *Criminal Procedure Code. Código de Processo Penal*. DECRETO-LEI Nº 3.689.

Brazil. (1967). *Constituição do Brasil. National Congress, Chamber of Deputies*. Retrieved August 30, 2022 from: https://www2.camara.leg.br/legin/fed/consti/1960-1969/constituicao-1967-24-janeiro-1967-365194-norma-pl.html

Brazil. (1988). *Constitution of the Federative Republic of Brazil* (3rd rev. ed.). The Federal Senate, Undersecretariat of Technical Publications.

Brazil. Federal Supreme Court STF. (2019a). *Ação Penal No. 937/2019*. Retrieved on August 30, 2022, from: https://portal.stf.jus.br/processos/detalhe.asp?incidente=4776682

Brazil. Federal Supreme Court STF. (2019b). *Portaria GP No. 69 de 14 de março de 2019*. Retrieved on August 30, 2022, from: https://www.conjur.com.br/dl/comunicado-supremo-tribunal-federal1.pdf

Brazil. Federal Supreme Court STF. (2020). *Regimento Interno: Atualizado até a Emenda Regimental n. 57/2020*. The Federal Supreme Court, Secretariat of high studies, research, and information. Retrieved August 30, 2022, from: https://www.stf.jus.br/arquivo/cms/legislacaoRegimentoInterno/anexo/RISTF.pdf

Bresson, E., & Stern, J. (2001). Efficient Revocation in Group Signatures. In K. Kim (Ed.), Lecture Notes in Computer Science: Vol. 1992. *PublicKey Cryptography. PKC2001*. Springer. doi:10.1007/3-540-44586-2_15

Bright, S. B. (1994). Counsel for the Poor: The Death Sentence Not for the Worst Crime But for the Worst Lawyer. *The Yale Law Journal, 103*(7), 1994. doi:10.2307/797015

Broadhead, S. (2018). The contemporary cybercrime ecosystem: A multi-disciplinary overview of the state of affairs and developments. *Computer Law & Security Report, 34*(6), 1180–1196. doi:10.1016/j.clsr.2018.08.005

Brown, S. (2019, September 9). *Presidential hopeful Bill de Blasio wants a tax on robots*. CNET. Retrieved September 28, 2022, from https://www.cnet.com/science/presidential-hopeful-bill-de-blasio-new-york-mayor-wants-tax-on-robots/

Brown, C. E. (2016). Article: LegalZoom: Closing the justice gap or unauthorized practice of law? *North Carolina Journal of Law & Technology, 17*, 219.

Brown-Heidenreich, S. (2018, December 14). Kiwibot catches fire outside Martin Luther King Jr. student union. *The Daily Californian*. Retrieved October 6, 2022, from https://www.dailycal.org/2018/12/14/kiwibot-catches-fire-outside-mlk-student-union

Brown, J., & Kurzweil, M. (2017). *The complex universe of alternative postsecondary credentials and pathways*. American Academy of Arts and Sciences. doi:10.1146/annurev-orgpsych-041015-062352

Brussee, V. (2022, Feb 11). *China's social credit score – Untangling myth from reality*. https://merics.org/en/opinion/chinas-social-credit-score-untangling-myth-reality

Bulos, U. L. (1997). *Mutação constitucional*. Saraiva.

Bulos, U. L. (2011). *Curso de direito constitucional* (6th ed.). Saraiva.

Bulrush, J. (2022, July 7). *Autonomous vehicles are a dream for drug smugglers — and the law must prepare*. Retrieved August 19, 2022, from https://thenextweb.com/news/law-enforcement-arent-ready-for-autonomous-vehicle-faciliated-crime-syndication

Bünz, B., Bootle, J., Boneh, D., Poelstra, A., Wuille, P., & Maxwell, G. (2018). Bulletproofs: Short proofs for confidential transactions and more. *2018 IEEE Symposium on Security and Privacy (SP)*, 315-334. 10.1109/SP.2018.00020

Buolamwini, J., & Gebru, T. (2018). Gender shades: Intersectional accuracy disparities in commercial gender classification. *Conference on fairness, accountability and transparency*, 77.

Bures, O., & Carrapico, H. (Eds.). (2018). Security privatization: How non-security-related private businesses shape security governance. Springer International Publishing. doi:10.1007/978-3-319-63010-6

Busato, P. C. (2010). De magistrados, inquisidores, promotores de justiça e samambaias: Um estudo sobre os sujeitos no processo em um sistema acusatório. *Sequência (Florianópolis), 31*(60), 133–161.

Cabarcas, D., Demirel, D., Göpfert, F., Lancrenon, J., & Wunderer, T. (2015). An Unconditionally Hiding and Long-Term Binding Post-Quantum Commitment Scheme. IACR Cryptol. ePrint Arch., 628.

Cai, C. (2018). Global Cyber Governance: China's Contribution and Approach. *China Quarterly of International Strategic Studies, 4*(1), 55–76. doi:10.1142/S2377740018500069

Caldwell, M., Andrews, J. T. A., Tanay, T., & Griffin, L. D. (2020). AI-enabled future crime. *Crime Science, 9*(1), 14. doi:10.118640163-020-00123-8

California Bar Association. (2019). *Legal Guide Pamphlets*. The State Bar of California. https://www.calbar.ca.gov/Public/Free-Legal-Information/Legal-Guide-Pamphlets

California Department of Justice. (2018, October 15). *California Consumer Privacy Act (CCPA)*. State of California - Department of Justice - Office of the Attorney General. https://oag.ca.gov/privacy/ccpa

Camenisch, J., Drijvers, M., & Dubovitskaya, M. (2017). Practical UC-secure delegatable credentials with attributes and their application to blockchain. *CCS '17: Proceedings of the 2017 ACM SIGSAC Conference on Computer and Communications Security*, 683–699. 10.1145/3133956.3134025

Campbell, D. A., & Lambright, K. T. (2020). Terms of Engagement: Facebook and Twitter Use among Nonprofit Human Service Organizations. *Nonprofit Management & Leadership*, *30*(4), 545–568. doi:10.1002/nml.21403

Campbell, D. A., Lambright, K. T., & Wells, C. J. (2014). Looking for Friends, Fans, and Followers? Social Media Use in Public and Nonprofit Human Services. *Public Administration Review*, *74*(5), 655–663. doi:10.1111/puar.12261

Canadian Broadcasting Corporation. (2014) *Turing test passed by computer*. https://www.cbc.ca/news/science/turing-test-passed-by-computer-1.2669649

Capez, F. (2022). *Sistema acusatório e garantias do processo penal*. Conjur. Retrieved September 13, 2022, from https://www.conjur.com.br/2021-out-07/controversias-juridicas-sistema-acusatorio-garantias-processo-penal#_ftnref2

Carnelutti, F. (1994). *Derecho procesal civil y penal*. Editorial Pedagógica Iberoamericana.

Carpenter, G. C. (2021). A Solution Hidden in Plain Sight: Closing the Justice Gap by Applying to Legal Aid the Market Incentives That Propelled the Pro Bono Revolution. Chap. L. Rev., 25(1).

Ceccorulli, M., & Lucarelli, S. (2013). Conceptualizing multilateral security governance. In The EU and multilateral security governance (pp. 11–25). doi:10.4324/9780203095430

Chainalysis Team. (2022, August 29). *2 Million and Counting: How Dozens of Pro-Russian Groups Are Using Cryptocurrency Donations to Fund the War in Ukraine*. Chainalysis.

Chainalysis Team. (2022, May 20). *North Korean hackers have prolific year as their unlaundered cryptocurrency holdings reach all-time high*. Chainalysis. Retrieved October 18, 2022, from https://blog.chainalysis.com/reports/north-korean-hackers-have-prolific-year-as-their-total-unlaundered-cryptocurrency-holdings-reach-all-time-high/

ChangK.HobbsW.RobertsM.Steinert-ThrelkeldZ. (2021). *Crisis is a Gateway to Censored Information: The Case of Coronavirus in China*. https://ssrn.com/abstract=3756577

Chen, S. (2018). Forget the Facebook leak: China is mining data directly from workers brains. *South China Morning Post*. https://www.scmp.com/news/china/society/ article/2143899/forget-facebook-leak-china-mining-data-directly-workers-brains

Chen, J., Pan, J., & Xu, Y. (2016). Sources of Authoritarian Responsiveness: A Field Experiment in China. *American Journal of Political Science*, *60*(2), 383–400. doi:10.1111/ajps.12207

Chesler, J. (2020, April 6). Kosovo's UBT to Create 20,000 Lessons Using AVR Platform. *EON Reality*. https://eonreality.com/ubt-kosovo-avr-platform/?lang=es

CheungA. S. Y.ZhaoY. (2013). *An Overview of Internet Regulation in China.* doi:10.2139/ssrn.2358247

Chevalier, M., & Lu, P. X. (2009). *Luxury China: Market Opportunities and Potential.* John Wiley & Sons.

Chiu, A. (2022, June 15). Limiting crypto helped the Texas power grid weather a heat wave. *The Texas Tribune.*

Cho, T. (2013). Voter migration and the geographic sorting of the American electorate. *Annals of the Association of American Geographers, 103*(4), 856–870. doi:10.1080/00045608.2012.720229

Christensen, C. M., Anthony, S. D., & Roth, E. A. (2004). *Seeing what's next: Using the theories of innovation to predict industry change.* Harvard Business Press.

Christensen, C. M., Baumann, H., Ruggles, R., & Sadtler, T. M. (2006). Disruptive innovation for social change. *Harvard Business Review, 84*(12), 94. PMID:17183796

Christensen, C. M., & Raynor, M. (2003). *The innovator's solution.* Harvard Business School Press.

Christensen, C., Raynor, M., & McDonald, R. (2015). What is Disruptive Innovation? *Harvard Business Review,* (December Issue), 44–53. PMID:17183796

Chudinovskikh, M., & Sevryugin, V. (2019). Cryptocurrency regulation in the brics countries and the Eurasian economic union. *BRICS Law Journal, 6*(1), 63–81. doi:10.21684/2412-2343-2019-6-1-63-81

Clandinin, D. J. (2007). *Handbook of narrative inquiry: Mapping a methodology.* SAGE Publications, Inc., doi:10.4135/9781452226552

Clarke, C. P. (2017, July 25). *ISIS Is So Desperate It's Turning to the Drug Trade.* RAND.

Clarke, C. P. (2016). Drugs & Thugs: Funding Terrorism through Narcotics Trafficking. *Journal of Strategic Security, Vol. 9, No. 3. Special Issue: Emerging Threats, 5,* 1–15. doi:10.2307/26473335

Classen, R. L., Magleby, D. B., Monson, J. Q., & Patterson, K. D. (2013). Voter confidence and the Election-Day voting experience. *Political Behavior, 35*(2), 215–235. doi:10.100711109-012-9202-4

Clautice, T. (2019). *Nation State Involvement in Cryptocurrency and the Impact to Nation State Involvement in Cryptocurrency and the Impact to Economic Sanctions Economic Sanctions Nation State Involvement in Cryptocurrency and the Impact to Economic Sanctions.* https://digitalcommons.lasalle.edu/ecf_capstones

Clayton, R., Murdoch, S. J., & Watson, R. N. M. (2006). Ignoring the Great Firewall of China. In Privacy Enhancing Technologies (pp. 20–35). Springer Berlin Heidelberg. doi:10.1007/11957454_2

Clements, R. (2018). Assessing the Evolution of Cryptocurrency: Demand Factors, Latent Value, and Regulatory Developments. *Michigan Business & Entrepreneurial Law Review.* Retrieved December 2, 2022, from https://repository.law.umich.edu/cgi/viewcontent.cgi?article=1079&context=mbelr

College, U. B. T. (2018, December 6). *Qendra Këshilluese, mbështetje kryesore psikologjike për studentët e UBT-së (Video).* UBT – Higher Education Institution. https://www.ubt-uni.net/sq/qendra-keshilluese-mbeshtetje-kryesore-psikologjike-per-studentet-e-ubt-se/

College, U. B. T. (2020, November 9). *UBT partner i projektit VTECH, mbajnë trajnimin e parë për Realitetin Virtual në Shqipëri.* UBT – Higher Education Institution. https://www.ubt-uni.net/sq/ubt-partner-i-projektit-vtech-mbajne-trajnimin-e-pare-per-realitetin-virtual-ne-shqiperi/

College, U. B. T. (2021, March 16). *Through "Nokia" UBT brings the 5G network to Kosovo*. UBT – Higher Education Institution. https://www.ubt-uni.net/en/through-nokia-ubt-brings-the-5g-network-to-kosovo/

College, U. B. T. (2022a). *Alumni Association*. UBT – Higher Education Institution. https://www.ubt-uni.net/en/study/alumni/alumni-association/

College, U. B. T. (2022b, September 10). *Suksesshëm UBT përmbyll një nga organizimet më madhështore të ceremonisë së gradimit të studentëve*. UBT – Higher Education Institution. https://www.ubt-uni.net/sq/suksesshem-ubt-permbyll-nje-nga-organizimet-me-madheshtore-te-ceremonise-se-gradimit-te-studenteve/

Collins, J. (2022, September 27). *Food delivery robot barges through crime scene, has no time for our human nonsense*. LADbible. Retrieved October 4, 2022, from https://www.ladbible.com/news/food-delivery-robot-barges-through-crime-scene-20220916

Commission of the European Communities. (1992). Amended Proposal for a Council Directive on the Protection of Individuals with Regard to the Processing of Personal Data and on the Free Movement of Such Data, *COM* (92) 422 final—SYN 297.

Conger, K., Fausset, R., & Kovaleski, S. F. (2019). San Francisco bans facial recognition technology. *The New York Times, 14*.

Consumer Data Protection Act. (2021, February 4). Virginia's Legislative Information System. https://lis.virginia.gov/cgi-bin/legp604.exe?211+ful+SB1392ES1

Contissa, G., Lagioia, F., & Sartor, G. (2017). The Ethical Knob: Ethically-customisable automated vehicles and the law. *Artificial Intelligence and Law*, *25*(3), 365–378. doi:10.100710506-017-9211-z

Corriere della Sera. (2013). La laurea senza mai seguire una lezione Bossi jr indagato a Tirana con altri 9 italiani. *Corriere della Sera*. https://www.corriere.it/cronache/13_dicembre_02/tirana-laure-false-indagato-renzo-bossi-5d3c80a4-5b51-11e3-bbdb-322ff669989a.shtml

Couger Team. (2021, April 21). *What Countries and States That Have Trialled Blockchain Voting Learned*. Medium.

Cowper, T. J. (2018, February 13). *Autonomous Vehicles- How Will They Challenge Law Enforcement?* Retrieved August 19, 2022, from https://leb.fbi.gov/articles/featured-articles/autonomous-vehicles-how-will-they-challenge-law-enforcement

Cox, C. (2022, September 20). *Treasury warns against Russia's efforts to evade sanctions with cryptocurrencies*. CNBC.

Craddock, P. (2022, September 19). EU Cyber Resilience Act: Cybersecurity Obligations for Connectable Hardware and Software Products Including IoT. *The National Law Review*. https://www.natlawreview.com/article/eu-cyber-resilience-act-cybersecurity-obligations-connectable-hardware-and-software

Crawford, S. (2018). Why Universities Need 'Public Interest Technology' Courses. *Wired Magazine*. https://www.wired.com/story/universities-public-interest-technology-courses-programs/

Creemers, R., & Webster, G. (2021, August 20). *Translation: Personal Information Protection Law of the People's Republic of China - Effective Nov. 1, 2021.* DigiChina. https://digichina.stanford.edu/work/translation-personal-information-protection-law-of-the-peoples-republic-of-china-effective-nov-1-2021/

Creemers, R. (2017). Cyber China: Upgrading Propaganda, Public Opinion Work and Social Management for the Twenty-First Century. *Journal of Contemporary China, 26*(103), 85–100. doi:10.1080/10670564.2016.1206281

Crosby. (2016). *BlockChain Technology: Beyond Bitcoin.* Academic Press.

CryptoWallet. (2021). What is SHA-256? *CryptoWallet.* Retrieved December 6, 2022, from https://cryptowallet.com/glossary/sha-256/

D'Arienzo, M. C., Boursier, V., & Griffiths, M. D. (2019). Addiction to Social Media and Attachment Styles: A Systematic Literature Review. *International Journal of Mental Health and Addiction, 17*(4), 1094–1118. doi:10.100711469-019-00082-5

D'ignazio, C., & Klein, L. F. (2020). *Data feminism.* MIT Press. doi:10.7551/mitpress/11805.001.0001

Dagres, H., & Slavin, B. (n.d.). *How Iran Will Cope with U.S. Sanctions.* https://www.atlanticcouncil.org/blogs/

Dahsyla, E. (2020, October 4). Të diplomuarit në kolegjet private ndihen të diskriminuar nga Komuna e Prishtinës në konkursin për mësimdhënës [Newspaper]. *Raporto Korrupsionin! KALLXO.Com.* https://kallxo.com/gjate/te-diplomuarit-ne-kolegjet-private-ndihen-te-diskriminuar-nga-komuna-e-prishtines-ne-konkursin-per-mesimdhenes/

Dall-E. (n.d.). https://openai.com/dall-e-2/

David, N., Justice, J. B., & McNutt, J. G. (2018). Smart Cities, Transparency, Civic Technology and Reinventing Government. In Smart Technologies for Smart Governments. Transparency, Efficiency and Organizational Issues. Springer.

David, N., Justice, J. B., & McNutt, J. G. (2015). Smart Cities are Transparent Cities: The Role of Fiscal Transparency in Smart City Governance. In P. R. B. Manuel (Ed.), *Transforming City Governments for successful Smart Cities. Empirical Experiences.* Springer. doi:10.1007/978-3-319-03167-5_5

Davies, J. L. (2001). The Emergence of Entrepreneurial Cultures in European Universities. *Higher Education Management, 13,* 25–43. doi:10.1787/hemp-v13-2-en

Davis, A. (2020). The Future of Law Firms (and Lawyers) in the Age of Artificial Intelligence. The Professional Lawyer, 27(1). doi:10.1590/2317-6172201945

Davis, K. (2019, October 23). *Food Delivery Robots pulled from Pitt campus after backlash about mobility.* 90.5 WESA. Retrieved October 4, 2022, from https://www.wesa.fm/identity-justice/2019-10-22/food-delivery-robots-pulled-from-pitt-campus-after-backlash-about-mobility

Dawes, S. S., & Helbig, N. (2010, August). Information strategies for open government: Challenges and prospects for deriving public value from government transparency. In *International Conference on Electronic Government* (pp. 50-60). Springer. 10.1007/978-3-642-14799-9_5

Dawes, S. S., Vidiasova, L., & Parkhimovich, O. (2016). Planning and designing open government data programs: An ecosystem approach. *Government Information Quarterly, 33*(1), 15–27. doi:10.1016/j.giq.2016.01.003

De Clercq, D., Danis, W. M., & Dakhli, M. (2010). The moderating effect of institutional context on the relationship between associational activity and new business activity in emerging economies. *International Business Review, 19*(1), 85–101.

De Cosmo, L. (2022, July 12). Google Engineer Claims AI Chatbot Is Sentient: Why That Matters. *Scientific American*. https://www.scientificamerican.com/article/google-engineer-claims-ai-chatbot-is-sentient-why-that-matters/# Emergency Medical Treatment and Active Labor Act, 42 U.S.C. § 1395dd (1986)

de Sio, F. S., & van den Hoven, J. (2018). Meaningful Human Control over Autonomous Systems: A Philosophical Account. *Frontiers in Robotics and AI*, *5*, 15. doi:10.3389/frobt.2018.00015 PMID:33500902

de Vries, A., Gallersdörfer, U., Klaaßen, L., & Christian, S. (2022). Revisiting Bitcoin's carbon footprint. *Joule*, *6*(3), 498–502. doi:10.1016/j.joule.2022.02.005

Deane. (2018). *The New Game: Cryptocurrency Challenges U.S. Economic Sanctions*. Academic Press.

Denhardt, R. B., Denhardt, J. V., Aristigueta, M. P., & Rawlings, K. C. (2018). *Managing human behavior in public and nonprofit organizations*. CQ Press.

Deto, R. (2020, December 2). *Pennsylvania legalizes autonomous delivery robots, classifies them as pedestrians*. Pittsburgh City Paper. Retrieved October 1, 2022, from https://www.pghcitypaper.com/pittsburgh/pennsylvania-legalizes-autonomous-delivery-robots-classifies-them-as-pedestrians
/Content?oid=18482040

di Pierro, M. (2017). What Is the Blockchain? *Computing in Science & Engineering*, *19*(5), 92–95. doi:10.1109/MCSE.2017.3421554

DigiChina. (2021, June 29). *Translation: Data Security Law of the People's Republic of China (Effective Sept. 1, 2021)*. DigiChina. https://digichina.stanford.edu/work/translation-data-security-law-of-the-peoples-republic-of-china/

Dion-Schwarz, C., Manheim, D., Johnston, P. B., & Rand Corporation. (2019). *Terrorist use of cryptocurrencies : technical and organizational barriers and future threats*. RAND.

DiPippa, J. M. A. (2016). Peter Singer, drowning children, and pro bono. *West Virginia Law Review*, *119*, 113.

Dixon, J., Hong, B., & Wu, L. (2021). The Robot Revolution: Managerial and employment consequences for firms. *Management Science*, *67*(9), 5586–5605. doi:10.1287/mnsc.2020.3812

DoNotPay. (n.d.). *DoNotPay: About Us*. https://donotpay.com/about/

Donyéa, T. (2022). New Jersey legislature moves to make state's cryptocurrency rules 'industry-friendly'. *WHYY*. Retrieved November 14, 2022, from https://whyy.org/articles/new-jersey-legislature-industry-friendly-cryptocurrency-rules/

Dorloff, A. (2017). Sozialkredit-System. China auf dem Weg in die IT-Diktatur. *Deutschlandfunk*. https://www.deutschlandfunk.de/sozialkredit-system-china-auf-dem-weg-in-die-it-diktatur-100.html

Dorokhova, E., Belykh, T., & Koren'kova, G. (2021). Economic and Legal Aspects of Cryptocurrency Usage in Russia. In *Strategic Latency Unleashed: The Role of Technology in a Revisionist Global Order and the Implications for Special Operations Forces* (pp. 327–332). 10.2991/aebmr.k.210710.055

Dragu, T., & Lupu, Y. (2021). Digital Authoritarianism and the Future of Human Rights. *International Organization*, *75*(4), 991–1017. doi:10.1017/S0020818320000624

Dressel, J., & Farid, H. (2018). The accuracy and limits of predicting recidivism. *Science Advances*, *4*(1), 4. doi:10.1126ciadv.aao5580 PMID:29376122

Duan, D. (2019). On authoritarian political representation in contemporary China. *Politics and Governance*, *7*(3), 199–207. doi:10.17645/pag.v7i3.2119

Dudely, S. (2021). Cryptocurrency: Will the Digitization of Currency Allow Malign Actors to Achieve Strategic Effects? In Strategic Latency Unleashed (pp. 269–282). Academic Press.

Dumas, B. (2022, September 22). *Delivery Robots emerge to bridge gaps as labor shortage drags on*. Fox Business. Retrieved October 7, 2022, from https://www.foxbusiness.com/technology/delivery-robots-emerge-bridge-gaps-labor-shortage-drags

Dunaway, J., & Lawrence, R. G. (2015). What Predicts the Game Frame? Media Ownership, Electoral Context, and Campaign News. *Political Communication*, *32*(1), 43–60. doi:10.1080/10584609.2014.880975

Du, P., Yang, D., & Yu, S. (2018). *The development of E-governance in China: Improving cybersecurity and promoting informatization as means for modernizing state governance*. Springer.

Dupuis, D., & Gleason, K. (2021). Money laundering with cryptocurrency: Open doors and the regulatory dialectic. *Journal of Financial Crime*, *28*(1), 60–74. doi:10.1108/JFC-06-2020-0113

Durbin, D. (2021, November 2). *Robots hit the streets as demand for food delivery grows*. AP News. Retrieved September 26, 2022, from https://apnews.com/article/coronavirus-robot-food-delivery-technology-business-5b7dbb782f89ef3d586eefb9a33d5e1c

Dutton, W. (1982). Technology and the Federal System. *Proceedings of the Academy of Political Science*, *34*(4), 109–120. doi:10.2307/3700973

Earl, J., & Kimport, K. (2011). *Digitally enabled social change*. MIT Press. doi:10.7551/mitpress/9780262015103.001.0001

Eckhouse, L. (2017, Feb. 10). Big data may be reinforcing racial bias in the criminal justice system. *The Washington Post*.

ECLI:EU:C:2020:559, Case C-311/18, Data Protection Commissioner v Facebook Ireland and Maximillian Schrems ("Schrems II"), 2020

Editors, A. (2017). *Potential impacts of self-driving cars*. Access Science.

Edwards, D. (2022, July 26). *Embark Trucks completes autonomous vehicle testing in snowy conditions*. Retrieved September 10, 2022, from https://roboticsandautomationnews.com/2022/07/26/embark-trucks-completes-autonomous-testing-in-snowy-conditions/53461/

Edwards, S. B., III. (2019). The Right to Privacy Is Dying: Technology Is Killing It and We Are Letting It Happen. In Cyber Law, Privacy, and Security: Concepts, Methodologies, Tools, and Applications (pp. 830-853). IGI Global.

Egan, M. (2022). Elizabeth Warren unveils bipartisan bill to crack down on crypto money laundering. *CNN*. Retrieved December 14, 2022, from https://www.cnn.com/2022/12/14/business/elizabeth-warren-bipartisan-crypto-crackdown/index.html

Egan, M. (2022). First on CNN: 'greed and deception.' Elizabeth Warren demands Sam Bankman-fried and FTX turn over a trove of Records. *CNN*. Retrieved December 3, 2022, from https://www.cnn.com/2022/11/17/investing/ftx-sbf-elizabeth-warren-crypto/index.html

Elazar, D. J. (1972). *American federalism: A view from the states*. Crowell.

Elazar, D. J. (1994). *The American mosaic: The impact of space, time, and culture on American politics*. Westview Press.

Election Assistance Commission. (2022a). *About the EAC*. https://www.eac.gov/about-the-useac

Election Assistance Commission. (2022b). *Statewide Voter Registration Systems*. https://www.eac.gov/statewide-voter-registration-systems

Emergency Medical Treatment and Active Labor Act, 42 U.S.C. § 1395dd (1986).

Endeshaw, A. (2004). Internet Regulation in China: The Never-ending Cat and Mouse Game. *Information & Communications Technology Law*, *13*(1), 41–57. doi:10.1080/1360083042000190634

Engels F. (1965). *Kazoku・Shiyuzaisan・Kokka no kigen: Lewis H. Morgan no kenkyu ni kanren shite* [The origin of the Family, Private Property, and the State: About the study of Lewis H. Morgan] (S. Tobara, Trans.). Iwanami Shoten. (Original work published 1884)

Ennis, P. J. (2021, Sep 14). Ethereum's Political Philosophy Explained. *CoinDesk*. https://www.coindesk.com/markets/2021/07/09/ethereums-political-philosophy-explained/

Ethereum.org. (n.d.) What is web3 and why is it important? *ethereum.org*. Retrieved December 6, 2022, from https://ethereum.org/en/web3/

Etzkowitz, H., & Leydesdorff, L. (2000). The Dynamics of Innovation: From National Systems and "Mode 2" to a Triple Helix of University–Industry–Government Relations. *Research Policy*, *29*(2), 109–123. doi:10.1016/S0048-7333(99)00055-4

EU. (2021). *The AI Act*. https://artificialintelligenceact.eu/wp-content/uploads/2021/08/The-AI-Act.pdf

Eubanks, V. (2018). *Automating inequality: How high-tech tools profile, police, and punish the poor*. St. Martin's Press.

EUCCC. (2019). *European Union Chamber of Commerce in China and Sinolytics. The Digital Hand. How China's Corporate Social Credit System Conditions Market Actors*. https://www.sinolytics. de/wp-content/uploads/2019/08/Sinolytics_The-Digital-Hand-How-Chinas-Corporate-Social-Credit-System-Conditions-Market-Actors.pdf

Euronews Albania. (2020, September 24). Korrupsion me provimet, SPAK arreston pedagogun dhe sekserin. *Euronews Albania*. https://euronews.al/korrupsion-me-provimet-arrestohet-pedagogu-dhe-sekseri/

European Commission. (2022, September 28). *New Liability rules on products and AI to protect consumers and foster innovation* [Press release]. https://ec.europa.eu/commission/presscorner/detail/en/IP_22_5807

FACTI. (2021). *Financial Integrity for Sustainable Development*. Report of the High Level Panel on International Financial Accountability, Transparency and Integrity for Achieving the 2030 Agenda, 25 February 2021. Available from https://www.factipanel.org/docpdfs/FACTI_Panel_Report.pdf

Fan, W. (2021). Peng Shuai: Who is the Chinese Tennis Star, and Where is She? *Wall Street Journal*. www.wsj.com/articles/peng-shuai-what-to-know-about-the-missing-chinese-tennis-star-11637324061

Fannin, R. (2022, May 21). *Where the billions spent on autonomous vehicles by U.S. and Chinese giants is heading*. Retrieved May 25, 2022, from https://www.cnbc.com/2022/05/21/why-the-first-autonomous-vehicles-winners-wont-be-in-your-driveway.html#aoh=165336535384 39&csi=0&referrer=https%3A%2F%2Fwww.google.com&_tf=From%20%2 51%24s&ampshare=https%3A%2F%2Fwww.cnbc.com%2F2022%2F05%2F21% 2Fwhy-the-first-autonomous-vehicles-winners-wont-be-in-your-driveway.html

FDIC. (2021, December 17). *How America Banks: Household Use of Banking and Financial Services*. Federal Deposit Insurance Corporation.

Felitto, G. (2021, June 7). *Is Law Enforcement Ready for Self-Driving Cars?* Retrieved August 21, 2022, from https://thecrimereport.org/2021/06/07/1333055/

Feng, L., Zhang, H., Chen, Y., & Lou, L. (2018). Scalable dynamic multi-agent practical byzantine fault-tolerant consensus in permissioned blockchain. *MDPI*. Retrieved December 20, 2022, from https://www.mdpi.com/2076-3417/8/10/1919#:~:text=Although%20 the%20PBFT%20algorithm%20is,low%20consensus%20speed%20and%20 scalability

Fernandes & Nelson. (2014). O Supremo Tribunal Federal e o ativismo judicial: (re)analisando o dogma do "legislador negativo". *Revista Eletrônica Direito e Política, Programa de Pós-Graduação Stricto Sensu em Ciência Jurídica da UNIVALI, 9*(1).

Ferrer, E., Jiménez, E., Lopez-Presa, J. L., & Martín-Rueda, J. (2022, April). Following Leaders in Byzantine Multirobot Systems by Using Blockchain Technology. *IEEE Transactions on Robotics*, *38*(2), 1101–1117. doi:10.1109/TRO.2021.3104243

Figenschou, T. U., & Fredheim, N. A. (2020). Interest Groups on Social Media: Four Forms of Networked Advocacy. *Journal of Public Affairs*, *20*(2). Advance online publication. doi:10.1002/pa.2012

Fillion, S. (2020, August 25). If Estonia figured out online voting 15 years ago, why can't the US? *Forbes*. Retrieved December 9, 2022, from https://www.forbes.com/sites/stephaniefillion/2020/08/24/if-estonia-figured-out-online-voting-15-years-ago-why-cant-the-us/?sh=6301f0f015ab

Finkelstein, S. (2020). Article. *FEATURE:GPSolo*, *37*, 12.

Finklea, K. (2017). *Dark Web*. www.crs.gov

Flynn, J. (2022). *18+ Food Delivery Statistics [2022]: Online Ordering Industry Numbers You Need To Know*. Zippia.com. Feb. 2, 2022, https://www.zippia.com/advice/food-delivery-industry-statistics/

Foley, S., Karlsen, J. R., Putniņš, T. J., Goldstein, I., Jiang, W., Karolyi, A., Weber, M., & Easley, D. (2019). *Sex, drugs, and bitcoin: How much illegal activity is financed through cryptocurrencies?* https://www.emcdda.europa.eu/attachements.cfm/att_194336_EN_TD3112366ENC.pdf

Follow My Vote. (2021). Blockchain comparison, a closer look at Bitcoin, BitShares, and Ethereum. *Follow My Vote*. Retrieved December 10, 2022, from https://followmyvote.com/blockchain-comparison-a-closer-look-at-bitcoin-bitshares-and-ethereum/

Foot, P. (1967). The problem of abortion and the doctrine of the double effect. *Oxford Review, 5.*

Foucault, M. (1975). *Surveiller et punir.* Gallimard.

Frank, D.-A., Chrysochou, P., Mitkidis, P., & Ariely, D. (2019). Human decision-making biases in the moral dilemmas of autonomous vehicles. *Scientific Reports, 9*(1), 13080. doi:10.103841598-019-49411-7 PMID:31511560

Frankenfield, J. (2022a, October 13). What is proof of work (POW) in blockchain? *Investopedia.* Retrieved December 9, 2022, from https://www.investopedia.com/terms/p/proof-work.asp

Frankenfield, J. (2022b, October 15). Understanding double-spending and how to prevent attacks. *Investopedia.* Retrieved November 17, 2022, from https://www.investopedia.com/terms/d/doublespending.asp

Frankenfield, J. (2022c, October 20). What does proof-of-stake (pos) mean in crypto? *Investopedia.* Retrieved December 11, 2022, from https://www.investopedia.com/terms/p/proof-stake-pos.asp

Frankenfield, J. (2022d, September 16). What are smart contracts on the blockchain, and how do they work? *Investopedia.* Retrieved November 15, 2022, from https://www.investopedia.com/terms/s/smart-contracts.asp

Frost, L. (2020, May 19). ISIS's $300 million war chest is hidden in Bitcoin, says think tank. *Decrypt.*

Fuchsbauer, G., & Pointcheval, D. (2009). Anonymous Consecutive Delegation of Signing Rights: Unifying Group and Proxy Signatures. In V. Cortier, C. Kirchner, M. Okada, & H. Sakurada (Eds.), Lecture Notes in Computer Science: Vol. 5458. *Formal to Practical Security.* Springer. doi:10.1007/978-3-642-02002-5_6

Fukushima S. (2022). FATF対日相互審査の教訓 [Lessons Learned from the FATF Mutual Review of Japan]. *Journal of the Japanese Institute of International Business Law, 50*(2).

Gale, M. (2022, April 12). *Meet Zoox, The 100% EV And Autonomous Human Transportation Company From Amazon.* Retrieved May 25, 2022, from https://www.forbes.com/sites/michaelgale/2022/04/12/meet-zoox-the-100-ev-and-autonomous-human-transportation-company-from-amazon/?sh=2ce3af5a3cd0

Gan, N. (2022). China's bank run victims planned to protest. Then their health codes turned red. *CNN.* https://edition.cnn.com/2022/06/15/china/china-zhengzhou-bank-fraud-health-code-protest-intl-hnk/index.html

Gapski, H., & Packard, S. (2021). *Super-Scoring? Datengetriebene Sozialtechnologien als neue Bildungsherausforderung.* Kopaed Verlag.

Garay, A. F. (2013). *La doctrina del precedente en la Corte Suprema.* Abeledo Perrot.

Garnaut, L. S., Yao, Y., & Wang, X. (2012). *Private Enterprise in China.* ANU Press.

Garnett, R. W. (2006). Debate: Judicial activism and its critics. *University of Pennsylvania Law Review, 155,* 112.

Gauer, Kueshner, & Updegrave. (2022). *The 2022 Crypto Crime Report.* Academic Press.

Gazeta Liberale. (2022, October 4). UET udhëheq në fushën e ndërkombëtarizimit/ Fiton 7 projekte të reja Erasmus+ [Newspaper]. *Gazeta Liberale.* https://liberale.al/uet-udheheq-ne-fushen-e-nderkombetarizimit-fiton-7-projekte-te-reja-erasmus/

Geiger, R. L. (1986). *Private Sectors in Higher Education: Structure, Function, and Change in Eight Countries.* University of Michigan Press. doi:10.3998/mpub.7381

Geyser, W. (2021). *What is TikTok? The fastest growing social media app uncovered.* Influencer Marketing Hub. Retrieved from https://influencermarketinghub.com/what-is-tiktok/

Giesecke, H. C. (2007). Gaining Legitimacy: A Continuum on the Attainment of Recognition. In S. Slantcheva & D. C. Levy (Eds.), *Private Higher Education in Post-Communist Europe: In Search of Legitimacy* (pp. 95–118). Palgrave Macmillan. doi:10.1057/9780230604391_5

Gil de Zúñiga, H., & Chen, H.-T. (2019). Digital Media and Politics: Effects of the Great Information and Communication Divides. *Journal of Broadcasting & Electronic Media*, *63*(3), 365–373. doi:10.1080/08838151.2019.1662019

Gillibrand, K. E. (2021, June 17). *S.2134 - 117th Congress (2021-2022): Data Protection Act of 2021 (2021/2022)* [Legislation]. https://www.congress.gov/

Gilman, H. R. (2016). *Participatory budgeting and civic tech: The revival of citizen engagement*. Georgetown University Press.

Goff, P., Rahn, C., & Diamond, S. (2016). *Predictive policing: The role of crime forecasting in law enforcement operations*. Routledge.

Goggin, S. N., & Byrne, M. D. (2007). An Examination of the Auditability of Voter Verified Paper Audit Trail (VVPAT) Ballots. *EVT*, *7*, 10–10.

Gogoll, J., & Müller, J. F. (2017). Autonomous cars: In favor of a mandatory ethics setting. *Science and Engineering Ethics*, *23*(3), 681–700. doi:10.100711948-016-9806-x PMID:27417644

Gohdes, A. R. (2020). Repression Technology: Internet Accessibility and State Violence. *American Journal of Political Science*, *64*(3), 488–503. doi:10.1111/ajps.12509

Goldsmith, M. (1989). State Policies and Programs for Encouraging Innovation and Technology Development. In *State and Local Initiatives on Productivity, Technology, and Innovation: Enhancing a National Resource for International Competitiveness.* Advisory Commission of Intergovernmental Relations.

Gomez-Ibanez, T. (n.d.). *Implications of Self Driving Vehicles on Public Policy with Tony Gómez-Ibáñez.* Retrieved July 13, 2022, from https://www.hks.harvard.edu/faculty-research/policy-topics/cities-communities/implications-self-driving-vehicles-public-policy

GoodallN. J. (2020). More than trolleys: plausible, ethically ambiguous scenarios likely to be encountered by automated vehicles. doi:10.31219/osf.io/bzkas

Goode, L. (2015). Anonymous and the political ethos of hacktivism. *Popular Communication*, *13*(1), 74–86. doi:10.1080/15405702.2014.978000

Goodfellow, I., Bengio, Y., & Courville, A. (2017). *Deep learning*. MIT Press.

Gordon, E., & Mihailidis, P. (Eds.). (2016). *Civic media: Technology, design, practice.* MIT Press. doi:10.7551/mitpress/9970.001.0001

Gould-Davies, N. (2020). Russia, the West and Sanctions. *Survival*, *62*(1), 7–28. doi:10.1080/00396338.2020.1715060

Grace. (2022, July 22). Top Metaverse use cases & applications. *LimeChain*. Retrieved November 18, 2022, from https://limechain.tech/blog/top-metaverse-use-cases

Greco, R. (n.d.). *Jurista Rogério Greco alerta para STF 'completamente dominado ideologicamente': 'Não se pode exercitar o seu direito sagrado de opinião.'* Retrieved September 4, 2022, from https://www.folhapolitica.org/2022/09/jurista-rogerio-greco-alerta-para-stf.html

Greenwood, J. (2022, March 7). *Anonymous claims Yandex data breach affecting 150K Users*. The Cybersecurity Times. Retrieved October 4, 2022, from https://www.thecybersecuritytimes.com/anonymous-claims-yandex-data-breach-affecting-150k-users/

Gregory, N., & Tenev, S. (2001). *The Financing of Private Enterprise in China*. Finance & Development. https://www.elibrary.imf.org/view/journals/022/0038/001/article-A005-en.xml

Griffiths, J. (2019). *The Great Firewall of China: How to Build and Control an Alternative Version of the Internet*. Zed Books Ltd. doi:10.5040/9781350225497

Guenduez, A. A., Mettler, T. & Schedler, K. (2020). Citizen Participation in Smart Government: A Conceptual Model and Two IoT Case Studies. *Public Administration and Information Technology,* 189-209.

Guerrero, M., & Marozau, R. (2022). Assessing the influence of institutions on students' entrepreneurial dynamics: Evidence from European post-socialist and market-oriented economies. *Small Business Economics*. Advance online publication. doi:10.100711187-022-00600-1

Guerrero, M., & Urbano, D. (2019). A research agenda for entrepreneurship and innovation: The role of entrepreneurial universities. In *A Research Agenda for Entrepreneurship and Innovation* (pp. 107–133). Edward Elgar. doi:10.4337/9781788116015.00012

Guerrero, M., Urbano, D., Cunningham, J., & Organ, D. (2014). Entrepreneurial universities in two European regions: A case study comparison. *The Journal of Technology Transfer*, *39*(3), 415–434. doi:10.100710961-012-9287-2

Guidry Jeanine, P. D., Waters Richard, D., & Saxton Gregory, D. (2014). Moving Social Marketing beyond Personal Change to Social Change: Strategically Using Twitter to Mobilize Supporters into Vocal Advocates. *Journal of Social Marketing*, *4*(3), 240–260. doi:10.1108/JSOCM-02-2014-0014

Gulasekaram, P., & Karthick Ramakrishnan, S. (2015). *The new immigration federalism*. Cambridge University Press. doi:10.1017/CBO9781316282410

Habermas, J. (1973). *Koukyosei no kozo tankan* [The Structural Transformation of the Public Sphere: An Inquiry into a Category of Bourgeois Society] (S. Hosoya & M. Yamada, Trans.). Miraisha. (Original work published 1962)

Hachigian, N. (2001). China's Cyber-Strategy. *Foreign Affairs*, *80*(2), 118–133. doi:10.2307/20050069

Hale, K., Montjoy, R., & Brown, M. (2015). *Administering elections: How American elections work*. Springer. doi:10.1057/9781137388452

Hamilton, J. (2022, July 18). US Treasury open to nonbanks issuing Stablecoins, official says. *CoinDesk Latest Headlines RSS*. Retrieved November 19, 2022, from https://www.coindesk.com/policy/2022/07/18/us-treasury-open-to-nonbanks-issuing-stablecoins-official-says

Hamilton, A., Madison, J., & Jay, J. (2012). *The Federalist Papers*. Van Haren Publishing.

Hamilton, S. D. (2001). A ortodoxia do sistema acusatório no processo penal brasileiro: Uma falácia. *Revista Da Faculdade De Direito De Campos*, *2*(2), 247–273.

Harrison, T. M., Pardo, T. A., & Cook, M. (2012). Creating open government ecosystems: A research and development agenda. *Future Internet*, *4*(4), 900–928. doi:10.3390/fi4040900

Hart, H. L. A. (1976). *Hou no gainen* [The concept of law] (M. Yazaki, Trans.). Misuzu-shobo. (Original work published 1961)

Hartley, J. (2006). *Innovation and its contribution to improvement: A review for policymakers, policy advisers, managers and researchers.* Department for Communities and Local Government.

Hasbi, A. H., & Mahzam, R. (2018). *Cryptocurrencies: Potential For Terror Financing?* www.rsis.edu.sg

Hastie, T., Tibshirani, R., & Friedman, J. (2009). *The elements of statistical learning: Data mining, inference, and prediction* (2nd ed.). Springer. doi:10.1007/978-0-387-84858-7

Hayek, F. (2020). *Kahei hakkou jiyuuka ron kaiteiban – Kyousou tsuuka no riron to jikkou ni kansuru bunseki* [Denationalisation of Money: The Argument Refined] (A. Murai, Trans.). Nikkei BP Classics. (Original work published 1976)

Hayes, A. (2021, August 31). *The Unintended Consequences of Self-Driving Cars.* Retrieved July 13, 2022, from https://www.investopedia.com/articles/investing/090215/unintended-consequences-selfdriving-cars.asp

Hayes, A. (2022, October 4). Stablecoins: Definition, how they work, and types. *Investopedia.* Retrieved December 4, 2022, from https://www.investopedia.com/terms/s/stablecoin.asp#:~:text=Stablecoins%20are%20cryptocurrencies%20that%20attempt,a%20commodity%20such%20as%20gold

Hayes, A. (2022a, September 28). Who is Satoshi Nakamoto? *Investopedia.* Retrieved December 5, 2022, from https://www.investopedia.com/terms/s/satoshi-nakamoto.asp

Hayes, A. (2022b, December 19). Blockchain facts: What is it, how it works, and how it can be used. *Investopedia.* Retrieved February 15, 2023, from https://www.investopedia.com/terms/b/blockchain.asp

He, L. (2022, October 31). *US curbs on microchips could throttle China's ambitions and escalate the tech war.* Retrieved December 10, 2022, from https://www.cnn.com/2022/10/31/tech/us-sanctions-chips-china-xi-tech-ambitions-intl-hnk

Healy Consultants Group. (2023). *Estonia cryptocurrency/bitcoin business set up in 2023.* https://www.healyconsultants.com/estonia-company-registration/cryptocurrency-solution/

Hébert, M. K. (2014). Come Hack with Me: Adapting Anthropological Training to Work in Civic Innovation. *Practical Anthropology*, *36*(2), 32–36. doi:10.17730/praa.36.2.405j1uvvn8584768

Hegstad, M. (2017). Networks Seek To Archive, Protect Environmental Information From Trump. *Inside EPA's Water Policy Report, 26*(3), 25-27.

Heikkillä, M. (2021). *China backs UN pledge to ban (its own) social scoring. UNESCO member countries endorse plan for 'ethical AI,' calling for clear red lines.* https://www.politico.eu/article/china-artificial-intelligence-ai-ban-social-scoring-united-nations-unesco-ethical-ai/

Henderson, W. D. (2018). *Legal Market Landscape Report.* (https://board.calbar.ca.gov/docs/agendaItem/Public/agendaitem1000022382.pdf)

Hendrickson, K. A. (2003). National Traffic and Motor Vehicle Safety Act. In S. I. Kutler (Ed.), *Dictionary of American History* (3rd ed., Vol. 5, pp. 561–562). Charles Scribner's Sons.

Hershkoff, H., & Miller, A. R. (2021). Article: Courts and civil justice in the time of covid: Emerging trends and questions to ask. *New York University Journal of Legislation and Public Policy*, *23*, 321.

Hertel-Fernandez, A., & Kashin, K. (2015). Capturing business power across the states with text reuse. Academic Press.

Hertel-Fernandez, A. (2014). Who passes business's "model bills"? policy capacity and corporate influence in U.S. state politics. *Perspectives on Politics*, *12*(3), 582–602. doi:10.1017/S1537592714001601

Hjerm, M. (2003). National sentiments in Eastern and Western Europe. *Nationalities Papers*, *31*(4), 413–430. doi:10.1080/0090599032000152933

Hodson, S. (2018). Applying WTO and FTA Disciplines to Data Localization Measures. *World Trade Review*, *18*(4), 579–607. doi:10.1017/S1474745618000277

Hoffman, S. (2017). Programming China. *Merics China Monitor*, *44*, 1–12.

Holt-Lunstad, J. (2015). Loneliness and Social Isolation as Risk Factors for Mortality: A Meta-Analytic Review. *SAGE Journals*. https://journals.sagepub.com/doi/10.1177/1745691614568352?url_ver=Z39.88-2003&rfr_id=ori:rid:crossref.org&rfr_dat=cr_pub
%20%200pubmed

Hong, Y., & Thomas Goodnight, G. (2020). How to Think about Cyber Sovereignty: The Case of China. *Chinese Journal of Communication*, *13*(1), 8–26. doi:10.1080/17544750.2019.1687536

Hooson, M. (2022, September 28). Bitcoin vs. Ethereum. *Forbes*. Retrieved December 5, 2022, from https://www.forbes.com/advisor/investing/cryptocurrency/bitcoin-vs-ethereum/

Hou, Y., & Lampe, C. (2017, June). Sustainable hacking: characteristics of the design and adoption of civic hacking projects. In *Proceedings of the 8th International Conference on Communities and Technologies* (pp. 125-134). ACM. 10.1145/3083671.3083706

Howard, P. N. (2010). *The Digital Origins of Dictatorship and Democracy: Information Technology and Political Islam*. Oxford University Press. doi:10.1093/acprof:oso/9780199736416.001.0001

Howson, C., Beyer, M. A., Idoine, C. J., & Jones, L. C. (2018). *How to Use Data4Good to Impact Society*. New York: Gartner. https://www.gartner.com/doc/3880666/use-data-good-impact-society

Huo, Y. (2022). *The Effect of Government Policies on Cryptocurrency Market*. Academic Press.

Hurtado, P. (n.d.). *Implications of Self-Driving Vehicles*. Retrieved August 21, 2022, from https://www.policechiefmagazine.org/implications-of-self-driving-vehicles/

Iansiti, M., & Lakhani, K. R. (2017). The truth about blockchain. *Harvard Business Review*. Retrieved November 15, 2022, from https://hbr.org/2017/01/the-truth-about-blockchain)

Iansiti, M., & Lakhani, K. R. (2017, January–February). The Truth About Blockchain. *Harvard Business Review*. https://hbr.org/2017/01/the-truth-about-blockchain

IBC-M. (2022). *International Bisiness College Mitrovica*. https://www.ibcmitrovica.eu/about/welcome-to-ibc-m/

IBM. (2020). *What is Artificial Intelligence (AI).* https://www.ibm.com/cloud/learn/what-is-artificial-intelligence

Iddon, P. (2022, November 15). *How Iran is cashing in on cryptocurrencies to evade U.S. sanctions.* Arab News.

Iglesia, A. D. L. (2022, April 18). Decentralizing what? When blockchain's perspectives clash. *The P2P Model.* https://p2pmodels.eu/decentralizing-what-when-blockchains-perspectives-clash/

Il Sole 24 ORE. (2012). Dopo la laurea del Trota in Albania ora tocca all'ex tesoriere Belsito: Diploma e la laurea in odore di falso. *Il Sole 24 ORE.* https://www.ilsole24ore.com/art/notizie/2012-05-04/dopo-laurea-trota-albania-173028.shtml?uuid=Abi2XkXF

Inoue T. (2012). *Sekai Seigi Ron* [World Justice Theory]. Chikuma--shobo.

INQUÉRITO 4781. (n.d.). *Portal Do Supremo Tribunal Federal.* Retrieved August 30, 2022, from https://portal.stf.jus.br/processos/detalhe.asp?incidente=5651823

Interesse, G. (2022). China's Cybersecurity Industry: A Market Analysis. *China Briefing News.* October 26, 2022. https://www.china-briefing.com/news/chinas-cybersecurity-industry-a-market-analysis/

International, S. A. E. (2021, April 30). *J3016_202104: Taxonomy and definitions for terms related to driving automation systems for on-road motor vehicles.* SAE International. Retrieved February 15, 2023, from https://www.sae.org/standards/content/j3016_202104/

Irani, L. (2015). Hackathons and the Making of Entrepreneurial Citizenship. *Science, Technology & Human Values, 40*(5), 799–824. doi:10.1177/0162243915578486

Irwin, A. S. M., & Milad, G. (2016). The use of crypto-currencies in funding violent jihad. *Journal of Money Laundering Control, 19*(4), 407–425. doi:10.1108/JMLC-01-2016-0003

Ito, M. (2018, March). *Kumikomi blockchain no shogeki (Jou)* [The impact of embedded blockchain (Volume one of the two)]. Takaoka Toko Co., Ltd. https://uquest.tktk.co.jp/embedded/outline/outline10.html

Ivey, A. E., Ivey, M. B., & Zalaquett, C. P. (2018). *Intentional Interviewing and Counseling: Facilitating Client Development in a Multicultural Society.* Brooks Cole.

Izdebski, K. (Ed.). (2019). *alGOVrithms—State of Play.* ePaństwo Foundation. Available at: https://epf.org.pl/en/projects/algovrithms/

Jablecka, J. (2007). Legitimation of Nonpublic Higher Education in Poland. In S. Slantcheva & D. C. Levy (Eds.), *Private Higher Education in Post-Communist Europe: In Search of Legitimacy* (pp. 179–200). Palgrave Macmillan. doi:10.1057/9780230604391_9

Jackson, E. (2021, November 18). *History of food delivery and how it's changed.* Thistle. Retrieved September 10, 2022, from https://www.thistle.co/learn/thistle-thoughts/history-of-food-delivery-and-how-its-changed

Jafar, U., Aziz, M. J. A., & Shukur, Z. (2021, August 31). Blockchain for Electronic Voting System-review and open research challenges. *Sensors.* Retrieved November 11, 2022, from https://www.ncbi.nlm.nih.gov/pmc/articles/PMC8434614/

Jahanian, F. (2018). *4 ways universities are driving innovation*. World Economic Forum. https://www.weforum.org/agenda/2018/01/4-ways-universities-are-driving-innovation/

Jansa, J. M., Hansen, E. R., & Gray, V. H. (2018). Copy and paste lawmaking: Legislative professionalism and policy reinvention in the States. *American Politics Research*, *47*(4), 739–767. doi:10.1177/1532673X18776628

Japan Association of New Economy. (2019, July 30). Blockchain no shakai jissou ni muketa teigen ~Ango shisan no shinpo kaisei wo ukete [Proposals for the social implementation of Blockchain - Following the revision of the new law on cryptographic assets]. *Naikakufu of Japan*. https://www8.cao.go.jp/kisei-kaikaku/kisei/meeting/wg/toushi/20191220/191220toushi03.pdf

Japan Blockchain Association. (2016, October 3). "Blockchain no teigi" wo koukai shimashita ["Definition of Blockchain" is now available] *JBA Nihon blockchain kyokai*. https://jba-web.jp/news/642

Jeffries, S. (n.d.). *Ethereum vs. bitcoin: Which crypto is better?* Nasdaq. Retrieved December 8, 2022, from https://www.nasdaq.com/articles/ethereum-vs.-bitcoin%3A-which-crypto-is-better

Jemielniak, D., & Przegalinska, A. (2020). *Collaborative Society*. MIT Press. doi:10.7551/mitpress/11587.001.0001

Jensen, E., Jones, N., Rabe, M., Pratt, B., Medina, L., Orozco, K., & Spell, L. (2021). *The chance that two people chosen at random are of different race or ethnicity groups has increased since 2010*. https://www.census.gov/library/stories/2021/08/2020-united-states-population-more-racially-ethnically-diverse-than-2010

Jia, L., & Liang, F. (2021). The Globalization of TikTok: Strategies, Governance and Geopolitics. *Journal of Digital Media & Policy*, *12*(2), 273–292. doi:10.1386/jdmp_00062_1

Jiang, Z. (2002). *Comprehensively build a society of moderate prosperity, open a new chapter in the pursuit of socialism with Chinese characteristics* [Speech to the XVI Chinese Communist Party Congress]. https://www.fmprc.gov.cn/web/ziliao_674904/zyjh_674906/t10855.shtml

Joh, E. E. (2019). Policing the smart city. *International Journal of Law in Context*, *15*(2), 177–182. doi:10.1017/S1744552319000107

Johnson, B. (2021, April 20). *ISIS Cyber Group Warns of Tracking Through Bitcoin Use*. Homeland Security Today.

Johnson, J. (2019). Legal research and artificial intelligence: A review of current applications and future possibilities. *Law and Technology*, *8*(1), 56–70.

Johnson, J. S. (2018). China's Vision of the Future Network-Centric Battlefield: Cyber, Space and Electromagnetic Asymmetric Challenges to the United States. *Comparative Strategy*, *37*(5), 373–390. doi:10.1080/01495933.2018.1526563

Johnson, M. S. (1996). Western Models and Russian Realities in Postcommunist Education. *Tertium Comparationis*, *2*(2), 119–132.

Johnson, R. (2018). AI in the criminal justice system: Opportunities and challenges. *Journal of Criminal Justice and Technology*, *5*(1), 34–40.

Jones, F. L., & Smith, P. (2001). Diversity and Commonality in National Identities: An Exploratory Analysis of Cross-National Patterns. *Journal of Sociology (Melbourne, Vic.)*, *37*(1), 45–63. doi:10.1177/144078301128756193

Jones, J. (2019). The use of artificial intelligence in the criminal justice system. *Journal of Criminal Justice*, *37*(2), 123–129.

Justice, J., McNutt, J. G., Melitski, J., Ahn, M., David, N., Siddiqui, S., & Ronquillo, J. C. (2018). The Civic Technology Movement: Implications for Nonprofit Theory and Practice. (89). In J. G. McNutt (Ed.), *Technology, Activism and Social Justice in a Digital Age*. Oxford University Press.

Kahneman, D., Slovic, P., & Tversky, A. (1982). *Judgment under uncertainty: Heuristics and biases*. Cambridge University Press. doi:10.1017/CBO9780511809477

Kahneman, D., & Tversky, A. (1984). Choices, values, and frames. *The American Psychologist*, *39*(4), 341–350. doi:10.1037/0003-066X.39.4.341

Kaloudis, A., Aspelund, A., Koch, P. M., Strand, Ø., Sørheim, R., & Aadland, T. (2019). *How Universities Contribute to Innovation: A Literature Review-based Analysis*. Norwegian University of Science and Technology. https://www.ntnu.edu/documents/1272711283/1276140112/Rapport_How+universities+contribute+to+innovation_web.pdf/86b6a699-0499-820e-0f52-35a7b7101de5?t=1574848729613

Kaplan, A. M., & Haenlein, M. (2010). Users of the World, Unite! The Challenges and Opportunities of Social Media. *Business Horizons*, *53*(1), 59–68. doi:10.1016/j.bushor.2009.09.003

Kassen, M. (2022). Open data governance as a theoretical concept: A stakeholder and institutional analysis. In M. Kassen (Ed.), *Open data governance and its actors: Theory and practice* (pp. 1–28). Springer International Publishing. doi:10.1007/978-3-030-92065-4_1

Katzman, K. (2022). *Iran Sanctions*. Academic Press.

Kaur, R., & Kaur, A. (2012). Digital signature. *Proceedings: Turing 100 - International Conference on Computing Sciences, ICCS 2012*, 295–301. 10.1109/ICCS.2012.25

Keatinge, T., & Danner, K. (2021). Assessing Innovation in Terrorist Financing. *Studies in Conflict and Terrorism*, *44*(6), 455–472. doi:10.1080/1057610X.2018.1559516

Kelleher, J. D., & Tierney, B. (2018). *Data science*. MIT Press. doi:10.7551/mitpress/11140.001.0001

Kethineni, S., & Cao, Y. (2020). The Rise in Popularity of Cryptocurrency and Associated Criminal Activity. *International Criminal Justice Review*, *30*(3), 325–344. doi:10.1177/1057567719827051

Kettl, D. F. (2017). *Little bites of big data for public policy*. CQ Press/Sage Publishing.

Khalid, A. (2017). *From Post-it Notes To Algorithms: How Automation Is Changing Legal Work*. NPR. https://www.npr.org/sections/alltechconsidered/2017/11/07/561631927/from-post-it-notes-to-algorithms-how-automation-is-changing-legal-work

Kharpal, A. (2017, June 2). *Bill Gates wants to tax robots, but the EU says, 'no way, no way'*. CNBC. Retrieved September 28, 2022, from https://www.cnbc.com/2017/06/02/bill-gates-robot-tax-eu.html#:~:text=Earlier%20this%20year%2C%20billionaire%20Gates,robot%20at%20a%20similar%20level.%E2%80%9D

Kiayias, A. (Ed.). (2011). Topics in Cryptology - CT-RSA 2011. In *The Cryptographers' Track at the RSA Conference 2011, San Francisco, CA, USA, February 14-18, 2011. Proceedings*. Springer Berlin Heidelberg. 10.1007/978-3-642-19074-2

Kilanko, V. (2021). The potential effects of Biden's infrastructure bill on the american economy. *International Journal Of Scientific Advances*, *2*(5). Advance online publication. doi:10.51542/ijscia.v2i5.1

Kim, S. J. (2021). *North Korea's Sanctions Evasion and Its Economic Implications*. Academic Press.

KimballD. (2013). Why Are Voting Lines Longer for Urban Voters? doi:10.2139/ssrn.2255009

Kim, S. (2021). Complete Analysis of Implementing Isogeny-Based Cryptography Using Huff Form of Elliptic Curves. IEEE Access. doi:10.1109/ACCESS.2021.3128515

King, B. A. (2017). Policy and precinct: Citizen evaluations and electoral confidence. *Social Science Quarterly*, *98*(2), 672–689. doi:10.1111squ.12303

King, B. A. (2020). Waiting to vote: The effect of administrative irregularities at polling locations and voter confidence. *Policy Studies*, *41*(2-3), 230–248. doi:10.1080/01442872.2019.1694652

King, G., Pan, J., & Roberts, M. E. (2013). How Censorship in China Allows Government Criticism but Silences Collective Expression. *The American Political Science Review*, *107*(2), 326–343. doi:10.1017/S0003055413000014

King, G., Pan, J., & Roberts, M. E. (2014). Political Science. Reverse-Engineering Censorship in China: Randomized Experimentation and Participant Observation. *Science*, *345*(6199), 1251722. doi:10.1126cience.1251722 PMID:25146296

King, G., Pan, J., & Roberts, M. E. (2017). How the Chinese Government Fabricates Social Media Posts for Strategic Distraction, Not Engaged Argument. *The American Political Science Review*, *111*(3), 484–501. doi:10.1017/S0003055417000144

King, S. F., & Brown, P. (2007, December). Fix my street or else: using the internet to voice local public service concerns. In *Proceedings of the 1st international conference on Theory and practice of electronic governance*. (pp. 72-80). ACM. 10.1145/1328057.1328076

Kinsella, B. (2021). Data Science for Social Good Volunteer Motivations and Limitations: An Exploratory Survey. *Chance*, *34*(3), W86–W95. doi:10.1080/09332480.2021.1981055

Kirby, M. (2002). *The Origins of Railway Enterprise: The Stockton and Darlington Railway 1821–1863*. Cambridge University Press.

Klaver, F. (2020, April). *The Economic and Social Impacts of Fully Autonomous Vehicles*. Retrieved July 13, 2022, from https://www.compact.nl/en/articles/the-economic-and-social-impacts-of-fully-autonomous-vehicles/

Klievink, B., van der Voort, H., & Veeneman, W. (2018). Creating value through data collaboratives: Balancing innovation and control. *Information Polity*, *23*(4), 379–397. doi:10.3233/IP-180070

Klineschmit, S. (2018). Policing Digital Sanctuaries: Exploring Environmental Advocacy through Technology-Enabled Monitoring and Enforcement Network Organizations. In Technology, Activism and Social Justice in a Digital Age. Oxford University Press.

Klosowski, T. (2021, September 6). The state of consumer data privacy laws in the US (and why it matters). *Wirecutter: Reviews for the Real World*. https://www.nytimes.com/wirecutter/blog/state-of-privacy-laws-in-us/

Knapp, M. (2022, January 12). *Nuro adds airbags (for people) to self-driving delivery vehicles*. PCMAG. Retrieved October 3, 2022, from https://www.pcmag.com/news/nuro-add-airbags-for-people-to-self-driving-delivery-vehicles

Knight, J., & Song, L. (2005). *Towards a Labour Market in China*. OUP Oxford. doi:10.1093/0199245274.001.0001

Koba, A. (2017). *Roma hou annai – Gendai no houritsuka no tameni* [Introduction of Roman law - for contemporary lawyers]. Keiso Shobo.

Koerner, A. (2013, July). Ativismo Judicial? Jurisprudência constitucional e política no STF pós-88. *Novos Estudos CEBRAP*, *96*, 69–85. doi:10.1590/S0101-33002013000200006

Koetsier, J. (2023, March 23). *GPT-4 Beats 90% of Lawyers Trying to Pass the Bar*. Forbes. https://www.forbes.com/sites/johnkoetsier/2023/03/14/gpt-4-beats-90-of-lawyers-trying-to-pass-the-bar/?sh=63c86be83027

Kosovo Accreditation Agency. (2022). Accredited Program for 20220-2023. *Agjencia e Kosovës për Akreditim*. https://akreditimi.rks-gov.net/

KostkaG. (2018). *China's Social Credit Systems and Public Opinion: Explaining High Levels of Approval*. https://ssrn.com/abstract=3215138 doi:10.2139/ssrn.3215138

Kourlis, R. L., & Gorsuch, N. M. (2020). Legal advice is often unaffordable. Here's how more people can get help. *USA Today*. https://www.usatoday.com/story/opinion/2020/09/17/lawyers-expensive-competition-innovationincrease-access-gorsuch-column
/5817467002/

Krahmann, E. (2003). Conceptualizing security governance. *Cooperation and Conflict*, *38*(1), 5–26. doi:10.1177/0010836703038001001

Kretzmann, J., & McKnight, J. (1993). Building communities from the inside out: A path toward finding and mobilizing a community's assets. Institute for Policy Research.

Krietzberg, I., & Lucas, A. (2022, July 17). *Restaurants are short-staffed, and that's taking a big toll on customers and workers alike*. CNBC. Retrieved October 7, 2022, from https://www.cnbc.com/2022/07/17/customer-service-suffers-at-short-staffed-restaurants-as-covid-takes-toll.html

Kroeger, M. (2015, May 28-30). *Plagiarizing policy: Model legislation in state legislatures*. Academic Press.

Kullenberg, C., & Kasperowski, D. (2016). What Is Citizen Science? – A Scientometric Meta-Analysis. *PLoS One*, *11*(1), e0147152. doi:10.1371/journal.pone.0147152 PMID:26766577

Kumar, A. (2019). Citizen-centric model of governmental entrepreneurship: Transforming public service management for the empowerment of marginalized women. *Transforming Government: People, Process, and Policy*, *13*(1), 62–75.

Kumar, A., & Mohan, J. (2020). Artificial intelligence for decision support in the public sector: A review. *Government Information Quarterly*, *37*(4), 554–564.

Kumar, R. (2018). Artificial intelligence and dispute resolution: A review of current applications and future directions. *Law and Technology*, *7*(2), 89–104.

Kuroda, H. (2019, December 4). *Kessai no innovation to chou ginko no yakuwari -Stable coin ga nagekaketa mondai* [Payments Innovation and the Role of Central Banks- the problems posed by stable coins]. Bank of Japan. https://www.boj.or.jp/about/press/koen_2019/data/ko191204a1.pdf

Kuvendi i Kosovës. (2011). *Ligji Nr. 04/L-037 për Arsimin e lartë në Republikën e Kosovës.* https://akreditimi.rks-gov.net/wp-content/uploads/2020/10/Ligji-per-Arsimin-e-Larte-ne-Kosove.pdf

Kuzio, T. (2002). The Myth of the Civic State: A Critical Survey of Hans Kohn's Framework for Understanding Nationalism. *Ethnic and Racial Studies*, *25*(1), 20–39. doi:10.1080/01419870120112049

Kwiek, M. (2007). The European Integration of Higher Education and the Role of Private Higher Education. In S. Slantcheva & D. C. Levy (Eds.), *Private Higher Education in Post-Communist Europe: In Search of Legitimacy* (pp. 119–134). Palgrave Macmillan. doi:10.1057/9780230604391_6

Lanctot, C. J. (2002). Conference on legal ethics: "What needs fixing?": Scriveners in cyberspace: online document preparation and the unauthorized practice of law. *Hofstra Law Review*, *30*, 811.

Lardy, N. R. (2014). *Markets Over Mao: The Rise of Private Business in China.* Columbia University Press.

Lassalle, F. (1942, January). On the Essence of Constitutions (Speech Delivered in Berlin, April 16, 1862). *Fourth International*, *3*(1), 25–31.

Lathrop, D., & Ruma, L. (Eds.). (2010). Open Government: Collaboration, Transparency, and Participation in Practice. O'Reilly.

Lavanderos, L. (2022). From cybersin to cybernet. Considerations for a cybernetics design thinking in the socialism of the XXI century. *AI & Society*, *37*(3), 1279–1292. doi:10.100700146-021-01354-2

LawGeex. (2017). Comparing the Performance of Artificial Intelligence to Human Lawyers in the Review of Standard Business Contracts. *LawGeex.* https://images.law.com/contrib/content/uploads/documents/397/5408/lawgeex.pdf

Lawtech UK. (2021). The contribution of lawtech to the UK economy. *LawtechUK* http://resources.lawtechuk.io/files/RPT-lawtech%20economic%20contribution-07-07-2021%20STC.pdf

LeCun, Y., Bengio, Y., & Hinton, G. (2015). Deep learning. *Nature*, *521*(7553), 436–444. doi:10.1038/nature14539 PMID:26017442

Lee, G. (2021). *Chinese Short Clip Video Case Study: Focusing on TikTok.* Korea Digital Publishing Society., doi:10.30580/kdips.2021.18.2.24

Lee, J. (2018). Legal chatbots and artificial intelligence: A review of current applications and future possibilities. *Law and Technology*, *7*(1), 34–48.

Lee, J. (2019). Chatbots and virtual assistants in government: An exploration of current uses and potential future developments. *Public Administration Review*, *79*(1), 120–129.

Lee, M. Y., & Edmondson, A. C. (2017). Self-managing organizations: Exploring the limits of less-hierarchical organizing. *Research in Organizational Behavior*, *37*, 35–58. doi:10.1016/j.riob.2017.10.002

Legal Information Institute. (n.d.). *Tort.* Legal Information Institute. Retrieved October 2, 2022, from https://www.law.cornell.edu/wex/tort

Legal Services Corporation. (2013). *Report of the Summit on the Use of Technology to Expand Access to Justice.* https://www.lsc.gov/sites/default/files/LSC_Tech%20Summit%20Report_2013.pdf

LegalZoom. (2001). *LegalZoom: About Us*. https://www.legalzoom.com/about-us

Lesaffre, C. (2019, June 18). Facebook va créer sa monnaie: "Nous allons demander des garanties", prévient Bruno Le Maire [Facebook will create its own currency: "We will ask for guarantees", warns Bruno Le Maire]. *Europe, 1*. https://www.europe1.fr/economie/facebook-va-creer-sa-monnaie-nous-allons-demander-des-garanties-previent-bruno-le-maire-3905215

Levendusky, M. S. (2009). *The partisan sort: How liberals became Democrats and conservatives became Republicans*. University of Chicago Press. doi:10.7208/chicago/9780226473673.001.0001

Levin, S. (2018, March 19). *Self-driving Uber kills Arizona woman in first fatal crash involving pedestrian*. Retrieved September 10, 2022, from https://www.theguardian.com/technology/2018/mar/19/uber-self-driving-car-kills-woman-arizona-tempe

Levy, D. C. (2004). *The New Institutionalism: Mismatches with Private Higher Education's Global Growth* [Working paper]. https://prophe.org/en/working-papers/the-new-institutionalism-mismatches-with-private-higher-education%E2%80%99s-global-growth/

Levy, D. C. (1986). *Higher Education and the State in Latin America: Private Challenges to Public Dominance*. University of Chicago Press.

Levy, D. C. (1999). When Private Higher Education Does Not Bring Organizational Diversity. In P. G. Altbach (Ed.), *Private Prometheus: Private Higher Education in the 21st Century* (pp. 15–43). Greenwood Press.

Levy, D. C. (2006). The Unanticipated Explosion: Private Higher Education's Global Surge. *Comparative Education Review*, *50*(2), 217–240. doi:10.1086/500694

Levy, D. C. (2007). Legitimacy and Privateness: Central and Eastern European Private Higher Education in Global Context. In S. Slantcheva & D. C. Levy (Eds.), *Private Higher Education in Post-Communist Europe: In Search of Legitimacy* (pp. 279–298). Palgrave Macmillan. doi:10.1057/9780230604391_14

Liang, B., & Lu, H. (2010). Internet Development, Censorship, and Cyber Crimes in China. *Journal of Contemporary Criminal Justice*, *26*(1), 103–120. doi:10.1177/1043986209350437

Liao, R. (2021, December 14). *California suspends Pony.ai driverless test permit after crash*. Retrieved September 10, 2022, from https://techcrunch.com/2021/12/14/pony-ai-suspension-driverless-pilot-california/

Library of Congress. (2021). *Regulation of Cryptocurrency Around the World: November 2021 Update*. http://www.law.gov

Liebkind, J. (2022, July 18). How blockchain technology can prevent voter fraud. *Investopedia*. Retrieved December 8, 2022, from https://www.investopedia.com/news/how-blockchain-technology-can-prevent-voter-fraud/

Li, H. (2016). The Weibo Celebrity Effect on Luxury Brands Marketing in China: An Analysis of Burberry's Weibo Marketing Strategy. *International Journal of Business and Commerce*, *5*(1), 1–11.

Lin, J. (2015). *Lin Junyue - Office of the Central Committee of the Communist Party of China for Network Security and Informatization ONSI*. http://www.cac.gov.cn/2015-10/28/c_1116956896.htm. (Chinese)

Lin. (2019). *Interview. Louvet, Sylvain. Monitored: seven billion under surveillance*. Arte docu.

Lindsay, J. R. (2015). The Impact of China on Cybersecurity: Fiction and Friction. *International Security, 39*(3), 7–47. doi:10.1162/ISEC_a_00189

Lin, K. J., Lu, X., Zhang, J., & Zheng, Y. (2020). State-Owned Enterprises in China: A Review of 40 Years of Research and Practice. *China Journal of Accounting Research, 13*(1), 31–55. doi:10.1016/j.cjar.2019.12.001

Linz, J. J., & Stepan, A. (1996, August 22). *Problems of Democratic Transition and Consolidation: Southern Europe, South America, and Post-Communist Europe*. The Johns Hopkins University Press.

Lippert, J. F., Furnari, M. B., & Kriebel, C. W. (2021). The Impact of the COVID-19 Pandemic on Occupational Stress in Restaurant Work: A Qualitative Study. *International Journal of Environmental Research and Public Health, 18*(19), 10378. doi:10.3390/ijerph181910378 PMID:34639678

Litan, R. E., Mitchell, L., & Reedy, E. J. (2007). The University As Innovator: Bumps in the Road. *Issues in Science and Technology, 23*(4). https://issues.org/litan/

Little, A. T. (2016). Communication Technology and Protest. *The Journal of Politics, 78*(1), 152–166. doi:10.1086/683187

Li, X. (2016). *Regulation of Cyber Space: An Analysis of Chinese Law on Cyber Crime*. International Journal of Criminal Justice Sciences., doi:10.5281/zenodo.56225

Li, Y., Chandra, Y., & Fan, Y. (2022). Unpacking Government Social Media Messaging Strategies during the COVID-19 Pandemic in China. *Policy and Internet, 14*(3), 651–672. doi:10.1002/poi3.282

Li, Y., Guan, M., Hammond, P., & Berrey, L. E. (2021). Communicating COVID-19 Information on TikTok: A Content Analysis of TikTok Videos from Official Accounts Featured in the COVID-19 Information Hub. *Health Education Research, 36*(3), 261–271. doi:10.1093/her/cyab010 PMID:33667311

Locke J. (2010). *Touchi niron kanyaku* [A complete translation of the Second Treatise of government] (Kato T. Trans.). Iwanami Shoten. (Original work published 1689)

Locke, J. (2013, May 13). *Second Treatise of Government*. Aziloth Books.

Lorentzen, P. (2014). China's Strategic Censorship. *American Journal of Political Science, 58*(2), 402–414. doi:10.1111/ajps.12065

Louvet, S. (2019). *Überwacht: Sieben Milliarden im Visier. Arte Doku*. https://www.arte.tv/de/videos/083310-000-A/ueberwacht-sieben-milliarden-im-visier/

Luhmann, N. (1997). *Die Gesellschaft der Gesellschaft* (Vol. 1). Suhrkamp.

Lupu, Y. (2015). Legislative Veto Players and the Effects of International Human Rights Agreements. *American Journal of Political Science, 59*(3), 578–594. doi:10.1111/ajps.12178

Lusk, K. (2022). *Public Interest Technology University Network: understanding the state of the field.* Boston University. https://open.bu.edu/handle/2144/44469

Lyngaas, S. (2022, July 10). *Here's how North Korean operatives are trying to infiltrate U.S. crypto firms*. CNN Politics.

MacArthur, B. D., Dorobantu, C. L., & Margetts, H. Z. (2022, August). Resilient government requires data science reform. *Nature Human Behaviour, 6*(8), 1035–1037. doi:10.103841562-022-01423-6 PMID:35941235

Macfarlane, E. (2021). Strengthening Sanctions: Solutions to Curtail the Evasion of International Economic Sanctions Through the Use of Cryptocurrency. *Michigan Journal of International Law, 42*(1), 199. doi:10.36642/mjil.42.1.strengthening

MacKenzie Sigalos. (2022, March 22). *Exxon is mining bitcoin in North Dakota as part of its plan to slash emissions.* CNBC.

Magill, K. (2022, August 12). *Aurora delays autonomous truck launch following supplier constraints.* Retrieved September 10, 2022, from https://www.supplychaindive.com/news/aurora-paccar-autonomous-truck-launch-delay-supply-constraints/629546/

Maharrey, M. (2018, August 6). *SWIFT and the Weaponization of the U.S. Dollar.* Foundation for Economic Education.

Mahdavieh, R. (2019). *Governments' Adoption of Native Cryptocurrency: A Case Study of Governments' Adoption of Native Cryptocurrency: A Case Study of Iran, Russia, and Venezuela Iran, Russia, and Venezuela.* http://library.ucf.edu

Maji, H. K., Prabhakaran, M., & Rosulek, M. (2011). Attribute-Based Signatures. In A. Kiayias (Ed.), Lecture Notes in Computer Science: Vol. 6558. *Topics in Cryptology – CT-RSA 2011. CT-RSA 2011.* Springer. doi:10.1007/978-3-642-19074-2_24

Makhlouf, F., & Selmi, R. (2022). *Do sanctions work in a crypto world? The impact of the removal of Russian Banks from SWIFT on Remittances Do sanctions work in a crypto world?* https://hal.archives-ouvertes.fr/hal-03599089

Malik, N. (2018). *How terrorists use encryption, the darknet, and cryptocurrencies.* Academic Press.

Manoharan, A. P., & McQuiston, J. (2016). Technology and pedagogy: Information technology competencies in public administration and public policy programs. *Journal of Public Affairs Education, 22*(2), 175–186. doi:10.1080/15236803.2016.12002239

Marian, O. (2013). Part of the Banking and Finance Law Commons, Criminal Law Commons, Internet Law Commons, and the Tax Law Commons Recommended Citation Recommended Citation Omri Marian, Are Cryptocurrencies Super Tax Havens? 112 MICH. *Michigan Law Review First Impressions, 112.* https://repository.law.umich.edu/mlr_fiAvailableat:https://repository.law.umich.edu/mlr_fi/vol112/iss1/2

Market Research Future. (2022, May 19). *Delivery robots market projected to reach valuation of USD 971.3 million by 2030, growing at a 31.3% CAGR - Report by Market Research Future (MRFR).* GlobeNewswire News Room. Retrieved September 28, 2022, from https://www.globenewswire.com/en/news-release/2022/05/19/2446711/0/en/Delivery-Robots-Market-Projected-to-Reach-Valuation-of-USD-971-3-Million-by-2030-Growing-at-a-31-3-CAGR-Report-by-Market-Research-Future-MRFR.html

Markets and Markets. (2021, April). *Delivery robots market size, share: Industry report (2021-2026).* Markets and Markets. Retrieved October 2, 2022, from https://www.marketsandmarkets.com/Market-Reports/delivery-robot-market-263997316.html

Markides, C., & Oyon, D. (2010). What to do against disruptive business models (when and how to play two games at once). *MIT Sloan Management Review, 51*(4), 25–32.

Marozau, R., & Guerrero, M. (2016). Conditioning factors of knowledge transfer and commercialisation in the context of post-socialist economies: The case of Belarusian higher education institutions. *International Journal of Entrepreneurship and Small Business, 27*(4), 441. Advance online publication. doi:10.1504/IJESB.2016.075014

Marozau, R., Guerrero, M., & Urbano, D. (2021). Impacts of Universities in Different Stages of Economic Development. *Journal of the Knowledge Economy, 12*(1), 1–21. doi:10.100713132-016-0359-7

Marr, B. (2021, December 10). *The Future of Delivery Robots*. Forbes. Retrieved September 26, 2022, from https://www.forbes.com/sites/bernardmarr/2021/11/05/the-future-of-delivery-robots/?sh=3bfc300d7337

Marr, B. (2022, October 12). Web3 and the future of Music. *Forbes*. Retrieved January 13, 2023, from https://www.forbes.com/sites/bernardmarr/2022/07/13/web3-and-the-future-of-music/?sh=6e0d96402a29

Martin, M. (2022, November 8). Crypto issues are a factor for 37% of US voters, study finds. *Forbes*. Retrieved November 15, 2022, from https://www.forbes.com/sites/mitchellmartin/2022/11/01/crypto-issues-a-factor-for-37-of-us-voters-study-finds/?sh=630c18
fd341d

Martinho, A., Herber, N., Kroesen, M., & Chorus, C. (2021). Ethical issues in focus by the autonomous vehicles industry. *Transport Reviews*, *41*(5), 1–22. doi:10.1080/01441647.2020.1862355

Martin, J. A. (2013). Closing gaps in international knowledge and participation: News attention, online expression, and the 2010 Haiti earthquake. *Mass Communication & Society*, *16*(3), 417–440. doi:10.1080/15205436.2012.713149

Martin, J. A. (2014). Mobile media and political participation: Defining and developing an emerging field. *Mobile Media & Communication*, *2*(2), 173–195. doi:10.1177/2050157914520847

Martin, J. A. (2015). Mobile News Use and Participation in Elections: A Bridge for the Democratic Divide? *Mobile Media & Communication*, *3*(2), 230–249. doi:10.1177/2050157914550664

Masterson, J. R. (2015). Social media in China: a double-edged sword. In S. B. Edwards & D. Santos (Eds.), *Revolutionizing the interaction between state and citizens through digital communications* (pp. 238–263). IGI Global. doi:10.4018/978-1-4666-6292-6.ch013

Masterson, J. R. (2020). Chinese Citizenry Social Media Pressures and Public Official Responses: The Double-Edged Sword of Social Media in China. In *Digital Transformation and Its Role in Progressing the Relationship Between States and Their Citizens* (pp. 139–181). IGI Global. doi:10.4018/978-1-7998-3152-5.ch007

Mauldin, M. D. (2016). No MPA left behind: A review of information technology in the master of public administration curriculum. *Journal of Public Affairs Education*, *22*(2), 187–192. doi:10.1080/15236803.2016.12002240

McAllister, K. (2019). To Raise Your Credit Score, Start by Understanding It. A beginner's guide to understandding how your credit score actually works. *Wall Street Journal*. https://www.wsj.com/articles/to-raise-your-credit-score-start-by-understanding-it-11548010800

McCormick, P. (2022). The Web3 debate. *The Web3 Debate*. Retrieved December 2, 2022, from https://www.notboring.co/p/the-web3-debate?s=r

McGavisk, T. (n.d.). *The positive and negative implications of GDPR in the workplace*. TDS. Retrieved September 22, 2022, from https://www.timedatasecurity.com/blogs/the-positive-and-negative-implications-of-gdpr-in-the-workplace

McGinnis, J. O., & Pearce, R. G. (2014). Colloquium: The legal profession's monopoly on the practice of law: the great disruption: How machine intelligence will transform the role of lawyers in the delivery of legal services. *Fordham Law Review*, *82*, 3041.

McGuinness, T. D., & Schank, H. (2021). *Power to the public: The promise of public interest technology*. Princeton University Press. doi:10.2307/j.ctv18b5dbz

McIlwain, C. D. (2019). *Black software: The Internet and racial justice, from the AfroNet to Black Lives Matter*. Oxford University Press.

McKeever, B., Greene, S., MacDonald, G., Tatian, P., & Jones, D. (2018). *Data Philanthropy: Unlocking the Power of Private Data for Public Good*. Urban Institute.

McKeever, E., Jack, S., & Anderson, A. (2015). Embedded entrepreneurship in the creative re-construction of place. *Journal of Business Venturing*, *30*(1), 50–65. doi:10.1016/j.jbusvent.2014.07.002

McKeown, M. (2008). *The truth about innovation*. Prentice Hall.

McKiernan, P., & Purg, D. (Eds.). (2013). *Hidden Champions in CEE and Turkey: Carving Out a Global Niche*. Springer Science & Business Media. doi:10.1007/978-3-642-40504-4

McNutt, J. G. (Ed.). (2018). Technology, Activism and Social Justice in a Digital Age. Oxford University Press.

McNutt, J. G., Brainard, L., Zeng, Y., & Kovacic, P. (2016). Information and Technology in and for Associations and Volunteering. In Palgrave Handbook of Volunteering and Nonprofit Associations. Palgrave Macmillan.

McNutt, J. G. (2019). Data justice and international development: An ethical imperative for policy and community practice. In *The Routledge Handbook of Social Work Ethics and Values* (pp. 280–286). Routledge. doi:10.4324/9780429438813-36

McNutt, J. G., & Goldkind, L. (2022). Intersectoral Collaboration and Social Justice Concerns in Artificial Intelligence Implementation at the Community Level. In J. Saura & F. Debasa (Eds.), *Handbook of Research on Artificial Intelligence in Government Practices and Processes* (pp. 38–54). IGI Global. doi:10.4018/978-1-7998-9609-8.ch003

McNutt, J. G., & Justice, J. B. (2016, September). *Predicting civic hackathons in local communities: Perspectives from social capital and creative class theory*. Presented at the *ISTR Conference*, Stockholm, Sweden.

McNutt, J. G., Justice, J. B., Melitski, M. J., Ahn, M. J., Siddiqui, S., Carter, D. T., & Kline, A. D. (2016). The diffusion of civic technology and open government in the United States. *Information Polity*, *21*(2), 153–170. doi:10.3233/IP-160385

McQuiston, J. M., & Manoharan, A. P. (2017). Developing E-Government coursework through the NASPAA competencies framework. *Teaching Public Administration*, *35*(2), 173–189. doi:10.1177/0144739417690582

Medlock v. LegalZoom.Com, Inc, 2013 S.C. Lexis, 362. (2013)

Mehrotra & Kharif. (2021, November 3). Ransomware H.Q.: Moscow's Tallest Tower Is a Cybercriminal Cash Machine. *Bloomberg*.

Meijer, A. J., Gil-Garcia, J. R., & Bolívar, M. P. R. (2016). Smart city research: Contextual conditions, governance models, and public value assessment. *Social Science Computer Review*, *34*(6), 647–656. doi:10.1177/0894439315618890

Meijer, A., & Bolívar, M. P. R. (2016). Governing the smart city: A review of the literature on smart urban governance. *International Review of Administrative Sciences*, *82*(2), 392–408. doi:10.1177/0020852314564308

Mello, M. A. (2022). *Dissenting vote. ADPF N° 572/DF*. Retrieved August 30, 2022, from https://redir.stf.jus.br/paginadorpub/paginador.jsp?docTP=TP&docID=754371407

Mergel, I. (2012). The Social Media Innovation Challenge in the Public Sector. *Information Polity*, *17*(3,4), 281–292. doi:10.3233/IP-2012-000281

Meshulam, D., Hausfeld, K., Kinney, J., & North, E. (2022, February 15). *U.S. Department of Justice, aided by cryptocurrency exchanges, seizes over US$3.6 billion in stolen Bitcoin*. DLA Piper.

Metinko, C. (2021). Legal Tech Makes Its Case With Venture Capitalists, Tops $1B In Funding This Year. *Crunchbase.* https://news.crunchbase.com/venture/legal-tech-venture-investment/

Millar, J. (2015). Technology as moral proxy: Autonomy and paternalism by design. *IEEE Technology and Society Magazine*, *34*(2), 47–55. doi:10.1109/MTS.2015.2425612

Miller, D., Sinanan, J., Wang, X., McDonald, T., Haynes, N., Costa, E., Spyer, J., Venkatraman, S., & Nicolescu, R. (2016). *How the World Changed Social Media.* UCL Press. doi:10.2307/j.ctt1g69z35

Mittiga, R. (2022). Political Legitimacy, Authoritarianism, and Climate Change. *The American Political Science Review*, *116*(3), 998–1011. doi:10.1017/S0003055421001301

Mollman, S. (2022). Oops. Sam Bankman-Fried's implosion took down Democrats' second-biggest donor with it as the party gears up to regulate crypto. *Fortune*. Retrieved December 11, 2022, from https://fortune.com/2022/11/10/sam-bankman-fried-ftx-joe-biden-democratic-party-second-biggest-donor/

Mondal, S. (2018). Artificial intelligence and the provision of government services to citizens. *Public Administration Review*, *78*(2), 283–292.

Mondal, S. (2019). Artificial intelligence and the management of public resources: A review of current applications and future possibilities. *Public Administration Review*, *79*(5), 845–855.

Monitor. (2021, September 5). Bie me 5% numri i studentëve në 2021, Universitetet që humbën studentë dhe ato që fituan. *Revista Monitor.* https://www.monitor.al/bie-me-5-numri-i-studenteve-ne-2021-universitetet-qe-humben-studente-dhe-ato-qe-fituan/

Monteiro, P., & Adler, P. S. (2022). Bureaucracy for the 21st century: Clarifying and expanding our view of bureaucratic organization. *The Academy of Management Annals*, *16*(2), 427–475. doi:10.5465/annals.2019.0059

Montesquieu. (2004). *The spirit of laws.* Barotche Books.

Montgomery, B. N., & Reifler, J. (2021). The effects of unsubstantiated claims of voter fraud on confidence in elections. *Journal of Experimental Political Science*, 1–16.

Moore, L., & Sawhney, N. (2019). *The West Virginia Mobile Voting Pilot - Under the Hood.* Retrieved December 20, 2022, fromhttps://sos.wv.gov/FormSearch/Elections/Informational/West-Virginia-Mobile-Voting-White-Paper-NASS-Submission.pdf

Moore, M. H. (1995). *Creating public value: Strategic management in government.* Harvard University Press.

Moraes, A. (2011). *Direito constitucional* (27th ed.). Atlas.

Moran, J. (2021, April 29). *S.1494 - 117th Congress (2021-2022): Consumer Data Privacy and Security Act of 2021 (2021/2022)* [Legislation]. https://www.congress.gov/

Morgan, S. (2021, April 27). Cybercrime to cost the world $10.5 trillion annually by 2025. *Cybercrime Magazine*. Retrieved October 4, 2022, from https://cybersecurityventures.com/cybercrime-damages-6-trillion-by-2021/

Morimura, S. (2020). *Houtetsugaku ha konnani omoshiroi* [Philosophy of law is so interesting]. Shinzan-sha.

Morita, T., & Managi, S. (2020). Autonomous vehicles: Willingness to pay and the social dilemma. *Transportation Research Part C, Emerging Technologies*, *119*, 102748. doi:10.1016/j.trc.2020.102748

Motamedi, M. (2018, December 6). *U.S. sanctions accelerate Iran's blockchain drive*. Al-Monitor.

Mulgan, G., & Albury, D. (2003). Innovation in the public sector. *Strategy Unit, Cabinet Office*, *1*(1), 40.

Muller, J. (2022, May 13). *Self-driving cars could be potential crime witnesses*. Retrieved August 19, 2022, from https://www.axios.com/2022/05/13/self-driving-cars-video-police-witness

Mustafa, A. (2020). Kosovo's social policy during self-management, UNMIK and independence: Persisting high inequality and social exclusion. *International Journal of Social Welfare*, *29*(1), 96–108. doi:10.1111/ijsw.12406

Mustafa, A. (2022). The causal mechanisms behind expansive basic universalism in Kosovo. *Global Social Policy*. Advance online publication. doi:10.1177/14680181221126029

Muzzey, D., Beard, C. A., & Beard, W. (1931). Review: The American Leviathan. *Political Science Quarterly*, *46*(1), 109–111. doi:10.2307/2143112

Nabilou, H. (2019). *How to Regulate Bitcoin? Decentralized Regulation for a Decentralized Cryptocurrency*. Academic Press.

Nadeem, M. A., Liu, Z., Pitafi, A. H., Younis, A., & Xu, Y. (2021). Investigating the Adoption Factors of Cryptocurrencies—A Case of Bitcoin: Empirical Evidence From China. *SAGE Open*, *11*(1). Advance online publication. doi:10.1177/2158244021998704

Nakamoto, S. (2009). Bitcoin: A Peer-to-Peer Electronic Cash System. *Satoshi Nakamoto Institute*. https://bitcoin.org/bitcoin.pdf

Nakanishi, T. (2019, July 17). Libra 'Kakkoku kisei ni shitagau' Bei kouchou kai de hihan ya kenen [Libra 'subject to national regulations,' criticisms and concerns at U.S. hearing]. *Nihon Keizai Shinbun Sha*. https://www.nikkei.com/article/DGXMZO47418380X10C19A7000000/

National Academy of Sciences. (2022), Cryptography and the Intelligence Community The Future of Encryption. The National Academies Press.

National Conference of State Legislatures. (2022). *Election Administration at State and Local Levels*. https://www.ncsl.org/research/elections-and-campaigns/electionadministration-at-state-and-local-levels.aspx

National Conference of State Legislatures. (2022). *Electronic Poll Books*. https://www.ncsl.org/research/elections-and-campaigns/electronic-pollbooks.aspx

National Cyber Security Centre. (2016, September 23). *Introduction to security governance*. https://www.ncsc.gov.uk/collection/risk-management-collection/governance-cyber-risk/security-governance-introduction

Nations, U. Convention on Road Traffic, 1042 § (1977). Retrieved from https://treaties.un.org/Pages/ViewDetailsIII.aspx?chapter=11&mtdsg_no=XI-B-19&src=TREATY

Neave, G. (2007). The Long Quest for Legitimacy: An Extended Gaze from Europe's Western Parts. In S. Slantcheva & D. C. Levy (Eds.), *Private Higher Education in Post-Communist Europe: In Search of Legitimacy* (pp. 27–54). Palgrave Macmillan. doi:10.1057/9780230604391_2

NershiK.CollinsM.ElzaynH. S.GrayJ.HeverlyR.HoffmanD.LeviM.LittrellC.MansfieldE. D.NielsonD.RobertsM.Rodine-HardyK.SharmanJ.SimmonsB.WerbachK. (2022). *How Strong Are International Standards in Practice? Evidence from Cryptocurrency Transactions.* https://ssrn.com/abstract=4186244 doi:10.2139/ssrn.4186244

Neto, K. T. (2022). *Inquérito do fim do mundo o apagar das luzes do direito brasileiro*. Editora E.D.A.

Neto, M. J. S. (2009). *Curso de Direito Constitucional* (4th ed.). Lumen Juris.

News. (2020, July 21). *The Political Case Against Self-Driving Cars*. Retrieved July 13, 2022, from https://mindmatters.ai/2020/07/the-political-case-against-self-driving-cars/#:~:text=in%20political%20philosophy%2C%20raises%3A,less%20harm%20to%20the%20environment

Newsome, G. (2013). *Citizenville*. The Penguin Press.

Nichols, G. (2017, December 12). *San Francisco bans delivery robots in most of the city*. ZDNET. Retrieved October 8, 2022, from https://www.zdnet.com/article/san-francisco-bans-delivery-robots-in-most-of-the-city/

Nicolescu, L. (2007). Institutional Efforts for Legislative Recognition and Market Acceptance: Romanian Private Higher Education. In S. Slantcheva & D. C. Levy (Eds.), *Private Higher Education in Post-Communist Europe: In Search of Legitimacy* (pp. 201–222). Palgrave Macmillan. doi:10.1057/9780230604391_10

NIST. (2022). *Call for Additional Digital Signature Schemes for the Post-Quantum Cryptography Standardization Process*. Available from the NIST homepage at https://csrc.nist.gov/csrc/media/Projects/pqc-dig-sig/documents/call-for-proposals-dig-sig-sept-2022.pdf

Noesselt, N. (2014). Microblogs and the Adaptation of the Chinese Party-State's Governance Strategy. *Governance: An International Journal of Policy, Administration and Institutions*, *27*(3), 449–468. Advance online publication. doi:10.1111/gove.12045

Nofer, M., Gomber, P., Hinz, O., & Schiereck, D. (2017). Blockchain. *Business & Information Systems Engineering*, *59*(3), 183–187. doi:10.100712599-017-0467-3

Noguchi, Y. (2017). *Blockchain Kakumei – Bunsan jiritsu gata shaki no shutugen* [Blockchain Revolution - Emergence of a decentralized and autonomous society]. Nihon Keizai Shuppan-sha.

Noguchi, Y. (2018). Kasou tsuuka ha dounaruka - Bubble ga owari, atarashii shinka ga hajimaru [What will happen to virtual currency? Bubble ends, new evolution begins]. *Diamond.*

Norry, A. (2017, October 11). *The History of Silk Road: A Tale of Drugs, Extortion & Bitcoin*. Blockonomi.

North, D. C. (1990). *Institutions, Institutional Change and Economic Performance*. Cambridge University Press. doi:10.1017/CBO9780511808678

Norton, K. L. W. (2020). The Middle Ground: A Meaningful Balance Between the Benefits and Limitations of Artificial Intelligence to Assist with the Justice Gap. *University of Miami Law Review*, *75*, 190.

Nott, G. (2022, April 12). *Robot Grocery Delivery Company starship investigating toddler 'hit and push' incident*. The Grocer. Retrieved October 5, 2022, from https://www.thegrocer.co.uk/technology-and-supply-chain/robot-grocery-delivery-company-starship-investigating-toddler-hit-and-push-incident/666529.article

Noveck, B. S. (2015). *Smart citizens, smarter state: The technologies of expertise and the future of governing*. Harvard University Press. doi:10.4159/9780674915435

Nozick, R. (1992). *Anarchy・Kokka・Utopia* [Anarchy, State, and Utopia] (I. Shimazu, Trans.). Bokutaku-sha. (Original work published 1974)

Nuñez, C. F. A. M., & Quintana, F. (2014, October). Repúblicas em conflito: A separação dos poderes made in America. *Revista de Informacao Legislativa*, *51*(204), 139–161.

Nyholm, S., & Smids, J. (2016). The Ethics of Accident-Algorithms for Self-Driving Cars: An Applied Trolley Problem? *Ethical Theory and Moral Practice*, *19*(5), 1275–1289. doi:10.100710677-016-9745-2

O'Donnell, G. (1988). Challenges to Democratization in Brazil. *World Policy Journal*, *5*(2), 281–300. https://www.jstor.org/stable/40209083

O'Leary. (2021, March 31). The bitcoin terrorists of Idlib are learning new tricks. *Wired.*

O'Neil, C. (2016). *Weapons of math destruction: How big data increases inequality and threatens democracy*. Broadway Books.

Oeste, R. (2022, August 18). *Hang rebate matéria de jornal sobre 'golpe' de empresários: "Fake news."* Revista Oeste. Retrieved August 23, 2022, from https://revistaoeste.com/politica/hang-rebate-materia-de-jornal-sobre-golpe-de-empresarios-bolsonaristas-fake-news/

Office of the Bureau of Public Roads. (n.d.). https://www.archives.gov/research/guide-fed-records/groups/030.html#30.1

Okamoto, T., & Takashima, K. (2020). Decentralized Attribute-Based Encryption and Signatures. IEICE Trans. Fundamentals, E103–A(1). doi:10.1587/transfun.2019CIP0008

Okamura, H. (2021). 個人情報保護法の知識 [Overviewing the Personal Information Protection Act] (5th ed.). Nikkei Bunko Publisher.

Oliveira, L. C. (2022, June 2). *O sistema acusatório no processo penal brasileiro e seus desdobramentos*. https://jus.com.br/artigos/98326/o-sistema-acusatorio-no-processo-penal-brasileiro-e-seus-desdobramentos

Olsen, J. P. (2008). The ups and downs of bureaucratic organization. *Annual Review of Political Science*, *11*(1), 13–37. doi:10.1146/annurev.polisci.11.060106.101806

Ooi, V., & Goh, G., (2022). Taxation of automation and artificial intelligence as a tool of labour policy. *eJournal of Tax Research*, *19*(2), 273-303.

Ora News. (2022). Programet që rrezikojnë mbylljen/ Kushi: Maturantët që i zgjedhin përjashtohen nga tarifa, mbështeten financiarisht. *Ora News*. https://www.oranews.tv/vendi/njoftim-i-rendesishem-nga-mas-tre-kategorite-e-studenteve-qe-perfitojne-bu-i1044380

Oregon State Bar. (2022). *Paralegal Licensing Proposal.* Oregon State Bar. https://www.osbar.org/lp

Osborne, D. (1988). *Laboratories of Democracy*. Harvard Business School Press.

Overton, M., & Kleinschmit, S. (2022). Transforming research methods education through data science literacy. Teaching Public Administration.

Pachuashvili, M. (2007). Legitimacy Sources and Private Growth in the Post-Communist Context. In S. Slantcheva & D. C. Levy (Eds.), *Private Higher Education in Post-Communist Europe: In Search of Legitimacy* (pp. 75–94). Palgrave Macmillan. doi:10.1057/9780230604391_4

Page, C. (2022, August 25). *Doordash hit by Data Breach linked to Twilio hackers*. TechCrunch. Retrieved October 4, 2022, from https://techcrunch.com/2022/08/25/doordash-customer-data-breach-twilio/

Palmer, M. (2019, September 4). *Europe's 6-wheeled delivery robots begin invasion of US campuses*. Sifted. Retrieved October 3, 2022, from https://sifted.eu/articles/starship-robot-delivery/

Palmer, N. A., & Perkins, D. D. (2012). Technological Democratization: The Potential Role of ICT in Social and Political Transformation in China and Beyond. *Perspectives on Global Development and Technology*, *11*(4), 456–479. doi:10.1163/15691497-12341236

Park, J. (2021). *The Lazarus group the cybercrime syndicate financing the North Korea state*. Academic Press.

Parker, E. (2018). Can China Contain Bitcoin? *MIT's Technology Review*, *121*(1), 61–65.

Park, J. (2016). Artificial intelligence and public safety: A review of current applications and future directions. *Public Administration Review*, *76*(6), 973–982.

Parks, E., & Zoga, D. (2020, September 11). *NBC 5 responds: What happens if you're in a crash with a robot?* NBC 5 Dallas-Fort Worth. Retrieved October 3, 2022, from https://www.nbcdfw.com/news/nbc-5-responds/nbc-5-responds-what-happens-if-youre-in-a-crash-with-a-robot/2442345/

Pasquale, F. (2019). Article: A Rule of Persons, Not Machines: The Limits of Legal Automation. *The George Washington Law Review*, *87*, 2.

Patel, M., Sotsky, J., Gourley, S., & Houghton, D. (2013). *The emergence of civic tech: Investments in a growing field*. The Knight Foundation.

Patterson, D. (2022). Explaining web3: From the Blockchain and crypto to nfts and the Metaverse. *CBS News*. Retrieved December 4, 2022, from https://www.cbsnews.com/news/web3-blockchain-crypto-nft-metaverse-explainer/

Patton, M. Q. (2014). *Qualitative Research & Evaluation Methods* (4th ed.). SAGE Publishing. https://us.sagepub.com/en-us/nam/qualitative-research-evaluation-methods/book232962

Payson, J. (2021). *When cities lobby: How local governments compete for power in state politics*. Oxford University Press.

Peach, S. (2021). *Why did Alibaba's Jack Ma disappear for three months?* https://www.bbc.com/news/technology-56448688

Pellert, A. (1999). *Die Universität als Organisation. Die Kunst Experten zu managen*. Böhlau.

Perrin, A. (2021, November 11). *16% of Americans say they have ever invested in, traded or used cryptocurrency*. Pew Research Center.

Perrow, C. (1970). *Organizational analysis: A sociological view*. Brooks/Cole.

Person, & Derby, M. S. (2022). Ny Fed official: Digital Dollar Could Speed Foreign Exchange Settlement. Reuters. Retrieved December 7, 2022, from https://www.reuters.com/markets/currencies/ny-fed-official-digital-dollar-could-speed-foreign-exchange-settlement-2022-11-04/

Person, & Godoy, J. (2022). FTX's Bankman-Fried, Tom Brady and other celebrity promoters sued by crypto investors. *Reuters.* Retrieved December 6, 2022, from https://www.reuters.com/legal/ftx-founder-bankman-fried-sued-us-court-over-yield-bearing-crypto-accounts-2022-11-16/

Peshkopia, R. (2008). In Search of the Private, Public, and Counterpublic: Modernity, Postmodernity, and Postsocialism. *New Political Science, 30*(1), 23–57. doi:10.1080/07393140701877684

Peshkopia, R. (2010). A Ghost from the Future: The Postsocialist Myth of Capitalism and the Ideological Suspension of Postmodernity. *Theoria (Pietermaritzburg), 57*(124), 23–53. doi:10.3167/th.2010.5712402

Peshkopia, R. (2014). Between Social Tutelage and Individual Responsibility: Dilemmas of Higher Education in Eastern Europe. *Policy Futures in Education, 12*(4), 542–556. doi:10.2304/pfie.2014.12.4.542

Peshkopia, R., Kristensen, L., & Logoreci, T. (2012). Agency and structure in Albanian post-socialist cinema: Lessons from three films. *Studies in Eastern European Cinema, 3*(2), 151–173. doi:10.1386eec.3.2.151_1

Pettigrew, S. (2017). The racial gap in wait times: Why minority precincts are underserved by local election officials. *Political Science Quarterly, 132*(3), 527–547. doi:10.1002/polq.12657

Pew Research Center. (2022, June 6). *Public Trust in Government: 1958-2022.* https://www.pewresearch.org/politics/2022/06/06/public-trust-in-government-1958-2022/

Pfaff, T. (2015). *The Definitive Guide to doing Data Science for Social Good.* https://www.kdnuggets.com/2015/07/guide-data-science-good.html

Phillips, B. (2022, April 14). Pizza Margherita may be fit for a queen, but was it really named after one? *National Geographic.*

Pickerill, J. M., & Bowling, C. J. (2014). Polarized parties, politics, and policies: Fragmented federalism in 2013–2014. *Publius, 44*(3), 369–398. doi:10.1093/publius/pju026

Pihlak, H. (2019). *I-voting – the future of elections?* e-Estonia. Retrieved November 19, 2022, from https://e-estonia.com/i-voting-the-future-of-elections/

Piquero, A. R., Jennings, W. G., Jemison, E., Kaukinen, C., & Knaul, F. M. (2021). Domestic violence during the COVID-19 pandemic - Evidence from a systematic review and meta-analysis. *Journal of Criminal Justice, 74*, 101806. doi:10.1016/j.jcrimjus.2021.101806 PMID:36281275

Pollitt, C., & Hupe, P. (2011). Talking about government: The role of magic concepts. *Public Management Review, 13*(5), 641–658. doi:10.1080/14719037.2010.532963

Polly. (2020, June 22). *6 Ways Autonomous Vehicles Will Impact Life As We Know It.* Retrieved July 13, 2022, from https://roboticsandautomationnews.com/2020/06/22/6-ways-autonomous-vehicles-will-impact-life-as-we-know-it/33326/

Popoveniuc, S., Kelsey, J., & Leontie, E. (2011). On the privacy threats of electronic poll books. In *Proceedings of the International Conference on Security and Cryptography* (pp. 281-286). IEEE.

Poreh, K. (2012). A Revolution in Communications: Democratization and the Growth of Cyberspace. *SAIS Review (Paul H. Nitze School of Advanced International Studies), 32*(2), 217–220. doi:10.1353ais.2012.0019

Porter, J. (2020, June 26). *Amazon is acquiring self-driving car startup Zoox*. Retrieved May 25, 2022, from https://www.theverge.com/2020/6/26/21304111/amazon-zoox-acquisition-self-driving-car-autonomous-vehicles-startup-robotaxi-bi-directional-vehicle

Powell, A. (2016). Hacking in the public interest: Authority, legitimacy, means, and ends. *New Media & Society*, *18*(4), 600–616. doi:10.1177/1461444816629470

Prado, G. (2002). *Sistema acusatório: a conformidade constitucional das leis processuais penais*. Editora Lumen Juris.

Protect Personal Data Privacy. SB21-190, Colorado General Assembly, 2021 Regular Session (2021). https://leg.colorado.gov/bills/sb21-190

Puussaar, A., Johnson, I. G., Montague, K., James, P., & Wright, P. (2018). Making open data work for civic advocacy. *Proceedings of the ACM on Human-Computer Interaction*, *2*(CSCW), 143. 10.1145/3274412

PYMNTS.com. (2022). *Crypto execs call for clearer regulations in wake of FTX collapse*. Retrieved December 15, 2022, from https://www.pymnts.com/cryptocurrency/2022/crypto-execs-call-for-clearer-regulations-in-wake-of-ftx-collapse/

Qian, Y. (2017). *How Reform Worked in China: The Transition from Plan to Market*. MIT Press. doi:10.7551/mitpress/8098.001.0001

Qin, A., Liu, J., & Chien, A. C. (2022). Police database in China was long left unsecured. *The New York Times*.

R7.com. (2020, June 18). *É um inquérito do fim do mundo, diz o ministro Marco Aurélio*. Retrieved September 17, 2022, from https://noticias.r7.com/brasil/e-um-inquerito-do-fim-do-mundo-diz-ministro-marco-aurelio-29062022

Rach.ipsa. [@rach.ipsa]. (2021). *rach.ipsa* [TikTok profile]. TikTok. Retrieved September 22, 2022, from https://www.tiktok.com/@rach.ipsa/video/7002708616255950085

Raguso, E. (2019, April 26). *Police: Berkeley Man, fed up with Kiwibot delivery robots, steals one*. Berkeleyside. Retrieved October 6, 2022, from https://www.berkeleyside.org/2019/04/26/police-berkeley-man-fed-up-with-kiwibot-delivery-robots-steals-one

Rawls, J. (2010). *Seigiron* [A Theory of Justice] (T. Kawamoto, Trans.). Kinokuniya Shoten. (Original work published 1971)

Reiff, N. (2020). Some US military overseas will vote via blockchain. *Investopedia*. Retrieved November 9, 2022, from https://www.investopedia.com/news/some-us-military-overseas-will-vote-blockchain/

Reiff, N. (2022a). Which industries will blockchain technology disrupt next? *Investopedia*. Retrieved November 1, 2022, from https://www.investopedia.com/tech/which-industries-will-blockchain-disrupt-next/

Reiff, N. (2022b). Bitcoin vs. Ethereum: What's the difference? *Investopedia*. Retrieved November 26, 2022, from https://www.investopedia.com/articles/investing/031416/bitcoin-vs-ethereum-driven-different-purposes.asp

Reisinger Walker, E. (2015). Mortality in Mental Disorders and Global Disease Burden Implications A Systematic Review and Meta-analysis. *JAMA Psychiatry*. https://jamanetwork.com/journals/jamapsychiatry/fullarticle/2110027

Reisz, R. D. (2007). Legitimacy Discourse and Mission Statements of Private Higher Education Institutions in Romania. In S. Slantcheva & D. C. Levy (Eds.), *Private Higher Education in Post-Communist Europe: In Search of Legitimacy* (pp. 135–156). Palgrave Macmillan. doi:10.1057/9780230604391_7

Reuters. (2020, December 8). Facebook's renamed cryptocurrency is still 'wolf in sheep's clothing': German Finance Minister. *Reuters*. https://www.reuters.com/article/uk-g7-digitial-facebook/facebooks-renamed-cryptocurrency-is-still-wolf-in-sheeps-clothing-german-finance-minister-idUKKBN28H202

Reverbel, C. E. D. (2009, April 30). Ativismo judicial e estado de direito. *Revista Eletrônica Do Curso De Direito Da UFSM*, *4*(1). Advance online publication. doi:10.5902/198136947028

Riley, J. (2021). The current status of cryptocurrency regulation in China and its effect around the world. *China and WTO Review*, *7*(1), 135–152. doi:10.14330/cwr.2021.7.1.06

RIT Kosova AUK. (2022). *About | RIT Kosovo | RIT*. https://www.rit.edu/kosovo/about#mission-and-vision

Ritchie, H. (2022). Leaked notes from Chinese health officials estimate 250 million Covid-19 infections in December: reports. *CNN*. https://edition.cnn.com/2022/12/23/china/china-covid-infections-250-million-intl-hnk/index.html

Roberts, M. E. (2020). *Censored: Distraction and Diversion Inside China's Great Firewall*. Princeton University Press.

Robinson, T. (2021, May 21). *How Iran Uses Bitcoin Mining to Evade Sanctions and "Export" Millions of Barrels of Oil*. Elliptic.

Rojas, H., & Puig-i-Abril, E. (2009). Mobilizers Mobilized: Information, Expression, Mobilization and Participation in the Digital Age. *Journal of Computer-Mediated Communication*, *14*(4), 902–927. doi:10.1111/j.1083-6101.2009.01475.x

Rold, A. D. (2012, May 8). Dopo la laurea in Albania, dubbi pure sulla maturità del Trota. *Linkiesta.it*. https://www.linkiesta.it/2012/05/dopo-la-laurea-in-albania-dubbi-pure-sulla-maturita-del-trota/

Rsakova, E. P., Inshakova, A. O., & Frolova, E. E. (2021). Legal Regulation of Internet Courts in China. In *Modern Global Economic System: Evolutional Development vs. Revolutionary Leap, 1515–21*. Springer International Publishing. doi:10.1007/978-3-030-69415-9_167

Rumbul, R. (2015). *Who Benefits From Civic Technology? Demographic and public attitudes research into the users of civic technology*. mySociety.

Saad, A. (2021, September 9). *O art. 142 da Constituição de 1988: Ensaio sobre a sua interpretação e aplicação* (Portuguese Edition). Independently Published.

Saito, K. (2014, May 16). *Bitcoin ni okeru transaction, sono tensei to eikyo* [Transactions in Bitcoin, their ability to exhibit and impact]. WIDE Technical-Report in 2014. http://member.wide.ad.jp/tr/wide-tr-ideon-bitcoin-transaction2014-00.pdf

Sanburn, J. (2013, May 18). Slices of History: Great Moments in Pizza Innovation. *Time: Business*. Retrieved September 10, 2022, from https://business.time.com/2013/05/18/slices-of-history-great-moments-in-pizza-innovation/

Sasaki, T., & Kim, T. C. (Eds.). (2002). *Koukyo Tetsugaku <6> Keizai kara mita koushi mondai* [Public philosophy <6> Public and private issues from the point of view of the economy]. Tokyo Daigaku Shuppankai.

Scheele, Losco, Crawley, & Vasicko. (2009). *Improving election administration with vote centers: Toward a national model.* Academic Press.

Schley, D. (2020). *Steam City: Railroads, Urban Space, and Corporate Capitalism in Nineteenth-Century Baltimore.* University of Chicago Press. doi:10.7208/chicago/9780226720395.001.0001

Schreurs, M. A., & Steuwer, S. D. (2015). Autonomous Driving - Political, Legal, Social, and Sustainability Dimensions. In M. Maurer, J. C. Gerdes, B. Lenz, & H. Winner (Eds.), *Autonomes Fahren* (pp. 151–173). Springer Berlin Heidelberg. doi:10.1007/978-3-662-45854-9_8

Schreurs, M. A., & Steuwer, S. D. (2016). *Autonomous Driving—Political.* Legal, Social, and Sustainability Dimensions.

Schrock, A. R. (2016). Civic hacking as data activism and advocacy: A history from publicity to open government data. *New Media & Society*, *18*(4), 581–599. doi:10.1177/1461444816629469

Schrock, A., & Shaffer, G. (2017). Data ideologies of an interested public: A study of grassroots open government data intermediaries. *Big Data & Society*, *4*(1). doi:10.1177/2053951717690750

Schwoerer, K. (2019). Individuals' Use of Twitter to Discuss Freedom of Information in the United States: A Social Network Analysis. *The Journal of Civic Information*, *1*(2), 45–59. doi:10.32473/joci.v1i2.119010

Scott, P. (2002). Reflections on the Reform of Higher Education in Central and Eastern Europe. *Higher Education in Europe*, *27*(1–2), 137–152. doi:10.1080/0379772022000003288

Seamans, R. (2022, March 9). *Tax not the Robots*. Brookings. Retrieved September 28, 2022, from https://www.brookings.edu/research/tax-not-the-robots/#footnote-1

SEEU Student Services Center. (2022). *Universiteti i Evropës Juglindore—Apliko!* Universiteti i Evropës Juglindore. http://www.seeu.edu.mk

Sensiba, J. (2022, May 21). *Ford-Backed Argo AI Begins Driverless Taxi Operations, & Why This Is Big News*. Retrieved May 25, 2022, from https://cleantechnica.com/2022/05/20/ford-backed-argo-ai-begins-driverless-taxi-operations-why-this-is-big-news/amp/

Serrano, M., Carlos, J., Papakyriakopoulos, O., & Hegelich, S. (2020). Dancing to the Partisan Beat: A First Analysis of Political Communication on TikTok. In *12th ACM Conference on Web Science* (pp. 257–66). Association for Computing Machinery.

Seth, S. (2022). Explaining the crypto in cryptocurrency. *Investopedia*. Retrieved November 27, 2022, from https://www.investopedia.com/tech/explaining-crypto-cryptocurrency/

Shark, A. R. (2016). The information technology gap in public administration: What we can learn from the certified public manager and senior executive service programs. *Journal of Public Affairs Education*, *22*(2), 213–230. doi:10.1080/15236803.2016.12002242

Sharma, R. (2022a). Non-fungible cryptocurrency (NFT): What it means and how it works. *Investopedia.* Retrieved November 24, 2022, from https://www.investopedia.com/non-fungible-cryptocurrencys-nft-5115211

Sharma, R. (2022b). Bitcoin's taproot upgrade: What you should know. *Investopedia.* Retrieved November 4, 2022, from https://www.investopedia.com/bitcoin-taproot-upgrade-5210039

Shaw, R. (1990). *Pioneer Canals and Republican Improvements. In Canals For A Nation: The Canal Era in the United States, 1790-1860*. University Press of Kentucky. https://www.jstor.org/stable/j.ctt5vkjxc.5

Sheehan, M. (2022, October 27). *Biden's Unprecedented Semiconductor Bet*. Retrieved December 17, 2022, from https://carnegieendowment.org/2022/10/27/biden-s-unprecedented-semiconductor-bet-pub-88270

Sheluchin, A. (2022, June 16). *A national digital currency has serious privacy implications*. The Conversation. Retrieved November 20, 2022, from https://theconversation.com/a-national-digital-currency-has-serious-privacy-implications-130520

Shirky, C. (2010). *Cognitive surplus: Creativity and generosity in a connected age*. Penguin.

Short, H. (2022, September 26). *Pizza Hut Canada launches door-to-door robot delivery service: Dished*. Retrieved September 26, 2022, from https://dailyhive.com/vancouver/pizza-hut-canada-robot-delivery-service

Shutsko, A. (2020). User-Generated Short Video Content in Social Media. A Case Study of TikTok. In *Social Computing and Social Media. Participation, User Experience, Consumer Experience, and Applications of Social Computing* (pp. 108–125). Springer International Publishing. doi:10.1007/978-3-030-49576-3_8

Sicular, T., Yang, X., & Gustafsson, B. (2022). The Rise of China's Global Middle Class in an International Context. *China & World Economy*, *30*(1), 5–27. doi:10.1111/cwe.12400

Simshaw, D. (2018). Ethical Issues in Robo-Lawyering: The Need for Guidance on Developing and Using Artificial Intelligence in the Practice of Law. *The Hastings Law Journal*, *70*, 174.

Singer, N. (2019). Top Universities Join to Push 'Public Interest Technology'. *The New York Times*. https://www.nytimes.com/2019/03/11/technology/universities-public-interest-technology.html

Singh, A. J., & Wang, B. (2021, May 31). *Impact of covid-19 on the hospitality industry and implication for operations and Asset Management*. Boston Hospitality Review. Retrieved October 7, 2022, from https://www.bu.edu/bhr/2021/05/31/impact-of-covid-19-on-the-hospitality-industry-and-implication-for-operations-and-asset-management/

Singh, C. (Ed.). (2019, September 13), France Finance Minister Calls Facebook Libra a Threat to 'Monetary Sovereignty'. *News18*. https://www.news18.com/news/tech/france-finance-minister-calls-facebook-libra-a-threat-to-monetary-sovereignty-2307541.html

Slantcheva, S. (2007). Legitimating the Difference: Private Higher Education Institutions in Central and Eastern Europe. In S. Slantcheva & D. C. Levy (Eds.), *Private Higher Education in Post-Communist Europe: In Search of Legitimacy* (pp. 55–74). Palgrave Macmillan. doi:10.1057/9780230604391_3

Smith, A. (2020). The role of artificial intelligence in the criminal justice system. *The Journal of Law and Technology*, *25*(2), 167–174.

Smith, A. D. (1991). *National identity*. Penguin.

Smith, A., & Johnson, R. (2018). Predictive policing and the use of artificial intelligence in law enforcement. *The Journal of Law and Technology*, *23*(1), 56–65.

Sorensen, J. E., & Sorensen, T. L. (1974). The conflict of professionals in bureaucratic organizations. *Administrative Science Quarterly*, *19*(1), 98–106. doi:10.2307/2391790

Soutar, G. N., & Turner, J. P. (2022). Students' preferences for university: A conjoint analysis. *International Journal of Educational Management*, *16*(1), 40–45. doi:10.1108/09513540210415523

Sowell, T. (2011). *Economic Facts and Fallacies* (2nd ed.). Basic Books.

Spencer, D. M., & Markovits, Z. S. (2010). Long lines at polling stations? Observations from an election day field study. *Election Law Journal*, *9*(1), 3–17. doi:10.1089/elj.2009.0046

Stackpole, T. (2022). What is web3? *Harvard Business Review.* Retrieved November 14, 2022, from https://hbr.org/2022/05/what-is-web3

Staff, B. B. C. (2022, June 27). *What are the sanctions on Russia and are they hurting its economy?* BBC News.

Stark, D., & Bruszt, L. (1998). *Postsocialist Pathways: Transforming Politics and Property in East Central Europe*. Cambridge University Press.

Starship. (2022, March 1). *$100m raised in last 30 days by Starship Technologies as demand for its autonomous delivery service triples in 2021* [Press release]. Retrieved February 15, 2023, from https://www.starship.xyz/press_releases/100m-raised-in-last-30-days

State Council. (2007). Stellungnahmen des Generalsekretariats des Staatsrates zum Aufbau des Sozialkredit-Systems. *Generalsekretariat des Staatsrates, 17.* http://www.gov.cn/zhengce/content/2008-03/28/content_1923.htm

State Council. (2014). *Bekanntmachung des Staatsrats: Grundriss des Aufbauplans für ein Sozialkredit-System (2014-2020).* http://www.gov.cn/zhengce/content/2014-06/27/content_8913.htm

State Statistics Office. (2022). *Enrolled undergraduate students.* State Statistics Office. https://www.stat.gov.mk/PrikaziPoslednaPublikacija.aspx?id=38

Statista Report: Cryptocurrencies. (2022). Statista.

Steenhuysen, J. (2022). New COVID model predicts over 1 million deaths in China. *Reuters.* https://www.reuters.com/world/china/new-covid-model-predicts-over-1-mln-deaths-china-through-2023-2022-12-17/

Stein, R. M., Mann, C., Stewart, C. III, Birenbaum, Z., Fung, A., Greenberg, J., Kawsar, F., Alberda, G., Alvarez, R. M., Atkeson, L., Beaulieu, E., Birkhead, N. A., Boehmke, F. J., Boston, J., Burden, B. C., Cantu, F., Cobb, R., Darmofal, D., Ellington, T. C., ... Wronski, J. (2020). Waiting to vote in the 2016 presidential election: Evidence from a multi-county study. *Political Research Quarterly*, *73*(2), 439–453. doi:10.1177/1065912919832374

Stepasiuk, T. (2014). Civic hacking: A Motivational Perspective. *New Visions in Public Affairs.*, *6*, 21–30.

Stephens, J. (2017). *Civic Technology: Open Data and Citizen Volunteers as a Resource for North Carolina Local Governments*. School of Government, The University of North Carolina at Chapel Hill.

Stetar, J., Panych, O., & Tatusko, A. (2007). State Power in Legitimating and Regulating Private Higher Education: The Case of Ukraine. In S. Slantcheva & D. C. Levy (Eds.), *Private Higher Education in Post-Communist Europe: In Search of Legitimacy* (pp. 239–256). Palgrave Macmillan. doi:10.1057/9780230604391_12

Stevenson, A. (2010). *Oxford Dictionary of English*. OUP Oxford.

Stewart & Ansolabehere. (2013). *Waiting in line to vote*. Academic Press.

Streck, L. L., Tassinari, C., & Obach Lepper, A. (2015, June 6). O problema do ativismo judicial: Uma análise do caso MS3326. *Revista Brasileira De Políticas Públicas*, *5*(2). Advance online publication. doi:10.5102/rbpp.v5i2.3139

Strickland v. Washington, 466 U.S. 668, 104 S. Ct. 2052, 80 L. Ed. 2d 674, 1984 U.S. LEXIS 79, 52 U.S.L.W. 4565 (Supreme Court of the United States May 14, 1984).

Sts Cyril and Methodius University. (2021). *Анализа за запишување студенти во академската 2021/2022 година, УКИМ-Одделение за настава, Скопје*. University Sts. Kirill and Metodiu. https://www.ukim.edu.mk/dokumenti_m/Analiza_2021-2022.pdf

Suciu, P. (2022). Is TikTok Really A National Security Threat? *Forbes Magazine*. https://www.forbes.com/sites/petersuciu/2022/11/18/is-tiktok-really-a-national-security-threat/

Suri, M. V. (2013). *From Crowdsourcing Potholes to Community Policing: Applying Interoperability Theory to Analyze the Expansion of "Open311"*. Berkman Center for Internet & Society at Harvard University.

Susha, I., Grönlund, Å., & Van Tulder, R. (2019). Data driven social partnerships: Exploring an emergent trend in search of research challenges and questions. *Government Information Quarterly*, *36*(1), 112–128. doi:10.1016/j.giq.2018.11.002

Susha, I., Janssen, M., & Verhulst, S. (2017). Data collaboratives as "bazaars"? A review of coordination problems and mechanisms to match demand for data with supply. *Transforming Government: People. Process and Policy*, *11*(1), 157–172.

Susha, I., Janssen, M., Verhulst, S., & Pardo, T. (2017, June). Data collaboratives: How to create value from data for public problem solving? In *Proceedings of the 18th Annual International Conference on Digital Government Research* (pp. 604-606). New York: ACM. 10.1145/3085228.3085309

Suspitsin, D. (2007). Between the State and the Market: Sources of Sponsorship and Legitimacy in Russian Nonstate Higher Education. In S. Slantcheva & D. C. Levy (Eds.), *Private Higher Education in Post-Communist Europe: In Search of Legitimacy* (pp. 157–178). Palgrave Macmillan. doi:10.1057/9780230604391_8

Su, Z., & Meng, T. (2016). Selective Responsiveness: Online Public Demands and Government Responsiveness in Authoritarian China. *Social Science Research*, *59*(September), 52–67. doi:10.1016/j.ssresearch.2016.04.017 PMID:27480371

T7 (Director). (2022, October 22). *Bie numri i studentëve në universitet publike, rritet në ato private*. T7. https://www.youtube.com/watch?v=TXWES4-YH_A

Taeihagh, A., & Lim, H. S. M. (2018). Governing autonomous vehicles: Emerging responses for safety, liability, privacy, cybersecurity, and industry risks. *Transport Reviews*, *39*(1), 1–26. doi:10.1080/01441647.2018.1494640

Takaragi, K., Kubota, T., Wohlgemuth, S., Umezawa, K., & Koyanagi, H. (2023). Secure revocation features in EKYC - Privacy protection in Central Bank Digital Currency. *IEICE Transactions on Fundamentals of Electronics, Communications and Computer Sciences*, *E106.A*(3), 325-332. Available from https://www.jstage.jst.go.jp/article/transfun/E106.A/3/E106.A_2022CIP0008/_pdf/-char/ja

Talbott, S. (2020, June 23). *The Political Economy Of Autonomous Vehicles*. Retrieved July 13, 2022, from https://www.forbes.com/sites/selikajosiahtalbott/2020/06/23/the-political-economy-of-autonomous-vehicles/?sh=74f629d4155
5

Tannenbaum, R., & Schmidt, W. H. (1973). How to Choose a Leadership Pattern. *Harvard Business Review, 51*, 162–180.

Tao, L. (2018). Jaywalkers under surveillance in Shenzhen soon to be punished via text messages. *South China Morning Post, 27.*

Tassev, L. (2022a, January 22). *Iran to Pilot 'National Cryptocurrency,' Considers Blockchain Tech for Stock Market.* Bitcoin.Com.

Tassev, L. (2022b, June 26). *Iran Aims to Roll Out Pilot Version of Crypto Rial Within 2 Months.* Bitcoin.Com.

Tate, K. (2014). History of A.I.: Artificial Intelligence. *Live Science.* https://www.livescience.com/47544-history-of-a-i-artificial-intelligence-infographic.html

Teichmann, F. M. J. (2018). Financing terrorism through cryptocurrencies – a danger for Europe? *Journal of Money Laundering Control, 21*(4), 513–519. doi:10.1108/JMLC-06-2017-0024

Teixeira, A. V. (2012, January). Ativismo judicial: Nos limites entre racionalidade jurídica e decisão política. *Revista Direito GV, 8*(1), 37–58. doi:10.1590/S1808-24322012000100002

telegrafi.com. (2018, March 22). *Studentët e UBT-së do të marrin pjesë në garën më të madhe të fakulteteve të drejtësisë në botë* [Newspaper]. Telegrafi. https://telegrafi.com/studentet-e-ubt-se-te-marrin-pjese-ne-garen-te-madhe-te-fakulteteve-te-drejtesise-ne-bote/

telegrafi.com. (2022, October 17). *UBT-së i ndahet vendi i dytë në botë dhe sjell medaljen e argjendtë në Kosovë nga olimpiada botërore e robotikës.* Telegrafi. https://telegrafi.com/ubt-se-ndahet-vendi-dyte-ne-bote-dhe-sjell-medaljen-e-argjendte-ne-kosove-nga-olimpiada-boterore-e-robotikes/

Telma. (2022, May 31). Топ Тема дебата за високото образование- Опаѓа интересот за студирање на универзитетите во Македонија. *Telma.Com.Mk.* https://telma.com.mk/2022/05/31/top-tema-debata-za-visokoto-obrazovanie-opagja-interesot-za-studiranje-na-univerzitetite-vo-makedonija/

Terbrack, J. (2021, March 31). *Starship vs. whips: Robots pose potential risk on streets.* BG Falcon Media. Retrieved October 3, 2022, from https://www.bgfalconmedia.com/news/starship-vs-whips-robots-pose-potential-risk-on-streets/article_f31331b2-91af-11eb-af2b-d7feb69985a3.html

The Debate Over the Judicial Branch. (n.d.). *Center for the Study of the American Constitution.* Retrieved September 17, 2022, from https://csac.history.wisc.edu/document-collections/constitutional-debates/judiciary/

The e-Governance Academy (eGA). (2018). *eGA.* https://ega.ee/about-us/

The e-Governance Academy (eGA). (2018, February 21). Let's get acquainted: Uuno Vallner, the father of e-Estonia. *eGA.* https://ega.ee/news/uuno-vallner-father-estonias-estate/

The Shift. (2022). An A.I.-Generated Picture Won an Art Prize. Artists Aren't Happy. *New York Times.*

The State Bar of California. (2018). *Rules of Professional Conduct.* https://www.calbar.ca.gov/Attorneys/Conduct-Discipline/Rules/Rules-of-Professional-Conduct/Current-Rules

The State Bar of California. (2020). *Task Force on Access Through Innovation of Legal Service*. https://www.calbar.ca.gov/Portals/0/documents/ATILS_Fact_Sheet.pdf

The United States Government. (2022, March 25). *United States and European Commission joint statement on Trans-Atlantic Data Privacy Framework*. The White House. Retrieved November 20, 2022, available from https://www.whitehouse.gov/briefing-room/statements-releases/2022/03/25/united-states-and-european-commission-joint-statement-on-trans-atlantic-data-privacy-framework/

The White House. (2016, December). *Preparing for the future artificial intelligence, of artificial*. The White House of President Barak Obama. Retrieved September 27, 2022, from https://obamawhitehouse.archives.gov/sites/whitehouse.gov/files/documents/Artificial-Intelligence-Automation-Economy.PDF

Thiel Fellowship. (n.d.). *Thiel Fellowship: Why*. https://thielfellowship.org/

Thiemann, A. (2021). *Cryptocurrencies: An empirical view from a tax perspective JRC Working Papers on Taxation and Structural Reforms No 12/2021*. https://ec.europa.eu/jrc

Thompson Reuters. (2017). Ready or not: Artificial intelligence and corporate legal departments. *Thompson Reuters*. https://static.legalsolutions.thomsonreuters.com/static/pdf/S045344_final.pdf

Thompson, S. (1925). *A short history of American railways, covering ten decades*. Appleton and Company.

Thomson, G. (2021). Iowa House approves bill to legally recognize blockchain smart contracts. *Cointelegraph*. Retrieved December 15, 2022, from https://cointelegraph.com/news/iowa-house-approves-bill-to-legally-recognize-blockchain-smart-contracts

Tocqueville, A. (2022, September 17). *Da democracia na América*. Principia.

Tomusk, V. (2003). The War of Institutions, Episode I: The Rise, and the Rise of Private Higher Education in Eastern Europe. *Higher Education Policy*, *16*(2), 213–238. doi:10.1057/palgrave.hep.8300017

Torfing, J. (2016). *Collaborative Innovation in the Public Sector*. Georgetown University Press.

Travaglino, G. A. (2019). Support for anonymous as vicarious dissent: Testing the social banditry framework. *Group Processes & Intergroup Relations*, *22*(2), 163–181. doi:10.1177/1368430217722037

Treasury, H. M. (2021, October 14). *G7 Public policy principles for retail central bank digital currencies and G7 Finance Ministers and Central Bank governors' statement on central bank digital currencies and digital payments*. Retrieved October 18, 2022, from https://www.gov.uk/government/publications/g7-public-policy-principles-for-retail-central-bank-digital-currencies-and-g7-finance-ministers-and-central-bank-governors-statement-on-central-bank

Triple, A. (2022). *Cryptocurrency across the world*. TripleA.

Tsui, L. (2003). The Panopticon as the Antithesis of a Space of Freedom. *China Information*, *17*(2), 65–82. doi:10.1177/0920203X0301700203

Tung, R. L., & Worm, V. (2001). Network Capitalism: The Role of Human Resources in Penetrating the China Market. *International Journal of Human Resource Management*, *12*(4), 517–534. doi:10.1080/713769653

Turak, N. (2021, May 26). *Iran bans bitcoin mining as its cities suffer blackouts and power shortages*. CNBC.

Turing, A. M. (1950). Computing machinery and intelligence. *Mind*, *49*(236), 433–460. doi:10.1093/mind/LIX.236.433

Turkel, E. (2020). *The Diffusion of Civic Technology: How Technology and Devolution Are Reshaping Civic Life in Urban America* [Ph.D. Dissertation]. The University of Delaware.

Turner, A., & Irwin, A. S. M. (2018). Bitcoin Transactions: A digital discovery of illicit activity on the blockchain. *Journal of Financial Crime*, *25*(1), 109–130. doi:10.1108/JFC-12-2016-0078

U.S. Department of Justice. (2020, August 13). *Global Disruption of Three Terror Finance Cyber-Enabled Campaigns Largest Ever Seizure of Terrorist Organizations' Cryptocurrency Accounts*. U.S. Office of Public Affairs.

Ueda, K. (2022). *Kinyu System no Keizaigaku* [Economics of Financial Systems]. Nihon Hyoronsha.

UN Human Rights Committee (HRC). (1988). *CCPR General Comment No. 16: Article 17 (Right to Privacy)*. The Right to Respect of Privacy, Family, Home and Correspondence, and Protection of Honour and Reputation.

UNESCO. (2021). *Report of the Social and Human Sciences Commission (SHS) 41 C/73 22 November*. https://unesdoc.unesco.org/ark:/48223/pf0000379920#page=14

United States Department of Justice. (2020). Global Disruption of Three Terror Finance Cyber-Enabled Campaigns.

United States Postal Service. (2020). *Secure Voting System* (Patent No. 2020/0258338 A1). United States.

Universiteti i Prishties. (2021). *Mbahet ceremonia e diplomimit të studentëve në Fakultet të Edukimit*. https://edukimi.uni-pr.edu/page.aspx?id=1,6,1256

University of Malaga. (2019, February 26). *Universum College International Week 2019*. University of Malaga. https://www.uma.es/relaciones-internacionales/noticias/universum-college-international-week-2019/?set_language=en

Universum Collge. (2022). Alumni Services. *Kolegji Universum*. https://universum-ks.org/alumni/

US DoT. (2021, January). *Automated Vehicles: A comprehensive plan*. U.S. Department of Transportation. Retrieved October 1, 2022, from https://www.transportation.gov/sites/dot.gov/files/2021-01/USDOT_AVCP.pdf

US Government. (2021). *National quantum initiative supplement to the president's FY 2022 budget*. Report by the subcommittee on quantum information science committee on science of the national science & technology council national quantum initiative supplement to the president's FY 2022 budget, United States Government, Dec. 2021. Available from https://www.quantum.gov/release-of-the-national-quantum-initiative-supplement-to-the-presidents-fy-2022-budget/

Van de Ven, A., Polley, D., Garud, R., & Venkataraman, S. (2008). *The innovation journey*. Oxford University Press.

Vasconcelos, P. C. B. (1996). Teoria geral do controlo jurídico do poder público. Lisbon: Cosmos.

Verhulst, S., & Sangokoya, D. (2015). *Data collaboratives: Exchanging data to improve people's lives*. https://medium. com/@ sverhulst/data-collaboratives-exchanging-data-to-improvepeople-s-lives-d0fcfc1bdd9a

Verhulst, S. G. (2021). The Value of Data and Data Collaboratives for Good: A Roadmap for Philanthropies to Facilitate Systems Change Through Data. In *Data Science for Social Good* (pp. 9–27). Springer. doi:10.1007/978-3-030-78985-5_2

Vermont Bar Association. (n.d.). *Self-Help Resources*. Vermont Bar Association. https://www.vtbar.org/public-resources/

Versprille, A. (2022). US midterm elections may have big crypto implications. *Bloomberg.com*. Retrieved December 1, 2022, from https://www.bloomberg.com/news/newsletters/2022-11-03/us-midterm-elections-may-have-big-crypto-implications

Vitória, M. C. (2021, December). A inconstitucionalidade do inquérito n. 4.781 à luz de uma perspectiva democrática. *Revista Jurídica do Ministério Público Catarinense, 16*(35).

von der Oelsnitz, D. (2014). Die frustrierende Organisation: Ungeschicktes Job Design und forcierte Entfremdung. In *Die auszehrende Organisation* (pp. 89–112). Springer Gabler. doi:10.1007/978-3-658-05307-9_5

Wang, J., Chu, J., & Huang, J. (2019). Automation of legal document processing in the criminal justice system: A review. *Journal of Law*.

Wang, Y.-H., Gu, T.-J., & Wang, S.-Y. (2019). Causes and Characteristics of Short Video Platform Internet Community Taking the TikTok Short Video Application as an Example. *2019 IEEE International Conference on Consumer Electronics - Taiwan (ICCE-TW)*, 1–2. 10.1109/ICCE-TW46550.2019.8992021

Wang, X., Xu, L. C., & Zhu, T. (2004). State-Owned Enterprises Going Public The Case of China. *Economics of Transition, 12*(3), 467–487. doi:10.1111/j.0967-0750.2004.00189.x

Warschauer, M. (2001). Singapore's Dilemma: Control versus Autonomy in IT-Led Development. *The Information Society, 17*(4), 305–311. doi:10.1080/019722401753330922

Weber, I. M. (2021). *How China Escaped Shock Therapy: The Market Reform Debate*. Routledge. doi:10.4324/9780429490125

Weber, M. (1947). *The theory of social and economic organization*. The Free Press.

Weimann, G. (2016). Terrorist Migration to the Dark Web. *Source: Perspectives on Terrorism, 10*(3). https://www.jstor.org/stable/26297596

Weingroff, R. (1996). From 1916 to 1939: The Federal-State Partnership at Work. *Public Roads, 60*(1).

Weiss, J., Anisimova, T., & Shirokova, G. (2019). The translation of entrepreneurial intention into start-up behaviour: The moderating role of regional social capital. *International Small Business Journal, 37*(5), 473–501. doi:10.1177/0266242619831170

Wells, M. (2022, March 4). *Trendy bakery unleashes robots in San Francisco parks*. SFGATE. Retrieved October 8, 2022, from https://www.sfgate.com/food/article/treat-vending-robots-at-SF-parks-16975139.php

Wertz, D. (2020). *Understanding U.S. and International Sanctions on North Korea*. www.ncnk.org

Westermeyer, G. (2013). *The impact of private actors on security governance: An analysis based on German ISR capabilities for ISAF*. Springer VS Wiesbaden., doi:10.1007/978-3-658-02230-3

Wevorce (n.d.). https://www.wevorce.com/

Whyte, C. (2018). Cryptoterrorism: Assessing the Utility of Blockchain Technologies for Terrorist Enterprise. *Studies in Conflict and Terrorism*. Advance online publication. doi:10.1080/1057610X.2018.1531565

Wicker, R. F. (2021, July 28). *S.2499 - 117th Congress (2021-2022): SAFE DATA Act (2021/2022)* [Legislation]. https://www.congress.gov/

Wild Dolphin Project. (n.d.). *Wild Dolphin Project: Goals*. https://www.wilddolphinproject.org/our-research/goals-and-objectives/

Wiley, K., Schwoerer, K., Richardson, M., & Espinosa, M. B. (2022). Engaging Stakeholders on TikTok: A Multi-level Social Media Analysis of Nonprofit Microvlogging. *Public Administration*, (April), padm.12851. Advance online publication. doi:10.1111/padm.12851

Williams, R. D. (2020, December 1). To enhance data security, federal privacy legislation is just a start. *Brookings*. https://www.brookings.edu/techstream/to-enhance-data-security-federal-privacy-legislation-is-just-a-start/

Williams, B. (2020). *Ikikata ni tsuite tetsugaku ha nani ga ieruka* [Ethics and the limit of philosophy] (Y. Morigiwa & K. Shimokawa, Trans.). Chikuma-shobo. (Original work published 1985)

Williams, J. (2017). Artificial intelligence in the courts: Opportunities and challenges. *The Journal of Law and Technology*, *22*(2), 123–129.

Wilson, B., & Chakraborty, A. (2019). Planning smart (er) cities: The promise of civic technology. *Journal of Urban Technology*, *26*(4), 29–51. doi:10.1080/10630732.2019.1631097

Winston, C. (2020, August 25). *Autonomous vehicles could improve policing, public safety, and much more*. Retrieved August 21, 2022, from https://www.brookings.edu/blog/techtank/2020/08/25/autonomous-vehicles-could-improve-policing-public-safety-and-much-more/

Winston, P. (1992). *Artificial intelligence* (3rd ed.). Addison-Wesley.

Winter, P., & Lindskog, S. (n.d.). *How the Great Firewall of China Is Blocking Tor*. https://www.usenix.org/sites/default/files/conference/protected-files/winter_foci12_slides.pdf

Woesler, M. (2022). Digitalisierung und Kybernetik in China: Das Sozialkredit-System. In T. Klenk, F. Nullmeier, & G. Wewer (Eds.), *Handbuch Digitalisierung in Staat und Verwaltung*. Springer VS. doi:10.1007/978-3-658-23669-4_101-1

Woesler, M., & Warnke, M. (Eds.). (2023). *Sozialkybernetik in statu nascendi. Die Entstehungsgeschichte des chinesischen Sozialkreditsystems – Quellentexte zum chinesischen SCS*. Matthes & Seitz.

Woesler, M., Warnke, M., Kettner, M., & Lanfer, J. (2019). The Chinese Social Credit System: Origin, political design, exoskeletal morality and comparisons to Western systems. *European Journal of Chinese Studies*, *2*, 7–35.

Wolford, B. (2018, November 7). *What is GDPR, the EU's new data protection law?* https://gdpr.eu/what-is-gdpr/

Woods, D. (2022, October 19). *Why the US cut China off from advanced chips*. Retrieved November 17, 2022, from https://www.npr.org/2022/10/19/1129990818/why-the-us-cut-china-off-from-advanced-chips#:~:text=On%20October%207%2C%20though%2C%20just,or%20buy%20advanced%20semiconductor%20chips

World Economic Forum. (2022a). Slow moving regulatory decision making for cryptocurrency not economically favourable, New World Economic Forum Study finds. *World Economic Forum*. Retrieved December 9, 2022, from https://www.weforum.org/press/2022/07/slow-moving-regulatory-decision-making-for-cryptocurrency-not-economically-favourable-new-world-economic-forum-study-finds

World Economic Forum. (2022b). The macroeconomic impact of cryptocurrency and stablecoins. *World Economic Forum*. Retrieved November 8, 2022, from https://www.weforum.org/whitepapers/the-macroeconomic-impact-of-cryptocurrency-and-stablecoins

Wronka, C. (2021). Digital currencies and economic sanctions: The increasing risk of sanction evasion. *Journal of Financial Crime*. Advance online publication. doi:10.1108/JFC-07-2021-0158

Wu, X. (2007). *Chinese Cyber Nationalism: Evolution, Characteristics, and Implications*. Lexington Books.

Xi, C. (2022). *The End of the War or the Commencement of Battle?* The Chinese University of Hong Kong Faculty of Law Research Paper No. 2022-15.

Xiang, Yu. (2019). User-Generated News: Netizen Journalism in China in the Age of Short Video. *Global Media and China*, *4*(1), 52–71. doi:10.1177/2059436419836064

Xu, X. (2021). To Repress or to Co-opt? Authoritarian Control in the Age of Digital Surveillance. *American Journal of Political Science*, *65*(2), 309–325. doi:10.1111/ajps.12514

Yamamoto, M., Kushin, M. J., & Dalisay, F. (2018). How Informed Are Messaging App Users about Politics? A Linkage of Messaging App Use and Political Knowledge and Participation. *Telematics and Informatics*, *35*(8), 2376–2386. doi:10.1016/j.tele.2018.10.008

Yamazaki, S. (2016). *Blockchain gijutu no shikumi to kanosei* [How blockchain technology works and its potential]. Internet hakusho 2016 20 nen tokubetsu kinen gou [Internet White Paper 2016 20 Year Special Anniversary Issue]. Impress A&D. https://iwparchives.jp/files/pdf/iwp2016/iwp2016-ch04-03-p243.pdf

Yin, H. H. S., Langenheldt, K., Harlev, M., Mukkamala, R. R., & Vatrapu, R. (2019). Regulating Cryptocurrencies: A Supervised Machine Learning Approach to De-Anonymizing the Bitcoin Blockchain. *Journal of Management Information Systems*, *36*(1), 37–73. doi:10.1080/07421222.2018.1550550

Yuan, Z., Wang, G., & Khor, N. (2011). The Rise of the Middle Class in the People's Republic of China. SSRN *Electronic Journal*. https://studycrumb.com/alphabetizer doi:10.2139/ssrn.1806769

Yuan, D., & Guo, L. (2018). Video-based social media platforms and censorship circumvention: The case of China. *China Media Research*, *14*(3), 1–13.

Yudken, J. S., Croft, T., & Stettner, A. (2022, April 27). *Revitalizing America's manufacturing communities*. The Century Foundation. Retrieved September 16, 2022, from https://tcf.org/content/report/revitalizing-americas-manufacturing-communities/?agreed=1

Zahirah, R., & Ridwan, M. (2018). Kelompok Studi Mahasiswa Pengkaji Masalah Internasional The Utilization of Cryptocurrencies by the Terrorist Group as an Alternative Way of Hawala for Illicit Purposes Mahasiswa Ilmu Hubungan Internasional, Fakultas Ilmu Sosial dan Ilmu Politik. *Jurnal Sentris KSMPMI, 2*. https://dictionary.cambridge.org/dictionary/english/cryptocurrency

Zaltman, G., Duncan, R., & Holbek, J. (1973). *Innovations and organizations*. Wiley.

Zavrsnik, A. (2020). Criminal justice, artificial intelligence systems, and human rights. *ERA Forum*, *20*, 567-583.

Zeqiri, A. (2020, August 22). Komuna e Prishtinës diskualifikoi kandidaten për mësimdhënëse pasi ka nivelin Bachelor në kolegj privat [Newspaper]. *Raporto Korrupsionin! KALLXO.Com*. https://kallxo.com/lajm/komuna-e-prishtines-diskualifikoi-kandidaten-per-mesimdhenese-pasi-ka-nivelin-bachelor-ne-kolegj-privat/

Zgaga, P., Klemenčič, M., Komljenovič, J., Miklavič, K., Repac, I., & Jakačić, V. (2013). *Higher education in the Western Balkans: Reforms, developments, trends*. Faculty of Education, University of Ljubljana. http://pdc.ceu.hu/archive/00006816/01/higher-education-balkans_2013.pdf

Zhang, F., & Feeney, M. K. (2020). Engaging through technology: The role of administrative culture and mandates. *Public Management Review*, *22*(10), 1423–1442. doi:10.1080/14719037.2019.1630475

About the Contributors

Sam Edwards, J.D., LL.M, Associate Professor of Legal Studies, joined the faculty at Quinnipiac University in August 2019. He has been teaching graduate and undergraduate students since 2002. From 2002-2007 as an Associate Professor, he taught at Nagoya University's Graduate School of Law in Nagoya Japan. From 2007-2019 as an Associate and then Professor he taught environmental law at Green Mountain College in Vermont. Sam has an LL.M. in international environmental law from Nagoya University Graduate School of Law and a J.D. from Lewis & Clark Law School. Sam is admitted to the bar in California, The Territory of Guam, The Commonwealth of the Northern Mariana Islands, and Yap State in the Federated States of Micronesia. His research focuses on the intersection of technology and law and stems from his work in Micronesia, Japan, and Africa.

James R. Masterson is MPA Dir. and Assoc. Professor of Political Science at Morehead State University. He recently served as the Dir. of International Education. He earned his Ph.D. in Political Science from the University. of Cincinnati. His research focuses on IR theory and research methodology with specialization on IPE. His research interests include the IR of E Asia in general and analyzing the nexus of PRC foreign and economic relations with the world in particular. His publications examine the effects economic factors have on PRC political relations with its neighbors from 1987-2001, and the world from 1990-2005. He co-authored a piece in E-IR examining the effectiveness of China's air quality plan on pollution levels in Beijing. He published 2 book chapters on the social media in China and how its use affects government response to events. He has co-authored articles in the Cont. Journal of Anthropology and Sociology and Education About Asia. He is co-authoring manuscripts advancing our understanding of the influence that bilateral meetings with PRC's leaders have on its foreign relations.

* * *

Euller Tacito Dias de Almeida Andrade Filho obtained his law degree in 2017 from Dom Bosco College - Brazil, and has been a barrister lawyer ever since, acting mostly in real estate law. Mr. Andrade has published some articles throughout his career, specially about Judicial Activism, which has been his primary research interest.

Abdulla Azizi is Professor and Dean of Faculty of Contemporary Social Sciences at Southeast European University. He received his MA in International Law and Politics and his PhD in European Social Policy with the Faculty of Law, University St. Cyril and Method, Skopje, North Macedonia. In 2001-2002, he pursued specialization studies at the Ecole National d'Administration (ENA) in Paris.

His research interest is in Public Administration, International Organizations and European Union, and quality in higher education. He has served as Director of the Center for Research in Public Administration at SEEU, Member of the University Senate, and Faculty Coordinator for the Erasmus + Program.

Jay S. Bagga, Ph. D. is a Professor of Computer Science and Co-Director of the Voting System Technical Oversight Program (VSTOP) at Ball State University. He has published numerous articles and books and won several faculty awards. He has extensive experience in Election Systems Testing and Certification and has collaborated in developing testing and certification protocols for voting systems and electronic poll books.

Caroline Ban is Assistant Professor and BSW Program Director for the Social Work Program at Valparaiso University. She has taught a variety of classes on topics such as policing, community, policy/government, advocacy, human rights and field education. Ban redesigned the program's government/lobbying and advocacy program to give students direct experience lobbying for social issues. Ban is a 2019 Indiana Humanities Action Fellow for her work in policing and social work. Prior to moving to Indiana, Ban was a consultant and the Manager of Government Affairs for Beyond Housing, a nationally-recognized community development non-profit in St. Louis. Ban facilitated and staffed meetings with mayors and police chiefs from North St. Louis County, directly south of Ferguson. She helped local governments attract or save a combined total of $3 million through grant writing and/or municipal efficiencies. Ban was the lead author of a successful $750,000 multi-jurisdictional Federal Department of Justice grant to strengthen police and youth relations. Her efforts also helped set the table for the consolidation of seven local police departments and the first successful municipal merger in St. Louis Boundary Commission history. She also worked with the board, staff and community members to set and implement the organization's state advocacy agenda. Ban received her bachelor's degree in psychology at Carleton College, her MSW at Washington University in St. Louis, and she is currently enrolled in a DSW program at St. Thomas/St. Kate's.

Fernando Barrio joined Queen Mary University of London in 2018 as lecturer in both the School of Business and Management and the School of Law, after moving from Argentina, where he was Ordinary Professor of Cyberlaw and Director of Law at the Universidad Nacional de Rio Negro. He is currently Senior Lecturer in Business Law at SBM. Fernando carried out his undergraduate studies in International Relations at the Universidad de Belgrano, Buenos Aires, Argentina and Law at the University of London. He has a Masters in International Cooperation Studies (Legal System of International Cooperation) from the Graduate School of International Development of Nagoya University and a Ph.D. in International Cooperation from Nagoya University, Japan, as well as a Certificate in Learning and Teaching in Higher Education from the London Metropolitan University, a Postgraduate Professional Diploma in Peace Studies and Conflict Resolution from Chulalongkorn University, Thailand and a Postgraduate Certificate in Global Business and Politics from the Yale School of Management. The main topic that underlines his scholarship and public engagement experience is the global regulation of information technology, intellectual property, human rights and their application to different industries and business. Currently, Fernando is focused on Artificial Intelligence agri-food and agri-business regulation, particularly legal and policy issues of smart farming and biotechnology, climate change and sustainability related business regulatory framework, and community resilience policies. In Argentina was principal researcher at the Public Policy and Government Research Institute of the Universidad Nacional de Rio Negro and

Researcher Category 1 of the National System of Incentives to Academic Research of the Argentine Ministry of Education. Fernando was Japanese Ministry of Education and Science Scholar, the 2007 E. Desmond Lee Visiting Professor for Global Awareness at Webster University, HEA 2008 UK Law Teacher of the Year, and 2012 Rotary Peace Fellow. Before entering full-time academia, he was Technical Advisor to the Governor of Rio Negro Province, Chief of Non Judicial Negotiations at Piatigorsky Lawyers, and was an adjunct member of the faculty at the School of Law and Social Sciences at the Universidad de Belgrano, all in Argentina.

Bryan D Byers, Ph. D. is a Professor Emeritus of Criminal Justice and Criminology and Co-Director Emeritus of the Voting System Technical Oversight Program (VSTOP) at Ball State University. He has published over thirty-five refereed articles and book chapters on such topics as hate crime, crisis intervention, elder abuse, and death and dying. He has eight books to his credit.

Thiago Allisson De Jesus is a Lawyer. Post-Douter in Criminal Sciences by the Pontifical Catholic University of Rio Grande do Sul. Post-Douter by the Global Inequalities and Social Justice Program (Capes/Internationalization Program) by the Faculty of Law of the University of Brasília and by the Latin American Faculty of Social Studies (FLACSo). Doctor and Master in Public Policy from the Federal University of Maranhão. Doctorate in the Rule of Law and Global Governance (Universidad del Salamanca/Spain). Bacharel em Direito by UFMA. Adjunct Professor I at Graduation and Post-Graduation in Human Rights at the State University of Maranhão and current Director of the International Relations Course at UEMA. Member of the Institutional Committee for Research and Extension of Uema. Permanent Professor of the Professional Master's Degree in Law and Affirmation of Vulneráveis and Graduation in Law of the University of Ceuma. Scientific Initiation Counselor (FAPEMA and CNPQ). Member of the National and International Corps of Consultants of the National Council for Research and Post-Graduation in Law (CONPEDI). Leader of the Nucleus of Studies in Criminal and Contemporary Processes (NEEPPC/UEMA/DGP-Capes) and of the Nucleus of Studies in State, Public Security and Society (NEESS/CEUMA/DGP-Capes). Lecturer, Author and organizer of various works. Catholic Christian. It occupies Cadeira 27 of the Athenian Academy of Letters and Arts / ATHEART with patron Nauro Machado.

Victor Kilanko is an international economist with research interests in development economics, economics of education, health, crime, technology, and public policy.

Chad J. Kinsella, Ph. D. is an Associate Professor of Political Science, Co-Director of the Voting System Technical Oversight Program (VSTOP), and Managing Director of the Bowen Center for Public Affairs at Ball State University. He primarily teaches state and local government, campaigns, and public administration. His research focuses on electoral geography but he also examines state and local government and politics and he has written several articles and book chapters in these areas.

Takashi Kubota obtained LLB and LLM degree from University of Tokyo in 1990 and 1993, LLM degree from Harvard Law School in 1996, and Ph.D. degree from Osaka University in 2003. In 1990, he joined the Bank of Japan, the central bank, and served as as an economist and legal advisor. In 1998, he shifted to academics and has been dedicated to researching and teaching in Nagoya University (1998-2004) and Waseda University (2004-current). He edited and wrote English books such as "Cyberlaw for

Global E-Business: Finance, Payment and Dispute Resolution," Information Science Reference (2007), and Japanese books such as "Practice, policy and law of blockchain," Chuokeizai-sha Inc. (2018), and "Lectures of International Business Law, 3rd Edition," Chuokeizai-sha Inc. (2021).

Jin Kyung Lee is a Ph.D. student in the School of Public and International Affairs at the University of Cincinnati. Jinkyung's research interests specifically concern the relationship between valuable resources in technology and large political systems. Based on her academic and work experiences, she has developed an interest in international data resource agreements and their effect on global relationship development. Especially, Jinkyung is pursuing she will be able to enhance her interests in cybersecurity policies, conflicts between countries in science and technology, and the process of establishing related systems.

|Benjamin James Alexander Leo is a second year doctoral student studying International Relations and Comparative Politics. His research interests center on cryptocurrencies, and how their use is related to, or otherwise results in, the occurrence of various forms of political violence such as protests, terrorist attacks, and conflict onset. Currently he is investigating how anti-state groups are using cryptocurrencies to both facilitate donations and distribute funds within their organization. Also, I am researching how states are responding to cryptocurrency use within their borders, as well as how they are leveraging them to circumvent international sanctions. Previously he earned his B.A. from the University of Cincinnati, Summa Cum Laude, where he majored in Political Science and International Affairs.

Alfred Marleku is Lecturer of Political Science at the University for Business and Technology, Kosovo. He received his PhD in International Relations from Southeast European University, North Macedonia. He teaches methods and pedagogy in Political Science, whereas his research interests are in International Relations, Political Analysis and Higher Education. He has extensive experience in the fields of higher education quality assurance and curricula development.

John G. McNutt is Professor Emeritus in the Joseph R. Biden, Jr. School of Public Policy and Administration at the University of Delaware. Dr. McNutt is a specialist in the application of high technology to political and social engagement. His work focuses on the role of technology in lobbying, e-government and e-democracy, political campaigning and deliberation, organizing and other forms of political participation. He has conducted research on professional associations, child advocacy groups, consumer and environmental protection groups, social action organizations and legislative bodies. His most recent work looks at Web 2.0 political change technology and e-government and fiscal transparency. Dr. McNutt has co-edited or co-authored eight books and many journal articles, book chapters and other publications. He regularly presents at National and International conferences and sits on several editorial boards.

Ridvan Peshkopia is Lecturer of Political Science at the University for Business and Technology, Kosovo. He received his PhD in Political Science from the University of Kentucky; he also spent a year as a postdoctoral fellow with the George Washington University. His main research areas research are International Relations, Political Behavior, and Migration Studies. Before entering the academic life, he has been active in politics, and has served two terms as Representative in the Kuvendi i Shqipërisë (Albania's legislative).

Diogo Santos obtained his Bachelors of Laws in 2003 from the Federal University of Maranhão, in Brazil, Diogo Santos was granted a Japanese Ministry of Education (MEXT) scholarship to pursue a masters of law degree at the Graduate School of Law in 2007 and a PhD in International Development at the Graduate School of International Development in 2010, both at Nagoya University. Since returning to Brazil, Dr. Santos has served in various academic positions, such as Director of the School of Law of the State University of Maranhão, professor of International Law, economics and Constitutional Theory at the Dom Bosco University in São Luís, Brazil, and professor of International Law in the Federal University of Maranhão, professor of Constitutional Law and Department Chair at Estacio College, professor of Law ate Ceuma University undergraduate and graduate School of Law (PPGDir-UNICEUMA) and professor of International Relations at the State University of Maranhão. Diogo has also published a number of articles in the fields of Intellectual Property Law, Environmental Law, Democracy, Legitimacy and Presidential decrees in various international journals.

Kazuo Takaragi obtained BS and MS degree from Kyushu Institute of Technology and Tokyo Institute of Technology in 1975 and 1977 respectively. In 1977, he joined Hitachi, Ltd. At Systems Development Laboratory in Hitachi, he engaged in the research and development of system safety and security. In 1886, he obtained a doctor degree in safety engineering from the University of Tokyo. In 2004, he became the senior chief scientist at Yokohama Laboratory, Hitachi, Ltd. During the fall semester in 2007, he served as a visiting professor for Albert-Ludwigs-Universität Freiburg in Germany. In 2012, he joined National Institute of Advanced Industrial Science and Technology (AIST), Japan as a deputy director of Information Technology Research Institute. Currently, he is a project manager of AIST. He also serves as a chief executive officer of HISAFE, LTD., and a guest professor of Chuo University. He wrote books such as "Practice, policy and law of blockchain," Chuokeizai-sha Inc., "Information Security," Kindai Kagaku Sha Co., Ltd., "Firewall," Shokodo Co., Ltd., and "Internet Security," Ohmusha, Ltd., all in Japanese.

Keisha Williams is a nonprofit fundraising consultant with an extensive background in the fundraising experience. She is an adjunct Professor of Business at Eastern Kentucky University, and eLearning Specialist at Berea College. Keisha Williams holds a B.A in Business Management and MBA in Business Administration from Morehead State University, and blockchain certifications earned from Codecademy and Hong Kong University.

Martin Woesler, PhD in Sinology from Ruhr University Bochum, Germany, Academian of the European Academy of Sciences and Arts, Salzburg. Research Associate at Witten/Herdecke University. "Xiaoxiang Scholar" Distinguished Professor, doctoral supervisor and EU Jean Monnet Chair Professor with the Foreign Studies College, Hunan Normal University. Research on the Social Credit System in China, data rights, artificial intelligence, intercultural multimodal communication, Chinese society in transition, cultural comparison, Chinese literature, comparative literature, and translation studies.

Jingwen Wu is a Ph.D. student in the Department of Political Science at the University of Kentucky. Her research interests currently focus on the influence of modern communication technology on political outcomes, as well as Chinese environmental and foreign policies. She received M.P. A. from Morehead State University and B.A. from Central University for Nationalities (China).

Kazushi Yamamoto is a Visiting Professor at Tashkent State University of Law, a Fellow of Central Asian Legal Research Fellowship, and an Editorial Board Member of Tashkent State University of Law. He earned a Bachelor of Law while also studying for the bar at Nagoya University, School of Law and subsequently completed a Master of Law at Nagoya University, GSL. He later visited the University of Oxford and the University of Cambridge on a private basis. He promoted the Global COE Project of Nagoya University, funded by the Govt. of Japan and conducted his int'l collaborative research with the Director of Kennedy institute of Ethics of Presidential Kennedy Family of Georgetown University, as a Visiting Researcher KIE and a University Researcher of Georgetown University itself. As a Special Lecturer at Nagoya University, GSL, he contributed to the establishment of a Law School, as well as the global Legal Assistant project of Japan. He stationed in the Republic of Uzbekistan at Ministry of Justice and taught as Lecturer of Japanese Law at Tashkent State University of Law and Nagoya University Education and Research Center for Japanese Law within TSUL. He has delivered many speeches at int'l symposiums such as those organized by the United Nations and has published works through int'l organizations like the Organization for Security and Co-operation in Europe (OSCE). Having worked on his doctoral dissertation, Yamamoto participated as an integrator in the COI Project of Govt. of Japan led by the President of the University of Tokyo. He is also actively involved in management roles in some corporations. He is now preparing to publish worldwide with publishers such as Cambridge University/ International Press, Routledge, TSUL, The Japanese Institute of International Business Law, and others. He is also scheduled to deliver speeches at int'l symposium held by F.C.I.Arb in India, the Ministry of Justice, int'l organizations, the United Nations, and so on.

Index

Recommended Reference Books

IGI Global's reference books are available in three unique pricing formats: Print Only, E-Book Only, or Print + E-Book.

Order direct through IGI Global's Online Bookstore at **www.igi-global.com** or through your preferred provider.

ISBN: 9781799877851
EISBN: 9781799877875
© 2022; 402 pp.
List Price: US$ **215**

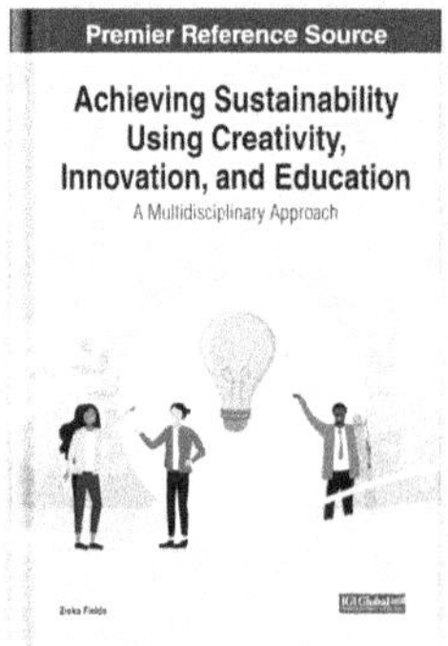

ISBN: 9781799879633
EISBN: 9781799879657
© 2022; 260 pp.
List Price: US$ **240**

ISBN: 9781799890904
EISBN: 9781799890928
© 2022; 307 pp.
List Price: US$ **195**

ISBN: 9781668437063
EISBN: 9781668437070
© 2022; 1,611 pp.
List Price: US$ **1,040**

ISBN: 9781799883395
EISBN: 9781799883418
© 2022; 327 pp.
List Price: US$ **215**

ISBN: 9781799868071
EISBN: 9781799868095
© 2022; 353 pp.
List Price: US$ **215**

Do you want to stay current on the latest research trends, product announcements, news, and special offers?
Join IGI Global's mailing list to receive customized recommendations, exclusive discounts, and more.
Sign up at: **www.igi-global.com/newsletters.**

Publisher of Timely, Peer-Reviewed Inclusive Research Since 1988

IGI Global
PUBLISHER of TIMELY KNOWLEDGE

www.igi-global.com | Sign up at www.igi-global.com/newsletters | facebook.com/igiglobal | twitter.com/igiglobal | linkedin.com/igiglobal

www.ingramcontent.com/pod-product-compliance
Lightning Source LLC
LaVergne TN
LVHW080630011026
849160LV00014B/1174
* 9 7 8 1 6 6 8 4 6 4 3 3 5 *